# The Avian Brain

## Ronald Pearson

Department of Zoology
University of Liverpool
Liverpool, England

1972

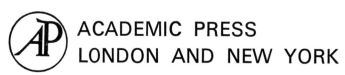

ACADEMIC PRESS
LONDON AND NEW YORK

ACADEMIC PRESS INC. (LONDON) LTD.
24/28 Oval Road
London NW1

*United States Edition published by*
ACADEMIC PRESS INC.
111 Fifth Avenue
New York, New York 10003

Library of Congress Catalog Card Number: 72-153526
ISBN: 0 12 548050 4

PRINTED IN GREAT BRITAIN AT
THE UNIVERSITY PRESS
ABERDEEN

# Preface

The work of embryologists, histologists, pharmacologists, physiologists and zoologists has now provided a wealth of information about the avian brain. Such studies form an interesting and constructive comparison for workers in many aspects of brain studies and no single reference source, providing a summary, exists. It was therefore a pleasure to accept the suggestion of Academic Press and attempt to fill this gap. If the book is somewhat idiosyncratic then this reflects, at least in part, the rather diverse field. Should it assist biologists to see both bird brains and other structures in perspective I will feel that my objectives have been attained.

I would like to express my personal appreciation to a number of people. To Dr. E. A. Salzen, now of the Department of Psychology, University of Aberdeen, with whom I worked for all too short a time ; to Professor P. M. Sheppard, F.R.S., of the Department of Genetics, and Professor A. J. Cain, Dr. C. L. Smith and Dr. J. W. Jones of the Department of Zoology in this University, without whose encouragement this book would not have been written. Needless to say they are in no way responsible for any inadequacies which it may have.

Finally I would like to acknowledge a far older debt to Professor E. N. Willmer, F.R.S., who long ago convinced me that a study of structure must precede a study of function and I would also like to thank the Production Department of Academic Press for their very kind help.

*University of Liverpool*
*January*, 1972

RONALD PEARSON

# Acknowledgements

I would like to acknowledge with gratitude the kind permission to reproduce illustrations from the following sources: Zoologischen Jahrb. Abt. Anatomie; "The Comparative anatomy of the cerebellum from Myxinoids through birds"; The Journal of Comparative Neurology; The Journal of Anatomy, The Royal Society, Proceedings B; Acta. Anat. and Bibliotheca anat.; The American Journal of Anatomy; The Journal of Experimental Zoology; Evolution; CIBA foundation "Hearing Mechanisms in Vertebrates"; Zeitschrift für wissenschaftliche Zoologie; Zeitschrift für Tierpsychologie; Anatomischer Anzeiger; Comparative Biochemistry and Physiology; The Journal of Zoology.

# Contents

## 1. Introduction

## 2. The Vascular System and Glial Cells

## 3. The Biochemistry of the Avian Brain

## 4. The Development of the Avian Brain

## 5. The Ear and Hearing

## 6. The Medulla and Afferent Systems

## 7. The Medullary Efferent Systems, The Olive, The Pons and Allied Structures

# 8. The Cerebellum

# 9. The Avian Eye and Vision

# 10. The Mesencephalon

# 11. The Diencephalon

1*

## 12. Hypothalamo-hypophyseal, Neuro-hypophyseal and Pineal Systems

## 13. Anatomy and Histology of the Forebrain Hemispheres and Related Systems

## 14. Experimental Studies on the Hemispheres and Related Structures

## 15. Electro-encephalographic Activity

## 16. The Relative Development of Different Brain Regions

## Envoi

# 1 Introduction

## I. GENERAL CONSIDERATIONS

In general terms both the gross and the detailed structure of the nervous system within a given taxon reflect the relative complexity of the morphological organization and behaviour. These are in themselves dependant upon the type of relationship which the animal has with its environment, and which its organ systems have with one another. The central nervous system of birds is no exception. Together with other amniotes, and more especially mammals, birds have a high degree of morphological differentiation. Furthermore, their behaviour patterns involve perceiving, and responding to, intricate visual, auditory and tactile cues. These facts pre-suppose the existence of a highly complex nervous system. The size of the brain itself exceeds that of reptiles, as can be seen from the often cited example of a lizard with a total body weight of 10 g. Such a specimen has a brain which weighs approximately 0·05 g. In contrast a small passerine bird of equivalent body weight has a central nervous system which weighs 0·5 g. Although this difference in size actually involves the entire brain it is particularly noticeable in terms of the cerebellum and forebrain hemispheres.

From a historical point of view it is interesting that the neuro-anatomical studies of these and other components of the avian brain, which were largely initiated during the closing decades of the last century, were somewhat interrupted in the latter part of the 1930s. This is of particular interest because it was at that very time that the complementary study of ethology was growing rapidly. Thirty or so years later the intellectual descendants of those early ethologists, and many behavioural psychologists, have now started to experiment widely upon avian brains. The particular interest of such workers lies in the possible neuronal pathways which may be implicated in the behaviour patterns that they observe. Although it would be ridiculous to suggest that there was not a continuing thread of neurological research on birds during the 1940s and 1950s, it is fair to say that at that time, when the work of ethologists was catching the imagination of both scientific and lay public alike, such research was spasmodic and largely confined to a limited number of isolated centres. During this period the information which did emerge was often an example of "spin-off", and was produced by interested individuals who considered that it would be informative to apply to the avian brain those

1

techniques which had been, or were currently being, developed by medically orientated research workers using mammalian material. It will be clear from subsequent chapters in this book that these techniques frequently fulfilled their user's expectations and produced fascinating results. If these results did not, perhaps, become as widely known as one might have expected, then this merely reflects the interests of the times—times during which such other subjects as molecular biology were beginning to compete successfully for the resources available to biologists.

It is also very instructive to notice how often the data which have emerged during the last twenty, and more especially the last ten years, corroborate at least some of the conclusions drawn by far older workers. These had painstakingly collated an immense number of anatomical results which are epitomized in the monumental works of Kappers (1921, 1947) and Kappers *et al.* (1936). The principal conclusions are represented in Fig. 140 of this book. Analogous compilations which represent a more physiological approach are those of Ten Cate (1936, 1965) and Buddenbrock (1953).

It is the intention of this present book to outline the extent of our knowledge on this whole subject, the anatomy, histology, physiology and overall integrative action of the avian brain. Naturally when doing so one necessarily indicates the extent to which its organization corresponds to that of vertebrate nervous systems in general. However, it is fair to say that, although a consideration of the brains of related species or classes frequently throws light upon the problems which particular research workers are encountering in the animals which interest them, this book is not written with the intention of providing an account of a type of brain which, although simpler, shares the characteristics of that found in mammals. It is clear that in the past a considerable proportion of the drive which produced descriptions of non-mammalian nervous material came from people whose primary interest lay in Man and mammals and who were seeking simpler, more easily investigated, but homologous, systems. Without any doubt at all this was both laudible and rewarding, but I would like to emphasize that in writing this book it has been my intention to consider the avian brain and sense organs *per se* and to emphasize that they are different from, but certainly not simpler than, mammalian brains. Where valid comparisons between avian and mammalian systems can be drawn, it is often beneficial to consider them. By doing so the answer to a particular problem may suggest itself. However, to go further than this and assume that just because a structure serves a particular function in one class it necessarily does so in another, is at best dangerous, and at worst ill-informed. Furthermore, to blithely assume that structures in different classes are homologous rather than analogous simply because they serve apparently related functions is also dangerous. In general, I have therefore limited this book to a consideration of the conditions which occur in birds.

Where, as in the chapter on brain biochemistry, the available data suggest that the situation is comparable with that in other vertebrates, I have indicated that this is the case, referred to general reviews, and restricted the text to the actual information which is available about avian material.

Wide ranging summaries of nervous systems and their function in numerous animals and experimental situations are provided by many works (see, for example, Bourne, 1968). Previous synoptic reviews of avian material include, besides those of Kappers *et al.* (1936) and Kappers (1947), those of Papez (1929), Portmann and Stingelin (1961) and Stingelin (1962). In recent decades the development of the Nauta–Gygax histological process (Nauta and Gygax, 1951, 1954) together with the stereotaxic atlases of both Tienhoven and Juhasz (1962) and Karten and Hodos (1967) promise great rewards in the future. Where no recent information on anatomical structures is available I have drawn on my own studies of avian material which are derived from specimens kindly provided by the Director of Chester Zoo and other local sources.

It is worth mentioning at this point the very interesting review of electrical stimulation studies on the brain of birds which was provided by von Holst and Saint Paul (1963). This has received a considerable amount of attention from ethologists and was discussed in detail by Hinde (1966). Sceptical of the existence of discrete and localized centres within the central nervous system of the hen, they concluded that:

1. A stimulation field may appear silent or active at different times;
2. Changing behaviour patterns may be evoked over a period of hours as the result of activating a particular stimulation field;
3. The same pattern or action can often be released by stimulating different fields.

Central to their interpretation is the concept of "mood shifts". They suggest that in life it is to be expected that stimulus fields remain silent in the presence of a dominant behaviour pattern such as fleeing or postural freezing which suppresses other activity. They conclude that, under these conditions, to say that the stimulus field is silent is incorrect. The activity which would normally result from such focal stimulation does not occur because it is being blockaded elsewhere. To explain the changing reactions which can be evoked at different times from one and the same field they invoke both the obvious and widely noted variations in the distance from the site of stimulation to which excitation may be propagated, and also the huge differences in individual behaviour thresholds which result from mood shifts and adaptation. Making the assumption that there are structures related to a number of neuromotor systems within the stimulus field and that all these are therefore excited, they concluded that the resultant behaviour is determined by the total dynamic situation which is currently prevailing, rather than by specific anatomical projections of the stimulus field. The fact that identical reactions can result from stimulating widely separated fields reflects, in turn, the fact that many movements are common to a variety of behaviour patterns. Locomotor unrest which may in one case be directed at an enemy may, in another, reflect thirst etc. Although to completely deny the existence within the avian brain of foci which are predominantly associated with at least certain sensory modalities and motor output would be to fly in the face of much of the evidence which is reviewed in the later chapters of this book, this highly critical assessment of many basic assumptions is a valuable background to any such considerations.

## II. AVIAN TAXONOMY

As birds constitute the most homogeneous class of the vertebrates it is not surprising that their systematic relationships are not easily stated with precision. It is certainly true that no single classificatory scheme is universally acceptable to ornithologists and zoologists. The particular difficulties relate to a limited number of orders and genera, are somewhat idiosyncratic, and reflect the preconceptions or predisposition of the relevant writer. Universally recognized as originating from the diapsid reptiles, from which they can be theoretically derived by relatively slight modifications of basic anatomy, there are some 25,000 known species or sub-species of living birds. Besides the exact interrelationships of these forms their degree of affiliation to fossil forms, such as the Jurassic sub-class Archaeornithes and the Cretaceous *Hesperornis*, are also difficult to establish with any great degree of confidence. A summary of these problems is provided in such works as Romer's "Vertebrate Paleontology", and Young's "Life of the Vertebrates". It is for this reason that the classificatory schemes which have appeared since the days of Linnaeus vary, as does their credibility. This variation not only reflects the items of knowledge which were either new, or considered to be critical, at the time they were produced, but also as was noted above the personal idiosyncracies of the particular workers. It is certainly clear that, as a result, the schemes which are considered acceptable today are markedly different from those which were put forward by for example Linnaeus himself, Buffon and Cuvier.

Broadly speaking, one can differentiate amongst living forms between the Ratites, Impennae and Carinates. The first group includes the Struthioniformes, or Ostriches; the Rheiformes, or rheas; the Casuariiformes, including the cassowary and emus; and the Apterygiformes or Kiwi. The second group comprises the penguins, and the third all other living birds.

By using zone electrophoresis Gysels (1970) showed that *Rhea, Casuarius* and the tinamiform genus *Crypturus* are clearly different from each other although showing similarities which suggest a closer relationship than with other birds. This would agree with the conclusions of Parkes and Clarke (1966). It is also interesting that, of the other avian orders, it is the Galliformes which have the closest similarities, in terms of lens protein reactions, with the ratites and tinamous. This corroborates the suggestions that tinamous are more closely associated with the ratite birds than with other carinates, and that the Galliformes are closer to the ancestral stock than are other carinates. On the other hand Sibley (1960) found that the tinamou pattern resembled neither the rheiform nor the galliform one. Stresemann (1927–34, 1959) favoured polyphyletic evolution.

## TABLE 1

*Systematic list of the orders of living birds based on Wetmore's classification (1934). The approximate numbers of species in certain orders after Mayr and Amadon.*

| | |
|---|---|
| Struthioniformes | Ostrich (1) |
| Rheiformes | Rheas (2) |
| Casuariiformes | Cassowaries, emus (5) |
| Apterygiformes | Kiwis (3) |
| Sphenisciformes | Penguins (16) |
| Tinamiformes | Tinamous (33) |
| Gaviiformes | Divers or loons (4) |
| Colymbiformes | Grebes (20) |
| Procellariiformes | Albatrosses, petrels, etc. (90) |
| Pelecaniformes | Cormorants, gannets, pelicans, etc. (54) |
| Ciconiiformes | Herons, storks, etc. (105) |
| Phoenicopteriformes | Flamingoes (6) |
| Anseriformes | Geese, Ducks, Screamers (148) |
| Falconiformes | Diurnal birds of prey (271) |
| Galliformes | Game-birds, etc. (241) |
| Gruiformes | Cranes, rails, cariamas, etc. (199) |
| Charadriiformes | Waders, gulls, auks, etc. (308) |
| Columbiformes | Sand-grouse, pigeons (308) |
| Psittaciformes | Parrots (316) |
| Cuculiformes | Turacos, cuckoo (147) |
| Strigiformes | Owls (134) |
| Caprimulgiformes | Nightjars, etc. (97) |
| Micropodiformes | Swifts, Humming-birds |
| Coliiformes | Mouse-birds (6) |
| Trogoniformes | Trogons (35) |
| Coraciiformes | Rollers, kingfishers, bee-eaters, hoopoes, hornbills, etc. (194) |
| Piciformes | Puffbirds, barbets, woodpeckers, toucans, etc. (381) |
| Passeriformes | Perching and singing birds (5072) |

Apart from the conjectural relationships of the orders grouped together under the heading of Ratites, and the degree of independence between the three groupings themselves, the inter-relationships of the neognathous orders produce a number of difficulties. Wetmore's classification is generally used and this is summarized in Table 1 but it has some inadequacies, and I have, for personal preference, followed the somewhat more anarchic system which is particularly favoured by some French speaking authors. In particular this affects the auks, gulls, waders and humming birds and throughout this book they are ascribed to the independent orders Alciformes, Lariformes, Charadriiformes *sensu stricto* and Trochiliformes. It is also worth emphasizing that some inconvenience awaits the unwary in the differing usage which occurs on the two side of the Atlantic for the ordinal names of the divers, grebes, herons, storks and rails. A comparison of these terms is provided by Table 2. Furthermore the status of the flamingoes also varies. They are usually relegated to a separate order by British, if not all European ornithologists, although Wetmore included them as a sub-order of his Ciconiiformes.

## TABLE 2

*A comparison of the principal differences in nomenclature on each side of the Atlantic.*

| Wetmore's nomenclature | British nomenclature | Common names |
|---|---|---|
| Gaviiformes | Colymbiformes | Divers or loons |
| Colymbiformes | Podicipidiformes | Grebes |
| Ciconiiformes | Ardeiformes | Herons and storks |
| Gruiformes | Ralliformes | Cranes, rails, cariamas, etc. |

*Note:* Wetmore's order Ciconiiformes differs from the Ardeiformes as it includes the Phoenicopteriformes or Flamingoes.

## III. DISSECTION OF THE BRAIN TO SHOW ITS GENERAL CHARACTERISTICS

Detailed guides for the dissection of the avian brain occur in most textbooks of practical vertebrate zoology but for completeness one can briefly outline the principal considerations at this point. They serve as an introduction to the overall anatomy and enable one to highlight the particular characteristics which are peculiar to birds. Prior to attempting any such gross dissections the brain should have been hardened in alcohol. It is situated very close to the roof of the skull and it is therefore necessary to remove the bone very carefully. When this has been done and the dorsal surface of the brain is exposed it can be viewed *in situ*. The olfactory bulbs are of variable size but often very small, and lie at the extreme anterior end. The forebrain has a pair of large hemispheres which have a relatively smooth surface in the chick, but which bear various surface structures in other birds. These grooves and prominences, together with their position, reflect the relative degree of development of the internal components such as the neostriatum, and the ventral, dorsal and accessory hyperstriatum which are discussed in detail in Chapter 13.

Lying between the anterior hemispheres at their posterior end is the small pineal body. The front part of the cerebellum is immediately adjacent to the hemispheres in this region; consequently, the optic lobes are not visible at this point although they can be seen to the side of it. Posteriorly the cerebellum projects backwards as far as the hind end of the roof of the fourth ventricle and covers the front part of the medulla oblongata. The medulla itself is wide and merges with the spinal cord. In such a dissection as this it becomes clear that the avian brain, as represented by, say, the pigeon, is bent and the axis of the brain is not merely a forward projection of the main axis of the spinal cord. The principal point of flexure is in the region of the midbrain or mesencephalon, but although the results of this flexure can be seen in dorsal view its actual location, in terms of the brain areas, necessitates at least a gross section.

After exposing the anterior part of the spinal cord the brain can be removed by transecting the cord near to the medulla, raising the brain and severing the various nerve roots as they are exposed. On the ventral surface (see Fig. 1) the olfactory lobes can be seen anteriorly, and either behind or above them, depending on the species of bird, lies the ventral surface of the hemispheres. The optic chiasma is a well-defined and large structure which is situated in the mid-line. The optic tracts can be seen leaving it and passing outwards to run dorsally and reach the optic lobes. Behind the chiasma the infundibulum projects into a small depression but unless considerable care is exercised the hypophysis is usually detached from the tuber cinereum when the brain is removed from the cranial cavity. In this case it will be found still remaining in the sella turcica which is a pit in the floor of the cavity. Behind the infundibulum is the large medulla which has a broad band of pontine fibres in its anterior region. With two exceptions the cranial nerves originate from regions which are comparable to those of other vertebrates.

If one makes a parasagittal section through the brain at a transverse level close to the mid-line of one forebrain hemisphere a number of further characteristics emerge. The hind part of the forebrain hemispheres overlaps the optic lobes dorsally and the ventricle of the forebrain is a very narrow cavity. This is predominantly restricted to the dorsal region and has a relatively thin roof. A detailed discussion of the components which surround it is

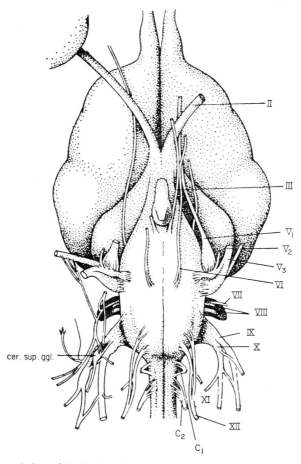

FIG. 1. A ventral view of the brain of the goose showing the cranial nerves (Portmann and Stingelin, 1961). $V_{1,2,3}$, branches of the trigeminus; $C_{1,2}$, cervical nerves; cer. sup. ggl., superior cervical ganglion.

again contained in Chapter 13, but it can be seen that there is an extensive development of telencephalic nuclear material. A vertical sagittal section which passes along the line separating the hemispheres shows that their posterior regions completely cover the mid-brain. Indeed it is as a result of the development of the forebrain on the one hand, and the cerebellum on the other, that the optic tectum is displaced laterally and assumes the characteristic avian position. Intervening between the roof of the mid-brain and the optic chiasma is the cavity of the third ventricle. The cerebellum is large, there is no cerebellar ventricle, and the medulla has a thick floor. The arrangement of the white and grey matter in the cerebellum gives rise to a tree-like appearance of the white matter which is known as the cerebellar arbor vitae. In such a section as this the flexures undergone by the axis of the brain are clearly detectable. In the pigeon the primary flexure at the level of the mid-brain results in the axis of the brain behind this point lying nearly at 90° to the more anterior axis. A second, or pontine flexure, below the cerebellum and at the front end of the medulla, bends the brain in the opposite direction to both this primary flexure and also to a nuchal flexure which makes the medulla take up an angle with the spinal cord.

## IV. THE BRAINS OF SOME FOSSIL GENERA

### A. The Brain of *Archaeopteryx*

The unusual morphological characteristics of the Jurassic genera *Archaeopteryx* and *Archaeornis* are almost universally interpreted as representing a condition which is intermediate between that of the ancestral diapsid reptiles on the one hand and those of more recent avian genera on the other. On *a priori* grounds one would therefore expect that the brain of *Archaeopteryx* might provide a comparable intermediate picture of one of the stages through which the avian brain has passed during phylogeny. A review of the brain of this genus was provided by Edinger (1926) whose conclusions were based on the apparent structure of the osseous material. Clearly such conclusions have to be accepted cautiously but certain features are of considerable significance.

Although the form of the olfactory lobes is not detectable Edinger concluded that the forebrain had narrow, elongated and somewhat curved hemispheres. Behind these the roof of the mesencephalic region seems to have been situated on the dorsal side of the brain. It therefore resembled the situation in reptiles and not that in birds where it undergoes a considerable amount of lateral displacement during development. The absence of any information about the morphological medial region made it impossible to decide whether the cerebellum actually penetrated between the two tectal components and hence foreshadowed the gross displacement characteristic of modern forms. With this in mind Edinger concluded that the overall plan of the brain, unlike that of the Cretaceous genera, resembled that of reptiles. However a number of characters which are associated with birds are present. The brain appears to have completely filled the cavity of the brain case, whereas, leaving aside the pterosaurs where the meninges and inter-meningeal spaces take up more room, casts of the cranial cavities of reptiles do not usually have such a definite brain-like appearance. Furthermore, in *Archaeopteryx* the axis of the brain does not appear to follow that of the cranium, but instead forms an angle with it. However, this also occurs in many fossil reptilian genera, and, in any case, the interpretation rests on the assumption that the skull of *Archaeopteryx* has been embedded in a similar position in relation to the body, as has that of *Archaeornis*.

## B. The Brains of the Cretaceous Genera

### (i) *The genus* Hesperornis

In view of the interesting information which Edinger obtained from the fossil *Archaeopteryx* material it is clearly important to consider that of the Cretaceous genera. Following the appearance of Marsh's (1880) description of the endocranial casts of these genera a very definite conception of their brain structure persisted for some seventy years. As a result of his studies Marsh concluded that there had been a progressive increase in the size of avian brains during post-Cretaceous time and, in view of the material which was represented in his figures and descriptions (1880, 1883) he considered that the brain of *Hesperornis* was quite small and more reptilian in character than that of any other adult bird. Furthermore that of *Ichthyornis* was remarkably small and also of a strongly reptilian nature. D'Arcy Thompson in a rather later paper (1890) made additional comments on the remains and again emphasized the small size of the brains. Apart from this small brain size he suggested that the birds were broadly comparable in overall appearance to modern *Colymbiformes* (≡ Gaviiformes). Unlike much of the neuro-anatomical work which dates from the end of the last century and is based on studies of modern birds these conclusions appear, in the light of more recent investigations, to be very wide of the mark.

According to Marsh's original figures the brain of the genus *Hesperornis* is approximately 64 mm long. That of *Ichthyornis* is only about half this size and measures 31 mm. The skull lengths have a slightly different size ratio since that of the former measures 257 mm and the latter 100 mm. Edinger (1951) pointed out that since the overall size of *Hesperornis* was apparently much greater than that of the pigeon-sized *Ichthyornis*, the latter had a relatively far larger brain. In view of this, and also of the entrenched conceptions which had grown up around Marsh's work, Edinger reviewed the neurological data which could be derived from a study of both the holotype and the plesiotype of *Hesperornis*, and also the *Ichthyornis* remains.

From a consideration of the holotype it became clear that the brain of *Hesperornis* could not have had the proportions which Marsh had ascribed to it. He had stated that the telencephalon was of very moderate size, was proportionately much shorter than that of any living genus of bird, and that the two lobes were narrow and sub-ovate in outline. Edinger conceded that it would certainly be fair to say that no extant bird has a telencephalon which is so short that the ratio of its length to that of the entire brain is 1 : 2·6. However, the fossil specimen does not have any features which would actually indicate such a short forebrain. Similarly it was equally clear that this part of the brain was by no means as slender as Marsh had suggested. Some of the illustrations which were provided by Marsh suggest that the cerebro–cranial relationships, and the point of maximum width of the "cerebral capsule", are quite disparate from those of all other birds including *Archaeopteryx*. In contrast Edinger concluded that the brain was actually much more avian in appearance.

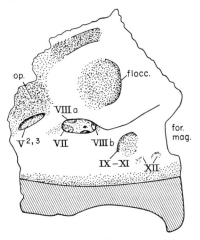

Fig. 2. The fragmentary remains of the cranial cavity of *Hesperornis regalis* showing the conclusions derived from it by Edinger (1951). Note part of the optic lobe pocket, floccular fossa, posterior foramen of trigeminal, internal auditory meatus with foramen for facial nerve and foramina VIII, IX, XI and pits round small XII.

The cranial roof had its highest vault between the post-orbital processes suggesting that as in other birds the comparable brain parameter was also located in this region. The olfactory bulbs were situated above the posterior region of the inter-orbital septum but the specimen does not reveal the size or form of their chambers. Clearly the reconstruction presented by Marsh is spurious. Exact information about the actual appearance of particular parts of the brain is actually only presented by the occiput of the plesiotype. This suggests that there was both a large corpus cerebelli and also a prominent flocculus reminiscent of that in recent genera. The floccular fossa is as wide and also as deep as that of modern forms and the medulla oblongata is

narrower than the cerebellum. Edinger contrasted this latter condition with
that which one finds in reptiles. However, of far greater significance, especially
in view of their apparently medial position in *Archaeopteryx*, was the lateral,
or even latero-basal position assumed by the optic tectum of each side. It
was in this region that the preserved portion of the cranial cavity appeared
to have its greatest width. Allowing for the inherent difficulties which are
involved in the interpretation of such fossil material, this is certainly very
suggestive, since behind the forebrain of modern genera the broadest transverse plane again occurs in the hind part of the optic lobe region.

## (ii) *The genus* Ichthyornis

In view of this summary of Edinger's conclusions about the *Hesperornis*
material it will not come as a surprise to anyone to find that the same author
considered that no data exist which substantiate Marsh's assertions about the
reptilian character of the brain of *Ichthyornis*. The cranial fragment of the
species *Ichthyornis dispar* gives more potential neurological information than
that of *Ichthyornis victor*. Like the fossil remains of other small bird skulls the
cranium had become embedded whilst it was lying on one side. As a result
of this it had been flattened laterally during the process of fossilization, and
the appearance of the major brain divisions is preserved. The most striking
avian characteristic is the considerable size of the optic lobes which are
represented by the large oval depression situated between the orbit and the
occiput, and comprising the chamber that housed the left lobe.

The process of crushing also revealed two other distinct areas of the endocranial cavity. Above and behind the chamber of the optic lobe lies that of
the cerebellum. This part of the skull had thick walls and both its shape and
position suggest that a large vaulted corpus cerebelli occupied, and moulded
the form of, the rear end of the brain case. In view of the apparently lateral
position assumed by the optic lobes and also the condition in Recent genera,
it is particularly interesting that Edinger concluded that a dent in the dorsal
outline of the cranium disclosed the existence of an area of contact between
the forebrain hemispheres and the cerebellum. She considered that the largest
diameter of the cerebellum itself could also be measured approximately by
using the distance between this dorsal dent and the dorsal rim of the foramen
magnum. The length of this oblique axis is 13–14 mm and its relationship
to the total brain length corresponds to the situation in Recent birds but not
to that in reptiles.

In front and above the crushed chamber of the optic lobe lies that of the
forebrain. As is the case in extant avian genera it has a somewhat funnel
shaped appearance. What is more the widest part in *Ichthyornis* also seems
to be a "temporal lobe", but neither the original width of this, nor the total
length of the hemisphere, were detectable. However, the hemisphere is clearly

marked for 10 mm in front of the cerebellum. In front of this region a 2 mm long, horizontal and tapering cone appeared to Edinger to represent the left olfactory bulb. This would correspond to the relative size and position of the supra-orbital bulbus of modern birds. She concluded that the apparently small size of the hemispheres, together with the fact that they only overlap half the length of the optic lobe, might be a result of the crushing which took place during the process of fossilization. Clearly the opposite may also be the case and these characters may be a true reflection of the condition in life.

It need hardly be said that such interpretations of fossil material are difficult and reflect very favourably on the perception, comprehension and intuition of the observer. Nevertheless, one has to maintain some scepticism over the finer details of the identification of individual nerve foramina. I have therefore restricted this account to the gross characters of the brain. One can certainly agree with Edinger's conclusion that by the Cretaceous period, and prior to the time that the major placental diversification began in the Cenozoic Era, the principal morphological characteristics which one associates with modern bird brains had been achieved at least in the mid- and hind-brain regions. Furthermore the forebrain of *Ichthyornis* was distinctly avian in such characters as its relationship with the position of the head, and the existence of a region of possible posterior contact with the cerebellum. In addition there are some indications that the olfactory lobes were reduced. From a consideration of modern brains it can be deduced that with the progressive increase in the importance of vision during avian phylogeny the relative size and importance of the avian olfactory centres became reduced (see Chapter 13). In this respect the data from these two Cretaceous genera would conform to such phylogenetic speculations, and it is interesting that in the phylogeny of two short lived families of Cenozoic birds, the dinorthids and aepyornithids, this trend was reversed. The olfactory bulbs appear to have become enlarged as the optic lobes became relatively smaller (Edinger, 1942). However, in general the remains of the Cretaceous genera, together with those of the Cenozoic *Numenius* (*Limosa*) *gypsorum* from the Gypsum beds of Montmartre, and *Cryptornis antiquus*, an early hornbill or hornbill analogue, suggest that most of the modern characters were present in the avian brain in excess of 60 million years ago, and that certainly by 40 million years ago the brain was modern in appearance.

# References

Bourne, G. H. (1968). "The structure and function of nervous tissue", 2 vols. Academic Press, London and New York.
Buddenbrock, W. von (1953). "Vergleichende Physiologie." II. Nervenphysiologie, pp. 396. Birkhauser, Basle.

Edinger, T. (1926). The brain of *Archeopteryx*. *Ann. Mag. Nat. Hist.* **9**, 151–156.

Edinger, T. (1942). L'encephale des Aepyornithes. *Bull. Acad. Malgache* **24**, 25–50.

Edinger, T. (1951). The brains of the Odontognathae. *Evolution* **5**, 6–24.

Gysels, H. (1970). Some ideas about the phylogenetic relationships of the Tinamiformes based on protein characters. *Acta Zool. Pathol. Antverpiensia* **50**, 3–13.

Hinde, R. A. (1966). "Animal behaviour; a synthesis of ethology and comparative psychology", pp. 534. McGraw-Hill.

Holst, E. von and Saint Paul, U. von, (1963). On the functional organisation of drives. *Anim. Behav.* **11**, 1–20.

Kappers, C. U. A. (1920–21). "Vergleichende Anatomie des Nervensystems." Bohn, Haarlem. (See also 1947.)

Kappers, C. U. A. (1947). "Anatomie comparée du système nerveux", pp. 754. Masson, Paris.

Kappers, C. U. A., Huber, G. C. and Crosby, E. L. (1936). "The comparative anatomy of the nervous system in vertebrates including Man." MacMillan (1960 reprint, Hafner, New York, 3 vols.)

Karten, H. J. and Hodos, W. (1967). "A stereotaxic atlas of the brain of the pigeon (*Columba livia*)", pp. 193. Johns Hopkins University Press, Baltimore.

Marsh, O. C. (1880). "Odontornithes." A monograph on the extinct toothed birds of North America. *Mem. Peabody Mus. Yale Coll.* **1**, 201.

Marsh, O. C. (1883). Birds with teeth. *Ann. Rep. U.S., Geol. Survey* **3**, 45–88.

Nauta, W. J. H. and Gygax, P. A. (1951). Silver impregnation of degenerating axon terminals in the central nervous system. 1. Technique. 2. Chemical notes. *Stain Technol.* **26**, 5–11.

Nauta, J. W. H. and Gygax, P. A. (1954). Silver impregnation of degenerating axons, a modified technique. *Stain Technol.* **29**, 91–93.

Papez, W. J. (1929). "Comparative Neurology." Crowell and Co., New York.

Parkes, K. C. and Clark, G. A. (1966). An additional character linking Ratites and tinamous, and an interpretation of their monophyly. *Condor* **68**, 459–471.

Portmann, A. and Stingelin, W. (1961). The central nervous system. *In* "The Biology and Comparative Physiology of Birds" (Marshall, A. J. ed.), vol. 2, pp. 1–36. Academic Press, London and New York.

Sibley, C. G., (1960). The electrophoretic patterns of avian egg white proteins as taxonomic characters. *Ibis* **102**, 215–284.

Stingelin, W. (1962). Ergebnisse der Vögelgehirnforschung. *Verhandl. Naturf. Ges. Basle.* **73**, 300–317.

Stresemann, E. (1927–34). Aves. *In* "Hdb. Zool." Kukenthal-Krumbach, vol. **7**. pp. 900.

Stresemann, E. (1959). The status of avian systematics and its unsolved problems. *Auk* **76**, 269–280.

Ten Cate, J. (1936). Physiologie des Zentral Nervensystems der Vögel. *Ergeb. Biol.* **13**, 93–173.

Ten Cate, J. (1965). The nervous system of birds. *In* "Avian Physiology" (Sturkie, P. D., ed.), pp. 697–751. Baillière, Tindall and Cassell.

Thompson, D'Arcy W. (1890). On the systematic position of *Hesperornis*. *Stud. Mus. Zool. Univ. Coll. Dundee.* **1** (10), 1–15.

Tienhoven, A. van., and Juhasz, L. P. (1962). Chicken telencephalon, diencephalon, and mesencephalon in stereotaxic coordinates. *J. Comp. Neurol.* **118**, 185–197.

Wetmore, A. (1934). A systematic classification of the birds of the World. *Smithson. Misc. Collect.* **89**.

# 2  The Vascular System and Glial Cells

## I. THE ARTERIAL SYSTEM

### A. Introduction

The arterial pathways which are involved in supplying blood to the avian brain were described by Kitoh (1962) and in more general terms by Baumel (1967) and Gillilan (1967). The overall pattern of distribution is shown in Fig. 3 and the variations which occur on this basic theme are represented in Figs 4–6. Kitoh's paper is particularly useful as it provides a summary of the synonymies between the nomenclature used by Hofmann (1900), Shiina and Miyata (1932) and Kaku (1959). The capillary system is described by Stoeckenius (1964).

Following his account of the consistent asymmetries which one observes in the arterial supply to the pigeon brain Baumel (1962, 1967) reviewed the widespread occurrence of such asymmetrical conditions. These had previously been reported by Carss (1895), Hofmann (1900), Klinckowström (1890) and Hughes (1934–35). In addition Beddard (1905), Wingstrand (1951) and Vitums *et al.* (1965) discussed the basal artery which is usually derived asymmetrically from the caudal branch of the internal carotid. Craigie (1940) also outlined the area of distribution of some vessels.

### B. The Inter-carotid Anastomosis

The two internal carotid arteries of most birds communicate with one another by some sort of anastomosis which is situated either just behind, or within the sella turcica. The only cases where such anastomoses have not been reported are five species of the passeriform sub-order Tyranni (Wingstrand, 1951; Baumel and Gerchman, 1968). In other genera the variations which occur upon a clear cut H-type anastomosis comprise I, X and border-line X–H types. These are represented by groups of approximately equal size in the list of species which was studied by the above authors. Although such an anatomical classification is by no means clear cut the classification serves as a convenient method of describing most of the common conditions. An overall view of the inter-carotid anastomosis together with the points at which other arteries originate is given in Fig. 5. A detailed series of diagrams which

15

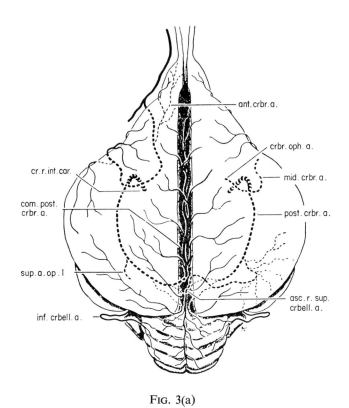

ant. crbr. a.

crbr. oph. a.

cr. r. int. car.

mid. crbr. a.

com. post.
crbr. a.

post. crbr. a.

sup. a. op . l

asc. r. sup.
crbell. a.

inf. crbell. a.

FIG. 3(a)

FIG. 3. (a) Dorsal view of the arteries supplying the avian brain; (b) Ventral view of the arteries supplying the avian brain (Baumel, 1962). Broken lines indicate vessels which are on a deeper plane. Branches of the superior artery of the optic lobe of the left side and the descending branch of the superior cerebellar artery are omitted. Note the inferior cerebellar artery coursing out of the cranial cavity into the optic region and then recurving into the cranial cavity to lie on the flocculus and side of the cerebellum.

Abbreviations: ant. crbr. a., anterior cerebral; bas. a., basilar; brr. sup. a. op. l., branches of superior artery of optic lobe; cau. r. int. car, caudal ramus of internal carotid; cr. r. int. car., cranial ramus of internal carotid; crbr. oph. a., cerebral ophthalmic; eth. a., ethmoid; inf. a. op. l., inferior artery of optic lobe; inf. crbell. a., inferior cerebellar; int. car. oph. a., internal carotid ophthalmic; mid. crbr. a., middle cerebral; post. crbr. a., posterior cerebral; asc. r. sup. crbell. a., ascending ramus of superior cerebellar artery; com. post. crbr. a., common posterior cerebral artery.

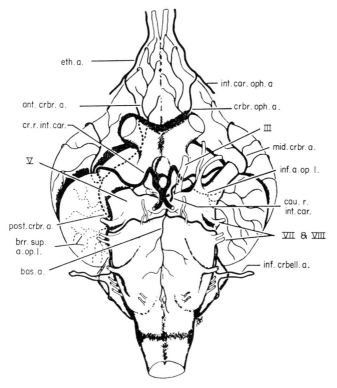

FIG. 3(b)

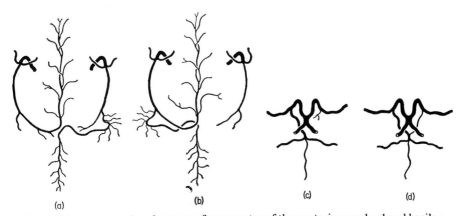

FIG. 4. Diagram showing the types of asymmetry of the posterior cerebral and basilar arteries supplying the pigeon brain (Baumel, 1962).

summarizes the situation in particular genera is contained in Fig. 6. In the few cases where several families from a particular order are represented in the available works there is very little evidence of a homogeneous distribution within the order as a whole. For example, within the Pelecaniformes the genus *Anhinga* has an H-type; *Phalacrocorax* both X and X–H types, and *Pelecanus* an I type. However there seems to be greater uniformity within the Passeriformes. Here representatives of five different families all possess H-type anastomoses although, as noted above, the Tyranni, with their unique, attenuated and anterio-posteriorly compressed hypophysis (Rahn and Painter, 1941; Wingstrand, 1951) seem to lack one altogether.

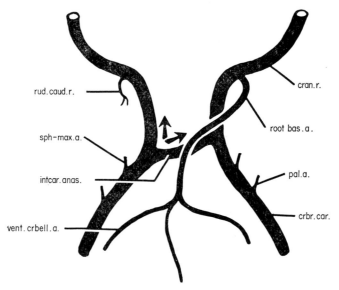

FIG. 5. An overall view of the avian inter-carotid anastomosis and the points of origin of the derivative arteries as seen in *Melopsittacus*. (Baumel and Gerchman, 1968).
Abbreviations: Cran. r., cranial ramus; crbr. car., cerebral carotid artery; intcar. anas., intercarotid anastomosis; pal. a., palatine artery; root bas. a., root basilar artery; rud. caud. r., rudimentary caudal ramus; sph-max. a., sphenomaxillary artery; vent. crbell. a., ventral cerebellar artery.

Baumel and Gerchman drew attention to two apparently unique patterns. In *Anhinga* the right half of the anastomosis is divided into two vessels and these join the right internal carotid independently of one another. Also the left internal ophthalmic artery springs directly from the anastomosis and not from the carotid as its right counterpart does. The second aberrant condition occurs in the trochiliform genus *Calypte*. In this humming bird the anastomosis is clearly doubled. One transverse vessel is situated in the usual post-

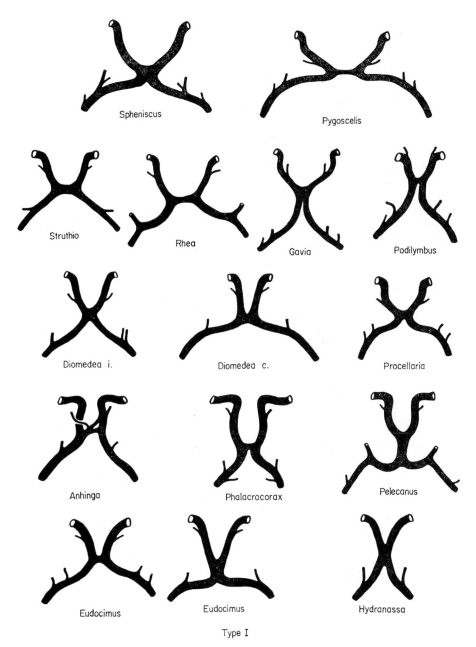

Type I

FIG. 6. The distribution of the three principal types (Types I, II and III) of inter-carotid anastomosis in a variety of genera (Baumel and Gerchman, 1968).

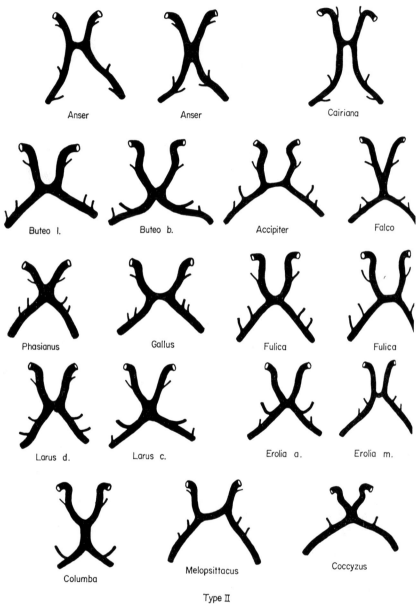

Anser   Anser   Cairiana

Buteo l.   Buteo b.   Accipiter   Falco

Phasianus   Gallus   Fulica   Fulica

Larus d.   Larus c.   Erolia a.   Erolia m.

Columba   Melopsittacus   Coccyzus

Type II

Fig. 6

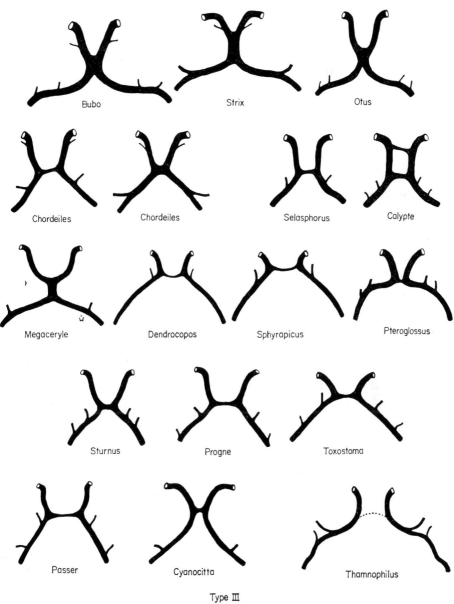

Type Ⅲ

Fig. 6

hypophyseal position but the other lies in front of the hypophysis. Baumel and Gerchman suggested that this unusual rostral component may reflect conditions which are closely analogous to those of the Tyranni. Although no trace of a functional anastomosis is visible in the adults of these birds there is, in all specimens, a vestigial, atrophied canal in front of the pituitary. Attached to the front of each cerebral carotid artery this is surrounded by a venous sinus which is confluent with other large sinuses that also lie in front of the hypophysis (see Section II below).

Barkow (1829) and Wingstrand (1951) provide additional information on both these and other species. Barkow described fine, plexiform, transverse anastomoses in both *Podiceps* and *Anas*. He also found plexiform communications together with a short transverse anastomosis in *Anser*, just a short transverse anastomosis in *Fulica*, *Gallus*, *Falco* and *Corvus*, and a short stem anastomosis in *Strix*, *Columba*, *Ciconia*, *Ardea*, *Cuculus* and *Oedicnemus*. Wingstrand has also described a long transverse anastomosis in passerines other than the Tyranni, and also in *Picus*, *Encognathus*, *Caprimulgus*, *Gallus*, *Sephanoides* and *Apus*. He considered that a lengthy longitudinal anastomosis characterized *Columba*, *Melanitta*, *Anser*, *Larus*, *Strix*, *Diomedea*, *Priocella* and *Procellaria* whilst there was a shorter one in *Pelecanoides* and *Milvagus*.

In order to assess the significance of these and allied structures it is as well to emphasize that birds do not, in general, possess the cerebral arterial circle, or circle of Willis, which is such an important and well-known feature of mammalian brains. This reflects the fact that the basilar artery is not usually formed by a union of both caudal branches of the cerebral carotids. As a result the inter-carotid anastomosis provides an interesting functional analogue of the circle. In at least some species it is large by comparison with the carotid derivatives themselves and may actually provide a somewhat better collateral communication than its mammalian analogue does. Even in the H-type, where the communication is smaller than is the case in I- or X-types, its bore is at least half that of the carotids themselves. The absence of such an anastomosis in the Tyranni is associated with, and compensated for by, the presence of symmetrical roots to the basilar artery. These arise on both sides from the posterior components of the carotids.

## C. The Cranial Branch of the Internal Carotid Artery

The arterial system in the region of the avian brain has a well established nomenclature. It is somewhat unfortunate that this does, however, frequently involve the use of the adjective "cerebral", which is clearly inappropriate in terms of birds. As each internal ($\equiv$ cerebral) carotid ascends out of the sella turcica it passes alongside the hypophysis and pierces the sellar diaphragm. In this way it reaches the hypothalamic region. At this level it divides to give rise to cranial and caudal branches (Baumel, 1967). The cranial branch then runs both forward and sideways, bends dorsally around the junction between optic tract and optic lobe, and gives off three major vessels at the level of the ventral part of the hemispheres. These are the posterior cerebral, medial cerebral and cerebro-ethmoidal arteries. Throughout the 44 species which were studied by Baumel this cranial branch shows one or other of two patterns of branching.

In most species it first of all bifurcates. This results in a backwardly directed posterior cerebral artery, and a forwardly directed common trunk that subsequently gives rise to the mid-cerebral and cerebro-ethmoidal vessels. However in *Diomedea, Columba* and *Melopsittacus*, the albatross, pigeon and budgerigar respectively, the final results of this division are achieved at a single level. Consequently in these genera there is no common trunk interposed between these vessels and the posterior cerebral artery.

## D. The Avian Basilar Artery

The avian basilar artery is usually asymmetrical and represents a prolongation, on one side only, of the posterior branch of the internal carotid. Descriptions of this apparently lop-sided condition include those of Beddard (1905), Hughes (1934–35), Wingstrand (1951), Vitums *et al.* (1965) as well as Baumel (1962). Baumel and Gerchman (1968) only recorded a bilateral and symmetrical origin in isolated specimens of *Procellaria aequinoctialis, Buteo buteo, Accipiter cooperi; Falco sparverius, Fulica americana,* the flycatchers *Sayornis phoebe* and *Elaenia martinica,* the manakin *Pipra mentalis,* and the tree-creeper *Dendrocincla homochroa.* However in a number of other cases there was a paired origin from both caudal rami but nevertheless one or other component was predominant. This was the case, for example, in specimens of the darter *Anhinga anhinga,* the duck *Cairiana moschata,* the coot *Fulica americana,* the toucan *Pteroglossus sanguineus,* the thrasher *Toxostoma rufum* and various lariform, falconiform and strigiform species such as *Larus delawarensis, Falco sparverius, Strix varia, Bubo virginianus* and *Otus asio.*

## E. The Anterior Choroidal Artery

In general the anterior choroidal artery is not a branch of the posterior cerebral but arises from the common trunk of the mid-cerebral and cerebro-ethmoidal vessels. It then runs towards the mid-line and typically exhibits a close association with the caudal part of the forebrain peduncle, lying in the deepest region of the transverse fissure. In most of the remaining species, which comprise one-third of the list studied by Baumel, it is a direct branch of the posterior cerebral. He suggested that in phylogenetic terms it may be in the process of shifting its origin from the cranial branch of the cerebral carotid to a point on the posterior cerebral system.

## F. The Posterior Cerebral Artery

The posterior cerebral artery itself supplies various parts of the fore-, mid- and hind-brain. The individual branches which serve these different regions were reviewed by Baumel (1967). Leaving aside the complications which arise in specimens which have a congenital bipartite condition, the principal branches are as follows:

1. The ventral branch which passes to the telencephalic hemispheres (ventral hemispheric ramus).

2. The branch to the occipital region of the hemispheres (occipital hemispheric ramus).

3. The vessels responsible for supplying the medial region of the hemispheres (inter, and dorsal hemispheric rami).

4. The vessel which runs to the dorsal part of the optic tectum.

5. The supply to the anterior and the dorsal regions of the cerebellum, the neighbouring parts of the tectum and the mid-brain generally (ascending and descending dorsal cerebellar vessels).

6. The vessel to the caudal part of the diencephalon adjacent to the optic lobe (dorsal diencephalic branch).

7. The anterior choroidal, pineal and posterior meningeal branches passing to the choroid plexi of the lateral and third ventricles, the epithalamic region, the pineal body and, finally, the dura mater which covers the occipito-parietal regions.

The major branches of the artery are all markedly asymmetrical and this asymmetry can assume one of five major patterns whose occurrence is summarized in Table 3. More specifically they can be described as follows:

*Type 1.* In this condition the unpaired, common, posterior cerebral artery arises from the posterior cerebral of one side. The unpaired dorsal cerebellar artery arises from its opposite partner and supplies both sides of the cerebellum. The common posterior cerebral sends branches to both hemispheres.

*Type 2.* This is a very skewed pattern in which both the unpaired, common, posterior cerebral artery and also the dorsal cerebellar arise from the posterior cerebral of the same side.

*Type 3.* There are two sub-divisions of this pattern which result in differing degrees of asymmetry.

(a) In some cases the common posterior cerebral arises from the posterior cerebral of one side whilst the dorsal cerebellar arteries are distributed symmetrically.

(b) This is broadly the converse of the former condition. Whilst the inter-hemispheric branches are both paired and symmetrical, the dorsal cerebellar artery is not paired. This condition is met with less frequently than the foregoing one.

*Type 4.* In this case the main branches of the posterior cerebral arteries are symmetrical and have a balanced distribution on the two sides of the brain.

## TABLE 3

*The differential occurrence of the five types of posterior cerebral artery in some avian species* (Baumel, 1967).

| | Type 1 | Type 2 | Type 3a | Type 3b | Type 4 |
|---|---|---|---|---|---|
| Sphenisciformes | | | | | |
| *Pygoscelis adeliae* | 1 | 1 | 1 | | |
| *Spheniscus demersus* | | | 1 | | |
| Struthioniformes | | | | | |
| *Struthio camelus* | 1 | 1 | | | |
| Rheiformes | | | | | |
| *Rhea americana* | | | | | 1 |
| Gaviiformes | | | | | |
| *Gavia immer* | | | 1 | | |
| Podicipitiformes | | | | | |
| *Podilymbus podiceps* | 1 | 1 | | | |
| Procellariiformes | | | | | |
| *Diomedea immutabilis* | | 1 | | | |
| *D. cauta* | | | 1 | | |
| *Fulmaris glacialis* | | | 1 | | |
| *Procellaria aequinoctialis* | | | 1 | | |
| Pelecaniformes | | | | | |
| *Pelecanus occidentalis* | | | 2 | | |
| *Phalacrocorax auritus* | 1 | | 1 | | |
| *Anhinga anhinga* | | | 1 | | |
| Ardeiformes | | | | | |
| *Hydranassa coerulea* | 1 | 1 | 1 | | |
| *Eudocimus albus* | | | 1 | | |
| *Bubulcus ibis* | 1 | | | | |
| Anseriformes | | | | | |
| *Cairiana moschata* | 5 | 2 | | | |
| *Anser anser* | | 1 | | | |
| Falconiformes | | | | | |
| *Falco sparverius* | | | 3 | | |
| *Buteo lineatus* | | | 1 | | 1 |
| *Accipiter cooperi* | | | | | 1 |
| Galliformes | | | | | |

## TABLE 3 (contd.)

|  | Type 1 | Type 2 | Type 3a | Type 3b | Type 4 |
|---|---|---|---|---|---|
| *Phasianus colchicus* |  | 1 |  |  |  |
| *Gallus domesticus* | 2 | 3 |  |  |  |
| Gruiformes |  |  |  |  |  |
| *Fulica americana* | 2 | 2 | 1 |  |  |
| Lariformes |  |  |  |  |  |
| *Larus californicus* |  | 1 |  |  |  |
| *L. delawarensis* |  | 1 |  | 1 |  |
| Psittaciformes |  |  |  |  |  |
| *Melopsittacus undulatus* | 1 |  | 4 |  |  |
| Cuculiformes |  |  |  |  |  |
| *Coccyzus americana* |  |  | 2 |  |  |
| Strigiformes |  |  |  |  |  |
| *Bubo virginianus* |  |  |  | 2 |  |
| *Strix varia* |  |  |  | 1 |  |
| *Otus asio* |  | 1 |  |  |  |
| Caprimulgiformes |  |  |  |  |  |
| *Chordeiles minor* | 4 | 2 |  |  |  |
| Trochiliformes |  |  |  |  |  |
| *Calypte costae* |  |  | 1 |  |  |
| *Selasphorus sasin* |  |  | 1 |  |  |
| Coraciiformes |  |  |  |  |  |
| *Megaceryle alcion* | 2 | 1 |  |  |  |
| Piciformes |  |  |  |  |  |
| *Sphyrapicus varius* |  |  | 1 |  |  |
| *Dendrocopus pubescens* |  |  | 1 |  |  |
| *Pteroglossus sanguineus* |  |  | 5 |  |  |
| Passeriformes |  |  |  |  |  |
| *Cyanocitta cristata* | 1 |  |  | 1 |  |
| *Toxostoma rufum* | 1 | 2 |  |  |  |
| *Progne subis* |  | 1 |  |  |  |
| *Sturnus vulgaris* | 1 |  |  |  |  |
| *Passer domesticus* |  | 1 |  |  |  |

The data which are summarized in Table 3 show the considerable amount of variation which occurs in the appearance of the posterior cerebral artery both within and between species. There is generally a reciprocal relationship between the areas of the telencephalon which are supplied by the posterior and middle cerebral arteries. In certain species the branches of the latter pass to those areas which are normally supplied by the posterior cerebral vessel. This is the case, for example, in the species of *Hydranassa, Bubolcus, Cairiana, Anser, Accipiter, Buteo* and *Larus*. In contrast the other extreme occurs in

*Pygoscelis, Falco, Bubo, Calypte* and *Selasphorus,* where the posterior cerebral vessel achieves its most extensive distribution. A similar reciprocal relationship is also found between those brain fields which are served by the dorsal and ventral cerebellar arteries. In *Anser, Coccyzus,* and all the Falconiformes and Procellariiformes which Baumel studied, the dorsal component is widespread on the dorsal, caudal and lateral cerebellar areas. In *Podilymbus* and *Pygoscelis* the reverse is the case and the ventral component supplies most of the area. In both the humming bird *Calypte* and all the Piciformes the ventral cerebellar artery virtually replaces the dorsal one.

## G. The Anterior Cerebral Artery

The anterior cerebral artery which arises, together with the ethmoidal, by a bifurcation of the cerebro-ethmoidal vessel, is also sometimes asymmetrical. Such asymmetries have been recorded in *Cairiana, Larus, Accipiter, Pteroglossus, Cyanocitta, Toxostoma,* and *Passer* (see Baumel *loc. cit.*) Kappers (1933) reported that the vessels of the two sides can actually anastomose but a number of authors have been unable to confirm this (Hofmann, 1900; Assenmacher, 1953; and Kitoh, 1962). However, more recently Baumel has described a definitive anterior cerebral anastomosis in *Rhea, Larus* and *Chordeiles,* and suggested that it may well be present in other genera.

It is therefore clear from the foregoing sections that a fair amount of variation exists in the cranial arterial system of avian genera. The overall condition differs from that in mammals in a number of ways and is essentially asymmetrical although this asymmetry can assume a number of forms.

## II. THE VENOUS SYSTEM

## A. Introduction

The classical monograph on the venous system of the avian brain is that of Neugebauer (1845). Gadow and Selenka (1891) followed his conclusions in their summary and this was widely accepted and used, for example, by Kaupp (1918) in his textbook. Further descriptions of the conditions which prevail in certain specific regions were provided by Corneliac (1935) and Wingstrand (1951); a developmental study of the fowl is contained in Hughes (1935) who did not, however, consider the adult condition, although in recent years this has been described by Richards (1968). There is no recent comparative study comparable with those on the arterial system. Neugebauer's results form the basis for all subsequent work. He concluded that the encephalic veins communicate with the general venous system by two pairs of ophthalmic veins, which lie in front of and behind the chiasma, and a pair of carotid

veins which surround the internal carotid arteries. He also described a series
of major venous sinuses within the dura mater. Briefly these were the longitudinal, occipital and occipito-foraminal, transverse semi-circular, petro-
sphenoidal, temporo-sphenoidal, basilo-annular and cerebellar sinuses.
Posteriorly a pair of occipital veins led out from the occipito-foraminal sinus.

Hughes summarized his embryological studies in terms of ventro-lateral
and mid-dorsal longitudinal veins. These are collective terms and under the
former heading, for example, he included the *v.v. capitis, medialis* and *lateralis*,
the primary head vein of Sabin, and both the primary and secondary stem
veins of van Gelderen. The discrepancies which exist between this ontogenetic account, that of Neugebauer (1845) on the turkey and that of Corneliac (1935) on the pigeon and chicken, are difficult to resolve. In view of the
variability of certain areas of the venous system within a single species
(Richards, 1968) and the specific variations of the arterial system (see Section
I) they may reflect real differences. Richards considered that the transverse
sinus in Neugebauer's account is possibly the anterior cerebral vein of Hughes
and seems to correspond to the cerebral anastomotic vein of Corneliac.

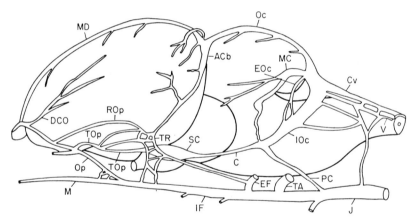

Fig. 7. A lateral view of the venous system draining the brain in the hen (Richards,
1968).

Abbreviations for Figs 7 and 8: ACb, anterior cerebral vein; C, carotid vein; Cv,
cervical sinus; DCO, dorsal cerebral ophthalmic vein; E, ethmoid vein; EF, external
facial vein; EOc, external occipital vein; IF, internal facial vein; IM, internal
mandibular vein; IOc, internal occipital vein; IOP, internal ophthalmic vein; J,
jugular vein; L, lingual vein, LOc, lateral occipital vein; M, maxillary vein; MC,
middle cerebral vein; MD, mid-dorsal sinus; MOc, median occipital vein; MOp,
median ophthalmic vein; Oc, occipital sinus; Op, ophthalmic vein; PC, posterior
cephalic vein; Pt, palatine vein; ROp, recurrent ophthalmic vein; Sc, sinus cavernosus; Sph, sphenomaxillary vein; TA, transverse anastomosis; TOp, temporal
ophthalmic vein; TR, temporal rete; V, vertebral vein; VCO, ventral cerebral
ophthalmic vein.

Again the vessel which Corneliac shows as an anterior extension of the peri-hypophyseal plexus in the pigeon is apparently mis-labelled the occipital vein and corresponds to the ophthalmo-temporal branch of Neugebauer's ophthalmic vein.

In his detailed account of the conditions in the fowl Richards (1968) laid emphasis on the dorso-ventral and cerebro-extracranial anastomoses. He demonstrated the complexity of the ophthalmic and occipital connections with vessels outside the cranium, and also that the principal outlet for blood from the cavernous sinus (Fig. 7) is provided by the carotid veins. The dorso-lateral evaginations of this sinus, which give off the internal ophthalmic veins, were considered by Wingstrand to represent the anterior part of Neugebauer's basilar venous ring. Unable to locate a significant area of anastomosis between this peri-chiasmatic ring and the brain sinuses he concluded that most of the blood from them must leave the cranium by way of the occipital region. The vertebral veins would therefore assume a far greater importance than is the case in mammals. Richards (1968) confirmed that this is certainly the case but showed the existence of several potential channels to the jugular veins. An indirect communication between both these and the brain sinuses exists in the temporal rete. He concluded that in general the sinuses of the fowl converge dorsally upon the prominent cervical sinus and that the occipital veins provide an extensive communication between this sinus, the vertebral veins dorsally and the jugular veins ventrally.

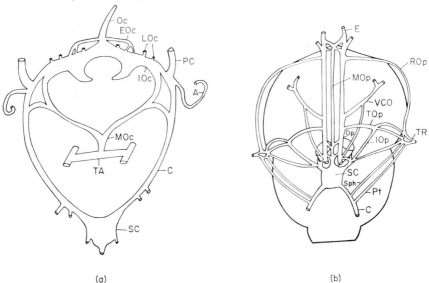

(a)                               (b)

FIG. 8. (a) Dorsal view of the occipital venous system; (b) ventral view of the cerebral-ophthalmic anastomotic veins in the fowl (Richards, 1968). Abbreviations as for Fig. 7.

**2***

The dorsal and ventral sinuses, together with the temporal rete and the extra-cranial veins are united by a complex ophthalmic system. The vertebral arteries provide the major venous outflow and the jugular veins are principally concerned with extra-cranial blood. These last are derived during development from superficially situated branches of the facial veins and that of the right side is substantially larger than that on the left. As in mammals the carotid veins envelop the internal carotid arteries and give rise to a bulbous cavernous sinus around the inter-carotid anastomosis.

## B. The Posterior Cephalic Vein

A semi-diagrammatic representation of the principal head veins and brain sinuses is provided by Fig. 7. The posterior cephalic is a major division of the jugular and emerges immediately behind the transverse anastomosis. This inter-jugular connecting vessel receives numerous highly variable collic veins from behind, and a complex pattern of veins from in front. The posterior cephalic receives much of the blood from the ventral brain sinuses and has three principal branches. These are the internal occipital, auricular and carotid veins. The first of these enters it from its dorsal position where the complex occipital system collects blood from the dorsal brain sinuses and passes it to the vertebral veins. The auricular vein enters the posterior cephalic at the same point as the internal occipital. Draining the inner ear through a series of fine labyrinthine vessels it maintains some connections via this network with the external facial vein. This last is a component of the anterior cephalic system and has cutaneous, temporal and palpebral contributions.

As in mammals the carotid artery is completely surrounded by the carotid vein which comprises the third principal tributary of the posterior cephalic. The carotid veins of each side, together with the sinus cavernosus which envelops the inter-carotid anastomosis, assume the characteristic H shape etc. which typifies the arterial system within the hypophysial region. The carotid veins themselves therefore comprise the principal outlet for the blood which is contained within the sinus since its connections with the orbital veins are both complex and variable.

## C. The Brain Sinuses

A convenient analysis of the intra-cranial venous system of fowl was suggested by Richards (1968) to consist of (i) the brain sinuses themselves; (ii) the occipital venous system and (iii) the ophthalmic venous system. The first of these three entities has itself three principal components which are each accompanied by various tributaries. The *mid-dorsal sinus* is a delicate vessel receiving branches from the olfactory area and communicating with the anterior part of the ophthalmic system. It drains the dorso-frontal area of the forebrain hemispheres and the choroid plexus by way of small, posterio-lateral branches. The *occipital sinus* runs above the cerebellum (see Fig. 7) is broad laterally and flattened in a dorso-ventral plane. Receiving two import-ant tributaries from the lateral surfaces of the cerebellum it also receives small and irregular side-branches from the dorsal regions. The first of the major tributaries is the anterior cerebral vein which drains the hemispheres via small anterior vessels and also receives blood from the area intervening between the forebrain and cerebellum. At least one branch, and usually the main trunk, enters the temporal rete thereby establishing contact with the extra-cranial venous return. The second tributary is the middle cerebral vein together with its branches. Entering the occipital sinus close to its hind end it drains the lateral cerebellar and floccular regions together with the posterior aspect of the tectum. Terminal branches anastomose with the carotid vein and labyrinth plexus as well as the auricular vein which was mentioned in associa-tion with posterior cephalic.

The third major sinus is the *cervical*. This is a posterior prolongation of the occipital sinus and arches over the spinal cord. As such it forms a series of connections between the vertebral veins of each side in the form of numerous lateral connecting vessels. Its overall location is again depicted in Fig. 7 which shows its relationships when viewed from the side.

## D. The Ophthalmic Venous System

The dorsal and ventral sinuses, the temporal rete and thereby the extra-cranial veins are united by a complex ophthalmic system. The predominant path of communication between the sinus cavernosus and the orbital veins is a vessel which passes lateral to the optic nerve and joins the temporal ophthalmic vein at its point of bifurcation. This vessel is the internal ophthalmic vein and it is closely associated with the artery of the same name. Indeed the basal region of the artery appears to be enveloped by the vein. Six other ophthalmic veins exist (Fig. 8b). The temporal ophthalmic is the first major branch of the ophthalmic ramus of the internal facial vein. It runs over the optic nerve behind and, dividing, gives rise to two vessels of more or less equal size. These both undergo extensive anastomoses with the temporal rete and also maintain contact with the brain circulation by way of the anterior cerebral vessel. The second major branch of the ophthalmic ramus of the internal facial vein is the recurrent ophthalmic. Posteriorly it joins with the more dorsal of the two divisions of the temporal ophthalmic. During its passage between the two, several small and irregular vessels enter it from the lateral regions of the brain.

A dorsal cerebral ophthalmic vein runs backwards from the junction between the mid-dorsal sinus and the ophthalmic branch of the internal facial. Distributed to the anterior and dorsal parts of the forebrain hemispheres it contributes by way of numerous tributaries to both the deep and superficial layers. A median ophthalmic vessel represents an anterior extension of the sinus cavernosus. Coming from the olfactory region it unites with the basal parts of both ethmoid veins and then passes backwards between the optic nerves to link up with the sinus. The ethmoids themselves drain the frontal regions and join the ophthalmic circulation at a common junction with the mid-dorsal sinus, the dorsal cerebral ophthalmic and the ophthalmic branch from the internal facial. In some 60% of his specimens Richards (1968) also detected a further ventral cerebral ophthalmic component which drained the anterio-ventral region, received a branch from the maxillary ramus of the internal facial, and joined with the ophthalmic ramus at a location more or less opposite to the point of entry on the temporal ophthalmic.

## E. The Occipital Venous System

The plexus which is located in the occipital region is mainly concerned with the inter-connections between the occipital branch of the posterior cephalic vein and the various components of the occipital and cervical sinuses; the origin and anastomoses of the vertebral veins; and the direct communications between the occipito-vertebral plexus and the jugular veins (see Fig. 8a). A large internal occipital vessel joins the ventral part of the occipital sinus to the posterior cephalic whilst several smaller branches enter the lateral occipital vein and there are two points of anastomosis with the external occipital.

Close to the base of the middle cerebral vein a small external occipital vessel loops ventrally to connect with the internal occipital by a number of branches. These last enter it at points opposite to the exits of the lateral occipital and vertebral veins. The lateral occipital itself arises from the internal occipital by a number of tributaries which are situated midway between the origin of the latter from the occipital sinus and its point of union with the posterior cephalic vein. Passing backwards for some way the lateral occipital gives off one branch to the ipsilateral vertebral vein and then divides into two. The median branch joins the vertebral vein whilst the lateral branch links up with the main trunk of the jugular just behind the point at which the posterior cephalic vein enters it. The median occipital is a vessel which arises from the mid-point of the inter-jugular anastomosis and then runs upwards for 5 mm prior to bifurcating. From this point of bifurcation branches of comparable size pass upwards to join the median aspect of the internal occipital vein almost opposite the site at which the external occipital enters it.

A small forwardly directed branch from the lateral component forms a direct connection between the median occipital and carotid veins.

# III. THE NEUROGLIA

## A. General Considerations

In view of the recent heightened interest in glial tissue, and the possible implication of certain components in the transport of nutritive material from the blood vessels to the brain cells, it is useful to include here a brief summary of the main literature. Although recent workers have not been using avian material to any great extent, neuroglial cells of all vertebrates have a number of characters in common. Fox (1964) has provided an extensive survey and discusses the confusion that had previously existed about the site of the blood-brain barrier, the size of the extra-cellular space and the status of both oedema and intra-cranial pressure. He suggested that the endothelial and glial cell membranes may be so close together that any pores which exist within them are obstructed. Diffusion may also be prevented by the molecules of other membranes, such as the sucker-feet of astrocytes. Stated tersely this amounts to the suggestion that whilst one membrane is permeable on its own, if two are closely apposed or adherent to one another, this permeability is impeded.

In fact the concepts which attempt to explain the function, or functions, of the glial cells were largely developed by speculations at the turn of the century. Their relative credibility and usefulness has however been extended by recent work. Such hypotheses are now largely concerned with the function of fibrous and protoplasmic astrocytes. Since oligodendrocytes, which are similar in both birds and mammals (Cemmermeyer, 1963), are implicated, together with the Schwann cells, in myelin formation they clearly have an important role in facilitating saltatory conduction.

The earliest idea, that astrocytes provide mechanical support within the nervous system, is probably a fair comment on one, possibly incidental, function of all glial components. The apparent ability of glial cells to continue to divide throughout life also implicates them in injury repair, and, following any neuronal disappearance, whether this is the result of injury or ageing, their place is taken by glial cells. This led Ramon y Cajal (1928) to postulate a dynamic inter-relationship between neurons and glial cells. Recently the use of electron microscopy has amplified our knowledge of the ability of both glial cells and Schwann cells to phagocytose parts of degenerating neurons. In the embryo, however, the normal neuronal development is apparently independent of that of neighbouring glial cells. Indeed both synaptic contact and axonal growth take place prior to the axons being surrounded by either glial cells or Schwann cells. Ramon y Cajal also suggested that glial components were necessary to prevent unwanted neuronal interaction. Although any such interaction seems unlikely today nevertheless the close association between astrocytes and synapses has been interpreted as providing a barrier to the spread of chemical transmitters.

The best known and longest enduring suggestion about glial function envisages that they fulfil a nutritive role. It was originally put forward by Golgi (1903). He drew attention to the fact that within the central nervous system the glial cells tend to occupy positions between the blood vessels and the neurons. Capillaries can in fact be practically covered by glial processes and, as these also have a close spatial relationship with neurones, he assumed that they were involved in the transfer of substances to these latter cells. In view of the immense histological documentation which he produced, this concept was widely accepted prior to the existence of any shred of experimental verification. Clearly all these suggestions lead one to infer special properties for the cells. The nutritive function pre-supposes special transport systems across the glial cell membranes. It would also suggest that any intercellular channels which one may observe are either of relatively little significance or, perhaps, closed. Hypotheses, such as those of Fox (1964), which were mentioned above and implicate the glial tissue in the blood-brain barrier, also imply the existence of properties that selectively prevent certain substances from reaching the neurons. The information which is at present available on these topics is summarized below. Some idea of the complexity of the avian glial tissue is provided by the description of the vascular organ of the lamina terminalis. Dellman (1964) described this as comprising four segments which are known respectively as the pre-chiasmatic, sub-commissural, pre-commissural and supra-commissural. Although there are only a few nerve cells, parenchyma cells are widespread and recognizable by their clear cytoplasm. Microglia is found predominantly in the outer zones of the organ and macroglia is largely represented by bipolar protoplasmic astrocytes. All such glial components were invariably found in the close vicinity of blood vessels.

## B. The Fine Structure of Glial Cells

Data exist which relate to the fine structure of astrocytes, oligodendroglia and glio-ependymal cells. As originally described, the fine structure of the first-named type suggested that, although branched, they possessed very little cytoplasmic differentiation. These rather negative results were subsequently found to reflect the method of preparation (Magnaini and Wahlberg, 1964). With improved methods of fixation various organelles were detectable and it was possible to differentiate between protoplasmic and fibrous varieties. Details of all these and other glial components were given by Glees and Meller (1968). Granular endoplasmic reticulum is scarce. There is a large number of free ribosomes; the Golgi complex is well-developed; osmiophilic bodies are numerous, but the mitochondria are of relatively small size. The vascular processes in which many of the osmiophilic bodies lie also contain glycogen granules and the presence of 60–100 Å thick filaments within both the perinuclear cytoplasm and the processes is typical of astrocytes.

Glees and Meller concluded that, in general, oligodendrocytes are more electron dense than the astrocytes. They have a predominantly round nucleus with a constant nucleolus. The ribosomes can be both free or attached to the endoplasmic reticulum, and the number of mitochondria varies. As the cells characteristically envelope the myelin sheath of medullated axons a direct contact with the vascular bed is rare. Other organelles which are found within them include dense bodies, multivesicular bodies, rods and granules. We have made some investigations of such structures elsewhere in the animal kingdom which show their enzymatic activities.

Finally some cells, such as the Muller cells in the eye (q.v.), are both ependymal and glial. The Muller cells stretch from the external to the internal limiting membrane of the retina and indeed may actually form these membranes, the internal by "end-feet" and the external by desmosomal thickenings in contact with receptor cells.

## C. The Activity of Glial Cells

When they are grown in tissue culture vertebrate glial cells have been reported to exhibit a resting potential of between 50 and 90 millivolts. Passing a depolarizing current through the cells whilst searching for impulse-like activity showed that even with 200 millivolt displacements of the trans-membrane potential glial cells remained passive. This clearly represents a basic difference from neurons. Furthermore current does not appear to undergo any significant spread from neurons to glial cells or vice versa. Even if the potential in one is changed by tens of millivolts that of the other remains unaffected. Most of the current flowing across the membranes is dispersed through the low resistance clefts which separate glial from nervous cells and this contrasts very markedly with the ease of current spread between individual glial cells which are inter-linked by tight junctions.

By using the membranes of glial cells and neurons as indicators of their ionic environment it was found that the inter-cellular clefts within the nervous system provide a rapid and effective pathway for the diffusion of such ions as $Na^+$ and $K^+$, and molecules such as sucrose, inulin, dextran and choline. The principal pathway for these substances would therefore seem to be around, rather than through, the glial components. However impulse activity in the non-myelinated axons of the optic nerves does appear to depolarize glial cells. Kuffler (1967) reported evidence for the release of $K^+$

which then accumulates within the intercellular spaces and leads to a reduction in the glial membrane potential. In such experimental situations the potassium ion concentration was estimated to undergo an increase from the normal 3, to 20 mequiv. per litre or more. Glial cell membrane depolarization which is analogous to this also seems to take place *in vivo* when the optic nerves are stimulated by incident light. The consequences of such $K^+$ fluctuations within the brain were discussed by Kuffler. The actual effect of $K^+$ depolarization on glial cells is not known but he suggested that as their membranes are good indicators of the local neuronal activity $K^+$ liberation may provide a signal for subsequent glial cell responses—possibly trophic in function. As the activation of groups of glial cells leads to current flow and to slow potential changes on the surface of the optic nerves, Kuffler also suggested that similar effects are likely to occur within the brain. He justifiably concluded from all this that it is probable that the ultimate conclusion will implicate glial structures in a variety of functions (see Chapter 3). It is also worth noting that Wardell (1966), who also investigated the glial response experimentally, concluded that it is not due to a regenerative, depolarization-activated increase of sodium permeability, but that it is mainly the result of an actual mechanical breakdown of the cell-membrane and can be produced by dielectric breakdown. He suggested that the "glial response" is in fact a general cellular property representing the behaviour of an electrically passive cell membrane under conditions of stress.

# References

Assenmacher, I. (1953). Étude anatomique du système artériel cervicocephalique chez les oiseaux. *Arch. Anat. Histol. Embryol.* **35**, 181–202.

Barkow, H. (1829). Anatomisch-physiologisch Untersuchungen vörzüglich uber das Schlagadersystem der Vögel. *Arch. f. Anat. Physiol.* **4**, 305–496.

Baumel, J. J. (1962). Asymmetry of encephalic arteries in the pigeon (*Columba livia*). *Anat. Anz.* **111**, 91–102.

Baumel, J. J. (1967). The characteristic asymmetrical distribution of the posterior cerebral artery of birds. *Acta Anat.* **67**, 523–549.

Baumel, J. J. and Gerchman, L. (1968). The avian inter-carotid anastomosis. *Amer. J. Anat.* **122**, 1–18.

Beddard, F. E. (1905). A contribution to the knowledge of the arteries of the brain of the class Aves. *Proc. Zool. Soc. London* **1**, 102–117.

Carss, E. (1895). "The brain of the turkey Meleagris gallopavo." Thesis, Cornell University.

Cemmermeyer, J. (1963). Similarities between oligodendrocyte and cerebellar ganglia cell nuclei in mammals and Aves. *Amer. J. Anat.* **112**, 111–139.

Corneliac, E. V. C. (1935). "Les rélations de la circulation labrinthique avec les circulations de l'hypophyse et de l'épiphyse." Iasi. Institutul de arte grafice.

Craigie, E. H. (1940). The area of distribution of the middle cerebral artery in certain birds. *Trans. Roy. Soc. Canada.* **34**, 25–37.

Dellman, H. D. (1964). Structure of the vascular organ of the lamina terminalis of the chicken. *Anat. Anz.* **115**, 174–183.

Fox, J. L. (1964). Development of recent thoughts on intra-cranial pressure and the blood-brain barrier. *J. Neurosurg.* **21**, 909–967.

Gadow, H. and Selenka, E. (1891). Vögel. *Bronn's Klass. Ordn. Tierreichs* **6**, 1–1008.

Gillilan, L. A. (1967). A comparative study of the extrinsic and intrinsic arterial blood supply to the brains of sub-mammalian vertebrates. *J. Comp. Neurol.* **130**, 175–196.

Glees, P. and Meller, K. (1968). Morphology of neuroglia. *In* "The Structure and Function of Nervous Tissue" (Bourne, G. H., ed.), pp. 301–325. Academic Press, London and New York.

Golgi, C. (1903). "Opera omnia", vol. **2**. Hoepli, Milan.

Hofmann, M. (1900). Zur vergleichenden Anatomie der Gehirn and Ruckenmark-sarterien der Vertebraten. *Z. Morph. Anthropol.* **2**, 247–322.

Hughes, A. F. W. (1934–35). On the development of the blood vessels in the head of the chick. *Phil. Trans. Roy. Soc. B.* **224**, 75–129.

Kaku, K. (1959). On the vascular supply in the brain of the fowl. *Fukuoka Acta med.* **50**, 4293–4306 (in Japanese).

Kappers, C. U. A. (1933). The forebrain arteries in plagiostomes, reptiles, birds and monotremes. *Proc. Kon. Akad. Wetensch, Amsterdam.* **36**, 52–62.

Kaupp, B. F. (1918). "The Anatomy of the Domestic Fowl." Saunders, Philadelphia.

Kitoh, J. (1962). Comparative and topographical anatomy of the fowl. XII Observations on the arteries and their anastomoses in and around the brain. *Jap. J. Vet. Sci.* **24**, 141–150 (In Japanese).

Klinckowström, A. (1890). Quelques récherches morphologiques sur les artères du cerveau des vertebrés. *Bih. Sven. Vet. Akad. Handl.* **15**, 1–26.

Kuffler, S. W. (1967). Neuroglial cells; physiological properties and a potassium mediated effect of neuronal activity on the glial membrane. *Proc. Roy. Soc. B.* **168**, 1–21.

Magnaini, E. and Wahlberg, F. (1964). Ultrastructure of neuroglia. *Ergebn. Anat. Entwick. Gesch.* **37**, 193–236.

Neugebauer, L. A. (1845). Systema venosum avium. *Nova Acta Leop. Carol.* **21**, 517–697.

Rahn, H. and Painter, B. T. (1941). A comparative histology of the bird pituitary. *Anat. Rec.* **79**, 297–312.

Ramon y Cajal, S. (1928). "Degeneration and Regeneration in the Nervous System." Oxford University Press.

Richards, S. A. (1968). Anatomy of the veins of the head in the domestic fowl. *J. Zool., Lond.* **154**, 223–234.

Shiina, J. and Miyata, D. (1932). Studies on the cerebral arteries of birds. I. The arterial supply on the brain surface of some kinds of birds. *Acta. anat. nippon.* **5**, 13–28 (In Japanese).

Stoeckenius, M. (1964). Die Kapillarisierung verschiedener Vögelgehirne. *Gegenbauers Morph. Jb.* **105**, 343–364.

Wardell, W. M. (1966). Electrical and pharmacological properties of mammalian neuroglial cells in tissue culture. *Proc. Roy. Soc. B.* **165**, 326–361.

Vitums, A., Michami, S. and Farner, D. S. (1965). Arterial blood supply to the brain of the White Crowned Sparrow. *Anat. Anz.* **116**, 309–326.

Wingstrand, K. G. (1951). "The Structure and Development of the Avian Pituitary." Gleerup. Lund.

# 3    The Biochemistry of the Avian Brain

## I. INTRODUCTION

A general review of the biochemistry of vertebrate brains was provided by Bartley *et al.* (1968). In general those data which are available for avian brains support their conclusions so that we may largely limit ourselves to the specifically avian information. The brain of homeotherms is exceedingly sensitive to any changes in the concentration of oxygen in the blood. Although there are possibly no marked overall variations in the oxygen consumption of a given brain in association with differing electro-encephalographic activity, no detailed studies of this problem have been carried out in birds. However, as mammalian cortical slices exhibit changes from a basal metabolic rate of 55 $\mu$ moles $O_2$/g/hour at rest, to 110 $\mu$ moles during periods of electrical stimulation, it is probable that comparable local changes do take place in avian brains and reflect the differential involvement of the various brain areas in a particular activity. A study of the metabolic changes which accompany the gradual morphological differentiation of the embryonic optic lobes in the chick (Gayet and Bonichon, 1961) certainly showed that such variations occur during development. As different topographical regions develop and become active in a definite temporo-spatial sequence, it is clear that marked changes must exist during this period. In the optic tectum they observed progressive increases in the levels of both oxygen consumption and succinic oxidase activity. There was a nineteen-fold increase in the oxygen uptake up to the moment of hatching and this was strikingly paralleled by the tissue nitrogen levels.

When studying the overall proton and electron transfer complexes Inouye *et al.* (1966) concluded that the electron transfer scheme which is depicted in Fig. 9 is applicable to avian brains. This is not particularly surprising because, although it was originally proposed for other organisms, we now know that the unique electronic constitution of the various components of the cytochrome series are fundamental to such activity. In FMN–FMNH$_2$ and NAD–NADH the oxidation-reduction activity is accompanied by an instantaneous redistribution of the energies of the molecular orbitals, and, in particular, those of the lowest empty, and highest filled orbitals. This occurs in such a way that in each case a particularly low-lying empty orbital is associated with the oxidized form, and a particularly high-lying filled orbital with the reduced form. The oxidized form has therefore got a natural inherent

38

tendency to accept electrons, and the reduced form to surrender them. Moreover this is associated with a particularly unusual characteristic of $FMNH_2$ in which the highest filled molecular orbital is an anti-bonding one. The sign of its energy coefficient is that which is generally associated with orbitals that are only occupied in an excited state. The occupation of such an orbital in a ground state, is, therefore, not unexpectedly a very unstable

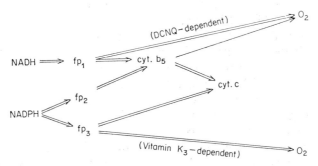

FIG. 9. A summary of the electron transfer scheme suggested by the microsome data of Inouye *et al.* (1966). DCNQ = 2,3-dichloro-1,4-naphthoquinone; fp = flavoprotein.

arrangement. $FMNH_2$ has therefore got a very strong inherent tendency to expel electrons which are located in this orbital. Such apparently unique orbital characteristics account for both its outstanding electron donor properties, and also the fact that it is particularly prone to auto-oxidization. This is not the case in the molecule of NADH which needs a system of higher potential to re-oxidize it. In nature this system is provided by a flavoprotein. For a further discussion of these orbital contributions to electron transfer within biological systems we can refer to Pullman and Pullman (1963). Inouye and his collaborators drew their conclusions following experiments which demonstrated the presence in the avian brain of a number of critical substances and activities. These may be summarized seriatim:

1. Cytochrome $b_5$;
2. NADH-, and NADPH-cytochrome *c* reductase activity which was mediated by cytochrome $b_5$;
3. Dichloronaphthoquinone-dependant NADH oxidase activity;
4. NADPH cytochrome *c* reductase which they assumed was identical with the napthoquinone-dependant NADPH oxidase in brain microsomes.

In more specific terms enzymatic activity was identified and compared with that in other vertebrates by Masai and Matsano (1961). They found that high levels of cytochrome oxidase were located in the avian choroid plexus. In the olfactory bulb the maximum activity for various oxidative enzymes was located in the glomeruli and there was less activity elsewhere. Clusters of granule cells lying between the synaptic glomeruli also show very low levels of oxidative enzyme activity so that there is a pattern which is strongly reminiscent of that which they and other workers observed in the cerebellar granular layer. Nevertheless, they concluded that there is no consistent, or at least easily detectable, correlation between cell density and the level of enzyme activity. A more overt topographical pattern emerged from the older work of Verzhbinskaya (1949) who investigated the distribution of carbonic anhydrase. Here the levels of activity increased in a roughly caudo–cranial order. There was a slight increase along the spinal cord and then successively higher levels in the medulla, pons and mesencephalon. The telencephalic hemispheres exhibited a level which was slightly less than that of the cerebellum. It is interesting that this appears to contrast rather strongly with the situation in the hog amongst mammals since in that species the levels of cerebral activity exceeded those observed in the cerebellum.

An indirect indication of the possible sites of massive, or chemically specific, enzyme activity within the hen brain was recently provided by Howell and Edington (1968). They administered sodium diethyldithiocarbamate to experimental birds and observed that lesions subsequently appeared within the cerebellum, medulla and spinal cord. Histologically these lesions comprised a marked degeneration of nerve fibres with a well-defined distribution. For example in the medulla one such tract occurred at the periphery and ran from there into the cerebellum. Since diethyldithiocarbamate inhibits many enzymes, particularly those which require sulphydryl groups for their activity, Howell and Edington concluded, on *a priori* grounds, that such inhibition had played a part in the production of the lesions. However, the list of enzymes which are inhibited in this way during *in vitro* experiments includes both succinic dehydrogenase and monoamine oxidase and it is of particular interest, if somewhat puzzling, that the areas in which large quantities of this last are known to occur were *not* affected. Clearly the exact mode of action and the initial site of inhibition remain to be established.

## II. FRACTIONATION OF BRAIN TISSUE

As a direct result of the marked heterogeneity of the basic cytological structure of brains the products of homogenization involve a somewhat wider range of particles than is the case in, say, liver. The presence of a large number of synaptic vesicles within neurons, coupled with the presence of the filamentous structures of glial cells, etc., renders the job of producing pure fractions rather difficult. To a certain extent this difficulty can be circumvented by using density gradient centrifugation. In this technique the cell free homogenate which is produced by low speed centrifugation is layered on to the top of a sucrose density gradient in a centrifuge tube. Subsequent high speed centrifugation distributes the particles as bands within this gradient. Each of these bands then represents a population of sub-cellular particles with the same density as the sucrose solution at that level. The plastic centrifuge tube is then either serially sectioned by a special cutter that automatically seals the bottom of each "slice" of liquid allowing it to be removed before the next section is cut, or, alternatively, the tube is pierced and the material from each band collected separately. A scheme for the isolation of different fractions using differential centrifugation is shown in Fig. 10.

Within such fractions, however prepared, Wahbe *et al.* (1961) concluded that the overall pattern of occurrence of mitochondria is broadly common to all vertebrates. The mitochondrial weight lies in the region of $1.69 \times 10^{-12}$ g wet weight, or $0.27 \times 10^{-12}$ g dry weight. The protein content is approximately

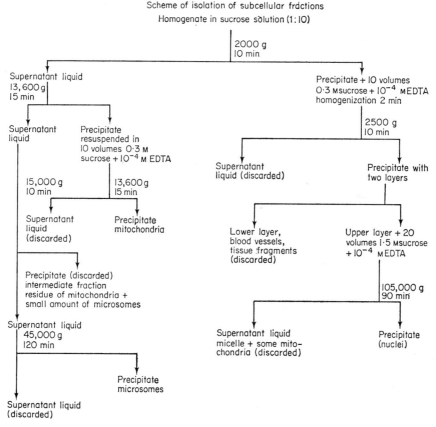

FIG. 10. A summary of some procedures for obtaining material for biochemical assays (Kreps *et al.*, 1964).

$0.11 \times 10^{-12}$ g. In the case of bird brains the total number of mitochondria in the M fraction after high speed centrifugation is $22.5 \times 10^{10}$ per g of tissue. This number is less than that which was obtained for the rat, and it far exceeded the values for both frog and turtle. Although this is suggestive it would be unwise in the present state of our knowledge to extrapolate from these results, which were obtained from some four species, and conclude that the mitochondrial concentration in this fraction is on the one hand homogeneous for all avian genera and on the other intermediate between that of reptiles and mammals. However, it is of importance to note from the functional point of view that a great deal of the hexokinase activity of brain material is always associated with such mitochondrial preparations whilst, using density gradient techniques, much of the acetyl choline, nor-adrenalin and 5-hydroxytryptamine activity can be shown to occur in a band which sediments above that which is enriched with mitochondria.

## III. THE BIOCHEMICAL ASPECTS OF BRAIN STRUCTURE

Once again it is important to emphasize that the internal heterogeneity of avian brains renders it most unlikely that the underlying sub-cellular biochemical composition is homogeneous. However, as Willmer (1960) has pointed out, tissues which are derived from similar germ layers frequently show certain similar biochemical tendencies. In the case of neural tube and

### TABLE 4

*The chemical composition of some bird brains in* mg/100 g *fresh weight.*

|                        | Chicken   | Duck     | Gull | Pigeon   |
|------------------------|-----------|----------|------|----------|
| Water (%)              | 79        | 78       | 81   | 79       |
| Total nitrogen (%)     | 1·1–1·2   | 1·7      | —    | —        |
| Total protein (%)      | 7·2–7·6   | 10·7     | —    | 45–50*   |
| Total lipid (%)        | 9·2–10·2  | 9·4      | —    | 4·2–5·4  |
| Phospholipid           | 4·2       | 3·7      | 4    | 4·4      |
| Cephalin (%)           |           |          | 2·3  | 2·4      |
| Lecithin               |           |          | 1·3  | 1·3      |
| Sphingomyelin (%)      |           |          | 0·4  | 0·7      |
| Cerebroside            |           |          | 1·4  | 1·3      |
| Cholesterol            | 1·6–2·0   | 1·8      | 1·3  | 1·4–2·0  |

* = dry weight. After Spector (1956).

neural crest material he drew attention to the predominance of tyrosin metabolism in one of several forms. Nevertheless in terms of the whole brain such gross similarities must be contrasted with the particular local topographical idiosyncracies. Unfortunately our data on these are as yet still very patchy and of varying degrees of credibility. A synopsis of the long-established gross composition of the brains of chickens, ducks, a species of gull and pigeons, expressed as mg/100 g fresh weight, is contained in Table 4. The lipid components such as sphingomyelins and cerebrosides are of a rather special nature and, in white matter at least, the sphingomyelins have a structural rather than a "biochemical" role.

## TABLE 5

*The composition of the embryonic avian brain.*

| Age in days | Wet weight (mg) | Lipid (mg) | Lipid phosphorus ($\mu$g) | Cerebrosides (mg) | Mucolipid (mg) |
|---|---|---|---|---|---|
| 5 | 38·6 | 0·58 | — | 0·031 | 0·016 |
| 8 | 136·8 | 2·31 | 62 | 0·132 | 0·189 |
| 9 | 246 | 3·78 | — | — | 0·316 |
| 10 | 167 | 3·92 | 108 | 0·259 | 0·247 |
| 13 | 339 | 8·57 | 231 | 0·615 | 0·627 |
| 15 | 557 | 12·9 | 352 | 1·032 | 1·125 |
| 18 | 794 | 25·2 | 373 | 1·640 | 2·045 |
| 20 | 860 | 37·9 | 980 | 2·690 | 3·59 |
| 20·5 | 803 | 34·5 | 933 | 3·84 | 1·86 |
| 22·5 | 814 | 40·1 | 1057 | 4·18 | 2·59 |

Garrigan and Chargaff (1963).

Comparable analytical data which refer to the composition of the brain during embryological development were provided by Garrigan and Chargaff (1963). These are summarized in Table 5 which demonstrates very clearly the gradual ontogenetic increase in the quantity of lipids, lipid phosphorus, cerebrosides and mucolipids that occurs during the pre-hatching period. These values are, however, not always easy to equate with adult values and a drop in the mucolipid content, at least, certainly appears to occur just prior to hatching. The actual distribution of such substances within the brain has been the subject of some further study but the difficulties which are encountered are illustrated, for example, by the fact that Singh (1964) found that intra-neuronal phospholipids can only be demonstrated by the controlled chromation method. Under these circumstances they appear as granular bodies in the cell perikaryon. However, in the case of *in vitro* studies Kreps *et al.* (1964), using chromatography on silicic acid impregnated paper, have extended our information on such subjects to the level of sub-cellular particles.

Using the centrifugation scheme which is represented in Fig. 10, Kreps and his co-workers showed that the phospholipid content is highest in the microsomes, somewhat lower (by 10–15%) in the mitochondria, and lowest of all

## TABLE 6

*The phospholipid content of two sub-cellular fractions, and also gross homogenates of avian brain tissue, in milligrams of phosphorus per g dry weight, per g of protein and as a percentage of the total phospholipids.*

| Phospholipid | mg P/g dry weight of fraction | mg P/g of protein | Phospholipid as a % of total phospholipids |
|---|---|---|---|
| Homogenate | | | |
| Diphosphatidyl inositol | 0·19 | 0·21 | 2·0 |
| Monophosphatidyl inositol | 0·43 | 0·47 | 4·5 |
| Sphingomyelin | 1·03 | 1·13 | 10·7 |
| Lecithin | 3·73 | 4·09 | 38·6 |
| Phosphatidyl serine | 1·21 | 1·38 | 12·5 |
| Phosphatidyl ethanolamine | 2·98 | 3·26 | 30·8 |
| Polyphosphatidyl glycerol | 0·09 | 0·07 | 0·9 |
| Mitochondria | | | |
| Diphosphatidyl inositol | 0·11 | 0·13 | 1·4 |
| Monophosphatidyl inositol | 0·42 | 0·51 | 5·1 |
| Sphingomyelin | 0·68 | 0·84 | 8·3 |
| Lecithin | 3·00 | 3·73 | 36·7 |
| Phosphatidyl serine | 1·20 | 1·49 | 14·5 |
| Phosphatidyl ethanolamine | 2·62 | 3·22 | 31·6 |
| Polyphosphatidyl glycerol | 0·20 | 0·25 | 2·4 |
| Microsomes | | | |
| Diphosphatidyl inositol | 0·11 | 0·13 | 1·1 |
| Monophosphatidyl inositol | 0·39 | 0·44 | 4·0 |
| Sphingomyelin | 0·96 | 1·27 | 10·0 |
| Lecithin | 4·60 | 5·49 | 47·5 |
| Phosphatidyl serine | 1·29 | 1·61 | 13·4 |
| Phosphatidyl ethanolamine | 2·31 | 3·00 | 24·0 |
| Polyphosphatidyl glycerol | 0 | 0 | 0 |

Kreps *et al.* (1964).

in the nuclei. The total composition of phospholipids in all these cell particles broadly conformed to that which was observed in whole cells but there were, however, some characteristic differences between the fractions. These are summarized in Table 6. In all cases lecithin is the most abundant molecule and its concentration exceeds that of the other two substances which occurred in rather high concentrations, viz. phosphatidyl ethanolamine and phosphatidyl serine. Sphingomyelin, phosphatidyl inositols and phosphatidyl glycerol occur in rather lower and variable concentrations. Reference to Table 6 will show that the microsome fraction is chiefly characterized by the absence of polyphosphatidyl glycerol which is always found in both the mitochondria and nuclei. There is also a difference in the quantitative ratio which exists between the phosphatides in the different fractions. Whilst microsomes are relatively richer in both sphingomyelin and lecithin than are mitochondria, in the latter organelles there is, on the contrary, a relatively higher total content of both ethanolamine and serine phosphatides.

As far as the concentration and occurrence of poly-unsaturated fatty acids of glycerophosphatides are concerned our knowledge is again limited, and is largely represented by the papers of Hanahan *et al.* (1960) and Miyamoto *et al.* (1966, 1967). It would appear that the principal types of poly-unsaturated fatty acids in the brain of embryonic chicks are the 20:4 and 22:6 derivatives. The yolk provided a continuous source of 18:2 but only trace amounts of w–3. Following injections of $(1-C^{14})$ acetate and other esters on the tenth day of incubation there was a 4–8 hour post-injection period of high specific activity in the brain phospholipids. This activity was 6 times as high as that registered for liver. The relative speeds of the incorporation of such labelled esters into various lipid fractions, and also into the fatty acids of glycerophosphatides, are shown in Table 7. In this table the resulting specific radioactivity of free fatty acids, of other saponifiable and non-saponifiable neutral lipids, and of the phospholipids, is expressed as a percentage of the radio activity of the total lipids. The quantity of acetate which was incorporated into the brain lipids was approximately twice that found when using the fatty acids, all of which exhibited a similar degree of labelling. As a result of this Miyamoto and his coworkers concluded that all the metabolic pathways which had previously been demonstrated in mammalian brains are also present in those of birds.

The concentration of both spermidine and spermine in the brain material of 18-day-old chick embryos was determined by Raina (1963). The former substance occurs in quantities of 33 m$\mu$M, and the latter at 462 m$\mu$M per g wet weight. Experimentally they are of importance in mitochondrial preparations because such polyamines are known to have a stabilizing effect on membranes. It has been suggested that this is a result of their high affinity for those cell constituents with acidic groups. Stevens (1970) has summarized their role in nature. All such polyamines bind strongly to both DNA and RNA. They stabilize the double helix of DNA, probably by forming a bridge across the narrow groove by involving electrostatic bonding with the phosphate group. They do not appear to alter the overall configuration of the DNA by doing this. Spermine enables single stranded RNA to fold into a more compact configuration, less susceptible to attack by ribonuclease. Both stimulate the DNA-primed RNA polymerase and facilitate the removal of RNA from the DNA–RNA–enzyme complex. Polyamines promote the association of ribosomal sub-units and also the binding of amino acyl transfer RNA to ribosomes. In certain organ systems there is a correlation between the concentration of spermidine and the rate of RNA synthesis.

## TABLE 7

*The relative rates at which* $(1\text{-}C^{14})$ *fatty acids are incorporated into the lipid fractions of chick brain, expressed as a percentage of the radioactivity of the total lipids.*

| Compound | Free fatty acids | Neutral lipids | | Phospholipids | |
| --- | --- | --- | --- | --- | --- |
| | | Saponifiable | Non-saponifiable | Glycero-phosphatides | Fatty acids of glycerophos-phatides |
| Acetate | 0·4 | 2·8 | 27·0 | 64·5 | 38·7 |
| Myristate | 0·5 | 4·0 | 25·6 | 65·0 | 44·5 |
| Palmitate | 0·3 | 4·5 | 19·2 | 76·0 | 68·5 |
| Stearate | 0·7 | 4·1 | 19·1 | 73·5 | 70·5 |
| Linoleate | 0·4 | 4·9 | 25·6 | 69·4 | 65·6 |
| Linolenate | 0·5 | 4·3 | 24·8 | 66·8 | 61·5 |

Miyamoto *et al.* (1967).

Other large molecules such as sterols are also only present in small quantities in the late embryonic and adult brain, although they may be more prominent during the earlier stages of its development. In fact, according to Grafnetter *et al.* (1965) the steroid concentration in the brain of chick embryos at the eighth day of incubation is in the region of 0–11 mg per g wet weight. By the time of hatching this value has diminished and the content of esterified sterols falls dramatically so that they are subsequently only present as trace molecules. It is not at all clear to what extent this is actually a function of the appearance of the blood brain barrier, although one would assume that it is. This is certainly at the root of the analogous temporal pattern of representation of the small ionic components. However Greenberg *et al.* (1948) long ago demonstrated the very marked difference between the high rate of incorporation of $C^{14}$ glycine into the proteins of foetal chick brain homogenates and the lower rates in the case of newly hatched birds.

Lajtha (1956) showed that, with increasing age, the chloride space of the brain is penetrated at a progressively slower rate by both chloride and thiocyanate ions. In the presence of a constant concentration of chloride in the blood that of the brain decreased by some 30% during the period from the fourteenth day of incubation until the first day after hatching. Subsequently it remained constant during the 40 days that it was observed. In contrast, as can be seen from Table 8, during the period in which the greatest drop in concentration occurs the water content only decreased by less than 10%. Assuming that all the chloride was in the extra-cellular compartment Lajtha calculated that this extracellular fluid phase of the chick brain was 23%. He also concluded that the decline in the uptake of chlorine-36 into this space was continuous and that the same general considerations applied to thiocyanate although the space which was involved was larger.

## TABLE 8

*The chloride and water content of the brain, and the chloride content of the blood, at various periods during the development of the chick.*

| Days Incubation | Serum Chloride (mequiv./kg) | Brain Chloride (mequiv./kg) | Water (%) |
|---|---|---|---|
| 10 | | | 91 |
| 13 | | | 90 |
| 14 | 119 | 48 | 89 |
| 18 | 110 | 37 | |
| 20 | 121 | 39 | 86 |
| After hatching | | | |
| 1 | 104 | 33 | |
| 3 | 108 | 31 | 84 |
| 5 | | | 84 |
| 7 | 121 | 34 | 82 |
| 12 | | | 81 |
| 26 | 111 | 28 | |
| 42 | 122 | 30 | |

Lajtha (1956).

The penetration of charged particles of greater size into the brain via the choroid plexus was studied by Klatzo et al. (1965) during their observations on the penetration of serum proteins into the central nervous system. The passage of protein conjugates into the choroid plexus was observed by utilizing isolated chick plexi which were incubated in vitro in various solutions of proteins (Smith et al., 1964). The viability of such preparations was ascertained by noting the presence of the characteristic and rapid ciliary movement, and the penetration of the protein conjugates was evaluated by autoradiography and/or fluorescence microscopy.

Their results showed that in the plexi of new-born chicks there is a conspicuous difference between the rates of penetration of fluorescein labelled albumin and similarly labelled γ globulin. Isolated plexi which had been derived from the lateral ventricles exhibited an intense accumulation of the albumin derivative when they were incubated for 15 min in a 0·1% solution made up in a balanced salt solution such as Tyrodes. There was no such uptake when the γ globulin conjugate was used. This ability to take up the fluorescein labelled albumin developed between the 9th and 13th day of incubation and it was inhibited by low temperatures, such as 4°C, and also by solutions which contained various metabolic inhibitors such as $10^{-4}$ M KCN; $10^{-2}$ M NaF; and $5 \times 10^{-3}$ M sodium azide. As a result they justifiably concluded that this reflected active transport. A rather weird effect was apparently seen when the incubating solutions contained both fluorescein labelled γ globulins and also unlabelled proteins. The fluorescein labelled globulin was taken up by the plexus material when it was incubated with unlabelled globulins but not when incubated with unlabelled albumin.

Although the transport of serum proteins by astrocytes and oligodendroglia was not actually demonstrated, Klatzo had earlier concluded that there was a migration of microglial cells laden with protein conjugates. The intense uptake of protein tracers by the microglial components in the sub-arachnoid space (Klatzo et al. 1965) was taken by them to indicate the supreme importance of this type of glia in relation to the transport and removal of soluble brain materials. In view of this one may note that in his wide ranging review of the blood-brain barrier Edström (1964) concluded that the choroid plexus is generally assumed to play a key role in the production of the cerebro-spinal fluid. However, other considerations are certainly relevant. These include both active and passive exchanges between the blood and the cerebro-spinal fluid; the dimensions of the extra-cellular space in the species or specimen under consideration; the glial tissue; and the rate limiting factors in the various compartments. In the case of active transport between the blood and the cerebro-spinal fluid it is certainly widely assumed that the choroid plexus plays a dominant role (see Pappenheimer et al., 1961; Davson et al., 1962). In this respect it is of considerable interest that at certain stages of development a brush border is reported on its cell edges, and also that the concentrations within it of various oxidative enzymes increase during chick embryology (Kaluza *et al.*, 1964).

# IV. RIBONUCLEIC AND DESOXYRIBONUCLEIC ACIDS

Many of the studies on the nucleic acid content of avian brains relate to the changes which take place during the process of neuronal differentiation. Such studies include those of Birge (1962), Fujita (1963, 1964), Gayet and Bonichon (1961), Grillo *et al.* (1964), Mandel *et al.* (1949) and Wegelin and Manzoli (1967). Birge observed that during maturation and differentiation ependymal cells can undergo characteristic fluctuations in their ribonucleic acid content. However, those ependymal components which are situated on the dorsolateral surfaces of the diencephalon, together with those which line the roof of both this region and the mesencephalon, display a high level of ribonucleic acid basophilia throughout the entire developmental period. As they retain certain primitive epithelial characteristics their cytological appearance conforms with their supposed function as secretory, possibly protein secretory, sources.

Fujita (1963) using auto-radiography observed that the extent of the cumulative labelling with tritiated thymidine indicated that the matrix cells of the neural tube in a 6-day-old embryo comprised a homogeneous population. It is not immediately clear to what extent such homogeneity extends beyond a comparable rate of thymidine uptake. He further (1964) suggested that the process of neuroblast formation from these primitive matrix cells in the region of prospective optic tectum, cerebellum and spinal cord only occurs at certain periods of development. Additional details of this process are provided in Chapter 4, Section II.

Biochemical measurements of the progressive increase which occurs in the ribonucleic acid concentration, the desoxyribonucleic acid concentration and the quantity of total protein during the period between the 10th and 19th day of incubation are provided by a number of these authors. Broadly speaking they are all consistent with the overt cytological changes. One can summarize the overall tendencies succinctly by saying that the concentration of desoxyribonucleic derivatives is generally exceeded by a much steeper increase in that of ribonucleic moieties. This last is itself paralleled by the changing levels of protein. Although the total amount of desoxyribonucleic acid per g of brain tissue actually decreases, that of the ribonucleic acid increases and these changes reflect the modifications which take place in the cell density. Hughes (1955), who used ultraviolet absorption methods on the developing spinal cord, concluded that from the 2nd to the 5th day of incubation high concentrations of ribonucleic acids occur in both the ventral horn and ependyma. After this period there is a temporary fall in the detectable levels which is subsequently followed by a further rise to reach a maximum at the fourteenth day. Wegelin and Manzoli (1967) have also demonstrated that the quantity of free nucleotides per mg of desoxyribonucleic acid increases up until the twelfth day. Thereafter it decreases until the time of hatching. The most interesting individual data were those for CTP. This had previously been shown to have a regulatory function in rats and to suppress protein synthesis; in their assays they found that it actually disappeared at the time of maximal protein levels.

## V. THE ENZYMES WHICH ARE INVOLVED IN CARBOHYDRATE AND AMINO ACID METABOLISM

There have been numerous studies on the developmental patterns which are exhibited by enzymes in avian brain homogenates. Most of these refer to the chick and reflect the use of easily cultured material by workers whose interests are varied. Emmanuelson and Helmer (1965) showed the presence of iso-citrate dehydrogenase, NAD-linked glutamate dehydrogenase, lactic dehydrogenase, phosphogluconate dehydrogenase, NADH-linked iso-citrate dehydrogenase and succinic dehydrogenase in the nervous system of 30–45 hour embryos. As might be expected, however, all such systems are complex and, for example, Doemel and Chilson (1969) found that the developmental pattern of adenosine 5′ phosphate aminohydrolase was influenced by numerous factors besides the rise in protein synthesis. Lowenthal et al. (1962) also demonstrated that overall similarities between the various vertebrate classes must be treated with caution. They separated the lactic dehydrogenase iso-

## TABLE 9

*The values for various enzyme types in the brains of pigeons and chickens expressed as the number of international units per g of tissue.*

| Enzyme type | Pigeon | Chicken |
|---|---|---|
| Lactate dehydrogenase | 4000 | 5000 |
| Phospho-hexose isomerase | 12,000 | 6000 |
| Aldolase | 400 | 400 |
| Malate dehydrogenase | 18,000 | 12,000 |

Dujovne et al. (1969).

enzymes electrophoretically and found that the "fingerprints" which were obtained from avian material differed conspicuously from those of both reptiles and mammals. Furthermore, as can be seen from Table 9, Dujovne *et al.* (1969) found that there is a fair degree of variation in the concentrations of such enzymes between various avian species. In view of the total temporo-spatial distribution of avian genera and the differences of developmental types it is clearly dangerous to indulge in unverified generalizations.

The topographical distribution of various enzymes in the striatal and brain stem structures was summarized by Baker-Cohen (1968). These are dealt with elsewhere. He concluded that the histochemical characteristics of the paleostriatum (*q.v.*) resembled those of the putamen-caudate complex in mammals. Prominent succinate dehydrogenase activity was detected in the analogue of Craigie's (1930) strio-tegmental tract nucleus. Both the neuron bodies and neuropile were very strongly stained. Such dehydrogenase activity also characterized the interpeduncular nucleus.

The pre-existing studies on the actual temporal sequence which one can observe in the developmental patterns of enzymes were summarized by Wang (1968). Although the categories which he suggested are somewhat wide, and therefore hide certain differences between different enzyme types, they provide a useful, if trite, synopsis of the observed events. Briefly he considered that there are two principal types of pattern:

1. Multiphasic sequences which exhibit more than one activity peak during development (Kato, 1959; Felicioli *et al.*, 1967);
2. Monophasic sequences where the maximal enzymatic activity occurs at a single well-defined point in development (Davidson, 1957; Knox and Eppenberger, 1966; Solomon, 1958; Walker and Walker, 1962; and Wang *et al.*, 1966).

Wang (1968) studied the levels of activity of eight enzyme types in avian brain material and compared the temporal patterns with those of the same enzymes in other tissues. The changing levels for glucose-6-phosphatase, malate dehydrogenase, isocitrate dehydrogenase, glucose-6-phosphate dehydrogenase, glutamic-oxaloacetic transaminase, glutamic-pyruvic transaminase and glutamate dehydrogenase are represented in Figs 11–17. By comparing these data with those for other organ systems Wang concluded

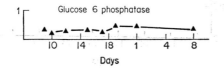

FIG. 11. Developmental patterns of glucose-6-phosphatase in chick brain. The enzymatic activity is expressed as $\mu$moles orthophosphate liberated/mg protein/hour. After Wang (1968).

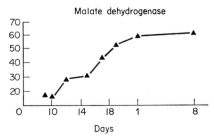

FIG. 12. Developmental patterns of malate dehydrogenase in chick brain. Enzymatic activity is expressed as $\mu$moles NAD reduced/mg protein/hour. After Wang (1968).

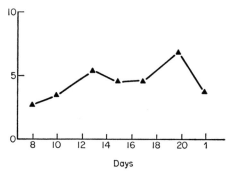

FIG. 13. Developmental patterns of iso-citrate dehydrogenase activity in chick brain. Activity is expressed as $\mu$moles NADP reduced/mg protein/hour. After Wang (1968).

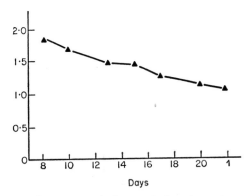

FIG. 14. Developmental patterns of glucose-6-phosphate dehydrogenase in chick brain. Activity is expressed as $\mu$moles NADP reduced/mg protein/hour. After Wang (1968).

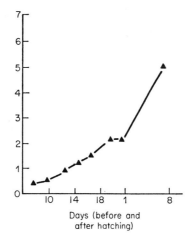

FIG. 15. Developmental patterns of glutamate dehydrogenase in chick brain. Enzymatic activity is expressed as $\mu$moles $NADH_2$ oxidized/mg protein/hour. (After Wang, 1968)

that there appeared to be two main trends. In the first of these there are either rapid increases or else gradual decreases until a day before hatching and then subsequent increases or decreases during the post-hatching period. In brain homogenates this applied to lactate dehydrogenase, malate dehydrogenase, glucose-6-phosphate dehydrogenase, glutamate dehydrogenase, glutamic-

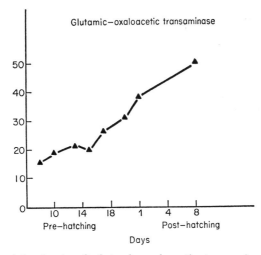

FIG. 16. The activity levels of glutamic-oxaloacetic transaminase in the chick. Enzymatic activity is expressed as $\mu$moles $NADH_2$ oxidized/mg of protein/hour. (After Wang, 1968).

3

oxalocetic transaminase, glutamic-pyruvic transaminase and glucose-6-phosphatase. In his second trend there are one or more peaks during development and, in the case of brain, this only applied to isocitrate dehydrogenase. It would be very interesting to have comparable data for a series of both nidicolous and nidifugous species.

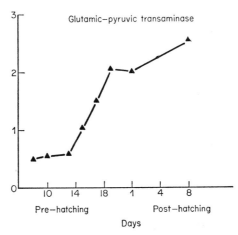

FIG. 17. The activity levels of glutamic-pyruvic transaminase in the chick. Activity expressed as in Fig. 16. (After Wang, 1968).

After considering these data Wang attempted to elucidate three problems. These were the possibility that low levels of enzymatic activity were due to the enzymes being bound to membranes and thereby rendered inert; that the varying levels of activity reflected the presence of either inhibitors or activators, or, indeed, both; and thirdly that a low overt level of activity was the result of substrate protection. In the case of the first possibility he found that the result of treating the homogenates with deoxycholate was to give a rise in the lactate dehydrogenase activity at the thirteenth and twentieth days of incubation in all tissues except the brain. The activity of homogenates of this organ system showed a drop in their activity levels which suggested some degree of inhibition. In the case of iso-citrate dehydrogenase such an increase in the observed level of activity occurred in material from the brain at the 10th and 15th days of incubation and also during the immediate post-hatching period. These were periods during which the enzymatic activity was otherwise rather low. In marked contrast to these rather sporadic effects in terms of the period of development, the activity of glutamic oxaloacetic transaminase was increased throughout all stages. Wang assumed that deoxycholate forms a complex at the surface of particles and the particle membrane is thereupon dispersed.

The possibility that fluctuations in the level of activity reflect the presence of inhibitors and/or activators was tackled quite straightforwardly. He combined homogenates from the brains of embryos which had reached different stages of development. If there was no deviation from the level of activity which was expected on the basis of the earlier experiments he concluded that both activators and inhibitors were absent. Finally, by using excess substrate he was able to satisfy himself that low activity, at least in the case of iso-citrate dehydrogenase, was not caused by either substrate protection or stabilization.

A comparison of Wang's results, which were obtained by using Leghorn embryos, with those which were obtained by Solomon (1958) using Rhode Island Red × Light Sussex hybrids shows some differences but a general measure of agreement. In the case of malate dehydrogenase and iso-citrate dehydrogenase, both of which are involved in the tricarboxylic acid cycle and hence in the major pathway of carbohydrate metabolism, the overall pattern is similar with higher malate dehydrogenase activity on the thirteenth day of incubation. After the 17th day this activity increases at a greater rate. A similar comparison of the glucose-6-phosphate dehydrogenase activity with that which was reported by Burt and Wenger (1961) shows that during the early period of development it is rather lower than is the case in New Hampshire Reds. Burt and Wenger had suggested that the high levels of activity which they detected reflected a rise in the activity of the pentose cycle responsible for producing the pentose needed for ribonucleic acid synthesis and also the $NADPH_2$ needed for various reductive syntheses associated with cellular and tissue differentiation. Rudnick and Waelsch (1955) had also reported that there is only minimal glutamine transferase activity in the brain during the early phases of both morphogenesis and normal growth. It begins to rise and achieves higher levels on the 11th or 12th day of incubation and, by the time of hatching, reaches levels which are similar to those observed in the liver.

## VI. FREE AMINO ACIDS AND THE METABOLISM OF GLUTAMATE AND AMMONIA

In general terms glutamate is the only amino acid which is used at an appreciable rate by vertebrate brain tissue. Bartley et al. (1968) have emphasized that it provides a limited source of energy, acts as a trap for ammonia and, if decarboxylated, is the immediate precursor of $\gamma$ amino-butyric acid which has an inhibitory effect on the activity of the central nervous system. Transamination between glutamate and oxaloacetate yields $\gamma$ oxoglutarate, which is a potential source of energy, and also aspartate (see Fig. 18).

In the avian brain Tsunoo et al. (1963a) showed the presence of a number of free amino acids. Starting with 10 kg of tissue they crystallized 0·4 g of histidine and an equivalent quantity of anserine, which were identified by salt formation, analysis and paper chromatography. By the use of column and paper chromatography they were also able to demonstrate the presence of aspartic acid, glutamic acid, serine, threonine, proline, valine, glycine, alanine, β-alanine, γ-amino butyric acid, phenylalanine, leucine, methionine,

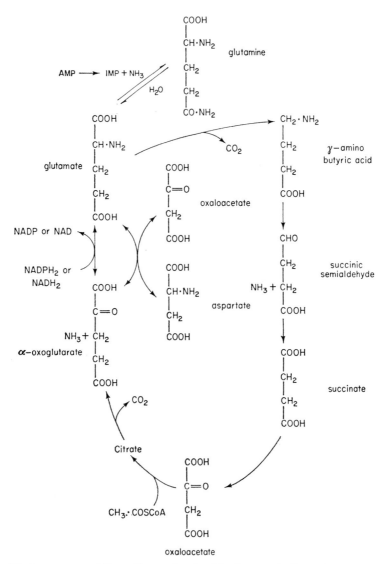

FIG. 18. A summary of the pathways involved in the metabolism of glutamate and ammonia in the brain. (After Bartley, Birt and Banks, 1968).

tryptophan, iso-leucine, lysine and arginine, all in the free form. An analogous situation was found in the avian eye by Tsunoo et al. (1963b) although in this latter case valine, methionine, tryptophan and isoleucine were absent from their list.

Recently Lajtha and Marks (1969) have given more detailed information. This included the concentration of the cysteine derivative taurine, together with that of 10 amino acids. In terms of the number of $\mu$moles per g of brain tissue these values were

| | | | |
|---|---|---|---|
| Glutamic acid | 12·2 | Arginine | 0·13 |
| Taurine | 3·8 | Valine | 0·07 |
| Aspartic acid | 2·8 | Leucine | 0·05 |
| Glycine | 0·42 | Tyrosine | 0·05 |
| Alanine | 0·70 | Methionine | 0·02 |
| Lysine | 0·07 | | |

The differential occurrence of one of these amino acids, glycine, in the different parts of the brain was detailed by Aprison et al. (1969). There appeared to be a marked increase in the concentration from the front to the back. The actual values in $\mu$moles per g wet weight are forebrain hemispheres, 1·21; diencephalon, 1·32; mesencephalon, 1·45; cerebellum, 1·18; pons, 3·89; and medulla, 4·79. It is interesting in this context that Galindo et al. (1967) reported that it had a potent and inhibitory effect on the neurons of the cuneate nucleus.

As was noted in Section III Greenberg et al. (1948) reported that the rate at which carbon-14 labelled glycine is incorporated into the proteins of embryonic chick brain homogenates is considerably in excess of that which can be observed in the case of newly hatched chicks. It is likely that the concentrations of the other amino acids undergo similar developmental fluctuations. Furthermore the effects of wound healing lead to other and comparable changes. Following the removal of one eye from a pigeon, Altman and Altman (1962) found that the uptake of glycine within the degenerating optic tract increases for a month or so. This change was contemporaneous with, and apparently related to, the phase of glial proliferation. No such effects occur if one eye is merely blindfolded.

As far as ammonia is concerned it is formed in the brain at a comparatively high rate by the deamination of adenosine monophosphate to inosine monophosphate; the hydrolysis of the amide group of glutamine; and, to a lesser extent, by the deamination of glutamate. It is a fairly widely applicable rule that any tendency for the concentration of ammonia to rise is counteracted by the conversion of $\alpha$-oxoglutarate to glutamate by glutamic dehydrogenase, and by the conversion of glutamate to glutamine by glutamine synthetase. The energy for the former reaction is provided by reduced coenzymes and for the latter by adenosine triphosphate. Any extensive drop in the ammonia concentration which may result from converting $\alpha$-oxoglutarate to glutamate results in the withdrawal of intermediates from the citric acid cycle. Such drops may be compensated for by the fixation of carbon dioxide using the malic enzyme and allowing pyruvate to be converted to malate.

# VII. THE INCORPORATION OF TRITIATED LYSINE AFTER IMPRINTING

An exceedingly interesting paper that reflects work which has been attempted, albeit unsuccessfully, by a number of other groups, was published by Bateson *et al.* (1969). They described what appear to be the effects of visual experiences in an imprinting situation on the incorporation of tritiated lysine into the protein constituents of the mid-brain, forebrain roof and forebrain base. They justifiably emphasized that such effects of a flashing light on the rate of incorporation of an amino acid into the forebrain proteins of early hatching chicks might be non-specific. It might reflect, for example, different amounts of visual stimulation *per se* although the observed results seemed to depend upon the maturational age, and, in this way at least, to parallel the imprinting process.

Prior to the start of the experiment six batches of Chunky chicks were hatched and then kept in an incubator. In view of the famous maturational effect on the imprinting process itself half of the birds were drawn from the early members of the hatch and the remainder from the late members. The former emerged some 6–9 hours before the peak period of hatching and the latter an equivalent time afterwards. Eighteen birds were taken from each of the batches and then distributed equally into three groups—experimental, light controls and dark controls. In an initial test the speed with which each experimental chick approached a flashing light was measured and recorded. Subsequently each bird was exposed alone to an identical flashing light for 60 min. At the end of this time it was weighed and given an intra-cardiac injection of 0·1 ml Locke solution containing 20 $\mu$Ci of lysine-4,5,T. A few chicks received faulty injections and were discarded.

Immediately after this injection the individual birds were again exposed to a flashing light for a further 45 min. After 40 min in a dark incubator the time which they now took to approach an identical flashing light was recorded. Bateson *et al.* concluded from other data that they could reasonably have been expected to have developed a preference for such a light, and, indeed, in the final test all but two of the experimental birds approached it more speedily than during the initial test. When this approach speed had been noted the birds were decapitated and the brains extracted. The specific activity for the various parts of the brain was then expressed as d.p.m./mg of protein and the results were corrected to a standard body weight of 50 g. The data refer to the incorporation of isotopically labelled amino acid into the acid-soluble substances which are largely protein. As a result they are a reflection of, but do not measure, the rate of protein synthesis.

In the early hatchers the mean value for the standardized specific activities of the forebrain roof was significantly greater than that for the controls. The values for the other regions were not. As the value for the forebrain roof exceeded that of its basal areas, which themselves had a higher value than the mid-brain, they noted that the observed effects of light on the rate of lysine incorporation were only detectable in the region where such incorporation was maximal. Further investigations on the effect of non-specific factors such as motor activity are not available at the time of writing.

## VIII. A GENERAL CONSIDERATION OF 5-HYDROXY-TRYPTAMINE, DOPAMINE AND NOR-ADRENALIN

There have been a number of studies on the occurrence of these substances in avian brains and not all of these are in obvious agreement with one another. As in the case of cholinesterase and alkaline phosphatase activity data which refer to specific brain areas are largely dealt with in the appropriate chapter elsewhere in this book. We may limit ourselves at this juncture to the more generalized considerations. In 1959, Nozdrachev suggested that the "inhibition" of flying activity which he observed in pigeons following serotonin injections, and also the adynamia of chickens under similar circumstances, both testify to damage to, or effects on, "sub-cortical" centres. The injection of small quantities resulted in a considerable improvement in the responses to conditioned stimuli. Following the injection of larger quantities of between 1–20 mg there was first of all an enhanced level of activity and vocalization which lasted some 5–7 sec. After this the preparation entered an immobile phase which persisted for a further 3–15 sec and during which there was widespread muscular relaxation. At no time did he observe any sign of the serotonin induced catalepsy which had previously been reported in frogs and dogs.

From beginnings such as these there have been, during recent years, a variety of biochemical, and biochemically based behavioural studies on these drugs. Whilst some are referred to here, others are featured in Chapters 14 and 15. By administering isotopically labelled tryptophan via the intraperitoneal and sub-arachnoid routes, and then comparing the results, Gal and Marshall (1964) showed that the two stage biosynthesis of 5-hydroxytryptamine can take place in the pigeon's brain.

A number of assessments of the actual brain concentrations exist. Correale (1956) gave values of 0·15 $\mu$g per g for the pigeon, and rather higher ones of

0·2 μg and 0·4 μg per g for hens and turkeys respectively. Brodie and Bog-danski (1964) gave even larger figures and compared them with the nor-adrenalin levels. These are contained in Table 10 whilst others, provided by Aprison and Takahashi (1965) and Aprison and Hingtgen (1965) are given in Tables 11, 15 and 51 below.

## TABLE 10

*The concentrations of serotonin and nor-adrenalin in μg per g of tissue.*

|         | Serotonin | Nor-adrenalin | Serotonin/ catecholamines |
|---------|-----------|---------------|---------------------------|
| Pigeon  | 0·78      | 0·37          | 2·1                       |
| Chicken | 1·0       | 0·60          | 1·7                       |

Brodie and Bogdanski (1964).

A recent comparative survey of the 5-hydroxytryptamine levels and the monoamine oxidase activity throughout the animal kingdom is presented in the paper by Baker and Quay (1969). Gunne (1962) concluded that chick brain contains 17% adrenalin by comparison with the values of 4–7·5% which had been obtained for various species of mammal. Basing his con-clusions on assays which utilized the ethylene diamine condensation method coupled with independant bio-assays he concluded that in amphibians, birds and mammals it is in fact the predominant catecholamine present in the brain. In fishes and reptiles he suggested that nor-adrenalin occupies this position. Comparative studies which deal with a number of such catecholamine assays have also been provided by Aprison and Hingtgen (1965), Pscheidt and Huber (1965), Pscheidt and Himwick (1963) and Juorio and Vogt (1967). These two last-named authors concluded that the highest dopamine concentrations occur in the basal nucleus and possibly also the endopeduncular nucleus. Fluorescence microscopy showed an intense diffuse fluorescence in this region comparable with that in the mammalian striatum which has recently received prominence in the press. It was also in this region that they obtained the highest values for endogenous 5-hydroxytryptamine. In contrast the highest values for nor-adrenalin were obtained in the hypothalamus (see Table 51) although relatively large amounts also occur in the cerebellum (Pscheidt and Himwick, 1963) and they again concluded that the concentra-tion of adrenalin in avian brain exceeds that in mammals.

In the pigeon brain reserpine-like drugs cause a drop in the monoamine concentration but this does not seem to be correlated with a corresponding rise in the levels of homovanillic acid. The administration of $\beta$-tetrahydro-naphthylamine leads to a moderate fall in all brain amines and their metabolites whilst injections of $\alpha$-methyl dopa results in the appearance of $\alpha$-methyl dopamine and $\alpha$-methyl nor-adrenalin. Unlike the situation which has been reported from some mammals the concentration of $\alpha$-methyl dopamine is, however, persistently low.

Pscheidt and Huber (1965) found that the autochthonous 5-hydroxy-tryptophan levels are highest in the telencephalon and, as reported by Juorio and Vogt (1967), the "basal" regions have concentrations which are in excess of those which can be observed elsewhere. Other brain regions showed low levels of dopamine. The levels of the decarboxylase activity towards both dopamine and 5-hydroxytryptophan appeared to decrease in the sequence pons-medulla; di- and mesencephalon; striatum; corticoid regions; cervical cord; thoracic cord; lumbar cord and cerebellum. From such data as are available it seems likely that specific variations occur and the situation is probably not totally homogeneous throughout all the Aves. For example in the chick there was twice as much activity in the pons-medulla by comparison with the situation in the pigeon. In general, however, there appeared to be a linear relationship between the different topographical levels of 5-hydroxy-tryptophan decarboxylase activity and the increase in the 5-hydroxytrypt-amine which was detectable following monoamine oxidase inhibition. The ratio of dopa decarboxylase to 5-hydroxytryptophan decarboxylase activity within the various regions varied between 4 and 6 but there was no direct relationship between the observable quantity of *endogenous* amines within the various regions and the intensity of the local decarboxylase activity. However, the direct relationship in the presence of monoamine oxidase inhibitors which was noted above agrees with the well-known theory that this oxidase regulates the oxidation of 5-hydroxytryptamine and that the decar-boxylase governs the rate of formation of this amine from 5-hydroxytrypto-phan.

In association with this information Aprison and Hingtgen (1965), Aprison *et al.* (1962) and Aprison and Takahashi (1965) have investigated the concentrations in the various brain regions during periods of drug-induced abnormal behaviour. Aprison *et al.* (1962) determined the rate of change of the 5-hydroxytryptamine concentration in the telencephalon, diencephalon + optic lobes, cerebellum and pons-medulla during the period of the complete behavioural effect (T) which followed an intra-muscular injection of 50 mg/ kg of DL5-hydroxytryptophan hydrate into trained pigeons. These birds had been trained to work on a multiple fixed ratio, fixed interval schedule of reinforcement. Only in the case of the telencephalon and the diencephalon did the concentration of 5-hydroxytryptamine pass through a maximum (at 30%

T) and then return to normal at the same time as the behaviour. The 5-hydroxytryptamine concentrations of the pons-medulla and cerebellum, together with those of the heart and lungs, certainly passed through maxima at about the same period (25–30 % T) but then returned to normal by 60 % T. The concentration in the liver, although following the same initial trend, remained high throughout the period of observation (128 % T).

## TABLE 11

*Mean dopamine and nor-adrenalin values in the brain regions of control pigeons and also those killed during and after behavioural disruption resulting from the administration of 5-hydroxyptryptophan.*

|                        | Telencephalon | Diencephalon + mesencephalon | Cerebellum | Pons-medulla |
|------------------------|:-------------:|:----------------------------:|:----------:|:------------:|
| Dopamine control       | 0·92          | 0·19                         | 0·08       | 0·21         |
| <100% T                | 0·93          | 0·09[b]                      | 0·01[b]    | 0·08[b]      |
| >100% T                | 1·05          | 0·17                         | 0·04       | 0·15         |
| Nor-adrenalin control  | 0·52          | 0·81                         | 0·24       | 0·95         |
| <100% T                | 0·52          | 0·65                         | 0·38[b]    | 0·71[b]      |
| >100% T                | 0·40          | 0·54[a]                      | 0·24       | 0·52[b]      |

[a] Significantly different from the control mean at 0·05 level.
[b] Significantly different from the control mean at 0·01 level.
Aprison and Hingtgen (1965).

Subsequently Aprison and Hingtgen (1965) extended this work which showed that the neurohumoural changes only paralleled the overt behaviour syndrome within the telencephalon and the diencephalon + optic lobes. Since it had been suggested that 5-hydroxytryptophan decarboxylase and dopa decarboxylase are one and the same enzyme they measured the dopamine and nor-adrenalin levels as well as those of 5-hydroxytryptamine following 5-hydroxytryptophan administration. The average period of depressed behaviour was determined for each bird. Then at various percentages of the total time (T) the preparations were decapitated and the four brain parts assayed. They could not detect any significant changes in the concentration of either dopamine or nor-adrenalin within the telencephalon but there were variations in the nor-adrenalin levels of the diencephalon + optic lobes, the pons-medulla and the cerebellum. They were inclined to discount any simple correlation between these last and the disturbed behaviour. Similar variations also occurred in the dopamine concentrations at these levels but it was only in the first two parts of the brain that such changes tallied with the temporal sequence of the behaviour pattern. They discussed these results, which are summarized in Table 11, on the basis that the dopamine is acting as a nor-adrenalin precursor or a transmitter, and concluded that although the depressed approach behaviour which follows 5-hydroxytryptophan adminis-tration can be most easily explained in terms of the action of free 5-hydroxy-tryptamine at appropriate sites, nevertheless the depression of dopamine levels within the mid-brain and pons-medulla must also be considered.

Masai *et al.* (1966) have summarized the known distribution of monoamine oxidase in the brains of vertebrates. In the avian genus *Uroloncha* a relatively strong positive reaction is found in the hippocampal, septal, preoptic, hypo-thalamic, habenular, paleostriatal and archistriatal areas, together with the peri-ventricular regions of the thalamus. In contrast, the hyperstriatal and neostriatal areas, together with most of the thalamic nuclei, including both the rotund and ovoidal nuclei, show relatively weak reactions. This can be compared with the analogous reactions for succinic dehydrogenase which are strongly positive in the hyperstriatal, neostriatal, paleostriatal, habenular and hippocampal regions and, within the thalamus, are only negative in the periventricular grey. With the exception of the habenula these grey areas are thought to be related to somatic function. Habenula, hippocampus and paleostriatum show active reactions for both monoamine oxidase and succinic dehydrogenase but the archistriatum, septal region, preoptic area and hypothalamus show weak reactions for succinic dehydrogenase. In general the degree of all such positive reactions is greater in birds than in mammals.

## IX. THE ACETYLCHOLINE ESTERASE, CHOLINE ACETYLASE AND ACETYLCHOLINE SYSTEMS

In view of the fundamental role which both acetylcholine and adrenalin play within the peripheral nervous system their presence, together with that of other amines such as $\gamma$-amino butyric acid, in synaptic vesicles of brain neurons is generally accepted as reflecting their functional role in such sites. In contrast to this neuronal distribution Brightman and Albers (1959) claimed that the principal butyrylcholine esterase systems of rooster brains are associated with the neuroglia, and more specifically the astrocytes. A number of overall synopses of both our knowledge of, and recent advances in, cholin-ergic physiology have been provided by recent symposia such as that arranged by the New York Academy of Sciences. The actual location of acetylcholine esterase systems within the sub-cellular fractions of pigeon brain, alongside data for both monoamine oxidase and choline acetylase are summarized in Table 12.

Alongside these data those of Table 13 give details of the variations which occur from one brain region to another and make a comparison possible with other data for choline acetylase, 5-hydroxytryptophan decarboxylase and monoamine oxidase.

## TABLE 12

*The relative activity of choline acetylase, monoamine oxidase and acetylcholine esterase in the sub-cellular fractions of pigeon brain.*

|  | Acetylcholine esterase (%) | Monoamine oxidase (%) | Choline acetylase (%) |
|---|---|---|---|
| Nucleus | 2 | 5 | 2 |
| Mitochondria | 32 | 83 | 26 |
| Microsomes | 60 | 12 | 28 |
| Supernatant | 6 | 0 | 44 |
| $M_1$ | 70 | 96 | 19 |
| $M_2$ | 26 | 4 | 15 |
| $M_3$ | 4 | 0 | 66 |

McCaman *et al.* (1965).

## TABLE 13

*The choline acetylase, acetylcholine esterase, 5-hydroxytryptophan decarboxylase and monoamine oxidase activities in four brain areas of the pigeon.*

| Brain area | Choline acetylase | Acetylcholine esterase | 5 HTP decarboxylase | Monoamine oxidase |
|---|---|---|---|---|
| Telencephalon | $2.26 \pm 0.26$ | $898 \pm 48$ | $0.26 \pm 0.02$ | $16.5 \pm 0.61$ |
| Diencephalon +optic lobes | $8.54 \pm 0.55$ | $2258 \pm 85$ | $0.26 \pm 0.02$ | $20.8 \pm 0.86$ |
| Cerebellum | $0.17 \pm 0.02$ | $1958 \pm 218$ | $0.03$ | $8.2 \pm 0.89$ |
| Pons-medulla | $6.80 \pm 0.46$ | $880 \pm 44$ | $0.19 \pm 0.01$ | $19.4 \pm 0.73$ |

Aprison *et al.* (1964).

Details of the interesting developmental data for acetylcholinesterase, which have been provided by Rogers and co-workers, are considered in the sections which deal with the specific brain regions. At this juncture we can note that it can be detected histochemically from 15 hours of incubation and prior to the full differentiation of the neuroblasts (Bonichon, 1958, 1960; Filogamo, 1960; Zacks, 1954). These data, which were summarized by Friede (1966), are not in themselves surprising in view of the fact that we now know that such esterase activity is detectable in a wide variety of non-nervous sites. It has also been known for a long time that the levels of activity rise prior to hatching (Nachmansohn, 1940) and this increase shows a well-marked topographical sequence. Proceeding in a caudo-cranial direction it has transitory peaks in some regions (Rogers *et al.*, 1960; Rebollo and Hekimian, 1955: Rogers, 1960). At an earlier date Wenger (1951) had also drawn attention to the correlation which exists between the rise in the acetylcholine esterase levels and the onset of overt electrical activity. Assays and histochemical studies on chick mesencephalon show this early appearance of activity prior to full neuroblast differentiation rather well. The maximal rate of increase occurs between 9–10 days and during this period it exceeds that of oxygen consumption. Wenger's original observations were, however, on chick spinal cord. Here the activity undergoes a five-fold increase prior to the appearance of electrical activity. In the spinal ganglion the esterase is detectable from the fourth day in neuroblasts and later large ganglion cells situated in a lateral position exhibit it. Subsequently it can then be found in smaller ganglion cells in the dorso-medial region (Strumia and Baima Bollone, 1964).

In their study of acetylcholine esterase and monoamine oxidase of adult brains Aprison *et al.* (1964) found that both were in large excess by comparison with the related synthetic systems. They also discovered that the activity levels of the enzymes of the cholinergic system exceeded that of 5-hydroxytryptophan decarboxylase. Indeed there appeared to be a direct linear relationship for two enzyme ratios in the pons, medulla, mesencephalon and telencephalon. These were the 5-hydroxytryptophan decarboxylase/monoamine oxidase, and the acetylcholine esterase/choline acetylase ratios.

As far as behavioural inclined data are concerned there is a considerable amount of experimental evidence from mammals which suggests that, when administered to pregnant specimens, chlorpromazine modifies the post-natal behaviour of the offspring. A comparable situation has been investigated in birds by Vernadakis (1969). Using White Leghorn chicks she measured the cholinesterase concentration within the forebrain, hypothalamus, cerebellum and spinal cord, together with the maximal electric shock seizures which could be obtained in preparations which had received chlorpromazine injections at the fifth and sixth days of incubation. Although the acetylcholinesterase activity was only slightly higher in the cord the duration of leg flexion was significantly shorter and that of extension longer. The result of these effects was that the chlorpromazine treated preparations exhibited more severe maximal electro-shock seizures than the controls.

In view of the importance of photo-period in determining sexual maturation and its concomitant phenomena during each annual cycle, it is also of considerable interest that Winget *et al.* (1967) found a very clear photic effect on the brain esterase levels. They studied the levels of activity in the diencephalon of chicks which had been maintained in differing photo-period régimes. Alongside the cholinesterase data they also cited comparable figures for both acid and alkaline phosphatase. These are all contained in Table 14. One can only speculate on the possibility that these reflect a direct photic effect on enzymatic synthesis and breakdown, or alternatively that they just reflect a lowered level of overall activity. Furthermore one can note the interestingly high values for alkaline phosphatase in view of the known occurrence of high values prior to the rise in acetylcholine esterase during embryonic development.

## TABLE 14

*The enzymatic activity levels in the diencephalon of chicks which had been kept under total darkness and others kept under diurnal light cycles. The units are the amount of phosphatase liberating* 1 *μmole of p-nitrophenol* $min^{-1}$ *from buffered p-nitrophenyl phosphate at* pH 10·5 *and* 4·8.

|  | 14 hours light<br>10 hours darkness | 0 hours light<br>24 hours darkness |
|---|---|---|
| Alkaline phosphatase | 2·33±0·17 | 4·50±0·50  IU |
| Acid phosphatase | 5·00±0·17 | 4·00±0·17  IU |
| Cholinesterase | 1·22±0·07 | 0·86±0·10 |

Winget *et al.* (1967).

The principal work which discusses other analogues of the cholinesterase system is that of Aldridge (1964). He concluded that within the chick brain there is evidence for two principal ali-esterases. The first of these is much the more sensitive to diisopropyl fluorophosphate and to Mipafox (= N. N. diisopropyl phosphodiamidic fluoride). It also hydrolyses phenylacetate at a faster rate than phenyl propionate. The other is considerably less sensitive to these inhibitors and hydrolyses phenyl propionate faster than phenylacetate. For a discussion of the inhibitory situation *vis a vis* butyrylthiocholine in the cerebellum see Chapter 8.

Before turning to acetylcholine itself we can interject a few comments on the available data referring to choline acetylase. The actual distribution of this enzyme is greatest in the mid-brain and pons-medulla. Within pigeon telencephalic material it was assayed by McCaman *et al.* (1965). Their data are summarized, together with comparable figures for acetylcholine esterase and monoamine oxidase, in Table 12. From that table it can be seen that the enzyme was predominantly associated with the nerve ending particles represented by the $M_3$ component of the sub-mitochondrial fraction. This component gave a value of 66% for the relative specific activity which can be defined as

$$\frac{\text{percentage of recovered enzyme}}{\text{percentage of recovered protein}}$$

This value can be compared with values of 7%, 15% and 6% obtained from the rat, rabbit and guinea pig respectively. In contrast the figures for the $M_2$ sub-mitochondrial fraction were necessarily lower than is the case in mammals. Comparing these results McCaman *et al.* concluded that they reflect different sub-cellular distributions within the two classes of animals. Although the choline acetylase is primarily within the axoplasm of birds it is possibly more closely associated with synaptic vesicles and their membranes in mammals.

## TABLE 15

*The variations of the acetylcholine, 5-hydroxytryptamine, dopamine and noradrenalin concentrations in the pigeon brain, expressed as* mμmoles/g.

| Brain region | Acetyl-choline | 5-hydroxy-tryptamine | Dopamine | Nor-adrenalin |
|---|---|---|---|---|
| Telencephalon | 15·5 | 5·85 | 6·52 | 3·07 |
| Diencephalon+ | | | | |
| optic lobes | 28·9 | 4·76 | 1·50 | 5·08 |
| Cerebellum | 1·6 | 1·30 | 0·85 | 1·48 |
| Medulla+pons | 18·1 | 6·64 | 1·70 | 5·67 |

Aprison and Takahashi (1965).

Clearly the results may also reflect the relative degree of binding in such a situation. If this is the case the "looser" binding in birds gives a greater ease of release. In this respect it is of interest that the other two enzyme categories which they studied, and which are also included in Table 12, had a similar distribution within the sub-cellular fractions of birds, rodents and lagomorphs.

Turning now to acetylcholine itself the data for the differential occurrence of this chemical in different parts of the brain are largely those of Aprison and his colleagues. Some comparison with 5-hydroxytryptamine, nor-adrenalin and dopamine can be derived from the figures of Table 15. This shows the concentration in m$\mu$moles/g for the four brain areas considered in Section VIII.

The actual paucity of acetylcholine data for specific points of the brain reflects the inherent difficulties of obtaining tissue for analysis. Although bioassays are sensitive enough to measure very small available quantities, the high rate at which it is metabolized, coupled with the ease with which it can change from the bound to the free form, make such assays subject to considerable error. However using the near freezing procedure which they had developed for use on rat brain Aprison and Takahashi (1965) determined the concentrations in the pigeon, and it is these that are contained in Table 15. That for the pons-medulla was 18 m$\mu$moles/g, and that for the mid-brain 29 m$\mu$moles/g. These values are higher than their analogues in the rat and probably represent the high choline acetylase activity which one can detect in these areas. The values of 1·6 and 15·5 m$\mu$moles/g for the cerebellum and telencephalon are comparable with those for the rat. Even a cursory examination of Table 15 will show that the acetylcholine is considerably in excess of the other molecular species. Aprison and Takahashi considered that this provided documentary evidence for the predominance and importance of cholinergic systems in the avian brain. Indeed they postulated that an excess of cholinergic neurons is present which follows a similar suggestion made by Whittaker (1953). They pointed out that if this is not the case then the obvious alternative would be that the available cholinergic neurons have a high storage capacity for acetylcholine.

A plot of the mean concentration within the four brain regions of any two of the neurohumours listed in Table 15 showed that in every case an increasing linear relationship exists within at least three regions. This suggested two questions. Why does it occur and, further, why is one point off the curve in every case? Surveying these problems Aprison and Takahashi pointed out that one might follow the suggestion that two physiological antagonistic systems exist—the ergotrophic and trophotropic systems. If acetylcholine and 5-hydroxytryptamine are related to one such system, and nor-adrenalin and dopamine to the other, then an increase in the components of one system would imply a concomitant increase in those of the other. If this was not the case a "balanced normal state" could never exist. The relationship between the two members of each of these pairs could result from dopamine being the precursor of nor-adrenalin on the one hand, and 5-hydroxytryptamine's postulated modulator action on the cholinergic system on the other.

In answer to the second question one naturally looks for local disparities in the concentration of the respective substances and these suggest themselves in the light of data from mammals. The deviant point in a plot of acetylcholine and dopamine is that for the telencephalon. It is well-known that dopamine sensitive neurons are in excess in the caudate and putamen of mammals where they provide the basis for dopamine therapy of Parkinsonian tremor. Such an excess in the pigeon would explain this deviant point. Similarly the odd site in the case of acetylcholine and nor-adrenalin is that for the pons-medulla. In mammals this region has very high concentrations of nor-adrenalin and in the dog only the hypothalamus has a higher value. Once again such local variations could account for the apparent disparity in the avian brain.

# References

Aldridge, W. N. (1964). A method for the characterisation of two similar esterases in the chicken central nervous system. *Biochem. J.* **93**, 619–623.

Altman, J. and Altman, E. (1962). Increased utilization of an amino acid and cellular proliferation demonstrated autoradiographically in the optic pathways of pigeons. *Exp. Neurol.* **6**, 142–151.

Aprison, M. H. and Hingtgen, J. N. (1965). Neurochemical correlates of behaviour. IV. Norepinephrin and dopamine in four brain parts of the pigeon during a period of atypical behaviour following the injection of 5-hydroxytryptophan. *J. Neurochem.* **12**, 959–968.

Aprison, M. H. and Takahashi, R. (1965). 5-hydroxytryptamine, acetylcholine, 3,4,dihydroxyphenylethylamine and nor-epinephrin in several discrete areas of the pigeon brain. *J. Neurochem.* **12**, 221–230.

Aprison, M. H., Takahashi, R. and Folkerth, T. L. (1964). Biochemistry of the avian central nervous system: I. 5-hydroxytryptophan decarboxylase, monoamine oxidase, and choline acetylase-acetylcholinesterase systems in several discrete areas of the pigeon brain. *J. Neurochem.* **11**, 341–350.

Aprison, M. H., Shank, R. P. and Davidoff, R. A. (1969). A comparison of the concentration of glycine, a transmitter suspect, in different parts of the brain and spinal cord in 7 different vertebrates. *Comp. Biochem. Physiol.* **28**, 1345–1355.

Aprison, M. H., Wolf, M. A., Poulous, G. L. and Folkerth, T. L. (1962). Neurochemical correlates of behaviour. III. Variation of serotonin content in several brain areas and peripheral tissues of the pigeon following 5-hydroxytryptophan administration. *J. Neurochem.* **9**, 575–584.

Baker-Cohen, K. F. (1968). Comparative enzyme histochemical observations on sub-mammalian brains. I. Striatal structures in reptiles and birds. II. Basal structures of the brain stem in reptiles and birds. *Ergebn. Anat. Entwicklungsgeschichte.* **40** (6), 7–70.

Baker, P. C. and Quay, W. B. (1969). 5-hydroxytryptamine metabolism in early embryogenesis—a review. *Brain Res.* **12**, 273–295.

Bartley, W., Birt, L. M. and Banks, P. (1968). "The Biochemistry of Tissues", pp. 163–223. John Wiley.

Bateson, P. P. G., Horn, G. and Rose, S. P. R. (1969). The effects of an imprinting procedure on regional incorporation of tritiated lysine into protein of chick brain. *Nature, Lond.*, **223**, 534–535.

Birge, W. J. (1962). A histochemical study of ribonucleic acid in differentiating ependymal cells of the chick embryo. *Anat. Rec.* **143**, 147–152.

Bonichon, A. (1958). 'L'acetylcholinesterase dans la cellule et la fibre nerveuse au cours du developpement. I. Differenciation biochimique précoce du neuroblast. *Ann. Histochim.* **3**, 85–93.

Bonichon, A. (1960). Development of cholinesterase in the mesencephalon of the chick embryo. *J. Neurochem.* **5**, 195–198.

Brightman, M. W. and Albers, R. W. (1959). Species differences in the distribution of extra-neuronal cholinesterase within the vertebrate central nervous system. *J. Neurochem.* **4**, 244–250.

Brodie, B. B. and Bogdanski, D. F. (1964). Biogenic amines and drug action in the nervous system of various vertebrate classes. *Progr. Brain Res.* **9**, 234–242.

Burt, A. M. and Wenger, B. S. (1961). Glucose 6 phosphate dehydrogenase activity in the brain of the developing chick. *Devel. Biol.* **3**, 84–95.

Correale, P. (1956). The occurrence and distribution of 5-hydroxytryptamine in the central nervous system of vertebrates. *J. Neurochem.* **1**, 22–31.

Craigie, E. H. (1930). Studies on the brain of the kiwi. *J. Comp. Neurol.* **49**, 223–357.

Davidson, J. (1957). Activity of certain metabolic enzymes during development of the chick embryo. *Growth* **21**, 287–295.

Davson, H., Kleeman, C. R. and Levin, E. (1962). Quantitative studies of the passage of different substances out of the cerebrospinal fluid. *J. Physiol.* **161**, 126–142.

Doemel, J. and Chilson, O. P. (1969). The developmental pattern of adenosine 5′ phosphate aminohydrolase in the brain of the embryonic chick. *Comp. Biochem. Physiol.* **29**, 829–833.

Dujovne, C. A., Levy, R. and Zimmerman, H. J. (1969). The correlation between serum and tissue levels of enzymes in six vertebrate species. *Comp. Biochem. Physiol.* **28**, 1193–1198.

Edström, R. (1964). Recent developments of the blood brain barrier concept. *Int. Rev. Neurobiol.* **7**, 153–190.

Emmanuelson, H. and Helmer, E. (1965). Dehydrogenase activities in early chick embryos. *Ark. Zool.* **18**, 449–459.

Felicioli, R. A., Gabrielli, F. and Rossi, C. A. (1967). The synthesis of phosphoenolpyruvate in gluconeogenesis. *Eur. J. Biochem.* **3**, 19–24.

Filogamo, G. (1960). Rapports entre l'activité acetylcholinesterasique et le dégré de differentiation des neurones de l'embryon de poulet. *C.R. Ass. Anat.* **46**, 251–255.

Friede, R. L. (1966). "Topographic Brain Chemistry", pp. 543. Academic Press, London and New York.

Fujita, S. (1963). Matrix cells and cytogenesis in the developing central nervous system. *J. Comp. Neurol.* **120**, 37–42.

Fujita, S. (1964). Analysis of neuron differentiation in the central nervous system by tritiated thymidine autoradiography. *J. Comp. Neurol.* **122**, 311–327.

Gal, E. M. and Marshall, F. D. Jnr. (1964). The hydroxylation of tryptophan by pigeon brain *in vitro*. *Progr. Brain Res.* **8**, 56–60.

Galindo, A., Krnjevic, K. and Schwartz, S. (1967). Micro-iontophoretic study on neurons in the cuneate nucleus. *J. Physiol.* **192**, 359–377.

Garrigan, O. W. and Chargaff, E. (1963). Studies of the mucolipids and the cerebrosides of chicken brain during development. *Biochim. Biophys. Acta.* **70**, 452–464.

Gayet, J. and Bonichon, A. (1961). Morphological differentiation and metabolism in the optic lobes of the chick embryo. *Regional Neurochemistry.* Proc. 4th Int. Neurochem. Symp. Varenna, pp. 135–150. Academic Press, London and New York.

Grafnetter, D., Gross, E. and Morganti, P. (1965). Occurrence of sterol esters in the chicken brain during pre-natal and post-natal development. *J. Neurochem.* **12**, 145–149.

Greenberg, D. M., Friedberg, F., Schulman, M. P. and Winnick, T. (1948). Studies on the mechanism of protein synthesis with radioactive carbon labelled compounds. *Cold. Spr. Harb. Symp. Quant. Biol.* **13**, 113–117.

Grillo, T. A. I., Okumo, G., Price, S. and Foa, P. P. (1964). The activity of uridine diphosphatase, glucose-glycogen synthetase in some embryos. *J. Histochem. Cytochem.* **12**, 275–281.

Gunne, L. M. (1962). Relative adrenalin content of brain tissues. *Acta Physiol. Scand.* **56**, 324–333.

Hanahan, D. J., Brockerhoff, H. and Barron, E. J. (1960). *J. biol. Chem.* 1917–1923.

Howell, J. McC. and Eddington, N. (1968). The neurotoxicity of sodium diethyl-dithiocarbamate in the hen. *J. Neuropath exp. Neurol.* **27**, 464–472.

Hughes, A. F. W. (1955). Ultraviolet studies on the developing nervous system of the chick. *In* "Biochemistry of the Developing Nervous System" (Waelsch, H., ed.), pp. 166–169. Academic Press, London and New York.

Inouje, A., Shinagawa, Y. and Y. (1966). Cytochrome $b_5$ and related oxidative activities in brain microsomes of fowl, toad and carp. *J. Neurochem.* **13**, 385–390.

Juorio, A. V. and Vogt, A. (1967). Monoamines and their metabolites in avian brain, *J. Physiol.* **189**, 489–518.

Kaluza, J. S., Burstone, M. S. and Klatzo, I. (1964). Enzyme histochemistry of the chick choroid plexus. *Acta Neuropath.* **3**, 480–489.

Kato, Y. (1959). The induction of phosphatase in various organs of the chick embryo. *Devel. Biol.* **1**, 477–510.

Klatzo, I., Wisniewsky, H. and Smith, D. E. (1965). Observations on the penetration of serum proteins into the central nervous system. *Progr. Brain Res.* **15**, 73–88.

Knox, W. E. and Eppenberger, H. M. (1966). Basal and induced levels of tryptophan pyrrolase and tyrosine transaminase in the chick. *Devel. Biol.* **13**, 182–198.

Kreps, E. M., Manukian, K. G., Patrikeeva, M. V., Smirnov, A. A., Chenkyaeva, E. I. and Cherkovskaia, E. V. (1964). Phospholipid distribution in hen brain. *Biokhimia.* **29**, 1111–1118.

Lajtha, A. (1956). The development of the blood brain barrier. *J. Neurochem.* **1**, 216–217.

Lajtha, A. and Marks, N. (1969). Dynamics of protein metabolism in the nervous system. *In* "The Future of the Brain Sciences" (Bogoch, S., ed.). Plenum Press.

Lowenthal, A., van Sande, H. and Karche, D. (1962). Fractionnement electro-phoretique de la déshydrogenase de l'acide lactique dans les proteins hydro-soluble du système nerveux central. *Arch. Int. Physiol. Biochim.* **70**, 420–422.

Mandel, P., Bieth, R. and Stoll, R. (1949). La répartition des diverses fractions lipidiques dans le cerveau de l'embryon de poulet durant la seconde partie de l'incubation. *C.R. Soc. Biol. Paris.* **143**, 1224–1226.

Masai, H., Kusunoki, T. and Ishibashi, H. (1966). Comparative studies of mono-amine oxidase and succinic dehydrogenase in vertebrates' forebrain. *In* "Evolution of the Forebrain" (Hassler, R., and Stephan, H., eds), pp. 271–275. Georg Thieme, Stuttgart.

Masai, H. and Matsano, S. (1961). Comparative neurological studies on respiratory enzyme activity in the central nervous system of sub-mammals. I. Birds. *Yokohama Med. Bull.* **12**, 265–270.

McCaman, R. E., Rodriguez de Lores Arnaiz, G. and Robertis, E. de (1965). Species differences in sub-cellular distribution of choline acetylase in the central nervous system. *J. Neurochem.* **12**, 927–935.

Miyamoto, K., Stephanides, L. M., and Bernsohn, J. (1966). *J. Lipid. Res.* **7**, 664.

Miyamoto, K., Stephanides, L. M. and Bernsohn, J. (1967). Incorporation of 1-$^{14}$C linoleate and linolenate into poly-unsaturated fatty acids of phospholipids of the embryonic chick brain. *J. Neurochem.* **14**, 227–237.

Nachmansohn, M. D. (1940). Cholinesterase in brain and spinal cord of sheep embryos. *J. Neurophys.* **3**, 396–402.

Nozdrachev, A. V. (1959). Concerning the effect of serotonin on the nervous system. *Dokl. Acad. Sci. U.S.S.R.* (Biol.) **125**, 454.

Pappenheimer, J. R., Heisey, S. R. and Jordan, E. F. (1961). Active transport of diodrast and phenolsulfonphthalein from cerebro-spinal fluid to blood. *Amer. J. Physiol.* **200**, 1–10.

Pscheidt, G. R. and Huber, B. (1965). Regional distribution of dihydroxyphenyl-alanine and 5-hydroxytryptophan decarboxylase and biogenic amines in the chicken central nervous system. *J. Neurochem.* **12**, 613–618.

Pscheidt, G. R. and Himwick, H. E. (1963). Chicken brain amines with special reference to cerebellar nor-adrenalin. *Life Science.* **7**, 524–526.

Pullman, B. and Pullman, A. (1963). "Quantum biochemistry", vol. 1, p. 867. John Wiley, New York.

Raina, A. (1963). Studies on the determination of spermidine and spermine and their metabolism in the developing chick embryo. *Acta Physiol. Scand.* **60**. suppl., **218**, 81.

Rebollo, M. A. and Hekimian, L. (1955). La fosfatasa alcalina el las primeras etapas del desarrollo del sistema nervioso central y periferico. *Acta. Neurol. Lat-amer.* **1**, 269–273.

Rogers, K. T. (1960). Studies on the chick brain of biochemical differentiation related to morphological differentiation. I. Morphological development. *J. exp. Zool.* **144**, 77–87.

Rogers, K. T., DeVries, L., Kepler, J. A., Kepler, C. R. and Speidel, E. R. (1960). II. Alkaline phosphatase and cholinesterase levels, and onset of function (see *ibid.* **145**, 49–60).

Rudnick, D. and Waelsch, H. (1955). Development of glutamo-transferase in the nervous system of the chick. *In* "The biochemistry of the developing nervous system" (Wallsch, H., ed). Academic Press, London and New York.

Singh, R. (1964). Some histochemical reactions for lipids in avian nervous tissue. I. The central nervous tissues. *J. Histochem. Cytochem.* **12**, 812–820.

Smith, D. E., Streicher, E., Milković, K. and Klatzo, I. (1964). Observations on the transport of proteins by isolated choroid plexus. *Acta Neuropath.* **3**, 372–386.

Solomon, J. B. (1958). Lactic acid and malic dehydrogenase in developing chick embryo. *Biochem. J.* **70**, 529–538.

Spector, W. S. (1956). "Handbook of Biological Data", pp. 584. Saunders.

Stevens, L. (1970). The biochemical role of naturally occurring polyamines in nucleic acid synthesis. *Biol. Rev.* **45**, 1–27.

Strumia, E. and Baima-Bollone, P. L. (1964). Ache activity in the spinal ganglia of the chick embryo during development. *Acta Anat.* **57**, 281–294.

Tsunoo, S., Horisaka, K., Kawasumi, M., Aso, K. and Tokue, S. (1963a). Free amino-acids in chick brain. Isolation and identification of histidine and anserine. *J. Biochem., Tokyo.* **54**, 355–362.

Tsunoo, S., Horisaka, K., Sasaki, S., Aso, K. and Tokue, S. (1963b). Free amino-acids in chicken eye. *J. Biochem. Tokyo.* **54**, 363–368.

Vernadakis, A. (1969). Sensitivity of chick embryos to chlorpromazine. *Brain Res.* **12**, 223–226.

Verzhbinskaya, N. A. (1949). Distribution of the enzyme carbonic anhydrase in the brain of vertebrates. *Izv. Akad. Sci. U.S.S.R. (Biol.)* **5**, 598–607.

Wahbe, V. G., Balfour, W. M. and Samson, F. E. Jnr. (1961). A comparative study on vertebrate brain mitochondria. *Comp. Biochem. Physiol.* **3**, 199–205.

Walker, M. S. and J. B. (1962). Repression of transaminase activity during embryonic development. *J. biol. Chem.* **237**, 473–476.

Wang, K-M. (1968). Comparative study of the development of enzymes involved in carbohydrate and amino-acid metabolism. *Comp. Biochem. Physiol.* **27**, 33–50.

Wang, K. M., Balzer, M. and Pascarella, M. (1966). ATPase patterns in tissues and their response to $Mg^{++}$ and $Ca^{++}$ in developing chick embryos. *Fed. Proc.* **25**, 450.

Wegelin, I. and Manzoli, F. A. (1967). Free nucleotides in chick embryo brain during development. *J. Neurochem.* **114**, 1161–1165.

Wenger, B. S. (1951). Cholinesterase activity in different spinal cord levels of the chick embryo. *Fed. Proc.* **10**, 268–269.

Whittaker, V. P. (1953). *Biochem. J.* **54**, 660.

Willmer, E. N. (1960). "Cytology and Evolution", 2nd ed. 1970, pp. 649. Academic Press, London and New York.

Winget, C. M., Wilson, W. O. and McFarland, F. Z. (1967). Response of certain diencephalic, pituitary and plasma enzymes to light in *Gallus domesticus*. *Comp. Biochem. Physiol.* **22**, 141–147.

Zacks, S. I. (1954). Esterases in the early chick embryo. *Anat. Rec.* **118**, 509–537.

# 4  The Development of the Avian Brain

## I. INTRODUCTION

For those people who are interested in the history of scientific ideas one can draw attention to a number of descriptions of the morphogenesis of avian brains which are of considerable antiquity. Baer (1837) was one of the first workers and gave an account of the situation in the chick. Subsequently during both the later part of the nineteenth century and also the early years of the present century a number of other workers made further contributions all of which influenced their successors to some extent. Some of these works are mentioned elsewhere in this book in sections which deal with particular brain regions. Others include those of Goronowitch, Henrich, Locy, McClure, Platt and Zimmermann who published in the closing decades of the last century; together with various workers such as Kamon (1906), Kupffer (1906), Meek (1907), Mesdag (1909), Biondi (1910), Tilney (1915), Palmgren (1921) and Schumacher (1928). These last were part of the great flowering of histological and embryological studies which characterizes the early part of the twentieth century and links it with the foregoing twenty years at the end of the nineteenth century.

Many of the works which were published around that time were particularly concerned with identifying the contributions made by different metameres. Carpenter's account gives details of histological structure but its content is largely restricted to a consideration of the development of the oculomotor and abducens components. However, various other brain regions and allied phenomena were described by some workers. Mesdag and Ingvar (1919), whose names are classics in terms of developmental studies, considered the whole brain and the cerebellum respectively, whilst both Tilney and Palmgren produced information about the diencephalon. An account of analogous detailed studies on the telencephalon can be omitted at this point, in view of the very considerable controversies which have raged over it, and left for the section on telencephalic structure. It is, however, important to emphasize that many of the ostensibly phylogenetic studies of the 1920s and 1930s included embryological data. For example, further information on the gradual development of various regions is provided by Durward (1934).

## II. CYTOLOGICAL STUDIES

The developmental aspects of neuronal cyto-differentiation have been summarized in recent years by Langman (1968) and Wechsler and Kleihues (1968). From their light and electron-microscopical investigations the latter workers concluded that there are three principal phases of neuronal development. These can be summarized briefly as:

(i)   The stage of undifferentiated matrix cells;
(ii)  A later stage of primordial determination and early cellular differentiation;
(iii) The final phases of differentiation and maturation.

The relationships envisaged between the developmental processes and cyto-differentiation are portrayed by Table 16, which deals in particular with the development of nerve cells and the other neuro-ectodermal constituents of the nervous system. More specific information on the cell lineage is contained in Table 17.

### TABLE 16

*The major phases of cytological development in the brain of* Gallus.

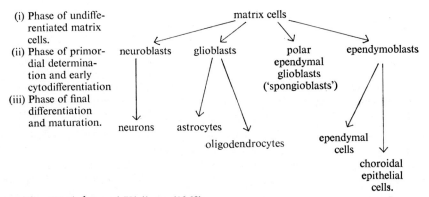

(i) Phase of undifferentiated matrix cells.
(ii) Phase of primordial determination and early cytodifferentiation
(iii) Phase of final differentiation and maturation.

After Wechsler and Kleihues (1968).

The archetypal matrix cells are characterized by a primordial bipolarity. Within the cytoplasm there are mitochondria together with a small Golgi-complex and also a few filaments and micro-tubules. The endoplasmic reticulum is barely seen and instead there are rosettes of free ribosomes. The gradual transformation of such neuro-ectodermal matrix cells into migrating neuroblasts is accompanied by a loss of the epithelial features. The number of free ribosomes and mitochondria increases but there is now a minor proliferation of the granular endoplasmic reticulum. As seen in explants of the spinal cord the final stages of both the maturation and differentiation of the definitive nerve cells are associated with an overt increase in the number of ribosomes bound to the endoplasmic reticulum. The proliferation of this granular endoplasmic reticulum then culminates in the formation of the ergastoplasm of the Nissl bodies. Even if the absolute number of free ribosomes increases during this period the ratio of free to bound ones is changed, and contrasts markedly with the situation in both matrix cells and primitive neuroblasts. Wechsler and Kleihues understandably suggest that such changes can probably be correlated with the progressive variations in the rate of protein synthesis which characterize cyto-differentiation.

## TABLE 17

*The probable cell lineages and relationships between different types of nerve cells in the avian central nervous system.*

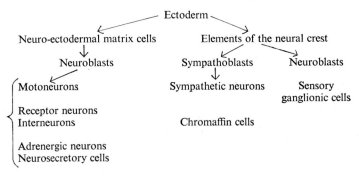

Using an auto-radiographic study of the rate of cumulative labelling which follows treatment with tritiated thymidine, Fujita (1963) showed that the matrix cells of the neural tube in a 6-day chick embryo become labelled at a linear rate. This labelling resulted in a value of 100% for the whole cell population after a period of 10 hours. As a result he concluded that the matrix cells which will subsequently give rise to various structures such as nerve cells, neuroglia and ependymal cells comprise a homogeneous cell population at this early stage. Klika and Jelinek (1965) who studied the areas membrancea ventriculi in chicks of comparable age by means of phase contrast and electron-microscopy suggested that its labyrinthine nature probably indicates involvement in electrolyte transport between the encephalic and amniotic fluids. It was for this reason that they postulated that it had an ionic regulatory function. It is interesting that this fits the hypothesis which was put forward by Willmer (1960), Pearson and Willmer (1963) for a bipartite classification of cell function. It is also worth drawing attention at this point to Willmer's theory of cell lineages which originally led him to direct attention to the predominance of tyrosin metabolism within the derivatives of the neural crest (Table 16). This has subsequently led to far-reaching investigations into the significance of tyrosin analogues in embryological potentiation.

Experimental determinations of the rates of both growth and maturation in chicken nervous tissue explants were carried out by Murray and Benitez (1968). They obtained the perhaps surprising result that substitution of deuterated water for the natural product caused accelerated growth and maturation together with some multiplication of differentiated neurons. In the case of cultures derived from the hypothalamus of 12- and 17-day-old embryos they found that the satellite cells did not participate in this observed cell division. Instead they remained relatively immobile and embracing a neuritic process which was never withdrawn. The optimal $D_2O$ concentration for these cells of hypothalamic origin was $12 \cdot 5\%$. At the same time that they presented these data Szentágothái (1968) pointed out that hypertrophy of the neuro-fibrillar apparatus similar to that which they observed is one of the standard structural reactions which one sees in a variety of pathological conditions. An extremely marked hypertrophy of this nature, including both the neurofilaments and neurotubules, is of widespread occurrence in such circumstances and comparable changes take place in some synaptic terminals during the early phase of cell degeneration.

Information on the morphological changes that occur during the later development of the choroid plexus, this time that of the telencephalon, was presented by Klatzo et al. (1966). After staining plexi derived from 8–20-day old embryos by Cowdry's method for mitochondria and also PAS, the most striking result was the observation of a change from pseudo-stratified to columnar epithelium at the 9th to 12th day. This was coupled with an overt increase in the number of mitochondria and the final distribution of these sub-cellular organelles was in close proximity to the brush border. In the chick this plexus has not descended into the developing ventricular cavity at the 9th day but has a superficial location on the dorsal surface between the telencephalon and diencephalon. At this stage one can just make out the branching of the third ventricular plexus which subsequently gives rise to the right and left ventricular plexi. By the 12th day the plexi are in the ventricular cavities. In an earlier paper Smith et al. (1964) had described the gradual changes in the rate of uptake of fluorescein labelled albumin which can be observed during this period. Uptake increases in a disto-proximal direction and indicates the progressive involvement of the emerging columnar cells as the mitochondria become aligned at the brush border. Up until the 12th day both this mitochondrial alignment and also FLA uptake are intermittent but they both become firmly established at that stage. Clearly such data substantiate the information on later embryos which was referred to in Chapter 3, Section III.

4*

# III. THE DEVELOPMENT OF THE CHICK BRAIN

At this point it is probably useful to include a brief review of the principal stages of the development of the most commonly studied avian brain, namely that of the chick. Further, and more specific details of, say, the cerebellum and telencephalon are given in the respective sections. We can therefore limit ourselves to the overall morphogenetic pattern. Surprisingly enough although isolated figures of stages older than 5 days of incubation are featured in such works as Hamilton (1952), Kappers (1947) and Kupffer (1906), a detailed description was not really produced prior to that of Rogers (1960). This presumably reflects the emphasis which was placed in both experimental work and undergraduate teaching on other organ systems. The finer points of brain development are frequently skated over with only a brief reference to the appearance of the main brain divisions. In his study Rogers used Barred Rock female × New Hampshire or Rhode Island Red eggs which had a hatching period of 21 days. At 4 days the major features of the brain are clearly demarcated, the sole exception being the absence of any great degree

Abbreviations used for Figs 19–23.

III–XII, cranial nerves
3rd ven., 3rd ventricle
4th ven., 4th ventricle
arb. vit., arbor vitae
cav. cbl., cavum cerebelli
cbl., cerebellum
cbl. roof, cerebellar roof
cer. h., forebrain hemisphere
chor. pl. III, choroid plexus III
chor. pl. IV, choroid plexus IV
com. ant., anterior commissure
com. inf., inferior commissure
com. pall., pallial commissure
com. post., posterior commissure
com. tec., tectal commissure
com. troch., trochlear decussation
corp. str., corpus striatum
ep., epiphysis
for. mon., foramen of Munro
h.b. roof, hind-brain roof
hyp., hypophysis (oral)
inf., infundibulum (neural)
isth., isthmus
iter, iter
lam. term., lamina terminalis
lat. floc., paraflocculus
lat. meso., lateral mesocoele

lat. ven., lateral ventricle
mam. tub., mammillary tubercle
med. ob., medulla oblongata
med. roof, medullary roof
mesen., mesencephalon
meten., metencephalon
meten. floor, metencephalic floor
meten. roof, metencephalic roof
myelen., myelencephalon
nuc. isth. nucleus isthmi
olf. l., olfactory lobe
olf. n. olfactory nerve
opt. chia., optic chiasma
opt. l., optic lobe
opt. n., optic nerve
opt. st., optic stalk
para., paraphysis
paren., parencephalon
pli. enc. ven., plica encephali ventralis
pons., pons
rec. opt., recessus opticus
synen., synencephalon
thal., thalamus
tor. trans., torus transversus
tub. post., tuberculum posterius
vel. trans., velum transversum

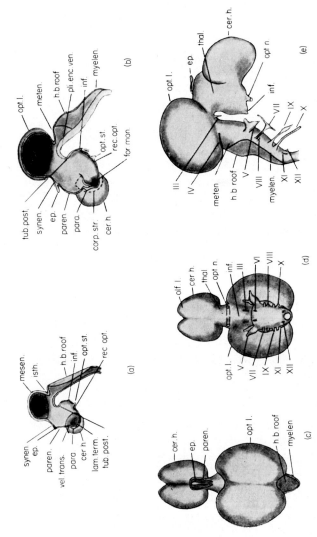

FIG. 19. (a) Median sagittal view of 96 hour chick brain. (b) Median sagittal view of 5 day brain. (c), (d) and (e) Dorsal, ventral and lateral views of 6 day brain. (Rogers, 1960).

of differentiation between the metencephalon and myelencephalon. A median sagittal section of such a brain is shown in Fig. 19(a). No portion of the roof is very greatly thickened but both the par-encephalic roof and also that of the hind-brain are particularly thin. The first appreciable thickening of the wall occurs on the 5th day. The hemispheres are almost round in section but traces of the olfactory lobes are already indicated at this time by ventro-lateral bulges. The foramina of Munro have become very narrow and form vertical passages. The basal forebrain components, which are discussed in more detail elsewhere, appear as thickenings which occupy about one-quarter of the width and some half of the height of the lateral ventricle. The lamina terminalis remains thin. Gradually the floors of the mid- and hind-brains increase in thickness and a thickened strip of optic tract fibres ascends each lateral thalamic wall. Most marked at points nearest to the optic stalk this tract extends about half way up the wall. The pontine flexure is still very obtuse and on the 5th day the cervical flexure is somewhat reduced by comparison with the condition 24 hours earlier. The diencephalon is still rather narrow in overall appearance with the epiphysis visible as a dorsal tube-like prominence. A median sagittal section of a brain on the 5th day is illustrated in Fig. 19(b) and enables an overt comparison to be made with the comparable section of Fig. 19(a). There is a clear but shallow constriction between the two optic lobes and the hind-brain is now more clearly differentiated into metencephalon and myelencephalon. All the cranial roots are present and those of V, VII, VIII, IX, X and XI undergo a gradual upward shift which reflects the considerable amount of growth that is taking place in both the pons and the ventral medullary region. The ensuing changes are most easily considered stage by stage.

## ca. Stage 29

During the 6th day the observable changes are not so extensive. Dorsal, ventral and lateral views of brains of this age are provided by Fig. 19(c, d, e). The walls continue to thicken and this is particularly true of the thalamic region where several grooves form on the inner surface. Anteriorly the fore-brain region has also increased in size and a thickening in the terminal lamina represents the transverse torus. In overall appearance the hemispheres are elongated and the olfactory lobes are clearly marked. As the hypophysis changes its position in relation to the diencephalic floor it begins to project out at right angles rather than lie parallel to it. The mamillary tubercles, together with the anterior, posterior, inferior and trochlear commissures are all visible. The median constriction between the optic lobes gradually deepens but the lateral mesocoeles are still in open contact and there is not yet any appreciable lateral displacement of the lobes themselves.

## ca. Stage 31

The gradual widening which will later result in the characteristic shape of the avian brain begins in earnest at the 7th day. The hemispheres begin to separate at their posterior poles and the same is true of the anterior regions of the optic lobes. As a result of this the tectal commissure linking these last-named lobes assumes a triangular form. On the metencephalon a distinct and rounded pouch by now represents the developing cerebellum and all the brain stem components become noticeably wider. Concomitant thickening of the walls also leads to a further reduction in the size of all the ventricles. In particular the median partition between the optic lobes constricts the underlying ventricle so that its cavity is reduced by about 50%. Elsewhere the thickenings in the parencephalic roof herald the gradual appearance of the choroid plexus in the region of the third ventricle. The progressive development of a secretory function in the constituent tissue was described in Section II. At the same time the thin part of the roof of the fourth ventricle is reduced in size, although not to the same extent as on the following day. Furthermore a thin area is still present in the floor of the region close to the optic recess. In front the remainder of the terminal lamina becomes considerably thickened and the olfactory bulbs are now somewhat elongated and taper at their distal ends. Posteriorly the pontine flexure, which has been in existence since the 5th day, undergoes a marked sharpening of its angle.

## Stage 34

By the 8th day this increase in the angle of the pontine flexure has been carried still further. Dorsal, ventral and lateral views of the brain at this stage, together with a median sagittal section, are contained in Fig. 20. The lateral displacement of the forebrain hemispheres and the lateral rotation of the optic lobes continues and is again accompanied by a widening of other brain parts. The lateral mesocoeles now communicate with each other, and also open into the iter, by a foramen whose size is only a quarter that of the 5th day. The nucleus isthmi is present and shows itself as an eminence in the hind part of the lateral mesocoele. Dorsally the epiphysis, which has been gradually elongating and shifting its position since day 5, is now, as in the adult, attached just in front of the posterior commissure.

In the forebrain the increased size of the telencephalic nuclear masses occludes most of the interior of the hemispheres, and the foramina of Munro are reduced to narrow slits which run from the paraphysis down to the transverse torus. As will be clear from Chapter 3, both of the definitive choroid plexi are also clear, and thickened cords of tissue within the parencephalic roof project forward through the foramen of Munro and into the lateral ventricle. The choroid plexus of the fourth ventricle is also indicated by

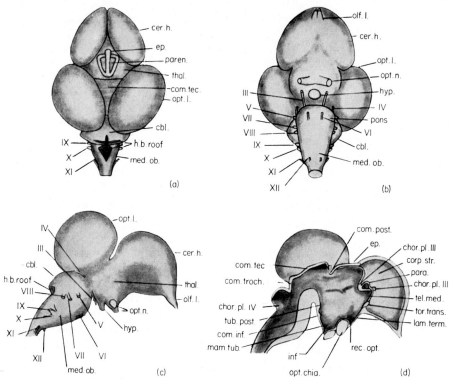

FIG. 20. (a–d). Dorsal, ventral, lateral and median sagittal views of an 8 day brain. (Rogers, 1960). For abbreviations see Fig. 19.

other thickened cords which grow up from each lateral wall towards the mid-dorsal line at the level of the constriction between the cerebellum and the medulla.

## *Stage 35*

During the 9th day of incubation the progressive rotation of the optic lobes is accelerated. Consequently they take up a lateral position and their longitudinal axes make an angle of 25° with the perpendicular. This progressive displacement increases the distance which separates them posteriorly. Furthermore the directional changes which take place in the region of the pontine flexure result in those medullary structures lying behind this point being aligned more or less in parallel with the axis of the forebrain. The choroid plexi continue to increase in size and that of the fourth ventricle now completely traverses the brain roof although thin striae alternate with thick ones and some thin, translucent roof still persists both in front of, and behind, the plexus tissue.

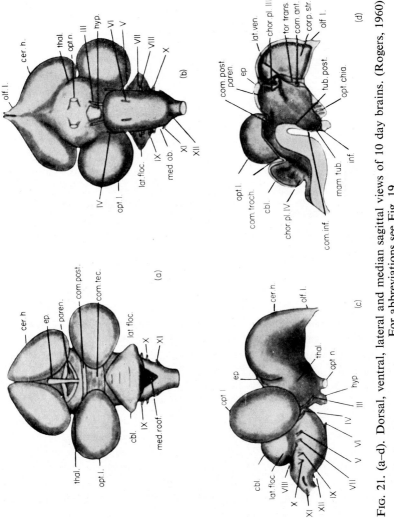

FIG. 21. (a–d). Dorsal, ventral, lateral and median sagittal views of 10 day brains. (Rogers, 1960). For abbreviations see Fig. 19.

## Stage 36

The optic lobes continue their rotation on the 10th day and their former anterior ends now have a definitively lateral position. However, the lateral displacement of the whole lobe also continues, indeed accelerates, and is associated with a downward tilting of the lateral regions. Dorsal, ventral, lateral and median sagittal views of the brain at this stage are contained in Fig. 21. The cerebellum has now attained a considerable bulk and begins to push forward between the hind part of the two tectal lobes. Lateral flocculi and paraflocculi, which were just present at 9 days are now definitive projections, whilst four transverse fissures are visible on the dorsal surface by 228 hours (see Section VII B below). Posteriorly the cerebellar component of the thin roof of the fourth ventricle now traverses the posterior face of the cerebellum itself although the development of the choroid plexus narrows it from behind. Anteriorly plexal growth causes a pronounced thickening of the parencephalic roof so that this projects into the ventricles as laminar plates which are expanded distally and overlie certain striatal components.

## Stage 37

At the 11th day the gradual expansion of the telencephalic hemispheres makes them press against the optic lobes and forces the outer borders of these backwards. By now 5 transverse fissures are visible on the dorsal surface of the cerebellum but the overall rate of growth of this region has now slowed down. Internally the cerebellar walls have thickened and they are now some four times their size on the previous day. Alongside this the cerebellar cavity is correspondingly reduced and only represents about one-third of the total cerebellar volume. When seen in section it becomes quite clear that there are seven visible fissures with the second to fifth penetrating deeply into the organ. The posterior wall also now projects out over the choroid plexus and presses down upon it so that the thin intervening roof is folded outwards. By the 12th day this has itself been transformed into plexus tissue although the area which lies behind the choroid plexus remains thin up until the 14th day. Anteriorly the development of the plexus in the region of the third ventricle results in its anterior projections completely filling the dorsal part of the foramina of Munro. Along with this the thickened walls of the optic lobes, together with the growing nucleus isthmi complex, continue to reduce the size of the lateral mesocoeles.

## Stage 38

During the 12th day it is the cerebellum which undergoes the most growth and its walls become so thick that the cavum cerebelli is very much reduced.

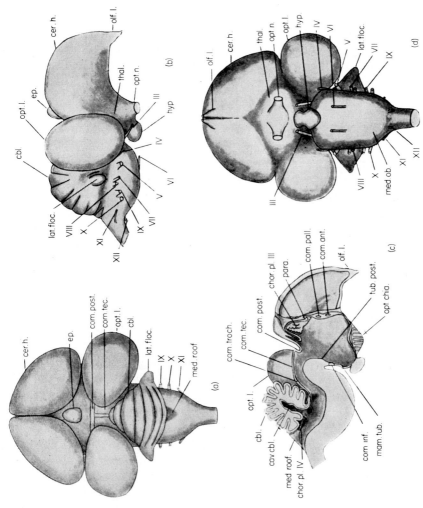

FIG. 22. (a–d). Dorsal, lateral, median sagittal and ventral views of 13 day brains (Rogers, 1960). For abbreviations see Fig. 19.

Six fissures are visible externally and a further 5 more can be detected when it is viewed in sagittal section. In front of this level the progressive increase in size of the isthmi and other nuclear complexes cuts down the cavity of the lateral mesocoeles even further. Dorsally the choroid plexus of the third ventricle now forms an almost complete covering for the diencephalic roof. There are indications of the habenular commissure between this plexus and the epiphysis, but it is rather difficult to determine the exact time at which these appear. There are also traces of a commissure within the transverse torus and just above the anterior commissure.

## Stage 39

Apart from a continuing increase in width throughout the whole length of the brain, together with a somewhat greater relative increase in the size of the cerebellum, the overall appearance of the chick brain on the 13th day of incubation is much the same as on the 12th. Figure 22 contains dorsal, lateral, median sagittal and ventral views of this stage which, with the foregoing qualifications, also depicts the situation on the previous day. The reduction in the volume of the cerebellar cavity has progressed to a stage in which it is only represented by a narrow vertical slit, and the continued growth of both the walls of the optic lobe and the nuclear complexes causes a similar reduction in the mesocoel of each side. These are by now just narrow, lateral cavities. In a comparable manner the choroid plexus of the third ventricle obliterates much of the cavity beneath the parencephalic roof. Those projections of the plexus which had previously penetrated the lateral ventricles are, however, pushed backwards by the development of the nuclear masses in the forebrain.

After 14 days of incubation the expansion of the hemisphere material in front, coupled with an analogous expansion of the cerebellum behind, finally crowds the optic lobes out and they become elongated laterally. Indeed the hemispheres have expanded so much that the epiphysis lies wedged between them. From this time until the 17th day there is also a progressive thickening of that part of the medullary roof situated just behind the choroid plexus, whilst within the cerebellum the definitive white fibre tracts of the arbor vitae begin to replace the cavum cerebelli. Externally further fissures are also visible so that a total of 13 can be counted with relative ease.

## Stage 41

At the 15th day both the breadth of the entire brain and the bulk of the cerebellum are even greater. This last is so large that it pushes strongly against the posterior end of the optic lobes. The arbor vitae within it is also so well developed that the final vestiges of the original cerebellar cavity are

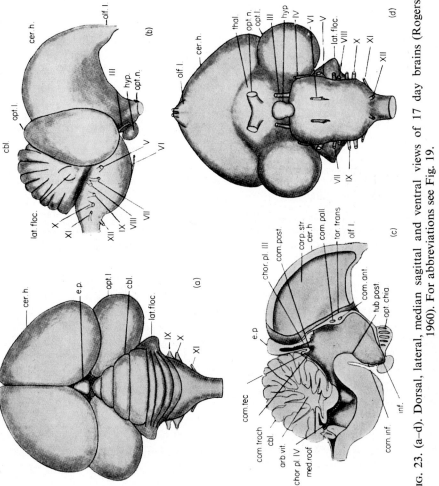

FIG. 23. (a–d). Dorsal, lateral, median sagittal and ventral views of 17 day brains (Rogers, 1960). For abbreviations see Fig. 19.

almost totally obliterated and only a small pocket can be seen remaining in a basal position.

### Stages 42 and 43

The growth of the cerebellum is, however, by no means complete. It continues throughout the 16th and 17th days and the organ projects further and further forward between the laterally displaced tectum of each side. By the end of the 17th day it has nearly reached the level of the epiphysis and this, reflecting the overall morphogenetic "pressures", gradually assumes a position in which it points backwards. Inside the cerebellum itself the proliferating arbor vitae now finally leads to a complete obliteration of the cavum cerebelli. Behind and below this the medullary roof is now so thick and expanded that it is folded into a convoluted mass. The more anterior roof of the mesencephalic and diencephalic regions is hidden from view because of the telencephalic and cerebellar enlargement. In Fig. 23 there are dorsal, lateral, ventral and median sagittal views of brains excised at this period and in which such characteristics can be detected. It will be clear that the principal features of the brain are therefore established by this time.

Although the gradual appearance of electrical activity in the brain is described in Chapter 15, we can note here a number of changes that take place during the stages which have just been described. Sedlaček (1966) found that the steady potential of the telencephalic hemispheres rose during incubation and maximum increases in the activity of the cerebellum and optic lobes occurred on days 16–17, and 20–21. The maximal levels for the forebrain were delayed by a further 2 days. Around the time of hatching the steady potential was ca. 14 mV. The depressant effect of $K+$ ions when applied to the hemisphere surface did not seem to manifest itself until the 18th day. Sedlaček and Maček (1966) found a low frequency impedance in hemisphere tissue at 400 cyc/sec, and high frequency impedance (at 10–40 megacyc/sec) in the hemisphere, cerebellum and optic lobes of 11–21-day-old embryos. Between the 17th and 21st day the low impedance rose from 1·5 to 3·5 K ohms, and the high impedance maximum rose from the range 1·5–2·5 K ohms, to a higher range of 7–8 K ohms. In all three regions a frequency dependant maximum of 25 megacyc/sec is reached on the 17th day and high frequency impedance increases in parallel.

## IV. THE DEVELOPMENT OF THE BRAIN OF
### STRUTHIO

As in the Kiwi, and the species of Section VI, our knowledge of the embryology of the brain of *Struthio camelus* reflects in large part the work of

Krabbe (1956). Kuenzi (1918) figured the adult brain and Ingvar (1918) the cerebellum of an embryo. Further studies of the adult condition by Craigie are considered in the chapters relating to the telencephalon but these do not contain any embryological information. Krabbe had 6 embryos at his disposal which measured 12, 16, 27, 35, 57, and 370 mm respectively. In terms of the 6 developmental stages which he established for mammalian material and which are referred to in Section VI these were:

(a) Intermediate between the first and second stage;
(b) Second stage;
(c) Third stage,
(d) Between the third and fourth stage:
(e) In a later and seventh stage.

Broadly speaking such stages represent

I. The presence of 4 distinct primordial regions.
II. Division of the telencephalon into hemispheres.
III. Division of the mesencephalon into tectal hemispheres.
IV. Pronounced flattening of the primordium and the appearance of an obtuse angle between anterior and posterior brain axes.
V. Rather greater flattening and more pronounced cerebellar development.
VI. More advanced than stage V but rather heterogeneous.

The earliest stage corresponds in its development to that of 5·5 mm *Melopsittacus* material and 8 mm *Pica* material. Already some differences occur. The entire anlage is larger than that of the budgerigar and magpie and about three times the size of the humming bird *Anisoterus*, a fact which can scarcely be considered surprising. It is, however, of considerable interest that the relative size of the forebrain is less than in the case of the budgerigar although the situation is very much the reverse in the case of the mesencephalon. The axis through the telencephalon, diencephalon and mesencephalon forms a right angle with the more posterior one that traverses both this last and the rhombencephalon. In relation to the total volume of the ventricles the walls are thin throughout but those of the basal mesencephalic and rhombencephalic regions are about twice as thick as those elsewhere.

The telencephalon is already divided into globular hemispheres the walls of which are in contact throughout their length. At the base two thin bundles of olfactory fibres penetrate the tissue but no definitive olfactory bulb exists. An optic fascicle located at each side of the telencephalic-diencephalic transition is hollow and communicates with the ventricle. The front part of the diencephalic region itself is formed by a rather large dorsal sac and from the hind border of this the thin stalk of the epiphysis emerges and enlarges distally into a pear shaped formation which is somewhat curved and directed forwards. Its external surface bears a few button-shaped excrescences. Behind this structure the diencephalon is roofed by a transverse metathalamic bridge at the base of which there is a prominence which separates it from the thalamic and hypothalamic *anlage*.

The mesencephalon is also distinctly separated from the thalamic area and has the form of a flattened globe with its basal walls thicker than the dorsal and lateral ones. The spacious ventricle within it is constricted behind prior to the expansion which gives rise to the even more capacious rhombencephalic ventricle. The walls in this narrowed isthmus region are rather thick. However it is the bulbo-pontine area with well developed trigeminal and stato-acoustic nerves which contributes the thickest part of the primordial brain, and dorsally a weak and rather indistinct cerebellar plate lies in front of the extensive ependymal roof.

In the 16 mm embryo the brain is somewhat straighter and the telencephalic–mesencephalic axis forms an obtuse angle with the hind axis. Furthermore, although the general proportions of each constituent region are very similar to those of the foregoing brain the diencephalon is somewhat smaller and the whole brain somewhat narrower. The forebrain hemispheres are rather compressed laterally. Their walls are thin and an inner densely nucleated sheet is surrounded by an outer, sparsely nucleated one of similar thickness. In the basal regions evidence of the developing striatum (see Section VIII) is provided by a densinuclear zone located near to the ventricular wall.

Dorsally a small sac with a large opening to the ventricle comprises a "paraphysis" with external button-shaped structures, and the diencephalic sac itself is of approximately equal size when viewed in both sagittal and transverse planes. Behind this, and arising from its posterior border, is the hollow epiphysis bearing vesicles on its external borders and some pigment granules within its cells. Further down the vertical axis thick walls indicate the developing thalamus. This is not clearly differentiated from the underlying hypothalamus, below which a collection of cavities represents the adenohypophysis. The mesencephalon is relatively larger than previously and, whilst somewhat flatter, is dilated both longitudinally and transversely. The walls remain thin by comparison with the ventricular volume and whilst there is no trace of tectal hemispheres, differentiation within the walls themselves leads to an outer and thinner layer provided with few nuclei, and an inner, thicker densi-nuclear zone. Posteriorly the ventricle is again constricted in front of the rhombencephalon which is now relatively smaller than was previously the case.

The 27 mm embryo is broadly comparable in its degree of differentiation with the 15 mm *Spheniscus*, 20 mm *Ardea*, 11 mm *Phasianus* and 14·5 mm *Astur* material of Table 18 although by now the actual size of the brain of *Struthio* exceeds that of these other genera. Other differences include the size of the mesencephalon, which is larger than the telencephalon in the penguin, etc. but not in the ostrich; the isthmus, which is not so markedly constricted in *Struthio*; and the optic fascicles, paraphysis and epiphysis which are relatively larger in *Struthio*. The pontine section is rather flattened and therefore resembles the condition in the pheasant. By the 35 mm stage the brain has developed considerably. The two brain axes are again obtuse, although at right angles in the intermediate stage, and the forebrain region is now enlarged in both lateral and vertical directions. The rhinencephalon forms a cone-shaped prominence whose central cavity communicates with the forebrain ventricle through a narrow canal. Its walls comprise a densely nucleated matrix situated close to the ventricle together with a more external layer that has moderately dense nuclei. This represents numerous large nerve cells and rather fewer small cells. A sheet of tissue which lies externally is continuous with the abres of the olfactory bundles.

In cross-section the telencephalic walls themselves are also more obviously stratified. There are three principal layers which differ in their nuclear density, and large striatal rudiments are situated basally. The diencephalon is similar in size to that of the 27 mm embryo but the mesencephalon is now divided into two discrete hemispheres by a median sagittal furrow. The walls of this region are also greatly thickened and equipped with marked cellular layers. Slightly broader than the telencephalon, it is sharply delimited from the rhombencephalon whose ependymal roof is now somewhat restricted and whose bulbo-pontine region shows a greater degree of differentiation

The fifth stage embryo is similar in most characters to those of *Spheniscus*, *Cygnus*, *Asio*, *Chlorostilbon* and *Melopsittacus*. The greatest differences concern *Asio* and *Chlorostilbon*. The angle between the two primary axes is now approximately 180°. The whole brain is very much larger and the telencephalic and mesencephalic hemispheres are in contact with each other. The cellular layers in the walls of the rhinencephalon are more clearly marked than those of the telencephalon, although when viewed in transverse sections the various nuclear rudiments of this last are now both large and prominent. An extensive choroid plexus is present within the lateral ventricles and further back the pars distalis of the hypophysis is much larger and more distinct than the pars nervosa. Below the dorsal diencephalic sac the thalami comprise two discrete blocks of tissue and the mesencephalon is now separated from both this region and the rhombencephalon. In fact this last-named region is now of a relatively small size by comparison with the more anterior brain components and Purkinje cells are detectable within the overlying cerebellum.

The seventh stage embryo has progressed further than any of the brains which are considered in Section VI and probably represents the degree of development that is attained within the immediate pre-hatching period.

The forebrain–mid-brain and mid-brain–hind-brain axes lie on a straight line, the total volume has increased considerably but nevertheless the degree of differentiation, other than that within the cerebellum, is not greatly increased. Such differences as do exist include the fissiform appearance of the lateral ventricles, the reduction in the size of the dorsal component of the diencephalon, the increased size of the tectal walls which exceed the volume of the remaining ventricle in the region, and the degree of differentiation within the pontine area.

As would be expected the cerebellum is by now divided into a series of thick folia which are separated from each other by narrow fissures and are often divided into sub-folia distally. Those in the hind part are thicker than the more anterior ones and in all of them there is a thin outermost layer of densely placed nuclei. Inside this a medullary layer with dispersed round nuclei is separated from an internal layer of intensely staining, densely packed, and small nuclei. Between these two layers there are numerous Purkinje cells. In conclusion one can note that at no stage does the whole brain primordium have a relationship to body size which would be comparable with that seen in the other genera, and, despite the overt differences which exist between ostriches and other birds, no great corresponding difference occurs in their brain development.

## V. THE DEVELOPMENT OF THE BRAIN OF *APTERYX*

A detailed description of an embryonic Kiwi brain was provided by Krabbe (1959). Apart from this account there are few others. In his classic description of the genus Owen (1841) did not describe the brain because of its state of decomposition. Craigie (1930) and Durward (1932) gave accounts of both the nuclei and axonic fascicles, but the only paper which deals with embryonic material is that of Parker (1891). The external features of the forebrain, particularly the large size of the olfactory bulb and the relatively long telencephalic hemispheres which have less dorsal curvature than those of other birds, have frequently been interpreted as primitive characters. However, the internal anatomy is essentially avian and the lateral ventricle is well-developed with the sub-divisions seen elsewhere. Perhaps the most interesting features of the adult are the presence of 5 distinct cell laminae situated just to the side of the so-called hippocampal fissure, together with traces of analogous laminations on another area of the lateral surface. These have been taken as indicating an origin from ancestors which already possessed a laminated neo-cortex similar to that of the most primitive living mammals. The occurrence of such laminae in this and other birds is discussed in Chapter 13. In other parts of the brain there are a number of unusual features. For example, the diencephalon and neighbouring regions are curiously distorted with the upper parts displaced backwards to a remarkable degree. As a result the optic chiasma lies at the front end of the third ventricle, the foramen of Munro appears directly over the habenular commissure, and the dorsal part of the diencephalon has been carried back to a position above the mid-brain. This means that a considerable amount of it lies directly above the optic tectum and is in fact fused with it. The various tracts and nuclei of these areas are therefore considerably displaced. Furthermore the habenulae are large, and the epiphysis is a mass of well developed glandular tubes. As might be expected in a flightless genus in which vision is not the predominant sensory modality the optic lobes are small, as is the cerebellum.

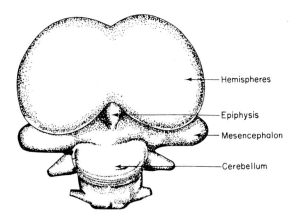

Hemispheres
Epiphysis
Mesencephalon
Cerebellum

FIG. 24. Dorsal view of the brain of an embryonic Kiwi. (After Krabbe, 1959).

The embryonic material which was studied by Krabbe was derived from a specimen which measured 6·6 cm from the root of the beak to the rudimentary tail. The brain anlage was clearly divided into telencephalon, diencephalon, mesencephalon and rhombencephalon (see Fig. 25), and it was the forebrain which had the largest volume. There was no distinct division between the metencephalon and myelencephalon but anteriorly a large presumed rhinencephalon was situated at the base of the rostral pole of the hemispheres. Histological examinations showed that this had a markedly stratified organization. An outer thick layer was probably myelinated and apparently devoid of nuclei. Inside this there was a thin layer provided with chromophilic nuclei that were rather larger than those elsewhere in the structure. Within this the bulk of the material was of uniform appearance and provided with small nuclei.

The hemispheres themselves were solid, pear-shaped masses in which a fissiform ventricle was visible. At the base of the caudal part of the fore-brain two optic fascicles issued out. These appear to be massive, slightly curved at their origin but straight distally, and within the brain they continue to a solid chiasma. In front of these fascicles there was a curious feature (Krabbe, 1957). This consisted of a low prominence which measured 1 mm in the longitudinal direction and 3 mm transversely. The hind part undergoes a gradual transition to the basal part of the telencephalon but it is well-demarcated by a furrow laterally. Rostrally it presents two lips which fuse behind.

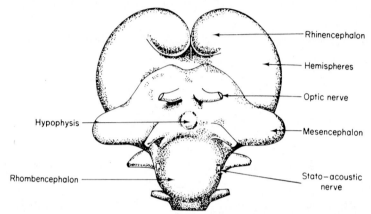

FIG. 25. Ventral view of the brain of an embryonic Kiwi. Note form of telencephalic hemispheres and the anterio-posterior compression. (After Krabbe, 1959).

Its internal cavity is continuous with the internal canal running through both the base of the forebrain and the median part of the diencephalon and which is continued towards the foramen of Munro. The tissue of which this structure is composed is stratified. A cuboidal ependyma lies in the vicinity of the large flat ventricular cavity and comprises 4–5 rows of cells with uniform nuclei. In some areas the nuclei give the impression of forming small chains. Particularly in the lip-shaped area there is a moderate degree of vascularization. Krabbe noted the morphological connection between this structure and the preoptic recess, named it the preoptic formation, but was unable to ascribe to it any particular function.

The diencephalic region presented no differentiation into hemispheres but was instead a massive structure separated from the telencephalon by external furrows and from the mesencephalon by further lateral ones. Its largest dimensions are its breadth and height and its longitudinal extent is small. Ventrally the hypothalamic region continues into an infundibulum which projects out as a funnel-like protuberance behind the optic tract. The neurohypophysis has the characteristic appearance described by Wingstrand in cases where there are 4 or 5 diverticula. (See Chapter 12, Section IV.)

The anterior commissure lies somewhat above the preoptic formation, is thick, runs transversely, seems to be principally composed of myelinated material with very few nuclei, and disappears within the bulk of the diencephalon. The epiphysis is rather large, has an irregular globular form, and its structure differed from that of Krabbe's *Struthio* material. In the ostrich it appeared to have a "coral-like" composition but in the Kiwi it seems to be less glandular and consist of shorter or longer tubes. However, as is pointed out in Chapter 12, Section VIII, variations in the appearance of the epiphysis are now known to take place during the life of individual birds and any definitive conclusions would be unwise.

The entire epithalamic region presents a strange feature which is comparable with that described by Craigie in the adult. It has the appearance of a sac with very thick lateral walls which are united at their base with a thin membrane. This possibly corresponds to the dorsal sac of reptiles. Basally the sac gives way to an abundant mass of choroid plexus and the upper part is filled with the basal part of the epiphysis. There was no trace of any parietal eye but Krabbe suggested that the mesenchymal tissue might represent a rudimentary paraphysis.

The mesencephalon can be divided into three parts and, although these do not have definite limits, they form a central "massif" and two lateral components which are continuous with it. It is these structures which, when taken together, make the widest area of the brain. Throughout the bulk of the region the tissue was fairly homogeneous and comprised cells with small round nuclei. The ventricle was a continuation of the dorsal part of the diencephalic ventricle, was dilated at the front and hind ends but was otherwise a long, narrow, sagittal fissure. There was no easily visible Reissner's fibre. The area of separation between this region and the rhombencephalon is marked by a deep furrow behind which are the cerebellar and ponto-bulbar components. The cerebellum itself is sub-divided into a number of posterior folia by a series of fissures but there are only a few shallow ones on its anterior region. Krabbe concluded that it was similar to those of other birds and not to those of either amphibians or reptiles. By comparison with that of *Struthio* embryos of comparable age it was, however, much smaller. Histologically there was an outermost densinuclear layer and within this a much thicker layer possessing few nuclei but Purkinje cells were not distinguished with any great certainty. Below and behind the cerebellum and ponto-bulbar region is massive and only traversed by a fissiform ventricle which is gradually closed posteriorly.

## VI. BRAIN DEVELOPMENT IN OTHER SPECIES

A summary of the development of the brain of the quail, *Coturnix*, which can be used for comparison with the chick as it is another representative of the order Galliformes, was provided by Padgett and Ivey (1960). They found that at 2 days of incubation all 5 brain divisions are clear, have acquired definite boundaries, and that the most prominent is the mesencephalon. In some embryos the size of the telencephalon approached that of the mesencephalon by the 3rd day but the latter usually remains prominent until the 6th day, whilst the relative size of the diencephalon diminishes at 96 hours. By 170 hours the mid-brain no longer protrudes very noticeably, the auditory meatus is clear, and the olfactory protuberances can be seen around the nasal pits on the beak.

Previously Krabbe (1952) had also provided a relatively detailed account of embryological material from thirteen genera representing the Casuariiformes, Sphenisciformes, Ardeiformes, Anseriformes, Falconiformes, Galliformes, Lariformes, Psittaciformes, Strigiformes, Micropodiformes, Trochiliformes and Passeriformes. These data provide a very valuable source of comparative information. As far as was possible he ascribed the material to one of the six stages which were referred to in Section IV. To do this he used criteria which he had initially developed while working with mammalian and reptilian embryos. A broad outline of the specimens in terms of these stages is contained in Table 18, although some additional *Anisoterus*, *Melopsittacus* and *Pica*

## TABLE 18

*A summary of the material used by Krabbe (1952) in terms of his stages of development (embryo size in mm).*

|  | Stage I | Stage II | Stage III | Stage IV | Stage V | Stage VI |
|---|---|---|---|---|---|---|
| *Dromiceius novaehollandiae* | 5·4 | 8·0 |  |  |  |  |
| *Spheniscus* sp. |  |  | 15·0 | 26·0 | 35·5 |  |
| *Ardea cinerea* | 4·5 | 11·0 | 20·0 |  |  | 55·0 |
| *Cygnus olor* |  | 10·0 |  | 30·0 | 44·0 |  |
| *Astur palumbarius* |  |  | 14·5 | 30·0 |  |  |
| *Phasianus colchicus* | 4·0 | 11·0 |  | 22·0 |  | 29·0 |
| *Larus ridibundus* |  | 7·0 |  | 20·0 |  |  |
| *Melopsittacus undulatus* |  | 8·0 |  |  | 19·0 | 23·5 |
| *Strix aluco* |  | 14·5 |  | 25·0 |  |  |
| *Asio otus* |  |  |  |  | 34·0 |  |
| *Chlorostilbon aureoventris* |  |  |  |  |  | 11 days |
| *Pica rustica* |  | 14·5 |  | 22·5 |  |  |
| *Parus* sp. | 5·0 | 8·5 |  | 15·0 |  |  |

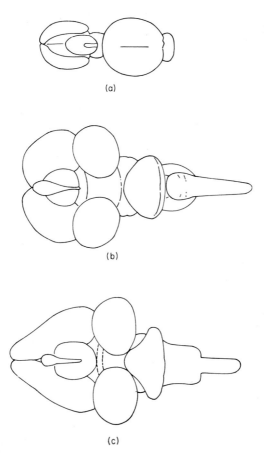

FIG. 26. Developmental stages of the penguin brain. (a) 15 mm embryo; (b) 26 mm; (c) 35·5 mm. (After Krabbe, 1952).

brains were in an intermediate phase between stages I and II. Outline drawings of the brains of *Phasianus*, *Ardea*, *Parus*, *Pica* and *Spheniscus* are represented in Figs 26–30.

## Stage I

On the whole Krabbe concluded that the primordial brains of *Dromiceius*, *Ardea*, *Phasianus* and *Parus* were similar in size to each other, to various primordial reptile brains, and also to those of mammals such as *Galeopithecus*, *Talpa* and *Vespertilio*. The four regions representing the telencephalon, diencephalon, mesencephalon and rhombencephalon were all sharply demarcated by transverse grooves. Only in *Ardea* was the situation somewhat different and this may reflect a difference in developmental stage rather than ontogenetic pattern. Here the groove which separates the telencephalon from the diencephalon was very deep and there was what Krabbe cryptically referred to as "a certain degree of metamerism". Furthermore the telencephalon had a pronounced transverse groove which divided it into a relatively narrow rostro-ventral, and a wider caudo-dorsal region. In contrast the diencephalon, which was also divided, had rostro-dorsal and caudo-ventral

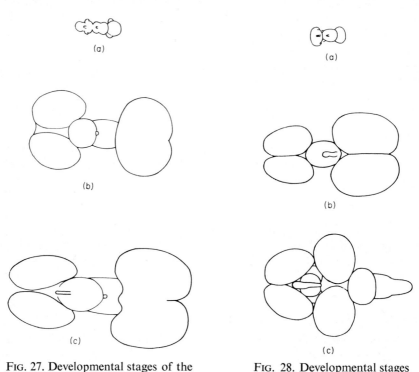

FIG. 27. Developmental stages of the heron brain. (a) 4·5 mm; (b) 11 mm; (c) 20 mm embryo. (After Krabbe, 1952).

FIG. 28. Developmental stages of the pheasant brain. (a) 4 mm; (b) 11 mm; (c) 22 mm embryo. (After Krabbe, 1952).

components. Some degree of this so-called metamerism was also evident in both the diencephalon and the rhombencephalon of *Dromiceius*, but not in either the telencephalon or diencephalon of *Phasianus* or *Parus* although some traces occurred in the rhombencephalon of the latter. It is by no means clear what exactly Krabbe had in mind when he used this term.

The situation in the material of *Anisoterus peretri* which was intermediate between stages I and II, was rather unusual. Here the hemispheres were very small whilst the mesencephalon was both large and flattened.

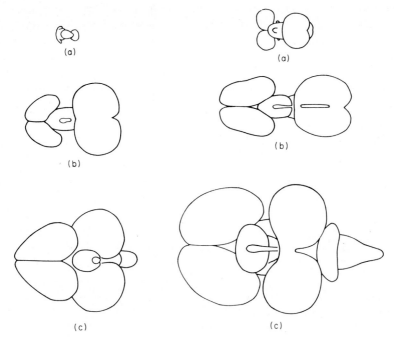

FIG. 29. Developmental stages of the brain of a titmouse. (a) 4 mm; (b) 8·5 mm; (c) 15 mm embryo. (After Krabbe, 1952).

FIG. 30. Developmental stages of the magpie brain. (a) 8 mm; (b) 14·5 mm; (c) 22·5 mm embryo. (After Krabbe, 1952).

## Stage II

This is characterized by the division of the telencephalon into two distinct hemispheres. The mesencephalon is either still unified, or else shows incipient tectal lobe formation by the presence of a faint groove which is situated over a small part of the median sagittal plane. The diencephalon is still fairly prominent and separates the posterior part of the hemispheres from the anterior part of the mesencephalon. Outlines of such brains are shown in Figs 27–30. The entire brain primordium is relatively narrow in *Melopsittacus* but wide in the other species, although there is a considerable amount of generic variation in the relative dimensions of the various parts. In particular this is true of the telencephalon and mesencephalon. Krabbe concluded that such variations do reflect specific and ordinal differences and are not discrepancies in his ascription of the brains to the appropriate developmental stages. The telencephalon is broader than the mesencephalon in *Dromiceius, Cygnus, Larus* and *Melopsittacus*. In *Parus* the opposite is true, whilst in *Ardea, Strix* and *Pica* both regions are of similar breadth. Furthermore although the total telencephalic volume is less than that of the mesencephalon in *Ardea, Larus, Strix* and *Parus*, the opposite is the case in *Cygnus* and *Melopsittacus*, and the volumes of both areas are comparable in *Dromiceius* and *Pica*.

There are, as one would expect, a number of other variations in various details. When seen in transverse section the hemispheres are generally globular or ovoidal. Those of *Ardea, Cygnus, Strix* and *Parus* are relatively shorter than those of *Dromiceius* and *Melopsittacus* and these last are also narrower. In *Pica* alone the lateral region is somewhat curved inwards and it was also only in this genus that Krabbe found that the optic cavity was closed. The diencephalon possessed a well-developed dorsal sac which was relatively broad in most genera but oblong in *Melopsittacus, Strix* and *Pica*. There was a small epiphysis in *Dromiceius* and *Ardea*, and a particularly tiny one in *Strix*. In contrast those of *Melopsittacus, Pica* and *Parus* were especially large. Ventrally the hypophyses also vary in size. The largest occurred in *Parus* and the smallest in *Strix* and in association with this the hypothalamic volume was also greatest in *Parus*. A small posterior sagittal groove indicates the beginnings of hemisphere formation in *Dromiceius, Ardea, Melopsittacus, Pica* and *Parus*, whilst the line of demarcation between this mid-brain region and the rhombencephalon behind it comprised a deep groove. As a consequence of this the isthmus of all species, particularly that of the heron, is very much reduced in size. In the hind-brain itself the cerebellar primordium forms an arch over the ventricle whose ependymal roof is usually extensive. The only exception to this occurred in the genus *Parus* where it was rather compressed. It is of interest that in this genus the trigeminal had a posterior position and was located in close proximity to the statoacoustic nerve. At this stage the overt appearance of the embryo's head region is very much affected by the mesencephalon. For example, in the budgerigar this conforms to the overall appearance of the brain and is narrow.

### Stage III

By the third stage of Krabbe's ontogenetic series the division of the mesencephalon into two tectal hemispheres is complete. The brain is relatively narrower in *Ardea* than is the case in *Spheniscus, Phasianus* and *Astur*. In all species the size of the mesencephalic region still exceeds that of the telencephalon. In *Ardea* and *Astur* but not in *Phasianus* and *Spheniscus* the rhinencephalon is clearly visible as a discrete structure on the basal region of the forebrain. In the last named genus the walls of the telencephalon are thicker than elsewhere and the whole structure has a rather aberrant shape.

In the diencephalic region there are a number of variations. The dorsal sac is conspicuously narrow in *Astur* whilst epiphysis development has given rise to a small structure in *Ardea* but a long and outwardly directed one in *Sphen-*

*iscus, Phasianus* and *Astur*. Complementarily, observation of the ventral surface shows that the hypophysis is larger in *Spheniscus* and *Ardea* than elsewhere. In the rhombencephalon the development of the cerebellar plate has not proceeded very much further than the condition in stage II, and the ependymal component is still very extensive. In *Phasianus* the pontine area is flattened so that the trigeminal and stato-acoustic nerves are relatively close together, whilst the compressed and curved form of this region in *Ardea* brings them into even greater proximity. Figures showing the overall appearance at this stage in *Spheniscus* and *Ardea* are contained in Figs 26 and 27.

## Stage IV

This stage is characterized by a pronounced flattening of the primordial brain coupled with the appearance of an obtuse angle between the anterior, or telencephalic–mesencephalic axis, and the posterior, or mesencephalic–rhombencephalic axis. As the relative length of the diencephalon is now reduced the occipital pole of the forebrain is closely apposed to the front part of the mesencephalon. Brains at this stage of development are shown in Figs 26, 28, 29 and 30. The marked lateral displacement of the tectal components in the mid-brain, which is so characteristic of the adult avian brain, is foreshadowed on the one hand by a slight degree of separation, and on the other by the appearance of a bridge between them. This is most pronounced in *Parus*. The actual proportional relationships which exist between the forebrain and mid-brain vary considerably. Although the forebrain is much the narrower in *Phasianus* and *Astur*, this difference is not so marked in *Larus*, *Cygnus*, *Spheniscus*, *Strix*, *Parus* and *Pica*. The long axis of the forebrain actually exceeds that of the mesencephalon in the species of the two first-named genera. As far as the respective volumes are concerned that of the telencephalon appears to exceed that of the mid-brain in *Spheniscus*, *Cygnus* and *Strix* but the two are comparable in *Astur*, *Phasianus*, *Larus*, *Pica* and *Parus*.

In overall appearance the forebrain again exhibits variations. When viewed from the side the top is convex in *Spheniscus*, *Astur* and *Strix*, and the basal regions are concave in *Cygnus*, *Phasianus* and *Strix*. When seen in dorsal view the hemispheres are elliptical and have a somewhat concave median side in *Spheniscus* and *Astur*. In the tit there is still no sign at all of the rhinencephalon and in the magpie it is rather small. In all the other species it is clearly visible on the ventral side and a more or less well-marked bundle of olfactory nerve fibres enters the smallest surface of the stub of the cone.

For the most part the hemispheres completely occlude any view of the diencephalon although the dorsal sac itself is still not covered. Small in *Strix*, the epiphysis in all other cases takes the form of a forwardly curved clavate body. The hypophysis is largest in *Larus*, *Pica* and *Parus* and projects forward to assume a unique appearance in the latter. In the mid-brain each lobe of the optic tectum is joined by a thin bridge to its contralateral partner and this bridge is shortest in *Phasianus*, *Strix* and *Parus*. Each tectal component is approximately globular in the penguin and *Pica*, but has an ovoidal shape in the remaining specimens. However, the overall shape of the region led Krabbe to construct two groupings. The first of these included *Phasianus*, *Astur* and *Strix*. Here the longest axis was in the transverse plane whilst in the other, which included *Larus* and *Cygnus*, the width was exceeded by the length. Behind this level the developing pontine area is now arched and both the breadth and volume of the cerebellum have increased considerably. In association with this there have also been corresponding changes in the relative extent of the ependymal roof of the rhombencephalon which is now very much reduced in extent and bears a deep transverse groove.

## Stage V

In Krabbe's (1952) descriptions stage V is not very different from the foregoing stage IV. The principal characteristics are a somewhat greater flattening of the entire brain primordium together with a rather greater degree of cerebellar development. This now begins to show traces of the adult avian structure particularly in the case of the penguin and swan material. It is less evident in the pheasant and budgerigar. The transverse axis of the mid-brain is much larger than the longitudinal one, but in four species the area is still larger than that of the forebrain. Indeed the hemispheres do not appear to have undergone much further development when their condition is compared with that in stage IV. The rhinencephalon has also remained of approximately similar size but is slightly further forward in the budgerigar. The ventral telencephalic region is outwardly convex in *Asio* and *Chlorostilbon* but remains concave in both *Cygnus* and *Melopsittacus*.

In the between brain the epiphysis is still relatively unmodified in *Chlorostilbon* but the size of the hypophysis exceeds that of other species. It is smallest in *Cygnus*, *Phasianus* and *Melopsittacus*. The developing cerebellum has increased in volume and in some species transverse grooves foreshadow the adult foliation. A brain of *Spheniscus* which was ascribed by Krabbe to this stage is shown in Fig. 26.

## Stage VI

The brains which fall into stage VI of Krabbe's scheme again vary in appearance. He justifiably intimated that in many respects their sole common developmental characteristic is that they are no longer in stage V. However, they do share a common flattened appearance, and, with the exception of *Ardea*, a conical rhinencephalon which has now assumed an anterior position and lies at the frontal pole of the hemisphere. The dorsal diencephalic sac is relatively reduced and the epiphysis, which is broader apically, lies within the groove separating the hemispheres. It is markedly larger than the hypophysis. The total breadth of the mesencephalon is still in excess of that of the telencephalon but the overall size of the region is now smaller than that of the forebrain in the heron and budgerigar. In *Phasianus* the opposite is true. In the budgerigar the bridge which unites the optic tectum of each side is by this time overlain by the developing cerebellum but it remains uncovered in *Ardea* and *Phasianus*. Nevertheless in all cases the cerebellum itself is now much larger and has a well-defined series of transverse fissures and folia.

# VII. THE DEVELOPMENT OF THE CEREBELLUM
## A. Development of Cerebellar Cells

The cerebellar plate is distinctly visible in the chick on the 4th and 5th days of development. It comprises a neuro-epithelial layer bordering the fourth ventricle, and a mantle layer containing neuroblasts. By the end of the 5th day the mantle layer situated in the lateral region develops both a deep and a superficial mantle zone. Herrick (1891) thought that the cells of the superficial zone arose mainly from the neuro-epithelium at the ventro-lateral angle of the fourth ventricle. He considered that they migrated from there to pass over the dorsal surface of the cerebellar plate. This concept was accepted by Forstronen (1963), Harkmark (1954a,b), Kershman (1938) and Kuhlenbeck (1950). Fujita (1964) showed that cerebellar neurons in the roof plate nuclei of the fourth ventricle are the earliest ones to differentiate and that they can be distinguished on the 3rd, 4th and 5th days of incubation. Furthermore, in the chick, most Purkinje cells appear during the 5th, 6th and 7th days, and those in the middle lobe appear first. Within the so-called uvular and nodular zones (see Table 31) the differentiation of some neuroblasts is delayed and in these regions Purkinje cells are still appearing up to the 13th day.

Using tritiated thymidine as a marker Hanaway (1967) studied the migration of cells at different stages of development by auto-radiography. He concluded that those cells which originate from the neuro-epithelial layer situated along the ventricular surface of the cerebellar plate migrate through previously formed layers of neuroblasts to reach the surface. As the development of the lateral portion of the cerebellar plate takes place approximately 24 hours prior to that of the medial region, the external granular layer appears there first. At no time was he able to detect a proliferation centre analogous to that reported by Herrick (1891), and located in the ventro-lateral angle of the fourth ventricle. He suggested that the early appearance of the granular layer in the lateral regions and the later appearance of such a layer in medial areas had led to the conclusion that cells migrate from the one to the other and had therefore given rise to the older suggestion.

Shortly after the formation of these early vestiges of the external granular layer its cells begin to actively proliferate. During this proliferative phase, which persists until day 15, relatively few cells leave the layer and migrate inwards. Thereafter there is a massive inward migration followed by differentiation giving the granular cells and also, according to Hanaway, possibly the glial cells. Within the internal granular layer they become intermingled with Purkinje cells and Golgi II neurones which originate in the neuro-epithelial layer between the 3rd and 6th day of incubation. Shortly after hatching, when the external granular layer has fulfilled its function in cell production, it disappears. The massive inwardly directed migration which results in its decrease in size takes place when the dendritic arborization of Purkinje cells has reached its fullest extent. Since the peripheral terminations of this arborization penetrate the inner zone of the external granular layer, it is possible that these processes are themselves in some way associated with the inward migration.

## B. Gross Cerebellar Development

As has been stated above, and was emphasized by Mesdag (1909), the rudiment of the cerebellum appears in the chick at 4·5 days of incubation. Ingvar (1918) observed a lateral swelling on each side of the rhombencephalon in 5-day chick embryos although these swellings had not yet undergone fusion in the mid-line of the roof. Jansen (Larsell, 1967) considered that such swellings correspond to the cerebellar part of the dorsal column of Bergquist and Kallén (1954). Following 8 days of incubation the cerebellum is tilted backwards and its bilateral rounded surfaces are separated by a median longitudinal furrow which penetrates to the ependymal region. A groove lying parallel to the margin isolates a slightly enlarged projection from the far more extensive medial mass. This groove, which represents the lateral component of the postero-lateral fissure, enables one to detect the rudiments of the flocculus. The, still paired, cell masses which are medial to it are the rudiments of the corpus cerebelli. Both Saetersdal (1956a) and Ingvar (1918) have described this situation in 7- and 5-day embryos respectively.

The ascending root of the Vth nerve, and, according to Jansen, possibly secondary trigeminal fibres, pass with spino-cerebellar fibres into the base of the developing cerebellum. In 9-day-old chick embryos some of these decussate across the mid-line in the anterior cerebellar region and form a cerebellar commissure which is related to the future corpus cerebelli. Vestibular root fibres follow the lateral and posterior cerebellar margins and one bundle decussates near to the postero-dorsal tip of the cerebellum. Between the two commissures which are formed in this way a thin zone of incipient cortex covers the ependymal layer medially.

The accounts of both Mesdag (1909) and Ingvar (1918) suggest that no fissures are visible on the 9th day. This is true for fissures which traverse the mid-line, although the postero-lateral depression separating the flocculus from the corpus cerebelli is by this time extending medially. At a slightly later stage the bilateral rudiments fuse and the median sagittal furrow disappears. The marginal zone has also extended backwards and inwards and those of each side exhibit enlargements which are the early signs of the developing nodular region and will subsequently fuse to form an unpaired cortical fold. The complete differentiation of the flocculo-nodular lobe from the corpus cerebelli is then complete.

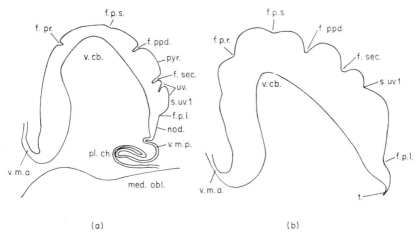

(a)                               (b)

FIG. 31. (a) The cerebellum of an 11·5-day-old duck embryo. (b) The cerebellum of a 65 mm *Phalacrocorax* embryo. (After Larsell, 1967).

Abbreviations: fpl, postero-lateral fissure; fppd, prepyramidal fissure; fpr, fissura prima; fps, posterio-superior fissure, fsec, fissura secunda; med. obl., medulla; nod, nodulus; s. uv. l., first uvular sulcus; v. m. a., anterior medullary velum; v. m. p., posterior medullary velum.

Our understanding of avian cerebellar anatomy is in large part the result of Larsell's researches (Larsell, 1948; and summarized posthumously by Jansen in Larsell, 1967), on material of *Anas*, *Gallus* and *Phalacrocorax* during the period of development which now follows. Amplified, qualified and extended to many other species by Saetersdal (1959a) they provide the basis for any understanding of the adult organization as referred to by subsequent authors. The corpus cerebelli of a 9·5-day chick embryo has four transverse furrows. Together with the postero-lateral fissure these correspond to the five fissures which were described by Saetersdal (1956a). The fact that Ingvar (1918) only found three may be a reflection of differing conditions during incubation. A considerable amount of confusion has arisen in the literature but Larsell concluded that the fissures which are present at this stage are the postero-lateral (*un* of Ingvar, 1918), secunda (*y* of Ingvar, 1918) the pre-pyramidal, prima (*x* of Ingvar, 1918) and intraculminary. The posterior superior appears in this scheme at 10 days and in some embryos its appearance is said to be more or less contemporaneous with that of a secondary fissure–uvular sulcus 1. It is this which Larsell thought corresponded to fissure *z* of Ingvar. A 65 mm cormorant embryo illustrated in Fig. 31(b) demonstrates all these components. The ontogenetic development of the duck is somewhat slower than that of the chick and as a result fissure appearance is appropriately retarded. Embryos of *Anas* have both cerebellar and vestibular commissures by the 10th day, whilst the vermal segment of the postero-lateral fissure, the fissura secunda, prepyramidalis and prima are visible at the 11th day. By 11·5 days these have been joined by the posterior superior fissure and uvular sulcus 1. A section of this stage in the cerebellar development of the duck is shown in Fig. 31(a). No indication of either intraculminary or preculminate fissures were thought to exist yet. Saetersdal (1956a) described all these components in the chick, pigeon and pheasant. Only the postero-lateral and secunda were present in the pigeon at the 10th day of incubation; the intraculminary appeared next, and was then followed by the prepyramidal and uvular sulcus 1. By the 13 day stage the fissura prima is fairly prominent and shallow indications of both the posterior superior and preculminate fissures are also visible.

The sequence of the appearance of these structures is more or less the same in the many different species which have been studied. There are, however, as one would expect, certain variations and also disagreements over the homo-logies. The work of Saetersdal (1959a) provided a survey of the developmental pattern in members of seven families of the Passeriformes, together with data relating to nine other orders. Indeed it was this work which finally elucidated a number of points that were incorrect in Larsell's account, and have been incorporated in Jansen's account (Larsell, 1967) and the foregoing paragraphs. He concluded that the first fissure to appear was always the pos-terolateral one, and that the corpus cerebelli is always primarily divided into two main divisions by the fissura secunda. It is important to emphasize this, although it has already been mentioned above, because it contrasts very markedly with the situation in mammals. In these animals it is the fissura prima, as its name suggests, that first appears on the corpus cerebelli, and the first fissures in the two classes are therefore not homologous. A series of sections illustrating the situation in embryos of differing ages from the genera *Anas, Mergus, Somateria, Ardea, Vannellus, Haematopus, Sterna, Larus, Corvus, Pica, Sturnus, Fringilla, Anthus, Motacilla, Phylloscopus, Turdus, Hirundo, Riparia, Chlorostilbon, Colibri,* and *Accipiter,* are repre-sented in Figs 32–37. Information which substantiates the conclusion that it is

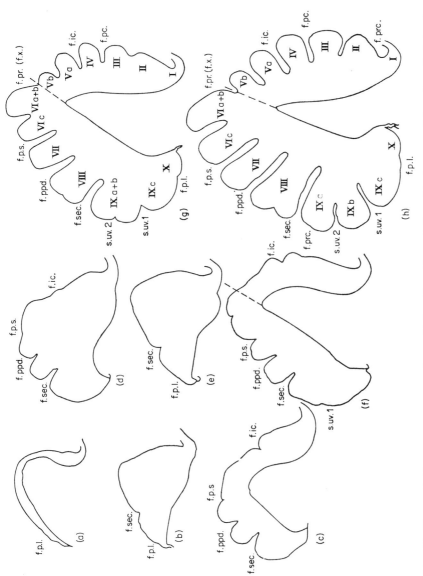

FIG. 32. The cerebellar structure as seen by sagittal sections through the mid-plane a, c, f–h and medio-lateral plane b, d, e of *Anas domesticus*. (a) 12 days; (b) 12 days; (c) 14 days; (d) 14 days; (e) 31 mm; (f) 35 mm; (g) 47 mm; (h) 53 mm (Saetersdal, 1959a).

Abbreviations: f. ic., intraculminary fissure; f. pc., preculminary fissure; f. ppd., prepyramidal fissure; f. pr., fissura prima; f. prc., precentral fissure; f. p. l., postero-lateral fissure; f. p. s., posterior superior fissure; f. sec., fissura secunda; s. uv. l., uvular sulcus 1; s. uv. 2, uvular sulcus 2.

the fissura secunda which forestalls the division of the corpus cerebelli by the fissura prima was obtained from *Sturnus vulgaris, Anthus spinoletta, Phylloscopus trochilus, Turdus pilaris* and *Riparia riparia.* Appearing first in the medio-lateral regions it then spreads both medially and laterally. Subsequently it is again usually the prepyramidal which precedes the intraculminary. For example, this was the case in *Pica pica, Fringilla coelebs, Motacilla alba* and *Phylloscopus trochilus,* but not, for example, in *Sturnus vulgaris* where the latter fissure is the first of the two to appear. In any event the result is the characteristic four-fold division of the entire cerebellar disc.

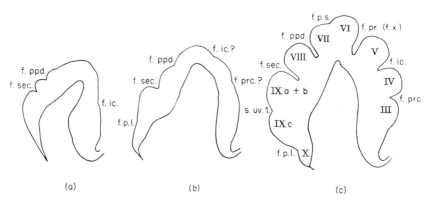

FIG. 33. Sagittal sections through the midplane of the cerebellum of *Somateria mollissima* (a) 32 mm and *Ardea cinerea* (b) 39 mm (c) 55 mm. (Saetersdal, 1959a).

Further development is then largely restricted to the posterior region. The primary uvular folium IX becomes sub-divided into two sub-lobes by the first uvular sulcus. More or less simultaneously, but often slightly before this, the posterior superior fissure is established between the appearance of what Larsell called the declive and that of the vermal tuber and folium (cf. *Sturnus vulgaris,* in Fig. 35; *Hirundo rustica* and *Riparia riparia* in Fig. 37). The establishment of these fissures is followed relatively quickly by that of the preculminate fissure in, say, *Riparia riparia, Turdus pilaris* and *Sturnus vulgaris.*

In view of the great emphasis which had been laid by other workers, and particularly both Ingvar (1918) and Larsell (1948) on fissure *x* (the fissura prima) as a cerebellar landmark, Saetersdal paid particular attention to it. In *Corvus cornix, Sturnus vulgaris, Turdus pilaris* and *Hirundo rustica,* he concluded that it first appears in a dorsal position and behind the intraculminary fissure which is already present. In species like *Pica pica* it may, as noted above, appear rather earlier but in any event it occupies the same relative position over a peak of the cerebellar disc and a little way in front of the cavum cerebelli. The further development of *Fringilla coelebs, Hirundo*

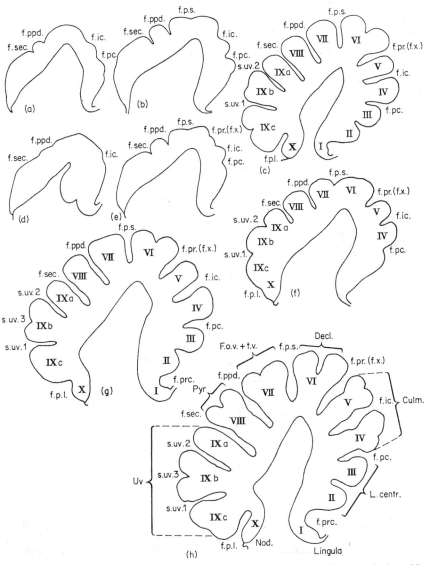

Fig. 34. Sagittal sections through the midplane a–c; e–h, and medio-lateral plane (d) of the cerebellum of: *Sterna hirundo;* (a) 25 mm, (b) 31 mm, (c) 41 mm; *Larus canus*, (d) 31 mm, (e) 35 mm, (f) 39 mm, (g) 42 mm, (h) 52 mm. (Saetersdal, 1959a).

*rustica* and *Riparia riparia* shows a rather interesting series of features. These are particularly marked in the members of the Hirundinidae. A comparison of the relative degree of development in 23 mm and 30 mm specimens of *Hirundo rustica* (Fig. 37) will show that the fissura prima does not increase in depth to the same extent that other areas undergo development during the intervening period. As a result of this it appears, in the oldest stage, to be very insignificant and comparable with the third uvular sulcus. Saetersdal also observed a similar tendency in both *Riparia riparia* and *Fringilla coelebs*. He concluded from this that, even in the adult brain of these Passeriformes, it ranks as one of the more insignificant cerebellar fissures. He also drew attention to the roughly uniform size of all the cerebellar fissures in those members of the Passeriformes which he was able to study.

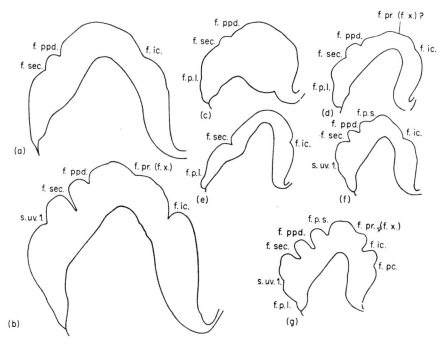

FIG. 35. Sagittal sections through the midplane a–b, d–g and mediolateral plane (c) of the cerebellum of *Corvus cornix*; (a) 34 mm; (b) 42 mm; *Pica pica*, (c) 27 mm; (d) 30 mm; *Sturnus vulgaris*, (e) 30 mm; (f) 34 mm (I); (g) 34 mm (II) (Saetersdal, 1959a).

In contrast to these varied passeriform species Saetersdal only had access to single specimens of the orders Psittaciformes, Strigiformes and Falconiformes. He concluded that amongst the Anseriformes the prepyramidal

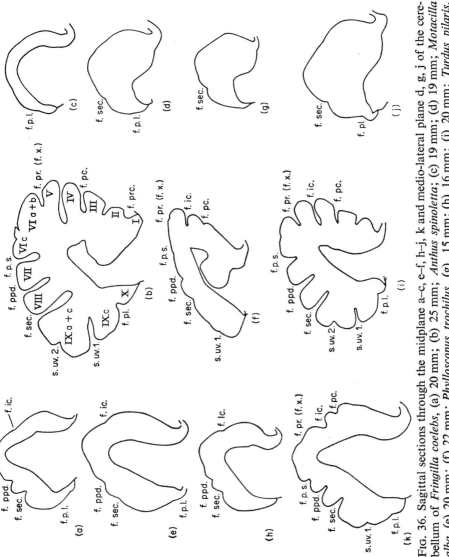

FIG. 36. Sagittal sections through the midplane a–c, e–f, h–j, k and medio-lateral plane d, g, j of the cerebellum of *Fringilla coelebs*, (a) 20 mm; (b) 25 mm; *Anthus spinoletta*; (c) 19 mm; *Motacilla alba*, (e) 20 mm; (f) 22 mm; *Phylloscopus trochilus*, (g), 15 mm; (h) 16 mm; (i) 20 mm; *Turdus pilaris*, (j) 23 mm; (k) 31 mm (Saetersdal, 1959a).

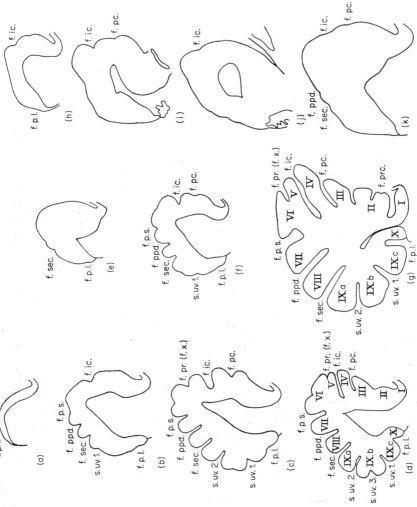

Fig. 37. Sagittal sections through the midplane (a–d, f–h, i (?)) and medio-lateral plane (e, j (?), k) of the cerebellum of *Hirundo rustica*, (a) 15 mm; (b) 20 mm; (c) 23 mm; (d) 30 mm, *Riparia riparia*, (e) 18 mm; (f) 20 mm; (g) 34 mm, *Chlorostilbon aureoventris* (h) and (i), *Colibri serrirostris*, (j); *Accipiter nisus*, (k) 25 mm. (Saetersdal, 1959a).

fissure was formed prior to both the intraculminary and posterior superior ones. This contrasts with the condition which Larsell described (1948, 1967) and in his study of *Mergus serrator* material Saetersdal (1959) concluded that the fissure which Larsell (1948) assumed to be the homologue of Ingvars fissure *x*, was in fact the homologue of the one situated immediately in front of it. He suggested that Larsell's specimens were too far apart in developmental time to enable him to detect the very considerable changes which occur during a relatively short period. Drawing attention to the 18-day specimen figured by Larsell (1948) he suggested that it was actually a 13-day specimen, and decided that fissure *x* lay in the declive (see Chaprer 8 below) between sub-lobes VIa, b, and VIc in view of its position relative to the cerebellar cavity. The sub-division of the so-called declive, which occurs rather later in *Anas* than in *Mergus*, is then the result of a fissure which is situated both higher up and further back on the cerebellar disc.

According to Saetersdal the situation in the three species of the Charadriiformes and Lariformes coincides very closely with what has now been described in the foregoing pages. The characteristic quartering of the corpus cerebelli by the secondary, prepyramidal and intraculminary (Saetersdal) fissures, certainly occurred in *Haematopus ostralegus*, *Vannellus vannellus* and *Larus canus*. He concluded that the intraculminary fissure (*sic*) appeared rather higher up on the cerebellar disc in *Larus* than elsewhere and appeared to have a rather similar location in both the Ardeiformes and Micropodiformes where the primary anterior fissure also runs high up on the disc. As a result of all these observations Saetersdal concluded with Larsell that the cerebellar disc was divided into a corpus cerebelli and a flocculonodular node by the postero-lateral fissure which penetrates to different depths in the varied species and is therefore occasionally hard to distinguish. However, it always seems to precede the other cerebellar fissures. The corpus cerebelli is then primarily divided by the secondary fissure which first appears in the medio-lateral region. During development it is very quickly followed by both the prepyramidal and intraculminary fissures, usually in that order. The principal exceptions to this sequence of events were *Chlorostilbon aureoventris* and *Colibri serrirostris* where the corpus cerebelli was first of all divided by the intraculminary fissure and where the three fissure stage was not available. Larsell (1967) concluded that in fact the relative size which is eventually attained by the folia adjacent to the fissures appears to be a factor which determines their order of appearance and relative depths. Furthermore the retarded appearance of his fissure *x* by comparison with the development of its supposed homologue in mammals may be due to its position on the strongly arched cerebellum. As the cortex develops and thickens sufficiently to become folded elsewhere, the tensions involved in rapid growth may prevent any folding along the line where it is to appear.

## C. The Causes of Cortical Fissure Formation

We have just seen that Larsell and Jansen (in Larsell, 1967) have implicated the general mechanical events which occur during cerebellar ontogeny in the causation of cerebellar fissures and their order of appearance. Saetersdal (1959b) drew attention to the various explanations which had previously been proffered. Larsell (1948) invoking the experiments of certain embryologists who had produced hyperplasia of amphibian central nervous systems by increasing the number of fibres which pass to a particular part suggested that similar processes had occurred during ontogeny. For example he suggested that the total number of proprioceptors and exteroceptors whose fibre paths impinge upon the cerebellum had increased during the evolution of large active vertebrates. In particular in the case of the avian cerebellum he postulated that there had been a process of inward thickening and folding along the zones of overlapping fibre systems which had also been accompanied by a simultaneous bulging of the cortex between the fissures.

Many other authors have correlated cerebellar fissure formation with the increasing maturity and growth of the cells of the cortex. As Saetersdal (1959b) emphasized this may be linked with a decrease in the thickness of, and a centripetal migration from, the external granular layer (see Section VIIA). Such suggestions have in the past been particularly favoured by Schaper (see Saetersdal, 1959a) and Berliner (1905) who worked with human foetal material. Saetersdal (1956b) had himself investigated this suggestion in relation to the avian brain and had found no evidence for a causal connection between the development of fissures and the rate at which the external granular layer decreases in thickness. Figure 38 contains a pictorial representation of a variety of data on cerebellar cortical development in *Gallus*. These include measurements of the mean thickness of the external granular, the molecular and the internal granular layers, together with information on the circumferences and areas of sagital sections and the relative mitotic rates.

Cortical folding within the region of the cerebellar disc begins, as we have already seen, with the appearance of the postero-lateral fissure at 7–8 days of normal incubation. Development within the corpus cerebelli itself does not begin until later, and more especially the 9th day of incubation. Cortical fissure formation in the fowl is then more or less complete by the 15th day by which time all the primary lobes and sub-lobes of the adult brain are established so that the foliar pattern is complete. Saetersdal pointed out that the data of Fig. 38 show that although the cerebellar circumference increases most rapidly between days 11 and 15 there is only a moderate increase in area at this time. He suggested that if the area measurements reflect changes in the total cerebellar volume, then the cortical development which is then taking place must be characterized by rapid longitudinal growth and accompanied

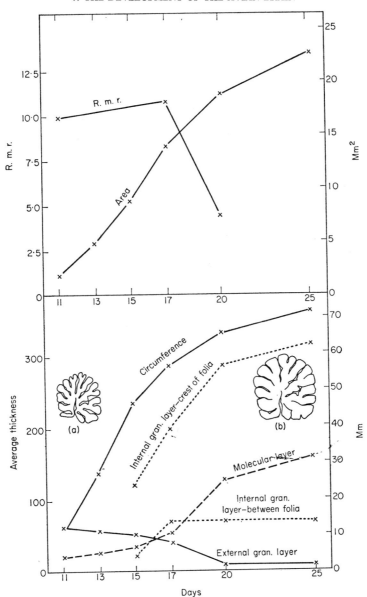

FIG. 38. Quantitative data on the cerebellar ontogenesis of *Gallus domesticus*. 1. Measurements of the average thickness of the cortical layers. 2. Measurements of the circumferences and 3. Areas of a series of sagittal sections. 4. Mitotic counts of the external granular layer expressed as relative mitotic rate during three successive stages of development (11, 17 and 20 days). Infolding phase indicated by heavy lines. Figures (a) and (b) represent sagittal sections of cerebella incubated for 15 and 25 days respectively (Saetersdal, 1959b).

by only a relatively slow increase in volume. It can be seen from the graphs that, as Hanaway (1967) has confirmed (see Section VIIA), the external granular layer retains almost its total thickness at the 15-day stage. The major period of vertical growth within the molecular and granular layers takes place later, and the reduction in the thickness of the external granular mantle is largely the result of cell migrations during days 17–20. These more advanced morphogenetic phases, which take place after day 15, are therefore characterized by such vertical growth.

Consequently Saetersdal (1956a, 1959b) concluded that one can distinguish two principal phases in cerebellar cortical development. There is, first of all, an *infolding phase* which is characterized by a high level of mitotic activity within the external granular layer, together with an intense longitudinal growth of the cortex. It is during this period that all the primary and secondary folia of the adult cerebellum are established, and fissure formation is virtually completed. Then, subsequently, there is a second or vertical growth phase. This is characterized by a drop in the levels of mitotic activity within the external granular layer, a rapid vertical growth or increase in thickness of the permanent cortical layers, and ultimately as a result, the complete disappearance of the external granular layer. In this respect it is interesting that Harman concluded that the increase in cerebral folding during mammalian ontogeny is characterized by rapid longitudinal growth but only slight changes in cortical thickness. From such results one may suggest that the dynamics of fissure formation are related to the proliferative activity of relatively undifferentiated cortical elements rather than to the subsequent differentiation and growth of cortical layers. Furthermore it may be related to considerable longitudinal growth in a relatively restricted volume.

## D. The Cerebellar Nuclei

Vertebrate cerebellar nuclei have been ascribed to various embryological sources and in the case of the chick and pigeon the cerebellar neural epithelium predominates (Baffoni and D'Ancona, 1958; Rudeberg, 1961). However, historically both the vestibular nuclei and the cerebellar cortex have been suggested. Rudeberg concluded that both the avian species cited conform to his fourth developmental category in which there are two successive cell migrations analogous to those described by Berquist and Kallén (1954). However as the migration periods observed by these latter authors occurred during phases of maximum mitotic activity and this was only the case for his first migration he considered that the succeeding ones in the thalamus and cerebellum are not homologous. Indeed both migrations may well be two stages of the same gross cell movement (see Section VIIA). The thin part of Migration A becomes incorporated into the superficial granular layer and the connections between this and the remainder of the cell-layer subsequently break.

The further development results in the appearance of the three definitive cerebellar nuclei (*q.v.*). The embryological data therefore seem to support the anatomical suggestions of Ramon y Cajal (1908) although Rudeberg was unable to detect his intercalated nucleus. Nevertheless it is important to emphasize the suggestions of Yamamoto *et al.* (1957). Studying adult material they decided that it was almost impossible to make a strict division of the cerebellar nuclear masses into discrete entities because they are very closely fused. They are certainly not independant masses as connections clearly exist between them (see Doty, 1946) but such bridges do not negate the existence of unified anatomical nuclei.

## VIII. TELENCEPHALIC DEVELOPMENT
### A. Introduction

Following the work of Kuhlenbeck (1938), which is fundamental to all subsequent interpretations of adult avian forebrain organization, the most recent widely-ranging and comparative work on telencephalic development is that of Haefelfinger (1958). Although the principal contributions in this last-named author's monograph refer to *Melopsittacus* and certain Passeriformes, he also gave brief accounts of the conditions in embryos of *Tyto alba, Micropus melba, Merops apiaster* and *Dromiceius novae-hollandiae.* In general he concluded that the developmental patterns in these species paralleled those which he described in the budgerigar. However, the differing

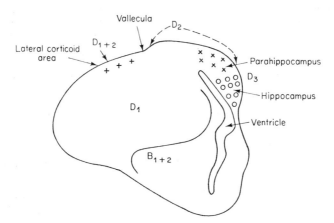

FIG. 39. Cross-section through the telencephalon of a chick embryo at 10 days of incubation. (After Kuhlenbeck, 1938).

degree of prominence assumed by various adult nuclear structures is reflected to a greater or lesser degree in ontogenetic variations. These adult variations are described in Chapter 13 but one may cite, for example, the massive appearance of the Wulst (the dorsal and accessory hyperstriatum) which characterizes the brain of the owl, *Tyto alba*, and which is foreshadowed by the early and preponderant development of these structures in the frontal region of the brain of the embryo. A summary of the available material for these species is contained in Table 19.

## TABLE 19

*A summary of the embryonic brain material which was used by Haefelfinger (1958).*

| Species | Days of development | Embryonic length (mm) | Hemisphere length (mm) |
|---------|---------------------|------------------------|-------------------------|
| *Tyto alba* | 3 | 12 | 0·43 |
|  | 8 | 37 | 3·42 |
|  | 13 | 50 | 3·94 |
| *Micropus melba* | 11 | 38 | 2·82 |
|  | 13 | 50 | 3·47 |
| *Merops apiaster* | 7 | 25 | 1·80 |
|  | 10 | 35 | 2·46 |
|  | 12 | 40 | 2·55 |
| *Dromiceius novae-* | 9 | 18 | 0·43 |
| *hollandiae* | 13 | 37 | 1·86 |
|  | 16 | 55 | 3·55 |

## B. Telencephalic Development in *Melopsittacus*

In *Melopsittacus undulatus* embryos which have been incubated for 2 days, and are 3 mm long, both the prosencephalon and rhombencephalon are distinct and the telencephalon has a length of 0·15 mm. Throughout the length of the brain the walls are of approximately uniform thickness and at its anterior end the medulla is completely closed although it still retains marked dorsal and ventral furrows. In the prosencephalon the width of the dorsal region exceeds that of the ventral side and as a result transverse sections appear heart shaped. By 3 days of development the differentiation of the mesencephalon has taken place and there are therefore three primary brain areas. When viewed in cross-sections the prosencephalon is now somewhat egg shaped in front but wider posteriorly, and, although thin dorsally, its walls are otherwise still the same thickness throughout. During the following day the dorsal part becomes broader and one can distinguish the paths towards the brain of those ectodermal cell migrations which will subsequently give rise to the olfactory tracts. At this time mitoses are very abundant within the ependymal cell layers.

## TABLE 20

*A summary of the notation used by avian embryologists in their studies of telencephalic development.*

| Early developmental phase | Intermediate developmental phase | Late developmental phase | Adult structure |
|---|---|---|---|
| Dorsal region of telencephalic zone | D3 | H1–H4 | Hippocampus and Parahippocampus |
| | D2 | 1. a.b.c.d. | Cortex |
| | | 2. Lateral corticoid region | |
| | | 3. Wulst primordium and nucleus diffusus | Accessory and dorsal hyperstriatum. Nucleus intercalatus. |
| | D1 | 1. Primordium of both neo-, and hyperstriatum | Ventral hyperstriatum, Neostriatum, Ectostriatum, |
| | | 2. Archistriatum | Archistriatum |
| Basal or ventral region of telencephalic zone | B1 | Primordium of paleostriatum | Paleostriatum augmentatum and primitivum |
| | B2 | | |
| | B3 | | Basal eminence |
| | B4 | | |

Haefelfinger (1958).

On the 5th day the hemispheres develop from the primary telencephalic region and measure 180 $\mu$ from the rostral pole to the interventricular foramen. Comparing this information with that which is contained in Section VI, it becomes clear that the brain is in stage II of Krabbe's classification. It is also comparable with the condition which is achieved by at least 96–120 hours in the chick (Section III). Histological examination reveals that the telencephalic walls are thick, and that this is especially true in the case of the ventromedian region. Here it is possible to distinguish the first traces of the adult paleostriatal components which are represented by $B_{1+2}$ of Table 20, and medial to these the median eminence ($B_3$). The dorso-lateral walls are still relatively unorganized. When seen in transverse sections the ependymal region is overlain by a matrix of chromatin-rich cells which are demarcated peripherally by a cell-free cortical zone. At the following, or 6th day, the hemispheres are markedly developed rostrally, appear oval in cross-section, and only their medial walls remain thin. On the lateral wall there is again a clearly defined limiting cortical zone but this is less obvious elsewhere. In the caudal region there is a comparably thick ventral layer which is associated with the developing, and by now more coherent, olfactory tract. Within the brain primordium itself there are also, by this stage, traces of the septo-mesencephalic tract which is so important in the adult brain and therefore referred to in a number of the subsequent chapters. On the dorso-medial wall the hippocampal region is thin, but laterally both the neostriatum and ventral hyperstriatum (q.v.) are foreshadowed by a marked swelling. A brief summary of what is now the customary alphabetical notation for the embryological components of the various regions is contained in Table 20.

The basal region continues to enlarge, particularly at the level of the interventricular foramen, and traces of the neo-paleostriatic fissure appear between the paleostriatal and neostriatal primordia so that they are now showing signs of incipient separation. The initial traces of the dorsal medullary lamina, which maintains this separation in the adult brain, are also visible in the lateral wall as a layer containing relatively few cell bodies. By the seventh day the ventral, lateral and dorsal walls are all thicker than the medial, and histologically one can begin to distinguish a number of distinct layers within them. Haefelfinger (1958) concluded that cell proliferation occurs predominantly within the ependymal layer and that there is an actual subsequent cell migration from this region towards the periphery. In general terms the various layers can be summarized as follows:

1. An ependymal layer in which both cells and nuclei are relatively large, and in which there is widespread mitotic activity;
2. A matrix of small and closely packed chromophilic cells;
3. A zone where the cells are more widely separated although they resemble those of the foregoing layer;
4. A peripheral or cortical layer which still lacks any overt evidence of organization and only contains a few cells.

Within the region in which the neostriatal and ventral hyperstriatal moieties are developing, a deep relatively acellular zone foreshadows the definitive appearance of the basal nucleus. Elsewhere the first traces of both the fronto-archistriatal, and the occipito-mesencephalic tracts are also detectable. As in the case of the septo-mesencephalic tracts which were mentioned above these are of great importance in the integrative action of the various parts of the adult brain.

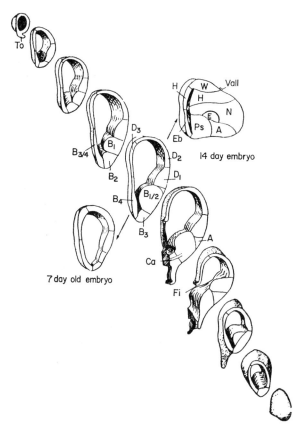

FIG. 40. Stereogram of the fore-brain of a 10 day embryo. After Haefelfinger (1958). For abbreviations, see Table 20.

Observation of later embryos, which have reached the 9th and 10th days of incubation, revealed that the olfactory bulb is now visible and lies in close apposition to the anterior pole of the hemispheres where it is closely associated with the olfactory tract. By now the stage of development has exceeded stage V in Krabbe's scheme. The *Vallecula*, a crucial landmark in the adult avian brain, is becoming clearly visible and delimits the area which will constitute the adult *Wulst*. In fact these actual limits are now demarcated throughout the relevant areas of the forebrain wall by a line which runs from the vallecula to the hyperstriatic sulcus on the ventricular side. This represents the area which intervenes between the earlier zones $D_1$ and $D_2$ of Table 20. Towards the front of the hemispheres it is situated about mid-way down the dorso-ventral plane but, as one progresses further backwards, it gradually assumes a more dorsal position and finally provides the definitive angle of the lateral ventricle. Within the Wulst itself there is a distinct area composed of darkly staining cells which probably represents the primordial accessory hyperstriatum.

By about this time the paleostriatum, which has been obvious as a distinct entity for some time previously, becomes firmly outlined by the now clearly visible dorsal medullary lamina. Towards the hind end the progressive formation of the neo-paleostriatic fissure is associated with a ventro-dorsal displacement of this lamina. The actual histological differentiation within the paleostriatal regions themselves has resulted in the appearance of the rudiments of both the paleostriatum primitivum and the bundles of the "brachium cerebri". Elsewhere the archistriatum is also a clearly defined nuclear mass which is situated somewhat laterally within the neostriatic-ventral hyperstriatic block. Fairly clearly divided into dorsal and ventral components it is joined to its contralateral homologue by the anterior commissure. Close to the inter-ventricular foramen the paleostriatum is actually relatively small and much of the latero-ventral area of the basal regions of the telencephalon is therefore formed from archistriatal tissue. A stereogram of a brain at this stage of development is contained in Fig. 40.

By the 12th day of incubation the overt adult appearance of the brain, together with the final disposition of the individual nuclear masses, is more or less attained. In the front the olfactory bulb is complete. The vallecula is clearly visible dorsally and circumscribes the definitive Wulst region, the predominant component of which is the accessory hyperstriatum. The cellular organization within both the neostriatum and the ventral hyperstriatum has progressed considerably and the two are separated by the constituent fibres of the hyperstriatic lamina. In the fore part of its extent the archistriatum is still situated laterally but further back its ventral component assumes a more ventral position on each side. As such it overlaps the paleostriatum and this results in a marked arching of the dorsal medullary lamina.

## C. Telencephalic Development in the Passeriformes

Following the work of Durward (1934) on the sparrow, Haefelfinger (1958) produced complementary information for *Corvus*, *Hirundo* and *Turdus*. The early phases of differentiation within the lateral wall parallel those of the budgerigar very closely. In *Passer* the primordium of the paleostriatum bulges into the ventricle by the 5th day of incubation. By the 7th day the neostriatic-ventral hyperstriatic primordium also comprises a thickening which lies within the lateral wall and towards the frontal pole. However the olfactory bulb is still not clearly defined and the olfactory tract comprises a thin band of fibres which runs beneath the hemispheres. Furthermore, the actual anterio-posterior extent of the hemispheres is still relatively small and, apart from the primordia, their walls remain of an almost uniform thickness with very little additional evidence of differentiation.

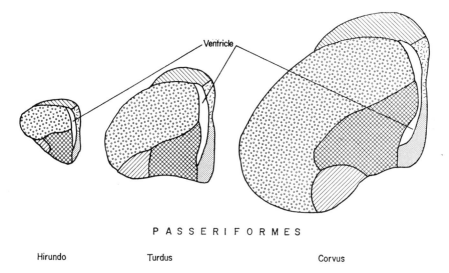

PASSERIFORMES

Hirundo                    Turdus                              Corvus

FIG. 41. A diagram showing the relative development of different hemisphere components. The developmental processes during ontogeny are modified accordingly. After Haefelfinger (1968). See also Figs 116–124.

By the 8th day of incubation transverse sections taken through the brain at a level which is close to the frontal pole show that the overall appearance is oval, and that there has been a general increase in the thickness of the walls. In front of the inter-ventricular foramen the primordium of the paleostriatum, together with that of the mutually inter-related neostriatum and ventral hyperstriatum, are very much enlarged. The actual limits of the Wulst are clear whilst, in the hind region, the basal eminence is present in a ventro-medial position. The fronto-archistriatic tract is visible, and, in the lateral region, the first traces of the dorsal medullary lamina are represented by a small cell-free zone.

These trends continue so that by the 9th day the thickening walls of the hemispheres have more or less totally occluded the anterior part of the ventricle. In the anterior region there are traces of the olfactory bulb ventrally, whilst the Wulst is even more well-defined dorsally. Further towards the hind-end the dorsal medullary lamina, which separates the paleostriatum from overlying structures, is showing traces of arching, and the nuclear mass itself is readily recognizable. This is also true of the anterior commissure which by now links the archistriatal components of each side. Further development results by the 10th day in the definitive form of the olfactory bulb which is a relatively insignificant structure in Passeriformes. At the same time the paleostriatal primordia can be differentiated into the discrete components of the augmentatum and primitivum. In the occipital region the principal component is the neostriatic–ventral hyperstriatic complex, which, like the Wulst in the frontal region, now occupies a position which is closely comparable with the final adult disposition. Haefelfinger drew particular attention to the differences which exist between the hemisphere indices of *Hirundo rustica*, *Turdus merula* and *Corvus corone*, which are respectively less than, equal to and greater than those of *Passer*. These are reflected by comparable differences in the appearance during ontogeny (see Fig. 41). Although a further and more detailed account of the actual differences between the hemispheres of the avian orders is more conveniently left for discussion in Chapter 13, one can cite, for example, the archistriatum which assumes a relatively ventral position below the lateral region of the paleostriatum in *Corvus*.

## References

Baer, K. E. von, (1837). *Uber Entwicklungsgeschichte der Thiere*. I (1828); II (1837).

Baffoni, G. M. and D'Anconi, G. (1958). Osservazioni sulla morfogenesi ed isto-genesi cerebellare in uccelli a prole inetta ed a prole precoce. *Rend. Acc. Maz. Lincei*. **8**, 456–463.

Berliner, K. (1905). Beitrage zur Histologie und Entwicklungsgeschichte des Kleinhirns, nebst Bemerkungen uber die Entwicklung der Funktionsstuchtigkeit desselber. *Arch. Mikr. Anat*. **66**, 220.

Bergquist, H. and Kallén, B. (1954). Notes on the early histogenesis and morpho-genesis of the central nervous system in vertebrates. *J. Comp. Neurol*. **100**, 627–659.

Biondi, G. (1910). Osservazioni sullo svilluppo e sulla struttura dei nuclei d'origine dei nervi oculomotore e trochleare nel pollo. *Riv. Ital. Neuropath*. **3**, 302–327.

Craigie, E. H. (1930). Studies on the brain of the Kiwi. *J. Comp. Neurol*. **49**, 223–357.

Doty, E. J. (1946). The cerebellar nuclear grey in the sparrow (*Passer domesticus*). *J. Comp. Neurol*. **48**, 471–499.

Durward, A. (1932). Observations on the cell masses in the cerebral hemisphere of the New Zealand Kiwi (*Apteryx Australis*). *J. Anat*. **66**, 437–477.

Durward, A. (1934). Some observations on the development of the corpus striatum of birds with especial reference to certain stages in the common sparrow (*Passer domesticus*). *J. Anat*. **68**, 492–497.

Forstronen, P. F. (1963). The origin and the morphogenetic significance of the external granular layer of the cerebellum as determined experimentally in chick embryos. *Acta Neurol. Scand*. **39**, 314–316.

Fujita, S. (1963). Matrix cells and cytogenesis in the developing central nervous system. *J. Comp. Neurol.* **120**, 37–42.

Fujita, S. (1964). Analysis of neuron differentiation in the central nervous system by tritiated thymidine autoradiography. *J. Comp. Neurol.* **122**, 311–327.

Haefelfinger, H. R. (1958). "Beitrage zur vergleichenden Ontogenese des Vorderhirns bei Vögeln." pp. 99. Helbing and Lichtenhahn, Basle.

Hamilton, H. L. (1952). "Lillie's Development of the Chick." Henry Holt, New York.

Hanaway, J. (1967). Formation and differentiation of the external granular layer of the chick cerebellum. *J. Comp. Neurol.* **131**, 1–14.

Harkmark, W. (1954a). Cell migrations from the rhombic lip to the inferior olive, the nucleus raphe and the pons. *J. Comp. Neurol.* **100**, 115–210.

Harkmark, W. (1954b). The rhombic lip and its derivatives in relation to the theory of neurobiotaxis. *In* "Aspects of Cerebellar Anatomy" (Jansen and Brodal, eds), pp. 264–284. Johan Grundt, Oslo.

Herrick, C. L. (1891). Architectonics of the cerebellum. *J. Comp. Neurol.* **1**, 5–14.

Ingvar, S. (1918). Zur Phylo-und Ontogenese des Kleinhirns. *Folia neurobiol.* **11**, 205–495.

Ingvar, S. (1919). "Zur Phylo- und Ontogenese des Kleinhirns nebst ein Versuch zu einheitlicher Erklarung der Zerebellaren Funktion und Lokalisation" Lunds.

Kamon, K. (1906). Zur Entwicklungsgeschichte des Gehirns des Huhnchens. *Anat. Hefte.* **30**, 562–650.

Kappers, C. U. A. (1947). "Anatomie comparée du système nerveux", pp. 754. Masson, Paris.

Kershman, J. (1938). The medulloblast and the medulloblastoma. *Arch. Neurol. Psych.* **40**, 937–967.

Klatzo, I., Wiśniewsky, H. and Smith, D. E. (1966). Observations on the penetration of serum proteins into the central nervous system. *Progr. Brain. Res.* **15**, 73–88.

Klika, E. and Jelinek, R. (1965). Area membranacea ventriculi in the chick. *Z. Zellforsch.* **63**, 950–959.

Krabbe, K. H. (1952). "Studies on the Morphogenesis of the Brain in Birds." pp. 110. Munksgaard, Copenhagen.

Krabbe, K. H. (1956). "Studies on the Morphogenesi of the Brain in *Struthio Camelus.*" Munksgaard, Copenhagen.

Krabbe, K. H. (1957). Sur une formation singulière du cerveau d'un embryon de Kiwi (*Apteryx Mantelli*). *L'encephale.* **5-6**, 612–622.

Krabbe, K. H. (1959). "Studies on the Brain Development of the Kiwi (*Apteryx Mantelli*)", pp. 19, Munksgaard, Copenhagen.

Kuenzi, W. (1918). Versuch einer systematischen Morphologie des Gehirns der Vögel. *Rev. Suisse Zool.* **26**, 17–112.

Kuhlenbeck, H. (1938). The ontogenetic development and phylogenetic significance of the cortex telencephali in the chick. *J. Comp. Neurol.* **69**, 273–301.

Kuhlenbeck, H. (1950). The transitory superficial granular layer of the cerebellar cortex. *J. Amer. Med. Women's Ass.* **5**, 347–351.

Kupffer, K. von. (1906). Die morphogenie des Centralnervensystems. *In* "Handbuch der vergleichenden und experimentellen Entwicklungslehre der Wirbeltiere." 2 vols.

Langman, J. (1968). Histogenesis of the central nervous system. *In* "The Structure and Function of Nervous Tissue." (Bourne, G. A., ed.), pp. 33–67. Academic Press, London and New York.

Larsell, O. (1948). The development and sub-divisions of the cerebellum of birds. *J. Comp. Neurol.* **89**, 123–190.

Larsell, O. (1967). "The Comparative Anatomy and Histology of the Cerebellum from Myxinoids Through Birds." (Jansen, J., ed.), pp. 291. University of Minnesota.

Meek, A. (1907). The segments of the vertebrate brains and head. *Anat. Anz.* **31**.

Mesdag, T. M. (1909). *Bijdrag tot de ontwikkelingsgeschiendenis van de structuur hersenen bij het kip.* Dissertation Groningen.

Murray, M. R. and Benitez, H. H. (1968). Action of heavy water ($D_2O$) on growth and development of isolated nervous tissues. In CIBA symposium "Growth of the Nervous System" (Wolstenholme, G. E. W. and O'Connor, M., eds), pp. 148–178.

Owen, R. (1841). On the anatomy of the southern *Apteryx* (*Apteryx Australis Shaw*). *Trans. Zoo. Soc. London* **182**, 257–309.

Padgett, C. S. and Ivey, W. D. (1960). The normal embryology of the *Coturnix* quail. *Anat. Rec.* **137**, 1–12.

Palmgren, E., (1921). Embryological and morphological studies on the midbrain and cerebellum of vertebrates. *Acta Zool.* **2**, 1–94.

Parker, T. J. (1891). Observations on the anatomy and development of *Apteryx*. *Phil. Trans. Roy. Soc. London* **182,** 25.

Pearson, J. L. and Willmer, E. N. (1963). Some observations on the actions of steroids on the metaplasia of the amoeba *Naegleria grubberi*. *J. Exptl. Biol.* **40,** 493.

Ramon y Cajal, S. (1908). Los ganglios centrales del cerebelo de las aves. *Trab. lab. biol. Madrid.* **6,** 195.

Rogers, K. T. (1960). Studies on the chick brain of biochemical differentiation related to morphological differentiation. I. Morphological development. *J. Exp. Zool.* **144**, 77–87.

Rudeberg, S. I. (1961). *Morphogenetic studies on the cerebellar nuclei and their homologisation in different vertebrates including man.* Tornblad Institute for comparative embryology. Lund. pp. 148.

Saetersdal, T. A. S. (1956a). On the ontogenesis of the avian cerebellum. I. Studies on the formation of fissures. *Univ. i Bergen Arb. Nat.* No. 2. 1–15.

Saetersdal, T. A. S. (1956b). On the ontogenesis of the avian cerebellum. II. Measurements of the cortical layers. *Univ. i Bergen Arb. Nat.* No. 3; 1–53.

Saetersdal, T. A. S. (1959a). On the ontogenesis of the avian cerebellum. III. Formation of fissures with a discussion of fissure homologies between the avian and mammalian cerebellum. *Univ. i Bergen Arb. Nat.* 1959. No. 3; 1–44.

Saetersdal, T. A. S. (1959b). On the ontogenesis of the avian cerebellum. IV. Mitotic activity in the external granular layer. *Univ. i Bergen Arb. Nat.* 4; 1–39.

Schumacher, O. (1928). Beitrage zur Entwicklungsgeschichte des Vertebratengehirns. IV. Die Entwickelungsgeschichte des Kiebitzgehirns. *Zeitschr. Anat. u. Entwicklungsges.* **87**, 139–251.

Sedlaček, J. (1966). Development of the steady potential of the brain of chick embryos. *Physiol. Bohem.* **15**, 111–116.

Sedlaček, J. and Maček, O. (1966). Development of brain impedance in chick embryos. *Physiol. Bohem.* **15**, 104–110.

Smith, D. E., Streicher, E., Milković, K. and Klatzo, I. (1964). Observations on the transport of proteins by isolated choroid plexus. *Acta Neuropath.* **3**, 372–386.

Szentágothái, J. (1968). Discussion of the action of $D_2O$ on nervous tissues. In CIBA symposium on "Growth of the Nervous System" (Wolstenholme, G. E. W. and O'Connor, M., eds), p. 175.

Tilney, F. (1915). The morphology of the diencephalon floor. *J. Comp. Neurol.* **25,** 213–282.

Wechsler, W. and Kleihues, P. (1968). Protein metabolism and cytodifferentiation in the nervous system. *In* "Macromolecules and the Function of the Neuron" (Lodin, Z. and Rose, S. P. R. eds), pp. 73–90. *Excerpta Medica.*

Willmer, E. N. (1960). "Cytology and Evolution", pp. 649. 2nd edition, 1970. Academic Press, London and New York.

Yamamoto, S., Ohkawa, K. and Lee, I. (1957). On the cerebellar nuclei of birds. *Acta hist. jap.* **13,** 129–139.

# 5　The Ear and Hearing

## I. INTRODUCTION

Sonic stimuli and specific behavioural responses to them are developed to a greater extent in birds than in any other class of vertebrates. Indeed the acoustic behaviour of birds is one of the oldest fields of biological research (see, for example, Busnel, 1963). Schwartzkopf (1968) and many earlier workers have emphasized that the ability to imitate sounds which were not previously contained in the species repertoire has been achieved at least twice, once in the Psittaciformes and once in the Passeriformes. Such imitative sound production is otherwise limited to the Hominoidea. A number of workers such as Konishi (1965) and also Schwartzkopf (1968) have suggested that certain calls are innately produced without ever having been heard by the singer, and Thorpe (1956, 1961) has reviewed the details of many Casper Hause experiments. Also all avian species are able to locate a fellow bird when it is calling, although this ability is generally not as precise as in many mammals. This difference is probably related to the supreme position which vision (q.v.) occupies amongst avian sensory modalities and it is certainly worth noting in this respect that those birds with the greatest "sonic acuity" are nocturnal, viz. members of the Strigiformes, Caprimulgiformes and *Collocalia*.

Various authors have demonstrated that only small differences exist between the song of normal blackbirds and that which is produced by birds which have been experimentally deafened at 18 days of age. Although such differences as do exist show that auditory feedback is certainly involved in monitoring the sounds produced, nevertheless it has been widely assumed that the feedback stimulation effect is a relatively minor one. The nature of avian vocal communication was reviewed, with special reference to the inheritance and development of song patterns by Thorpe (1961) and Hooker (1968). Although studies on the development of calls are fairly numerous relatively few of these works include detailed anatomical investigations as well. In this respect the work of Wurdinger (1970) holds out hope for many further papers. Other papers are summarized by Lanyon (1960), Marler (1960) and Hinde (1969).

Hooker considered the various functions of song and the possible explanations for such phenomena as duet singing and mimicry. She concluded that the precise form of the vocal behaviour in a given species reflects its ecology and suggested that strongly territorial and competitive birds such as those of the more temperate climates tend to be dependant upon very early auditory experience for fixing their subsequent vocal behaviour. Less territorial, and also tropical birds, possibly retain the ability to learn new song patterns for longer or unlimited periods. However, in view of the limited number of species which have currently been investigated, it would clearly be rather dangerous at this present juncture to place too much emphasis on such stimulating generalizations.

Konishi (1965) proposed a neurological model for the auditory monitoring system of the white crowned sparrow. This is illustrated in Fig. 42. The thick arrows indicate motor outflow from the central nervous system to the syrinx and other organs which are involved in motor control. Diagram 1 shows the situation in an intact bird which develops an abnormal song if it has not been exposed to a normal model during the early, formative, critical period. In diagram 2 the model is matched by the vocal output to the template which it acquired at this "imprinting" period. Deafening, or any other form of interference with the auditory feedback, causes an abnormal song regardless of whether a normal template has been acquired (1B) or not (1A). The template cannot be used in the absence of auditory feedback but, in the case of a specimen which has already performed the appropriate motor pattern which leads to normal song, non-auditory feedback alone can maintain the established vocalizations following deafening. However, one must conclude that, despite this established fact that both songs and less complex sound stimuli can be produced in the absence of further auditory control, a most important component in avian bio-acoustic behaviour is peripheral sonic analysis by the ear.

## II. FREQUENCY DISCRIMINATION AND THE AUDITORY THRESHOLDS

In recent years there have been a number of reviews which cover avian hearing (Pumphrey, 1961; Vinnikov and Titova, 1964; and Schwartzkopf, 1968). The summary provided here will permit a clearer appreciation of the medullary auditory centres.

With the development of recording frequency analysers which have a short time constant, it became apparent that certain avian auditory characteristics either compete with, or are better than, those of man. The frequency and intensity of variations which occur in the song of birds greatly exceed those which the human ear can detect (Potter et al., 1947). When imitating a song the mocking bird Mimus polyglottus undertakes very fast frequency changes which certainly lie outside our perceptual ability (Borror and Reese, 1956), and antiphonal calls (Thorpe 1963; Thorpe and North, 1965) demonstrate what are apparently very efficient temporal acoustic resolutions.

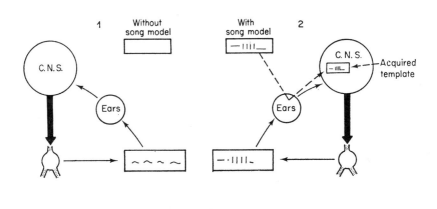

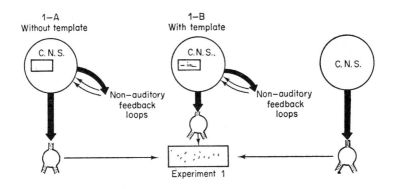

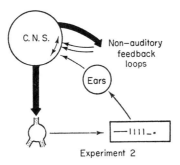

Fig. 42. Analysis of vocal control systems. (After Konishi, 1965.)

The actual discriminatory ability of birds has been widely discussed in relation to that of man and other mammals. Both *Steatornis caripensis* amongst the Caprimulgiformes, and *Collocalia* in the Micropodiformes, produce and detect sounds that are analogous to those of *Rousettus*. This is the only megachiropteran genus which is known to use echo-location (Griffin, 1953, 1958; Medway, 1959, 1962; Novick, 1959). In other genera Trainer (1946) has demonstrated threshold curves up to 7 and 8 kilocyc/sec. Vassil'iev (1933) claimed to be able to detect an upper limit of 12 kilocyc/sec whilst Brand and Kellogg (1939) extended this to 15 kilocyc with a lower limit of 200 cyc/sec. Wever and Bray (1936) recorded cochlear microphonics up to 11,500 cyc/sec but Edwards (1943) established a definite upper limit of 7,600 cycles, and a lower one at 60 cycles. Whitfield (1968) drew attention to the fact that Heise (1953) had tested pigeons over a range of frequencies up to 11,000 cycles but it is not clear whether he was able to obtain a response throughout his experimental range. It is probably significant that the threshold curve, which does not change by more than 10 decibels over the range 400–4,000 cyc/sec, shoots up abruptly when it reaches this latter level at a rate of over 70 decibels/octave. This clearly suggests that a break of some sort occurs in this region. Whitfield's own highest unit had a characteristic frequency of 4,130 cyc/sec, and, as its cut off was not infinitely steep, the correspondence appears to be quite good. Although Heise did obtain responses at 8 kilocycles these were only observed with thresholds of 80 decibels. It can therefore be seen that apart from Whitfield's data these results still leave the upper and lower limits only imprecisely defined. This is especially clear when one remembers that many of the determinations were made by using a Galton whistle which produces a broad spectrum of noise superimposed upon its computed frequency. Pumphrey (1961) doubted the existence of a useful sensitivity to frequencies in excess of 10 kilocyc/sec except in the case of owls and parrots, and even in these it is unlikely in view of our present knowledge. Measurements such as those of Schwartzkopf (1955) who used the cochlear potential of *Pyrrhula pyrrhula,* the bullfinch, have established that the band of maximum sensitivity lies between 2 and 4 kilocyc/sec, and that the sensitivity decreases as the extreme frequencies are approached. This is depicted in Fig. 43. Comparable and corroborative data for the budgerigar and the pigeon are shown in Fig. 44.

Knecht (1940) utilizing a training method demonstrated that the discriminative ability of the genera *Loxia* and *Melopsittacus* enabled them to differentiate between frequencies which only differed by 0·3–0·7%. He also concluded that the upper limit lay between 14 and 20 kilocyc/sec, a suggestion which clearly conflicts with the data of Fig. 44, and the lower limit at 40 cyc/sec. A summary of the variety of more or less probable suggestions which have previously been made for the upper and lower limits of a number of

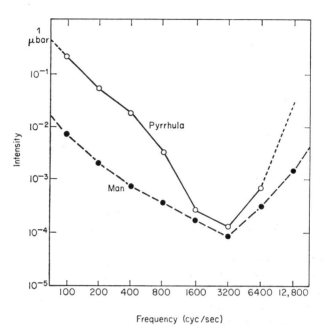

FIG. 43. Auditory threshold curves of *Pyrrhula pyrrhula,* the bullfinch, and man. (After Schwartzkopf 1968).

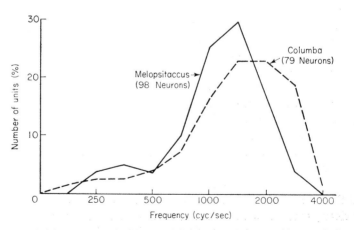

FIG. 44. Auditory response curves of *Melopsittacus* and *Columba.* (Schwartzkopf, 1968).

species are contained in Schwartzkopf (1955) and Kare (1965). As far as the frequency differentiation is concerned Pumphrey, noting the rather better values for the discriminative ability of man, concluded that this necessitated a sonic analyser of high quality. He considered that this must be a peripheral system as no obvious mechanism was available which would enable the central nervous system to perform such discrimination. It is, however, essential to observe at this point that in the special case of the owls which perform binaural localization very considerable medullary modifications exist which are considered in Chapter 6.

## III. DIRECTIONAL LOCALIZATION

Schwartzkopf (1955) measured the directional discrimination of *Pyrrhula* and discovered that this depends upon two factors:
  i. directional properties which ensure that sources out of the sagittal plane produce effects of different intensity at each ear;
  ii. the ability to recognize the temporal differences in the arrival of such sounds at both the proximal and distal ear.

Factors of this nature account for the fact that Granit (1941), who used pure tones, observed a directional ability which was 10 times less than that observed by Englemann (1928). The latter worker used the compound tones of chirping chicks as his stimuli. Marler (1955) has drawn attention to the considerable biological significance of such differences. The presence of transients in the songs of the oscine passerines renders their songs accurately detectable. In contrast, the pure tone of the hawk alarm cry, together with its gradual beginning and ending, are devoid of those elements which would provide an extensive time comparison. This makes it very difficult to detect either the direction or location from which it originates.

In the particular case of *Steatornis* echo-location is dependant upon emissions with a pulse of 7 kilocyc/sec, and the pulse repetition rate is high so that the interval separating them is only 2·3 msec. To obtain a discrimination which would compare with that of the majority of bats, the receptor elements would need to be some 10 times as highly damped and the speed of response must almost certainly exceed that of man. That the receptors of the avian ear are heavily damped was indeed suggested by Schwartzkopf amongst others.

Batteau (1967) has recently carried out an analysis of the functional significance of the mammalian pinna. In view of the results of Schwartzkopf (1955) on localization in *Pyrrhula*, it is useful to note his conclusions. He considered that the general function of the pinna is to introduce, by means of delay paths, a transformation of the incoming signal and that the inverse transform defines the location of the sound source. A relatively simple system of delays, attenuations and signed additions, which could be used to construct the

6

inverse transformations, could easily be realized by the central nervous system. These could be used for both binaural and monaural sound localization and also sound recognition. In the light of such conclusions one has to view the medullary organization of auditory nuclear masses with renewed interest.

The explanation of the functional significance of the overtly asymmetric ears, which were originally described in some strigiform genera by Pycraft (1898), dated initially from the paper by Pumphrey (1949), who considered the physical pre-conditions for three-dimensional directional analysis. Pumphrey's theory requires excellent monaural frequency discrimination which is met on *a priori* grounds by the extraordinary length of the basilar membrane. Nevertheless it has to be assumed that this long membrane is adapted for discriminating the highest tones. It seems probable that the ears differ in the frequency function of their directional characteristics. Recently Payne (in Schwartzkopf, 1968) has shown how, depending upon the direction of the sound source, the ears of *Tyto alba* may receive sound fractions of different pitch. Payne carried out experiments with stuffed heads in which the middle ear had been replaced by a microphone. Under these conditions the only auditory stimuli which are large enough to be of use appear to be restricted to frequencies which are in excess of 10 kilocyc/sec. Despite the suggestions of Pumphrey (1961) it is probable that in this range the owl's ear, in common with those of other avian genera, is relatively insensitive. Schwartz-kopf (1968) using an analogous method of investigation, but on anaesthetized specimens of *Asio otus*, studied the cochlear potentials as a function of the direction of the sound source. With frequencies between 3–4 kilocyc/sec, a range in which the owl's ear is very sensitive, only very small differences can be detected in the evoked results. As this conflicts with the performance of conscious animals he concluded that the system is only fully efficient when the bird operates the ear flaps voluntarily, the asymmetry being primarily a dynamic process which has only been secondarily developed as a static morphological adaptation. This means that any such experiments can only detect a fraction of the total capacity for directional location. If one adds to this the suggestions which were made by Batteau for man, and bears in mind the very considerable differences in relay time which can be observed experimentally at different levels of the ascending auditory pathway even on the ipsi-lateral side (see Chapter 14), this would certainly be a reasonable conclusion.

## IV. THE EXTERNAL EAR

On their path from the air to the sense organs sound stimuli pass through the sound collecting apparatus of the external ear, and the transforming components of the middle ear. The external auditory meatus is a relatively short tube which is usually slightly curved and, although its walls remain membranous, is limited by, for example, the occipital bone. It is usually associated externally with modified feathers whose form is subject to a fair amount of variation. However, in such orders as the Struthioniformes, Casuariiformes and Falconiformes it is naked. In *Asio* and certain other strigiform genera the cavity can be closed by two cutaneous folds. The anterior of these is under voluntary muscular control and this enables it to be used in the directional location of sounds which originate behind the bird. It is also worth drawing attention once again to the now long-established fact that, indeed, many of the owls which have a well-developed ability to practice acoustic location have a very considerable aural and even cranial asymmetry as in *Aegolius funereus*.

Cordier and Dalcq (1954), Pumphrey (1961) and Schwartzkopf (1968) have already reviewed much of the recent literature. The correlations which exist between the shape of the ear opening, the structure of the associated feathers and both ecological and behavioural characteristics have been demonstrated by Dementiev and Ilytchev (1963), Ilytchev (1960, 1961a,b, 1962a,b) Ilytchev and Izvekova (1961), Kartaschev and Ilytchev (1964). Pumphrey drew attention to the generalized functional significance of a structural feature such as the mammalian pinna. By increasing the area over which acoustic energy can be abstracted it gives directional locating properties and prevents the masking of signals by wind excited noise. The absence of such a structure from the avian ear, coupled with the high speed at which many species habitually fly, necessitates the almost universal feather covering over the meatus. This clearly reduces both the drag which is due to turbulence, and also the auditory masking. However such advantages are obtained at the expense of some absolute, and also some directional, sensitivity. A more recent consideration of the role which the pinna fulfils in mans sound localization is that of Batteau (1967) which was referred to in the previous section.

Kartaschcv and Ilytchev (1964) have demonstrated that there is also considerable variation in the actual site of the ear opening in relation to say the orbit amongst certain Alciformes. Their data are portrayed in Fig. 45. In general it is situated below and behind the eye but in *Aethia* the two are in closer proximity than is the case in other genera. The remainder have the ear further back but at differing vertical levels. In *Cephus grylle*, *Alca* and *Uria* it is nearer to the plane of the eye than in *Lunda* and *Fratercula*, whilst in the case of *Lunda cirrhata* the disparity is particularly noticeable and it is at a level which is markedly below that of the eye. Other work by the Soviet authors has also demonstrated that in the majority of species the variations in feather form result in the anterior aural tectrices being denser than the posterior ones, and, as can be seen from both Fig. 46 and Table 21, their outlines

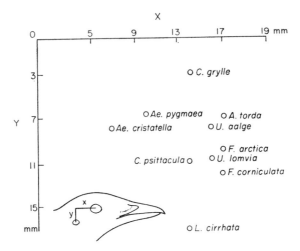

FIG. 45. A diagram showing the position of the ear opening relative to the orbit in some alciform birds. (After Kartaschev and Ilytchev, 1964).

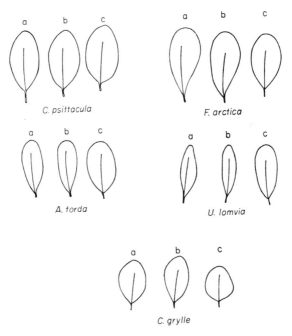

FIG. 46. The form of the tectrices in some alciform birds. (After Kartaschev and Ilytchev, 1964).

are more elongated. The inter-radial space is occluded by this dense arrangement and the first rows of the pterylae auris comprise a hydrodynamic surface. These adaptations are particularly pronounced in the definitive fish-eaters *Alca*, *Fratercula* and *Uria*.

## TABLE 21

*A comparison of the distance between the aural feathers in a number of species.*

| | Distance between feathers in mm. | |
| | Tectrices auris anteriores | Tectrices auris posteriores |
| --- | --- | --- |
| *Uria lomvia* | 0·62 | 0·87 |
| *Uria aalge* | 0·75 | 0·94 |
| *Alca torda* | 1·13 | 1·44 |
| *Fratercula arctica* | 0·44 | 0·69 |
| *Fratercula corniculata* | 0·37 | 0·69 |
| *Cephus grylle* | 0·89 | 0·89 |
| *Aethia cristatella* | 0·5 | 0·5 |
| *Aethia pygmaea* | 0·75 | 0·75 |
| *Cyclorrhynchus psittacula* | 0·75 | 0·5 |

Kartaschev and Ilytchev (1964).

The efficiency of the external ear in sound amplification has been measured by Ilytchev and Izvekova (1961). They replaced the tympanic membrane by an apparatus which measured the pressure of the sound. A comparison between the initial pressure and that transformed to the ear-drum suggested to them that in this respect the most efficient ears were those of hen harriers, members of the order Strigiformes and the *Corvidae*. In contrast the external ears of loons seemed to hardly amplify sounds at all. They also noted that in most cases the ears exhibited a definite tuning with respect to sounds of differing frequency and direction. In contrast to the conditions in the mammals that they studied they found that the maximal coefficient for directional sound in an eagle, gull, heron, loon and pigeon was at 135°, whilst in the case of the kite it was 90°.

## V. THE MIDDLE EAR

The general appearance and structure of the avian middle ear was summarized by Portmann (1950) and Stresemann (1934). It is an air-filled derivative of the first detectable visceral slit, is closed from the external meatus by

the tympanum, and opens to the buccal cavity by the Eustachian tube. Each
of these tubes open, in close proximity to each other, into a dorsal cavity and
this antrum communicates with the buccal cavity by a narrow median slit.
The average pressure on the two sides of the eardrum is therefore the same
but rapid changes in atmospheric pressure—sounds—are almost entirely
conducted to the drum by the meatus.

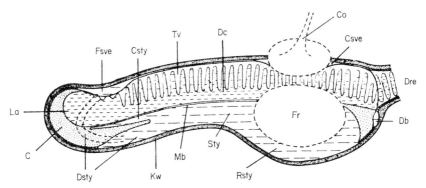

FIG. 47. Longitudinal section of the avian cochlea and lagena (Schwartzkopf, 1968).
    Abbreviations:

| | |
|---|---|
| C, Cartilage | Fr, Fenestra rotunda |
| Co, Columella | Fsve, Fossa scalae vestibuli |
| Csty, Cavum scalae tympani | Kw, Bony wall |
| Csve, Cisterna scalae vestibuli | La, Lagena |
| Db, Ductus brevis | Mb, Membrana basilaris |
| Dc, Ductus cochlearis | Rsty, Recessus scalae tympani |
| Dsty, Ductus scalae tympani | Sty, Scala tympani |
| Dre, Ductus reuniens | Tv, Tegmentum vasculosum. |

    The tympanic cavity is maintained in contact with the labyrinth by the two
fenestrae, the oval and rotund windows, which as can be seen from Fig. 47,
are situated about a quarter of the way along the length of the cochlea. The
oval window is in contact with the occluded scala vestibuli and the rotundum
with the scala tympani. Displacements of the tympanum which result from
sounds are communicated to the inner ear via the columella. This is a bone of
complex origin whose footplate nearly fills the oval window (Fig. 47). It is
held in position by an impervious but flexible annular ligament which links
its edges to those of the window. The tympanic cavity is continued into the
surrounding bone by a system of ramifying cavities which are lined by the
tympanic epithelium and seem to correspond to the other pneumatic cavities
of the skeleton. Sometimes one of these extensions even reaches to the mandibular articulation and enters the mandible. Elsewhere the cavity of the first
visceral slit is reduced to a small isolated vesicle which is situated close to the
squamosal. The origin of the innervation of this region is by no means clear.

Some authors consider that it is related to the facial whilst others attribute to it a component of the vestibular branch of the VIIIth nerve. The whole comprises the para-tympanic organ of Vitali.

The tympanum is fixed to a number of bones. These include the basisphenoid, squamosal and occipital complex. The distal cartilaginous regions of the columella connect it to the ossified stapedial region, the columellar footplate and the oval window. A supra-columellar process which is associated with a small bone runs to its upper posterior edge, and a longer, infra-columellar process, to its ventral edge. Together with Platners ligament these hold the columella in position and permit the transmission of sound to the oval window. Pumphrey (1961) emphasized that, although the actual mechanics of the transfer of sound power from tympanum to cochlea appear to be simpler in the avian ear than in that of mammals, they are nevertheless not clearly understood. Most discussions on the subject make the assumption that both the eardrum and the footplate are massless free pistons linked by a massless rod, the columella, which is normal to both. If this is the case then the transformation ratio will equal the ratios of the eardrum and footplate, and the sound pressures in the scala vestibular region on the one hand, and the auditory meatus on the other, will be in this ratio. A detailed series of measurements of the eardrum area and the columellar base area, together with the ratios of eardrum and footplate areas and eardrum and body surface areas, was provided by Schwartzkopf (1955). These are contained in Table 22.

Other workers have shown that the ratio of the ear-drum to footplate areas is low in the alciform genera listed below (Kartaschev and Ilytchev, 1964) and this has been taken as indicating a lower auditory sensitivity than is the case in many birds.

| | |
|---|---|
| *Uria lomvia* | 9·55 |
| *Fratercula corniculata* | 5·05 |
| *Lunda cirrhata* | 7·1 |
| *Cephus grylle* | 3·6 |
| *Aethia cristatella* | 12 |
| *Cyclorrhynchus psittacula* | 13·3 |

Since the columella is only partially ossified and lies at an angle to the eardrum the assumption that it is a uniform and rigid rod normal to the tympanum is clearly a gross oversimplification. Pumphrey pointed out that this results in a rocking motion of the columella which acts as a lever whose fulcrum is one edge of the oval window and whose short and long arms are the radius of the footplate and the whole length of the columella itself. This lever introduces an additional transformation factor which becomes predominant when, as in the Strigiformes, the angle of incidence of the columella on the ear-drum is very high. Indeed in the owls, Schwartzkopf noted that the inner aspect of the footplate is approximately hemispherical. He suggested that when the movement of the footplate is principally a rotation in the plane of the stapes about a point in the annular ligament there is turbulence and a consequent loss of power in the perilymph.

The values in Table 22 indicate the clearly demonstrable variation of the middle ear transformation ratios. One must interpolate the fact that they are not a measure of the actual range, which is probably much greater, because

## TABLE 22

*The area of the body surface, the ear drum (movable part) and the footplate in some avian ears, together with their ratios.*

| Species | Body surface | Relative size of ear drum (ear drum/ body surface) | Ear drum area cm² | Columella base area cm² | Area ratio ear drum footplate |
|---|---|---|---|---|---|
| *Phylloscopus collybita* | 4·0 | 0·020 | 0·078 | 0·0036 | 22 |
| *Phylloscopus trochilus* | 4·5 | 0·021 | 0·094 | 0·0034 | 28 |
| *Parus communis* | 5·0 | 0·018 | 0·089 | 0·0039 | 23 |
| *Parus caeruleus* | 5·1 | 0·016 | 0·084 | 0·0032 | 26 |
| *Hippolais icterina* | 5·7 | 0·015 | 0·086 | 0·0030 | 29 |
| *Sylvia atricapilla* | 6·6 | 0·019 | 0·126 | 0·0044 | 29 |
| *Hirundo rustica* | 7·4 | 0·010 | 0·071 | 0·0038 | 19 |
| *Fringilla coelebs* | 7·9 | 0·015 | 0·114 | 0·0041 | 28 |
| *Parus major* | 7·9 | 0·013 | 0·104 | 0·0042 | 25 |
| *Pyrrhula pyrrhula* | 9·0 | 0·013 | 0·117 | 0·0048 | 24 |
| *Passer domesticus* | 9·6 | 0·0094 | 0·091 | 0·0042 | 22 |
| *Turdus merula* | 20·9 | 0·0077 | 0·160 | 0·0073 | 22 |
| *Pica pica* | 35·5 | 0·0075 | 0·265 | 0·0116 | 23 |
| *Corvus corone* | 65·5 | 0·0053 | 0·347 | 0·0151 | 23 |
| *Asio otus* | 44·9 | 0·0107 | 0·480 | 0·0120 | 40 |
| *Strix aluco* | 66·4 | 0·0089 | 0·593 | 0·0198 | 30 |
| *Columba livia* | 47·8 | 0·0043 | 0·204 | 0·0116 | 14 |
| *Gallinula chloropus* | 41·7 | 0·0032 | 0·132 | 0·0078 | 16 |
| *Fulica atra* | 84·0 | 0·0025 | 0·209 | 0·0106 | 19 |
| *Anas platyrhynchos* | 82·5 | 0·0034 | 0·285 | 0·0109 | 26 |
| juv 10d | 15·2 | 0·0054 | 0·082 | 0·0055 | 15 |
| *Buteo buteo* | 86·1 | 0·0039 | 0·330 | 0·0180 | 18 |
| *Podiceps cristatus* | 86·0 | 0·0016 | 0·140 | 0·0095 | 16 |
| *Phasianus colchicus* | 113·0 | 0·0033 | 0·368 | 0·0133 | 28 |
| *Gallus domesticus* | 153·0 | 0·0019 | 0·291 | 0·0133 | 22 |
| juv. 40d | 27·6 | 0·0052 | 0·144 | 0·0083 | 17 |
| juv. 1d | 10·5 | 0·0066 | 0·069 | 0·0060 | 11 |
| *Grus grus* | 245·0 | 0·0017 | 0·418 | 0·0169 | 25 |

From Pumphrey (1961) after Schwartzkopf (1955).

of the differences in the angle of incidence of the columella on the tympanum. It is also probable that the oblique insertion of the columella, together with the ability of the cartilaginous extra-stapedial component to bend under excessive pressure on the ear-drum before the annular ligament is over-strained, are the reasons for the relative immunity of birds to trauma in the presence of loud noise.

## VI. THE COCHLEA

From a historical point of view it is important to note that the peripheral components of the avian vestibulo-cochlear system were studied at the close of the last century by Retzius (1884), Krause (1906) and Gray (1908). The bony cochlea is a finger-like tube through the centre of which runs the scala media or cochlear duct. An outgrowth of the membranous labyrinth, this contains endolymph and is held in position by a pair of cartilaginous shelves which project inwards from the bony cochlea, but lack the strong appearance which is to be seen in mammals. The floor comprises the basilar membrane and the tegmentum vasculosum. At its apical end is the macula lagenae which consists of a group of sensory hair cells associated with oto-conia. This is adherent to cartilage which is in its turn anchored by the trabeculae to the bony cochlea. The possible involvement of this structure in avian sound perception has been discussed by Pumphrey (1961).

As Gray (1908) pointed out the amount of pigment in the avian labyrinth rarely renders it visible to the naked eye. Present in the ostrich, rhea and a tinamou, he found that it was particularly abundant over the perilymph recess, the adjacent parts of the cochlea and on the upper surface of the vestibule, and that it was also present in the vicinity of nerves. He concluded that the ostrich has the largest labyrinth to be observed in birds but that as the cochlear component is small the length of this structure is scarcely longer than that of the gannet. The measurements which he made on a number of species are contained in Table 23 but it should be noted that these represent the distance between the anterior margin of the oval window and the cochlear tip. As the oval window is situated some quarter of the way along the avian cochlea the overall length would be somewhat in excess of these values. In the ostrich he found that the cochlear diameter at the level of this window is 2·5 mm as is the major axis of the window itself. He drew attention to the fact that in the case of the smaller cochlea of *Rhynchotus* (a tinamou) the relative size of the structure is actually greater than that of *Struthio* and *Rhea* when compared with gross external parameters. More recent comparative data which were provided by Schwartzkopf (1968) are presented in Fig. 48. From these values it can be seen that the avian cochlea is large by comparison with that of reptiles but it nevertheless lacks the complex coiled form which

6*

is so characteristic of mammals. It is worth noting in this context that the only functional explanation which has been advanced to account for the coiled structure in mammals is that of Békésy (1953) who suggested that the curvature of the tectorial membrane limits its bending, or at least its maximal deformation, to a fairly closely defined area (Whitfield, 1967).

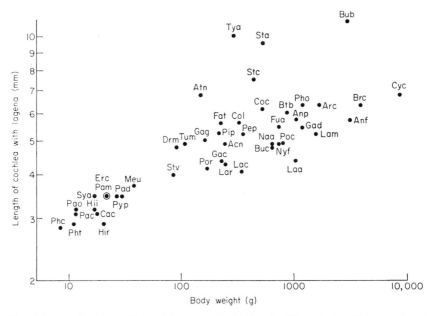

FIG. 48. Length of cochlea with lagena in birds of different size (Schwartzkopf, 1968).

Abbreviations:

Acn, *Accipter nisus*
Afn, *Anser fabilis*
Anp, *Anas platyrhynchos*
Arc, *Ardea cinerea*
Atn, *Athene noctua*
Brc, *Branta canadensis*
Btb, *Buteo buteo*
Bub, *Bubo bubo*
Buc, *Bucephala clanga*
Cac, *Carduelis cannabina*
Coc, *Corvus corone*
Col, *Columba livia*
Cyc, *Cygnus cygnus*
Drm, *Dryobates major*
Erc, *Fringilla coelebs*

Fat, *Falco tinnunculus*
Fua, *Fulica atra*
Gac, *Gallinula chloropus*
Gad, *Gallus domesticus*
Gag, *Garrulus glandarius*
Hii, *Hippolais icterina*
Hir, *Hirundo rustica*
Laa, *Larus argentatus*
Lac, *Larus canus*
Lam, *Larus marinus*
Meu, *Melopsittacus undulatus*
Naa, *Numenius arquata*
Nyf, *Nyroca fuligula*
Pac, *Parus caeruleus*
Pad, *Passer domesticus*

Pam, *Parus major*
Pao, *Parus communis*
Pep, *Perdix perdix*
Phc, *Phylloscopus collybita*
Pho, *Phasianus colchicus*
Pht, *Phylloscopus trochilus*
Pip, *Pica pica*
Poc, *Podiceps cristatus*
Pyp, *Pyrrhula pyrrhula*
Sta, *Strix aluco*
Stc, *Steatornis caripensis*
Stv, *Sturnus vulgaris*
Sya, *Sylvia atricapilla*
Tum, *Turdus merula*
Tya, *Tyto alba*

Inside the cochlea of freshly killed birds the massive gelatinous arch of the tegmentum vasculosum is in contact with the bony walls and almost totally occludes the space within the scala vestibuli. This therefore *lacks* the wide and open lumen that is to be seen in the mammalian structure. Careful investigations do, however, reveal the existence of "perilymphatic fissures" between the tegmentum and the endosteum. Below the columella these enlarge to form the cisternae scalae vestibuli as shown in Fig. 47 (Schwartzkopf, 1968).

## TABLE 23

*The measurements of certain cochlear and labyrinthine parameters according to Gray (1908).*

| Species | Length from the vertex of the superior canal to the tip of the cochlea (mm) | Cochlea Length from the front of the oval window (mm) | Major axis of oval window (mm) |
|---|---|---|---|
| *Struthio masai* | 18·0 | 6·0 | 2·5 |
| *Apteryx mantelli* | 10·5 | 2·5 | 1·0 |
| *Rhynchotus rufescens* | 9·5 | 2·5 | 1·0 |
| *Colymbus septentrionalis* | 13·0 | 4·5 | 1·75 |
| *Sula capensis* | 17·0 | 6·0 | 2·0 |
| *Phalacrocorax carbo* | 13·5 | 4·5 | 1·75 |
| *Nycticorax griseus* | 13·0 | 4·0 | 1·5 |
| *Anas boscas* | 11·0 | 4·0 | — |
| *Buteo vulgaris* | 15·5 | 4·5 | 2·0 |
| *Gallus domesticus* | 12·5 | 3·75 | 1·5 |
| *Lagopus scoticus* | 10·0 | 3·5 | 1·5 |
| *Balearica pavonina* | 15·0 | 4·5 | 1·5 |
| *Goura coronata* | 13·0 | 4·5 | 1·5 |
| *Licmetis nasica* | 11·5 | 5·0 | 1·5 |
| *Speotyto cunicularia* | 12·5 | 5·25 | 1·75 |
| *Corvus corone* | 13·0 | 3·5 | 1·75 |
| *Turdus musicus* | 8·5 | 3·5 | 1·0 |

Pumphrey (1961) observed that it looked as though in the design of the avian ear, and more especially the cochlea, minimization of viscous and turbulence resistance is of less importance than in mammals. However, he was, at the time, working on the assumption that the overall avian cochlear structure was more closely comparable with that of mammals than it is now known to be. The simplicity of the structure of the papilla basilaris contrasts with that of the organ of Corti, the fenestra rotunda and fenestra ovale lie a quarter of the way along its length, and the axis of the columella is approximately at right angles to the line of the cochlea. In the latter case it is only amongst the owls that the situation approximates to that in mammals (Schwartzkopf, 1968).

It is probable that like the stria vascularis in mammals the tegmental membranes serve, at least in part, a trophic function. There are numerous blood vessels within the perilymph of the scala tympani and also of course in the scala vestibuli. Pumphrey suggested that if the papilla basilaris is a regenerative sound amplifier, as was suggested for the organ of Corti by Gold (1948), then it would certainly need an adjacent energy source. In view of the absence of any obstruction in the mammalian ear comparable with the tegmentum vasculosum of birds several authorities have concluded with Pumphrey that such obstructions are of less consequence in avian hearing. The residual cisternae scalae vestibuli which, under these circumstances, correspond to the open scala of other classes, connect with the scala tympani by the ductus brevis at the basal end. Although the presence of a definitive helicotrema at the distal end has been denied, Schwartzkopf (1968) noted that one exists. It is situated between the lagena and the cochlea and at the end of a channel, the ductus scalae tympani, which penetrates the supporting cartilage of the inner ear (see Fig. 47).

The basilar membrane is both broader and shorter than that of mammals as would be expected from the overall cochlear dimensions. There are also differences between the two classes in the size of its constituent fibres. Although the transverse fibres increase in length towards the apex, this increase is less than 2 : 1 and contrasts with the fourfold changes that occur in mammals. The total number of fibres is 1200 in *Columba* and some psittaciform species. As can be seen from Fig. 48 the absolute length of the membrane is variable and naturally follows cochlear length. Leaving aside the owls, which are distinguished from most other birds by the length of their cochlea, the data of Schwartzkopf (1968) and Winters (1963) are of very different magnitude to those of Gray (1908), and suggest that the swan has a cochlear length of 6·8 mm. Although large by comparison with that of certain other avian genera this size compares unfavourably with that of a mouse. However, the presence of about 10 times as many sensory cells in a given transverse section of the papilla basilaris by comparison with the organ of Corti, suggests that the net effect may be comparable with the condition in mammals but that it is achieved in a different manner. With reference to the damping effect, which, as was mentioned in Section III, would be necessary for echolocation in *Steatornis*, Schwartzkopf suggested that this damping, when coupled with the short length of the basilar membrane, would constitute a limitation on any frequency analysis. On the other hand one would expect an enhanced temporal resolving ability comparable with that which one observes during song. It is certainly clear that although good frequency analysis does exist it is only within a band which is narrower than that of mammals.

Finally there is the possible involvement in hearing of the lagenar region. This is definitely an auditory organ in fishes and, in birds, its numerous afferent nerve fibres accompany those from the cochlea into the medulla. What is more, unlike the majority of the fibres from the sacculus they seem to enter the medullary auditory nuclei and not those associated with vestibular function (see Chapter 6). Pumphrey (1961) suggested that it may be an integral part of the mechanisms which are associated with the known auditory sensitivity of the unhatched young. However, in opposition to this, he also cited the design of the cochlear architecture which he considered would ensure that the endolymph in the proximity of the lagenar macula was stagnant and not affected by perilymph movements originating at the columella. In view of the true nature of the scala vestibuli and its restricted lumen, at least in the adult, these observations are difficult to interpret further.

## VII. THE HAIR CELLS

The sensory hair cells of the avian papilla basilaris comprise a stretched epithelial layer embedded between supporting cells. It is through these that they maintain their connection with the basilar membrane. As was mentioned above they are more numerous than their mammalian homologues and in the parrots some 40 have been counted in a single transverse section. The hair cell processes are anchored at their distal end within the overlying tectorial membrane. Each contains within it a polarized kinocilium and 40–50 stereocilia which are comparable with those seen elsewhere in labyrinthine and neuromast organs (Dijkgraaf, 1963; Vinnikov et al., 1965a). There is no tunnel comparable with that which in mammals separates the inner from the possibly more sensitive outer hair cells which generate the microphonic effects. Because there is, in fact, a widespread absence of those features which characterize the mammalian organ of Corti Schwartzkopf (1968) urged the use of the term papilla basilaris. This usage has been followed here in spite of the fact that the other term is by now deeply entrenched in the literature.

Despite these differences Wersall et al. (1965) together with Vinnikov et al. (1965a) suggest that the synaptic relationships of the papilla basilaris are comparable with those which can be detected in the organ of Corti. The nerve endings at the base of the hair cells seem to be of two distinct types. One contains large mitochondria and the other numerous synaptic vesicles. The endings form several small buttons at each hair cell so that they resemble the Type II synaptic connections of Wersall (1956), rather than the large nerve chalices of Type I that occur in the utricle. Cordier (1964) found a pre-synaptic body at the side of the hair cell and various mitochondria within the nerve ending in the case of afferent connections. The efferent fibres have numerous synaptic vesicles and there are additional membranes in the adjacent area of the hair cells.

Vinnikov and Titova (1964) in their monograph on the organ of Corti draw attention to the fact that in the avian ear the hair cells of the lagena macula acoustica are still taking the weight of the otoliths. In their evolutionary scheme they suggest that such structures have gradually been modified to give the definitive structure of the organ of Corti, free from otoliths and covered by the coarse fibred tectorial membrane. Hawkins (1965) using mammalian material discussed the cytoarchitectural features which are of importance in the vertebrate ear when it is functioning as a transducer of great dynamic range and is also a frequency analyser. Disturbance or disruption of the cellular pattern by the action of ototoxic antibiotics, or intense sound, indicates differences of susceptibility to these two forms of injury in the various hair cell rows. The accompanying reductions in cochlear microphonic potentials support the hypothesis that their origin lies in the outer hair cells.

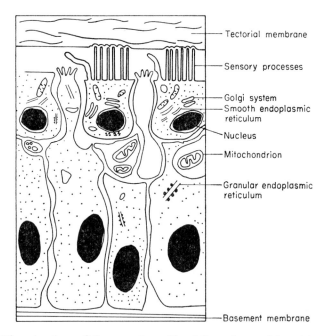

FIG. 49. Fine structure of the receptor cells of the avian cochlea. After Vinnikov (1969) and others.

The fine structure of the avian hair cells is shown diagrammatically in Fig. 49. They can bear up to 100 sensory hairs on their distal surface in contrast to the 8, 12, 30 or 40 which one sees in mammals. Vinnikov and Titova found that throughout the labyrinths of both the hen and the pigeon they have a high total protein content. They also contain large concentrations of glycogen, phosphorylase, phosphatases, and the enzymes of the succinic oxidase system, particularly cytochrome oxidase. Together with acetylcholine esterase high concentrations of these enzymes are also particularly prevalent at the synaptic endings and neurons of the vestibular and auditory ganglia. They appear to be lacking in the hair cells of the acoustic maculae in both the utricle and saccule, the cristae of the semi-circular canals, and in the region of the acoustic macula of the lagena. In contrast the sensory hair cells of the papilla basilaris have high specific activities. According to these Soviet workers the situation in birds is intermediate between that of reptiles and mammals. The concentrations of cholinesterases are greater than those in reptiles and more comparable with those in mammals. This is possibly a reflection of the common homeothermy. Unlike those of reptiles they are also detectable at the hair bearing pole and these Russian authors further concluded that within the avian auditory system there exists a third type of cholinesterase which is unique amongst vertebrates.

The processes which are involved in the perception of sound at the level of the avian hair cell have been studied by various workers. According to Mazo (1958) experimentally produced injury to different cochlear areas showed that the receptors for low tones are located closer to the apex or the lagena. Those for high tones, are, on the contrary, closer to its base although it is not clear whether it is the lagena or papilla basilaris that is involved. Vinnikov and Titova (1964) showed that 30–60 min exposures to sounds of both high and low frequencies results in an initial rise, and then a subsequent fall, in the activity of the enzyme systems of the hair cells within the papilla, the lagena, and the maculae of the utricle and saccule. Within the "organ of Corti" itself such enzymatic changes were also demonstrable in the actual hairs themselves. However it was not possible to detect a gradient of enzyme activity along the cochlea, in association with these low and high frequency stimuli, which would be comparable with the gross sensitivity gradient observed by Mazo. Indeed they obtained the distinct impression that in birds those sensory cells which are sensitive to high and low frequencies respectively are not spatially separated from each other.

In concurrence with the general suggestions of sensory physiologists such as Nachmansohn, Vinnikov and Titova (1964) intimate that the transmission of an impulse from an excited hair across the labyrinthine synapses is associated with an acetylcholine–acetylcholine esterase system. This would also agree with the specific suggestions of Derbyshire and Davis (1935a, b) that acetylcholine plays an integral part in the initiation of impulses within the eighth nerve. The application of the acetylcholine esterase inhibitor physostigmine certainly delays the appearance of cochlear potentials in the mammalian ear. In *Columba*, Dacha and Martini (1942) and Martini (1941) found that, after extracting perilymph from the ampulla of the posterior semicircular canal and then applying a sonic stimulus of 436 cyc/sec for 10 min, newly extracted perilymph contains acetylcholine. Gisselson (1950) was unable to confirm these results but he did detect considerable levels of acetylcholine esterase activity in the endolymph, and, to a lesser extent, in the perilymph. This has led to the suggestion that in the areas where the sensory hairs are free from otoliths and are subjected to endolymph movement the high concentration of acetylcholine is related to the mechanical bending of the hairs after which excitation is possible.

By considering the condition of the macula acoustica of the avian lagena Vinnikov and Titova postulated a scheme for the successive phases which may have occurred in the phylogenetic and physiological transition from a mechano-receptor to a specialized chemoceptor. This they think reflects the situation in the highly differentiated sound receptors of birds and mammals. They consider that in these sound receptors waves in the endolymph excite the hairs by both mechanical and chemical stimuli. They emphasize the chemical aspect and their suggestions therefore differ from many of the earlier speculations which had focused on mechanical deformation consequent upon either a standing or a variable moving wave in the basilar membrane or other constituent structures. The release of a unit quantity of acetylcholine was envisaged which produces a series of changes in the hair cells of a particular cochlear region. These changes are said to be detectable as overt morphological and chemical transformations. They cite in particular a rounding of the cytoplasmic body, "pulsation" of the nucleus, and changes in the deposition characteristics of vital dyes. These they interpret as reflecting a complex but reversible change in the protein components of the cytoplasm. Associated with this change there are biochemical processes which have a special topographical distribution within the cells and also occur in a strict temporal sequence. It is difficult to be other than somewhat sceptical about some of these suggestions in view of the very high degree of cytological localization which would have to be detected. These pre-suppose, at the very least, a very large number of serial sections viewed at the electron-microscope level. However, as a yet incompletely proven theory their suggestions are stimulating. They consider that the hair cells are "high energy antennae" which become depolarized by means of endolymphatic acetylcholine under the influence of shearing and travelling waves. The "energy of depolarization" activates the cell processes. During sonic exposure the conversion of the glycogen contained within the hair cell, its combination with protein, and its topographical relationship to phosphorylase activity led them to conclude that the energy of anaerobic metabolism is intimately involved in these processes. In itself this is obviously a reasonable *a priori* hypothesis and it would conform with data such as those of Bosher and Warren (1968) which are noted below.

# VIII. BIO-ELECTRICAL PHENOMENA

Schwartzkopf and Bremond (1963) described the microphonic and action potentials which can be detected at the round window. The overall characteristics of such bio-electric phenomena in the avian inner ear correspond to those which have been reported in both mammals and reptiles although there are some differences in detail by comparison with the supposed conditions of mammals. Even at a very low sound pressure there is not a directly linear increase of amplitude with intensity (Schwartzkopf, 1968). This lack of linearity is likely to result in distortion of the stimulating signal and Schwartzkopf suggests that the properties of the microphonics may be explicable in terms of a non-linear damping effect exerted through the tegmentum vasculosum.

The nervous components of the cochlear potentials represent the activity of the auditory nerve and its synapses. With suitable techniques Schwartzkopf and Bremond were able to detect stimulus synchronous discharges up to 3 kilocyc/sec. The action potentials are certainly influenced by metabolic changes and exhibit both adaption and latency variations as a function of stimulus intensity. In view of the lack of histological differentiation in the avian ear Stopp and Whitfield (1964) (see Chapter 6, Section V) understandably concluded that the summating potentials which they demonstrated are not explicable as in the case of mammals in terms of inner hair cell activity. Necker (see Schwartzkopf, 1968) has shown that they are almost certainly of complex origin, with two positive and one negative components. In some respects they appear to behave as microphonics but show a differential dependance upon the supply of oxygen, and again the kind of metabolic interference which one observes warns against any simple equation between the summating potentials of birds and mammals. However, a brief word on certain mammalian results is not out of place. Using flame spectrophotometry Bosher and Warren (1968) proved that the sodium and potassium concentrations in *rat* endolymph were 0·91 and 154 mequiv/litre respectively. In the perilymph the analogous values were 138 and 6·9 and the average endolymphatic potential was +92 millivolts. During anoxia this positive potential disappeared and was replaced by one of opposite sign which reached an average −42 millivolts after 4·5 min. The principal ionic changes were a progressive increase in the endolymphatic sodium concentration and an associated decrease in the potassium concentration. As a result Bosher and Warren concluded that the low sodium concentration of the endolymph is maintained by active transport of sodium and chloride out of, and potassium into, the cochlear duct. The mechanism for this probably involves the stria vascularis and the system, or systems involved are very dependant upon oxidative metabolism. In this connection one can also refer to the work of Schmidt and Fernandez (1962). Endocochlear d.c. potentials that are characteristic of both classes of homoiotherms and associated with a high metabolic rate were studied by Schmidt whilst Necker, again in Schwartzkopf, has extended the previous studies and demonstrated an improved resistance to metabolic stress. The voltages which he detected were normally of the order of +20 millivolts but dropped to −30 or −40 during short term nitrogen respiration. However, it is extremely interesting that, unlike the mammalian auditory nerve, that of birds can continue to discharge for some time after the endocochlear potential has turned negative.

# References

Batteau, D. W. (1967). The role of the pinna in human sound localization. *Proc. Roy. Soc. B.* **168,** 158–180.

Békésy, G. V. (1953). Description of some mechanical properties of the organ of Corti. *J. Acoust. Soc. Amer.* **25,** 770–785.

Borror, O. J. and Reese, C. R. (1956). Vocal gymnastics of the wood thrush. *Ohio J. Sci.* **56,** 177–182.

Bosher, S. K. and Warren, R. L. (1968). Observations on the electrochemistry of the cochlear endolymph in the rat. *Proc. Roy. Soc. B.* **171,** 227–247.

Brand, A. R. and Kellogg, P. P. (1939). Auditory responses of starlings, English sparrows and domestic pigeons. *Wilson. Bull.* **51.** Michigan.

Busnel, R. G. (1963). "Acoustic Behaviour of Animals," pp. 993. Elsevier.

Cordier, R. (1964). Sur la double innervation des cellules sensorielles dans l'organe de Corti du Pigeon. *C.R. hébd. Séance. Acad. Sci. Paris.* **258,** 6238–6240.

Cordier, R. and Dalcq, A. (1954). *Traité de Zoologie. Vertebrés généralités.*

Dacha, U. and Martini, R. (1942). Il riflesso di fullio nei piccioni con stimolazioni Sonora di diversa altezza. *Boll. soc. ital. biol. sper.* **17,** 331.

Dementiev, G. P. and Ilytchev, V. D. (1963). *Falke.* **10,** 123–125; 158–164; 187–191.

Derbyshire, A. J. and Davis, H. (1935a). Probable mechanism for stimulation of the auditory nerve by the organ of Corti. *Amer. J. Physiol.* **113,** 35.

Derbyshire, A. J. and Davis, H. (1935b). Action potential of the auditory nerve. *Amer. J. Physiol.* **113,** 476.

Dijkgraaf, S. (1963). The functioning and significance of lateral line organs. *Biol. Rev.* **38,** 51–106.

Edwards, E. P. (1943). Hearing ranges of four birds. *Auk.* **60,** 239–241.

Englemann, W. (1928). Untersuchungen uber die Schallokalisation bei Tieren. *Z. Psychol. Physiol. Sinnesorg.* **105,** 317–370.

Gisselson, L. (1950). Experimental investigation into the problem of humoral transmission in the cochlea. *Acta oto-laryngol.* Suppl. **82.**

Gold, T. (1948). Hearing. II. The physical basis of the action of the cochlea. *Proc. Roy. Soc. B.* **135**, 492–498.

Granit, O. (1941). Beiträge zur Kenntnis des Gehörsinnes der Vögel. *Ornis Fennica.* **18**, 49–71.

Gray, A. A. (1908). "The Labyrinth of Animals", 2 vols. Churchill, London.

Griffin, D. R. (1953). Acoustic orientation in the oil bird, *Steatornis. Proc. Natl. Acad. Sci.* **39**, 884–893.

Griffin, D. R. (1958). "Listening in the Dark", pp. 413. Yale.

Hawkins, J. E., Jr. (1964). Hearing. *Ann. Rev. Physiol.* **26**, 453–480.

Hawkins, J. E. Jr. (1965). Cytoarchitectural basis of the cochlear transducer. *Cold Spring. Harb. Symp. Quant. Biol.* **30**, 147–157.

Heise, G. A. (1953). Auditory thresholds in the pigeon. *Amer. J. Psychol.* **66**, 1–19.

Hinde, R. A. (editor) (1969). "Bird Vocalisations", pp. 394. Cambridge University Press.

Hooker, B. I. (1968). "Birds" *in "Animal Communication"* (Seboek, T. A., ed.), pp. 311–337. Indiana University Press.

Ilytchev, V. D. (1960). External part of the auditory analyser in birds. (This and following in Great Russian) *Zool. J.* **39**, 1871–1877.

Ilytchev, V. D. (1961a). External part of the auditory analyser in Impennae. *Scient. reports of the school of Biol. Sci.* **2**, 51–55.

Ilytchev, V. D. (1961b). Adaptations of the outer parts of the avian sound analysers to life in water. *Rep. Moscow Univ.* **1**, 22–25.

Ilytchev, V. D. (1961c). Some regularities in the evolution of the external ear in vertebrates. *Zool. Journ.* **12**, 1795–1806.

Ilytchev, V. D. (1962a). Avian aural feathers, their structure and function. *Dokl. Acad. Nauk. U.S.S.R.* **144**'s, 1185–1188.

Ilytchev, V. D. (1962b). On the evolution of the auditory compartments of the avian skull. *Dokl. Acad. Nauk U.S.S.R.* **144**, 934–937.

Ilytchev, V. D. (1966a). *Zh. vyssh. nerv. Deyat.* **16**, 480–488.

Ilytchev, V. D. (1966b). *Vest. Moskow Univ. Biol.* **4**, 32–36.

Ilytchev, V. D. and Dubinskaja, G. R. (1966). *Zool. Zh.* **45**, 1580–1582.

Ilytchev, V. D. and and Izvekova, L. M. (1961). *Zool. Zh.* **11**, 1704–1714.

Kare, M. R. (1965). The special senses. *In* "Avian Physiology" (Sturkie, P. D., ed.), pp. 406–446. Baillière, Tindall and Cassell.

Kartaschev, M. M., and Ilytchev, V. D. (1964). Uber das Gehörorgan der Alkenvögel. *J. Ornithol.* **105**, 115–136.

Knecht, S. (1940). Uber das Gehorsinn und die Musikalitat der Vögel. *Z. Vergleich. Physiol.* **27**, 169–323.

Konishi, M. (1965). Role of auditory feedback in the control of vocalisation in the white crowned sparrow. *Z. tierpsychol.* **22**, 770–783.

Krause, R. (1906). *Gehororgan; Handb. Entwickl. lehre Wirbeltiere* II. Part I.

Lanyon, W. E. (1960). The ontogeny of vocalisations in birds. *In* "Animal Sounds and Communication" (Lanyon, W. E. and Tavolga, W. N. eds), pp. 321–347. Amer. Inst. Biol. Science.

Marler, P. (1955). Characteristics of some animal calls. *Nature, Lond.* **176**, 6–8.

Marler, P. (1960). Bird songs and mate selection. *In* "Animal Sounds and Communication" (Lanyon, W. E. and Tavolga, W. N. eds), pp. 348–367. Amer. Inst. Biol. Science.

Martini, V. (1941). Liberazioni di sostanza acetilcolonosomile nel orecchio interno durante la stimolazione sonora. *Arch. Sci. Biol.* **27**, 94.

Mazo, I. L. (1958). *The study of the origin and function of the inner ear.* Abstr. Diss. Moscow, ear, nose and throat research inst.

Medway, Lord (1959). Echolocation among *Collocalia. Nature, Lond.* **184,** 1352–1353.

Medway, Lord (1962). The swiftlets of the Niah cave Sarawak. *Ibis.* **104,** 45–66, 228–245.

Novick, A. (1959). Acoustic orientation in the cave swiftlet. *Biol. Bull. mar. biol. Lab. Woods Hole.* **117,** 497–503.

Portmann, A. (1950). Oiseaux. Traité de zoologie. Masson, Paris.

Potter, R. K., Kopp, G. A. and Green, H. C. (1947). "Visible Speech". Van Nostrand, New York.

Pumphrey, R. J. (1949). The sense organs of birds. *Smithsonian rep.* 1948. 305–330.

Pumphrey, R. J. (1961). Sensory organs, hearing. *In* "Biology and Comparative Physiology of Birds" (Marshall, A. J. ed.), 2 vols. Academic Press, New York and London.

Pycraft, W. P. (1898). A contribution towards our knowledge of owls. *Trans. Linn. Soc.* **7,** 223–275.

Retzius, O. (1884). "Das Gehörorgan der Reptilien, Vögel und Saugetiere." Stockholm.

Schmidt, R. S. and Fernandez, C. (1962). Labyrinthine d.c. potentials in representative vertebrates. *J. Cell. Comp. Physiol.* **59,** 311–322.

Schwartzkopf, J. (1955). Schallsinnesorgane, ihre Funktion und biologische Bedeutung bei Vögeln. *Acta 11th Congr. Intern. Ornithol.* 1954, pp. 189–208.

Schwartzkopf, J. (1968). Hearing in birds. *In* CIBA symposium "Hearing in Vertebrates", pp. 41–59.

Schwartzkopf, J. and Bremond, J. C. (1963). A method for obtaining cochlear potentials in birds. *J. Physiol., Paris* **55,** 495–518.

Streseman, E. (1934). Aves. *In* "Hdb. Zool." Kukenthal-Krumbach. Vol. 7, p. 900.

Thorpe, W. H. (1956). "Learning and instinct in Animals." Methuen, London. (2nd ed. 1963).

Thorpe, W. H. (1961). "Bird song: the Biology of Vocal Communication and Expression in Birds." Cambridge University Press.

Thorpe, W. H. (1963). Antiphonal singing in birds as evidence for avian reaction time. *Nature, Lond.* **197,** 774–776.

Thorpe, W. H. and North, M. E. W. (1965). Origin and significance of the power of vocal imitation with special reference to antiphonal singing of birds. *Nature, Lond.* **208,** 219–222.

Trainer, J. E. (1946). Thesis, Cornell University.

Vasiliev, M. F. (1933). Uber das Unterscheidungsvermögen der Vögel fur die hohen Tone. *Z. Vergleich, Physiol.* **19,** 424–438.

Vinnikov, Y. A. (1969). The ultrastructural and cytochemical basis of the mechanism of function of the sense organ receptors. *In* "The Structure and Function of Nervous Tissue." (Bourne, G. H. ed.), pp. 265–392. Academic Press, London and New York.

Vinnikov, Y. A. and Titova, L. K. (1964). "The Organ of Corti, its Histophysiology and Histochemistry." pp. 253. Consultants Bureau. New York.

Vinnikov, Y. A., Osipova, I. V., Titova, L. K. and Govardovski, V. I. (1965a). Electron microscopic data on the avian organ of Corti. *Zh. obshch. Biol.* **26,** 138–159.

Vinnikov, Y. A., Titova, L. K. and Aronova, M. Z. (1965b). *Acta histochem.* **22,** 120–154.

Wersall, J. (1956). Studies on the the structure and innervation of the sensory epithelium of the cristae ampullares in the guinea pig. *Acta oto-laryngol.* suppl. **126.**

Wersall, J., Flock, A. and Lundquist, P. G. (1965). Structural basis for directional sensitivity in cochlear and vestibular sensory receptors. *Cold Spring Harb. Symp. Quant. Biol.* **30,** 115–132.

Wever, E. G. and Bray, C. W. (1936). Hearing in the pigeon as studied by electrical responses of the inner ear. *J. Comp. Psychol.* **22,** 353–363.

Whitfield, I. C. (1967). *The auditory pathway.* Edward Arnold. Pp. 209.

Whitfield, I. C. (1968). "Hearing Mechanisms in Vertebrates", p. 59. CIBA Society symposium.

Winters, P. (1963). Vergleichende qualitative und quantitative Untersuchungen an der Horbahn von Vögeln. *Z. Morph. Okol. d. Tiere.* **52,** 365–400.

Wurdinger, I. (1970). Erzeugung, Ontogenie und Funktion der Lautauserungen bei vier Gansearten (*Aner indicus, A. caerulescens, A. albifrons* und *Branta canadensis*) *Zeitschr. f. Tierpsychol.* **27,** 257–302.

# 6 The Medulla and Afferent Systems

## I. INTRODUCTION AND GENERAL RELATIONSHIPS

In the adult brain the embryonic rhombencephalon is represented by the medulla, pons and cerebellum. The general developmental history of these was described in Chapter 4 and the cerebellum itself is discussed in Chapter 8. In this and the following chapter we can consider the bulbo-pontine region. For convenience the principal afferent and efferent systems are considered separately although the afferent systems have been studied in rather greater detail by recent workers and much of the information on the efferent systems relates to neuro-anatomy. Since the afferent units are necessarily bound up with their peripheral sensory components these are also briefly described in the appropriate sections.

The medulla oblongata lies between the spinal cord and the mesencephalon and posteriorly it grades into the cord so that it is not always easy to define a precise medulla/spinal transition point. Broadly speaking one can, nevertheless, take this as situated at a level just in front of the root of the first cervical nerve. As the result of being situated in this important intermediate position between the point of entry of peripheral nervous components and the more anterior parts of the brain both the medullary and pontine regions are traversed by fibre tracts travelling to and from these areas. However, within the bulbo-pontine complex itself there are a number of nuclear moieties which are involved, amongst other things, with the control of the heart, the respiratory apparatus, the alimentary canal and also both the vestibular and cochlear regions. Thus there are three major neurological constituents comprising the nuclei of origin and termination of cranial nerves V to XII, the nuclear masses other than those directly associated with cranial nerves, and also those fibres that pass through without actually synapsing in the medulla itself. Scattered amongst these nuclei and tracts, and forming a very considerable part of the medullary structure, are the various components of the reticular formations. These comprise more or less discrete masses of both grey and white matter whose specific interconnections remain largely uncertain.

Kilmer *et al.* (1969) have summarized the importance of the reticular components at various levels of the brain in their article on a plastic concept of reticular function. It is this system that commits a vertebrate to one of fewer

154

than 30 incompatible overall modes of behaviour, say eating, drinking, copulating, flying, waking, sleeping, etc. It appears to be crucial to the animal behaving as a well-integrated unit instead of a loose connection of separate organs and brain regions. The reticular formation receives unprocessed information from all sensory and effector systems as well as from autonomic sources. It can be interpreted as sending out control signals that direct, tune and set filters on all inputs, and it also controls relays at many levels. The orders of magnitude for the number of reticular formation afferents, neurons and efferents are about the same for many vertebrates. Characteristically cells in the reticular formation rapidly cease to respond to repeated insignificant input. Multiple electrode studies disclose activity wheeling around reticular formation cells which may be a component of a reticular formation distributed "memory by reverberation". It is some such system as this which is generally suggested as the basis for reticular formation conditioning. A comparison of the contribution which the reticular formation makes to the medulla in various species is provided by Stingelin (1965).

The central canal of the spinal cord continues forward and widens out at the level of the calamus scriptorius to give the rhomboidal fourth ventricle. This is roofed over by both the choroid plexus and the cerebellum and is prolonged into the aqueduct of Sylvius. It is because of the overall rhomboidal shape of the ventricular components, at least during ontogeny, that the hind end of the vertebrate brain is referred to as the rhombencephalon. This shape seems to arise from mechanical effects resulting from the pontine flexure. As will be clear from the discussions of Chapters 3 and 4 the choroid plexus (*q.v.*) is not nervous and consists of highly vascular projections from the pia.

In general terms it is often stated that the efferent pathways and nuclei of the medullary region are rather more extensively developed than those of the ascending afferent systems but this clearly begs a few questions. The actual distribution and structure of the avian nuclei differ from those conditions which prevail in both reptiles and mammals although many generalized homologies are possible. These differences are usually ascribed to the individual peculiarities of both avian cranial structure and body posture. They are particularly marked in the static centres, the vagal and hypoglossal nuclei, and in the presence of an inferior olive. The actual function of this last is still far from clear although it has intimate, point to point, topological relationships with the cerebellum. Assumed by some authors to be intercalated in the spino-cerebellar system, it is implicated in ocular and pharyngeal control by others, and is dealt with in Chapter 7 to emphasize the apparently close functional relationship which exists with the cerebellum. Some idea of the distribution and nature of the medullary fibre tracts which were long ago established by neuro-anatomists can be obtained from Fig. 140. A more specific representation of both nuclei and tracts is contained in the transverse sections of Figs 50 and 57. In view of the clearly all important nature of the spinal cord in both afferent input and efferent output some discussion of this structure can conveniently precede the medulla itself.

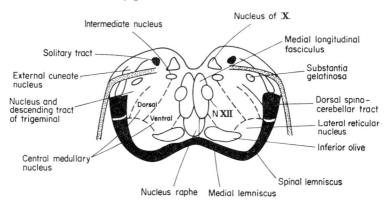

FIG. 50. Diagrammatic transverse section of the posterior region of the avian medulla.

## II. THE SPINAL CORD

### A. Neuro-anatomy

In all vertebrates the cord is a posterior prolongation of the medulla and in the five living classes it has a common overall appearance with dorsal sensory roots and ventral motor roots. A certain percentage of small unmyelinated efferent fibres also exit by way of the dorsal roots. The avian cord has, however, certain peculiar characteristics when it is compared with those of reptiles on the one hand and mammals on the other. There is an elongated cervical region, a reduced thoracic region, and, at the hind end, a variously called lumbo-sacral sinus, sinus rhomboidalis or glycogen body. This last is a large and gelatinous egg shaped structure which is situated between the diverging dorsal columns at the level of the sciatic nerve. Its development has been studied by Watterson (1949, 1952a,b) and histological investigations (Friede, 1966) show that it is composed of large and densely packed vesicular cells whose origin is controversial but may be of a glial nature. Meyer (1884), Gage (1917), Terni (1924, 1926), Olivo (1925), Doyle and Watterson (1949) and Masai (1961) have shown the very high glycogen content of the component cells, which, from micro-chemical data, seems to approach 75% of the lipid-free dry weight and represent 5–10% of the total body glycogen in chicks. It was for this reason that Terni suggested the name glycogen body. When cultured *in vitro* the cells exhibit glycogen storage at 15 days from explantation and during ontogeny such storage begins at the same time as the proliferation of the cells. The total amount of glycogen stored subsequently appears to be relatively independant of starvation. Lervold and Szepsonwold (1963) state that the rate of glycogen turn-over is rather slow.

Some idea of the variation of the segmental nervous components elsewhere in the cord can be obtained from the data of Table 24 which represents the cumulative results of a large number of the older investigations. In general there is also a marked difference in the size of the roots. The ventral ones are

## TABLE 24

*The variation in the distribution of segments within the various spinal regions.*

|  | Cervical segments | Thoracic segments | Lumbosacral segments | Coccygeal segments |
|---|---|---|---|---|
| *Struthio* | 15 | 8 | 19 | 9 |
| *Columba* | 12 | 8 | 12 | 6 |

Portmann (1950).

usually larger than the dorsal and this is accompanied by a corresponding difference in the relative sizes of the ventral and dorsal horns. This is particularly true in cursorial forms such as the Struthioniformes, Casuariiformes and Rheiformes. Within the cervical and lumbar regions there are also particularly marked enlargements which are related to the innervation of the wings and legs. Again, as one would expect, the lumbar enlargement predominates in the large cursorial forms, and the cervical in the much more numerous and typical volant orders. In these it usually involves the last 4–6 pairs of cervical roots. In the posterior region, and therefore in association with the glycogen body, there are three principal components. A lumbar plexus includes 2–4 roots, a sacral plexus of from 5–6 roots supplies the hind limbs and feet, and a rather poorly developed pudendal plexus supplies the caudal and pudendal regions.

Within the ventral horn itself the body musculature has long been considered to obtain its nerve supply from lateral or ventro-lateral cells. As would be expected from what has just been said the relative development of these cell groupings in the cervical and lumbar regions parallels the development of the ventral roots. The cervical enlargement is sometimes referred to in the literature as the flying centre. In a somewhat more anterior region, and close to the brain, the predominance of the ventral spinal aspect is sometimes reversed because of the presence of the spinal trigeminal nucleus dorsally.

Historically the distribution of tracts and fibres within the spinal cord was summarized in the work of Kappers *et al.* (1936). Its development was considered by Steding (1962). We can summarize the older conclusions and then briefly indicate the more recent contributions. An afferent spino-mesencephalic tract, possibly responsible for transmitting impulses derived from widespread tactile, thermal and nociceptive stimuli, was described in the ventro-lateral region at an early stage. Spinal ganglion cells have their central processes in the dorsal funiculi and the marginal zone. In birds these funiculi are of a relatively small size in comparison with those of reptiles. This has been ascribed, at least in part, to a reduced peripheral cutaneous sensibility in association with a covering of feathers. There are, however, numerous nociceptors over the body surface. More to the point, within the medulla the nucleus gracilis and the nucleus cuneatus are also small, and, since this reflects the lack of an accumulation of root fibres towards the anterior region, it has been suggested that the small size of both these nuclei and the dorsal spinal projections reflects the short course of ascending fibres within the cord. Such older descriptions also include details of an incipient medial lemniscus system; well-developed spino-cerebellar tracts occupying the entire area in the anterio-lateral regions and accompanied on their medial side by the analogous descending system; homolateral and contralateral vestibulo-spinal fibres; rubro-spinal fibres; and, in the ventral and ventro-lateral funiculi, tecto-spinal fibres. The overall pattern which results from such details and includes well-marked efferent pathways from the optic tectum, the cerebellar and the vestibular regions, would reflect the type of coordination which one would expect in such bipedal cursorial and flying forms in which vision is a predominant sensory modality. It is of great importance to emphasize the absence of any system which would correspond to the crossed pyramidal tracts of mammals, and also to note the apparently very large number of endogenous fibres many of which have a relatively short inter-segmental distribution.

Karten (1963) undertook a check of the ascending components of these avian spinal systems in pigeons. After performing spinal hemi-sections at the level of cervical roots 12–14, he used the Nauta-Gygax technique to follow the degenerating fibres, and hence the ascending pathways. With few exceptions all the terminations which he was able to detect were ipsilateral. Some fibres passed right through the medulla and terminated in more anterior brain regions, notably the cerebellum, thalamus and optic tectum. Within the bulbo-pontine area itself others ended in the cuneate and gracile nuclei; the posterior extremity of the inferior olive; the nucleus infima or commissural nucleus of the vagus-solitary complex; the nucleus of the solitary tract; the medial and lateral parts of the medial reticular nucleus of Cajal, more especially its medial to ventro-lateral zones; the raphe and paramedian nuclei. The actual location of such structures can be seen in Fig. 50. A comparison of the fibre composition within various roots which contribute to all these projections is contained in Table 28. Noback and Shriver (1966) summarized the similarities between sauropsidan pathways and mammalian lemniscal systems and they followed Karten in suggesting that direct spinothalamic fibres in the pigeon can be considered part of a lemniscal pathway.

## B. The Activity of the Spinal Tracts

Avian spinal reflexes, which can involve both single or multisegmental arcs, have been studied by many authors and the early works were reviewed by Ten Cate (1936). The tonic innervation of skeletal muscles was considered by von Boeke (1913), Hunter (1934) and Langelaan (1915). More specifically the sympathetic homeostatic corrections which occur during flight figured in Tiegs (1931) and van Dijk (1930). It has also been known for several centuries that decapitated birds can both run and flap their wings. Lamettrie (1751) recorded the fact for turkeys, Tosetti (1755) and Boerhaave (1795) for poultry, whilst analogous reports were made by Legallois (1812) and Flourens (1824), as well as the classical, *sensu stricto*, description by Herodianus. Specific details of spinal locomotory coordination have figured in a number of papers but are largely irrelevant here. They can be found by reference to the works of Gray (1968), Hinde (1966) and the older summary of Ten Cate.

The actual discharges which can be evoked in the ascending spinal tracts following the stimulation of muscles, skin and the nerves from legs or wing were investigated by Oscarsson et al. (1963). Using ducks they showed that the organization of the relevant tracts paralleled that which one finds in mammals. The dorsal tracts within the lateral funicular region undergo both mono-synaptic and poly-synaptic activation by the ipsilateral nerves. Ventral tracts, which are located in this same funiculus and also in the ventral funiculus, are only activated monosynaptically by contralateral nerves but poly-synaptic activation can come from both sides. For obvious reasons Oscarsson et al. concluded that the dorsal tracts are therefore uncrossed but that the last named or ventral ones have components which are crossed at the spinal level. Low threshold muscle afferents activate two such ascending tracts. One, a crossed tract, is associated with the legs, the other, an uncrossed one, with the wings. They also obtained evidence which suggested that the former, but not the latter, can be strongly inhibited by both skin and high threshold muscle afferents.

In Fig. 51 the uncrossed tracts at the cervical level lie above the broken line and those for the wing extend further towards the ventral side than do those for the leg. Oscarsson et al. concluded that the tract which is activated in this way corresponds to the ventral spino-cerebellar tract. The dorsal spino-cerebellar tract does not appear to exist in the duck, or, alternatively, it is not activated by low threshold muscle afferents. The anatomical descriptions of avian spino-cerebellar tracts which were referred to in Section IIA are somewhat fragmentary. Friedlander (1898), Sanders (1929), Kappers *et al.* (1936), Larsell (1948) and Whitlock (1952) describe separate dorsal and ventral ones. On the other hand Shimazono (1912) and Ingvar (1918) did not separate the spino-cerebellar system into two divisions. Kuhn and Trendelenburg (1911) claimed that the tract which arises at a lumbar level is crossed at the spinal level, and undergoes a gradual shift in position. Located ventrally in the

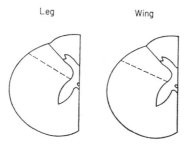

Leg                    Wing

FIG. 51. Diagram showing the distribution of tracts in the avian spinal cord.

lower thoracic segments it assumes a dorsal position by the time that the cervical region is reached. Clearly such a dorsal position corresponds to that described by Oscarsson *et al.* for the tract which is activated by low threshold muscle afferents in the leg nerves.

## C. The Spinal Ganglion Cells

Using both light and electron-microscopic techniques Pannese (1968) studied the changes which take place in the endoplasmic reticulum and ribosomes during the development of chick spinal ganglion cells. His results correspond to those of Wechsler and Kleihues (1968) which were referred to in Chapter 4, Section II. The rim of cytoplasm which is situated around the nucleus in undifferentiated cells contains numerous clusters of free ribosomes and only scanty traces of endoplasmic reticulum. The primitive neuroblasts are spindle shaped and their processes contain filamentous structures. In comparison with undifferentiated cells such primitive neuroblasts have many more free ribosomes and there is a prominent endoplasmic reticulum comprising long and inter-connected channels which are often clearly continuous with the nuclear envelope. This remarkable development of the endoplasmic reticulum is one of the main characteristics of differentiating ganglion cells. Sub-surface cisternae also become apparent for the first time at the primitive neuroblast stage.

Both the endoplasmic reticulum and the ribosomes of the intermediate neuroblasts are confined to the peripheral regions of the perikaryon. In contrast mitochondria, Golgi complexes, dense bodies, microtubules and neurofilaments are located more centrally. Units of smooth endoplasmic reticulum appear in the first intermediate neuroblast. In the pseudo-unipolar nerve cells the cisternae of the endoplasmic reticulum, together with the ribosomes, are mainly grouped in discrete areas. The rough endoplasmic reticulum becomes both more developed and more highly organized in the intermediate neuroblasts, and especially within the pseudo-unipolar cells, whilst in these latter entities there are somewhat unusual relationships between the endoplasmic reticulum and the Golgi complex. Both the free and the membrane attached ribosomes are arranged in clusters, polysomes, throughout all the developmental stages. The highest number of ribosomes in the membrane attached polysomes appeared to be at least 13. The overall development of spinal ganglion cells therefore clearly fits the pattern seen in other, for example, diencephalic derivatives.

## III. THE POST-OTIC AFFERENT SYSTEMS OF THE MEDULLA

### A. Vagal and Glossopharyngeal Afferents

The visceral and somatic afferent functions of vertebrate cranial nerves are to a great extent summarized by the chart of their segmental origins in

| Segment | Arch | Dorsal root | Ventral root |
|---|---|---|---|
| Pre–mandibular | Trabecula | R. op. profundus V | Oculomotorius III |
| Mandibular | Palato–pterygo–quadrate bar and Meckel's Cartilage | Rr. op superficialis maxillaris and mandibularis V | Trochlearis IV |
| Hyoid | Hyoid | Facialis VII Acousticus VIII | Abducens VI |
| 1st Branchial | 1st Branchial | Glossopharyngeus IX | (absent) |
| 2nd – 5th Branchial | 2nd–5th Branchial | Vagus X + Accessorius XI | Hypoglossus XII |

FIG. 52. Chart showing the cranial segmentation of vertebrates.

ontogeny and phylogeny. This well-known scheme is contained in Fig. 52. Since taste is relatively poorly developed and associated with the glossopharyngeal nerve the facial nerve has a small afferent input. Both the glossopharyngeal and the vagus nerves tend to have larger afferent roots than the facial and consequently the spinal and post-otic afferent systems are conspicuous components of the medullary sensory nuclei. This is particularly true in the case of the vagus which has often been described as incorporating the controversial avian spinal accessory. The work of Rogers (1965), which is referred to in Chapter 7, Section III, has, however, dispelled such suggestions. In the adult of *Gallus*, Rogers points to three possible sites for sensory cells associated with the nerve. He concluded that the most likely is on the course of the nerve just behind the jugular ganglion of the vagus where about 65 ganglion cells exist. Much evidence has been presented for peripheral migration of ganglion cells along nerves III, IV and VI (Rogers, 1957) and he suggested that an analogous migration may occur along XI. A second possible site is also suggested by the presence of such sensory cells in the root ganglion of the vagus (Hamilton, 1952), or, thirdly, it may lie in, or just in front of, the second cervical nerve.

The general relationships of the caudo-cranial nerves of the hen were described by Malinovsky (1962) and Bubien-Waluszewska (1968). In principle the latter concluded that no direct branches of the vagus enter the syringeal region, the pharynx or the oesophagus. Fibres of the pneumogastric which do enter these regions do so by way of an anastomosis which unites the pneumo-gastric and glossopharyngeal nerves. In the chick, but not in the pigeon, there is an anastomosis between the pneumogastric and the superior cervical ganglion of the sympathetic system. On the chick skull the orifices of the pneumogastric and glossopharyngeal are quite independent. That of the glossopharyngeal is situated in front of the jugular foramen, or vagus foramen, through which the pneumogastric and spinal nerves pass and she suggested that the glossopharyngeal orifice corresponds to the foramen lacerum of mammals. The nerve itself innervates the tongue, pharynx and syrinx, and, in addition to these, the upper part of the oesophagus. The pharyngeal branch which serves these three last-named topographical regions is considerably larger than the lingual.

Both the vagal and glossopharyngeal roots enter the medulla in a dorsal position and then proceed medially. On their way they pass through the descending root of the Trigeminus to which they are said to contribute cutaneous components. The overall relationships have been summarized by Cordier (1954), Kappers *et al.* (1936), Papez (1929), Portmann (1950), Port-mann and Stingelin (1961) and Stingelin (1965). The dorsal and medial position of the nucleus of the vagus is seen in Fig. 50. Visceral fibres which originate in the mucous membrane appear to either terminate in the grey matter of the medullary floor, or, and this seems to be the case for the majority, run posteriorly in the well-developed solitary fasciculus (Fig. 50). Ramon y Cajal (1909) described a pair of such fasciculi which disappear behind on each side in *Passer*. He concluded that the dorso-lateral partner received crossed fibres from the glossopharyngeal etc. and Kappers *et al.* (1936) concluded that the poorly developed avian gustatory modality would suggest that these fasciculi are associated with viscero-sensory impulses. The fibres within the fasciculus were recognized by Brandis (1894). Some 75% of them seemed to end in the grey matter of the commissura infima after decussating at that level. Besides the vagal and glossopharyngeal fibres other extra-fascicular axons paralleled the fasciculus to terminate in the viscerosensory area of the dorso-lateral medullary region. Others exhibit some ordinal variation, are easily visible as a plexus in the dorso-medial nucleus of *Casuarius*, but are very reduced in some genera, notably in the orders Sphenisciformes, Colymbiformes and Ardeiformes.

# B. Peripheral Sensory Structures

## (i) *Free Nerve Endings*

The sources of many of the afferent impulses which are conducted by both spinal and post-otic pathways are widely assumed to comprise the free nerve endings, Merkel's discs and corpuscles of Herbst. Free nerve endings are components of the so-called "protopathic" system, a term coined to describe the units which are concerned with "perceptions related to the preservation of life" (Head *et al.*, 1905). They are usually associated with nociception (Sherrington, 1906) but it would be an exaggeration to consider this as their sole function. In view of the distribution of free nerve endings within the vertebrate classes it has long been assumed that they are phylogenetically older than encapsulated endings.

The fibre, if myelinated, branches repeatedly at successive nodes of Ranvier and these arborizations lose their myelin sheaths and ramify throughout the terminal sensory area as telodendria. In birds the internal organs have numerous such structures which often form fine knobs, dendritic plexi or simple loops. There has been some dissension for a long time on the degree of functional independance which is exhibited by the final components from various nerves, at one time it having been suggested that they anastomose. Networks of some sort may traverse the zone between the dermis and the epidermis (Botezat, 1906). Schartan (1938) showed that analogous plexi of sympathetic origin are interspersed between them. Bing (1959) is one of the more recent authors to emphasize the importance of such peripheral networks. Like Sinclair (1955) he considered incorrect the general assumption, which is supported by electrophysiological research, that the capsular sensory structures mediate peripheral sensory modalities. Instead he placed massive emphasis on an undifferentiated nerve network in the skin. To what extent such devastating suggestions are correct, except in special areas like the cornea, is not clear.

## (ii) *Merckel's Discs*

In both the skin and the buccal cavity there are a large number of supposed tactile menisci or Merckel's discs. By analogy with the similar mammalian structures they have long been 'considered to be tango-receptors. They are formed by special tactile cells which are associated with cup-shaped networks of terminal nerve fibres. Around the whole structure there is a fine network which is derived from a small nerve fibre that has been considered to be part of the autonomic system. Similar but larger complexes derived from several tactile cells, and also corpuscles with two rows of component cells, were reported by Botezat. Clusters of dense terminal nets which are associated with secondary sensory cells to produce structures that resemble Krause's end-bulbs were also included alongside these discs by Portmann (1961).

## (iii) *Herbst's Lamellar Corpuscles*

As can be seen from Fig. 53 these are encapsulated sensory endings which appear to be analogous to the Vater-Pacinian corpuscles of mammals (Herbst,

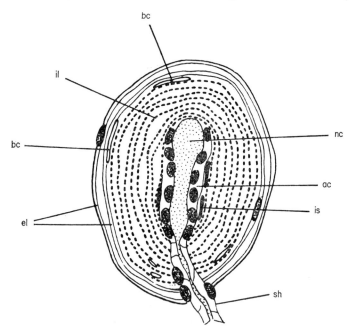

FIG. 53. Herbst corpuscle from the bill of a duck. Only the two innermost of the external lamellae are shown. Up to 12 may be present. (Portmann, 1961).

Abbreviations: ac, axial cells; bc, blood capillaries; el, external lamella system; il, internal lamella system; is, internal sheath cells; nc, ending of central nerve fibre; sh, sheath cells of the nerve fibre.

1848). They occur in a number of rather disparate foci which include the skin of the bill region, the tongue and palate of a variety of genera, and also elsewhere such as on the tibial region. Particularly large numbers are present on the bill of *Capella* and on the tongue of the Piciformes. Although clearly contributing to some pre-otic nervous imput they are included here for convenience since their distribution is widespread.

The lamellar capsule is derived from connective tissue cells and ensheaths a central structure. This is homogeneous in *Anas* but consists of numerous concentric lamellae in *Capella*. Within this there are two rows of nuclei and, centrally, a terminal nerve fibrillum. Occasionally such nervous components arborize into the adjacent cytoplasm and a second nerve fibre has also been described forming a network around the main one. Where they occur the outer lamellar structures have been interpreted as a system of hemispherical membranes which enclose a fluid filled space. Originally described as tango-receptors they were subsequently considered to react to hydrostatic variations in the blood and allied fluids. Then Schwartzkopf (1955) demonstrated that the well-known sensitivity of birds to distant explosions was probably not mediated by aural stimuli but possibly by those corpuscles of Herbst which are located between the tibia and fibula. He suggested that this would explain the apparently enhanced sensitivity which he said could be observed when the animal is in the resting position. Clearly as birds are rather more vulnerable under such conditions this may not be the end of the story.

In fact in their considerations of the morphology and histochemistry of cutaneous end organs in the chick and the hen, Winkelmann and Myers (1961) demonstrated that Vater-Pacini type corpuscles are found over the entire cutaneous surface. The sole exception is the plantar region. They could be found associated with feathers, or, alternatively, be situated close to the surface in areas of glabrous skin. It was particularly easy to study them in sites of the latter type, as, for example, in the cheek, the tarsal region or in peri-oral tissue. In the cheek region clusters of such examples were present around rudimentary feather follicles and in a typical ending the axial nerve fibre extends right to the apex terminating in a slight swelling. Similar organs were also present in the hard palate, but not the soft one, the tongue and the floor of the mouth. Others occurred in the eyelid in proximity to rudimentary feathers, but none were found in the conjunctiva. In all cases the cellular composition was quite distinct with two rows of cells at the wall of the inner bulb. By staining with toluidine blue, it was possible to show up a marked metachromasia in the capsular spaces surrounding the bulb, which was not greatly influenced by the application of hyaluronidase. Mayer's mucicarmine, alcian blue and also periodic acid Schiff staining all gave positive results showing the presence of acid mucopolysaccharides. Elastic tissue stains, such as the Gomori aldehyde-fuchsin procedure, gave faint positive reactions in the capsular substance. The use of both acetylthiocholine and butyrylthiocholine iodide gave positive reactions within the inner bulb but not the capsule. There was no recorded use of 62C47, iso-OMPA or other inhibitors so that one may merely conclude from this that either acetylcholine esterase or non-specific cholinesterases are present within the organ.

### (iv) Taste

The taste-buds of birds have been reviewed, together with other aspects of this avian sensory modality, by Kappers *et al.* (1936), Portmann (1961) and Duncan (1963). By comparison with those of other vertebrates they were only described relatively recently. At the beginning of the present century, Botezat (1904, 1910) and Bath (1906) found them in the buccal region of many species, especially on the basal region of the tongue and on the palate. They never appeared to be associated with discrete papillae as are those of mammals. Moore and Elliott (1946) added further precision to these older results. They found an average of 37 taste buds on the tongue of pigeons, with 27 and 59 as the upper and lower limits; 70% of these were situated dorsally and behind the dorsal fold and their average diameter was 30 $\mu$. Bath calculated that there might be 200 in the starling, and from 300–400 in parrots. Further values are cited by Duncan (1960) and are included in Table 25.

## TABLE 25

*The reported and calculated numbers of taste-buds in some birds.*

| Species | Number of taste-buds | Reference |
|---|---|---|
| Pigeon | 27–59 | Moore and Elliot (1946) |
| Parrot | 300–400 | Bath (1906) |
| Young bullfinch | 77 | Duncan (1960) |
| Adult bullfinch | 42–51 | Duncan (1960) |
| Starling | 200 (?) | Bath (1906) |

Their structure can be seen from Fig. 54. In many cases they are surrounded by a group of follicular cells which form a well-defined sheath which is not clear in other vertebrates (Bath, 1906) but they are otherwise comparable to those of other classes. In one form this sheath is limited to the basal region and the more distal part of the bud, which otherwise bears the taste pore, is naked. Others can have a complete sheath whilst follicular cells are totally lacking in species of the order Psittaciformes. The particular type of bud which is found in a given species seems to be uniform and the number of sensory cells within them can vary from 10–40. The lower values are recorded from passeriform species and the highest from parrots.

As in the mammals the individual buds also contain both gustatory cells, with their nuclei in the distal region, and supporting cells. The majority clearly lie within the area which is innervated by the glossopharyngeal nerve, not that of the facial, and this is corroborated by the available electro-physiological data. The nerve supply at the taste-bud itself was described by Botezat. According to him there are two distinct plexi. A dense, basal infra-gemmal plexus, and a larger intra-gemmal plexus whose components have a close relationship with the definitive gustatory cells.

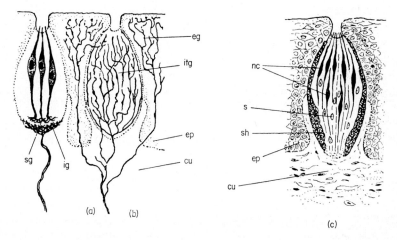

FIG. 54. The general structure of avian taste buds. (a) Sensory cells with sub-gemmal and infra-gemmal nerve supply. (b) intra-, and extra-gemmal nerves. (c) general structure (Portmann, 1961), cu, cutis; eg, extra-gemmal nerve fibres; ep, epidermal layer; ig, infra-gemmal plexus; itg, intra-gemmal nerve fibres; nc, sensory cells; s, sustentacular cells; sg, sub-gemmal plexus; sh, sheath cells.

Turning to functional studies Halpern (1961) assessed the effect of a variety of thermal, chemical and mechanical stimuli on the avian tongue by electrical monitoring of the multi-unit activity within the lingual branch of the glosso-pharyngeal nerve. More specifically Engelmann (1937) investigated the taste modality in hens. He concluded that there was qualitative discrimination be-tween NaCl/HCl, HCl/acetic acid and glycerol/$MgCl_2$. In each case the bird preferred the more dilute solution. $MgCl_2$ was preferred to both acids and salt. Duncan (1964) provided a review of the available data on the pigeon and these are also dealt with in his earlier papers (1960, 1962). He concluded that inorganic acids and bases produce a similar pattern of rejection when using the "one bottle" technique. However, at equal pH, the stimulatory efficiency of organic acids parallels their lipoid solubility and surface activity. He also suggested that the buffering capacity of saliva is probably of only minor importance. If sweetening agents were added to the acid solutions this did not increase their acceptability. No change occurred if saccharine was added, and the addition of 18 g/litre of glucose actually reduced the level of consumption, although not to a significant extent. Alongside this information he also discussed two birds which were apparently taste-blind. Kare and Pick (1960) had previously deduced that such cases probably reflected a reduced sensitivity. In Duncan's birds their established drinking pattern was, however, changed when they were presented with acid solutions which other specimens rejected. He therefore considered that they were at least able to detect the presence of an unpleasant stimulus. This being the case he went on to postulate that the apparent "taste-blindness" was the result of central effects rather than an overt deficiency in the gustatory sensory mechanisms. Clearly both diminished peripheral sensitivity and central nervous parameters may be involved.

From the available data he concluded that the feral pigeon's behavioural response to acid stimulation is not dissimilar to that observed in many mammals. Within the taste modalities of sour as well as salt the physiological mechanisms are essentially the same in the two classes. There are, however, certainly both inter-specific and inter-ordinal differences and such differences are also very clear in the olfactory modality which is considered in Chapter 13, Section II. The rejection threshold for acids is higher in ducks, hens (Engelmann, 1960) and bullfinches than in pigeons (Duncan, 1964). In contrast to these conclusions Duncan found no correlation between the sweet and bitter modalities in birds and mammals. Many bitter substances were readily accepted by both pigeons and chickens whilst, on the contrary, they can show a marked aversion to certain organic substances which man is not able to detect.

The results of Kitchell et al. (1959) are broadly in agreement with these conclusions although the situation of the bitter and sweet modalities is not clear. Using electro-physiological monitoring techniques these workers found that both lingual and pharyngeal taste receptors had response characteristics which are comparable with those of mammals. Impulses from these receptors, as would be expected from the neuro-anatomical evidence of Bubien-Waluszewska (1968) and older workers, reach the brain via the peripheral branches of the glossopharyngeal nerve. They could obtain no evidence for either trigeminal or facial involvement. The impulses in the chick indicated distinct responses to salt, bitter, acid and water, and none to either sucrose or saccharine. On the other hand both glycerine and ethylene glycol, which taste sweet to man, gave marked responses. In the pigeon there were comparable results for acid, salt and water, but, as Duncan found, no responses to bitter and only an equivocal one to sweet, since only 50% of the birds had a positive reaction to saccharine. There was also a very marked effect which appeared to be related to the ambient temperature. When this was maintained at a constant 39°C the high level of spontaneous activity within the lingual and laryngo-lingual nerves masked the effects of applying test solutions to the tongue. Warming the tongue itself reduced, but did not obliterate this activity, whilst a drop in temperature of 3°C caused an abrupt increase. They were of the opinion that, in the absence of warm fibres, the relative temperature of food may be determined on the basis of a drop in such activity.

## C. The Vagal Control of Respiration

Historically a very large number of workers have been interested in the vagal systems which are, or may be, involved in avian respiratory control. More especially this refers to the control of the ventilation rate. To date there are a number of results which indicate the direct involvement of vagal afferent impulses in maintaining a flow of information about the phase of the respiratory cycle to the central nervous system. Evidence for vagal motor control is still not incontrovertible and, in mammals, it is largely spinal components which supply this. Many of the relevant contributions are explicable in terms of afferent control. Some of these are of considerable age and date from the earliest days of experimental neurophysiology. These include Fredericq (1883), Francois-Franck (1906) and Siefert (1896) who considered the central foci. Possible labyrinthine involvement with such foci was suggested by the results of Fano and Masini (1894), Huxley (1913) and Paton (1913, 1927). Nervous control of respiration which implicates the vagus was also discussed in the works of Brown-Séquard (1860), Bert (1870), Knoll (1880), Couvreur (1892), Bourgeois (1896), Siefert (1896), Grunwald (1904), Stübel (1910) and Niko-laides (1914). The role of the carbon dioxide tension in the blood has also been compared on a number of occasions with the situation in mammals as far as it was understood at the relevant times. The principal contributions are those of Bordoni (1888), Baer (1896), Grober (1899), Nagel (1901), Trèves and Maiocco (1905), Foa (1911), Viet (1911) and Stehlik (1927), whilst all these factors were also considered by Ten Cate (1936).

There is now fairly general agreement that unilateral vagal transection results in a short and temporary reduction in the rate of breathing. In the duck, chicken and pigeon bilateral transection also leads to a comparable slowing (Bert, 1870; Grober, 1899; Grunwald, 1904; Orr and Watson, 1913; Graham, 1940; Hiestand et al., 1953 and Sturkie, 1965). In these birds the rate may decrease to 4–7 cyc/min and either return to normal after a variable period which ranges from 7 min to 3 days, or, alternatively, lead to death. This has been interpreted as suggesting that the respiratory control is subject to stimulatory effects from receptors other than those which are mediated by the vagus.

Sina (1958) concluded that bilateral vagotomy gives a transient respiratory standstill in either the expiratory or mid-position. Transection of one vagal trunk resulted in a reduction of the respiratory rate to either one-half or one-third normal. Although there was some subsequent recovery the original level was never attained again. If this hemilateral section was followed by transection of the other vagus there was a further rate reduction coupled with a prolongation of the expiratory phase and a marked diminution in amplitude. The complementary situation involving electrical stimulation of the vagus gave a variable response which appeared to be dependant upon the intensity of the stimulus. Weak to moderate levels gave increased respiratory rates but strong stimuli generally inhibited breathing in the inspiratory and, occasionally, the expiratory position.

Some quite definite species specific results have been recorded following vagal section. Panting was abolished in the fowl, duck and quail according to Hiestand and Randall (1942) and Richards (1968). This, however, was not the case in the pigeon (Saalfeld, 1936; Richards, 1968). After vagotomy, pigeon preparations did not start panting if the body temperature was increased but the rate per minute was less than in the intact bird. Appropriate afferent vagal stimulation gave both normal breathing and thermal panting in vagotomized fowl preparations. Richards considered that it was in fact quite possible that panting is subject to a greater degree of central control in some species than is the case in others. As thermal polynea can only be elicited and maintained by electrical stimulation in fowls he suggested that it is dependent upon vagal drive. On the other hand in the pigeon the panting centre, which was described by Saalfeld (1936) as located in the anterior dorsal mid-brain, together with impulses from the anterior hypothalamic region (Akerman et al., 1960), appear to control polynea by way of efferent medullary paths but not necessarily without the intermediation of thoracic receptors.

Fowle and Weinstein (1966) discovered that peripheral electric shocks cause a change in both respiratory frequency and tidal volume. They concluded from the persistence of this effect during barbital anaesthesia, that the electric shock was effective without being "consciously" perceived. Following the suggestions of Salmoiraghi and Burns (1960) they wondered whether the observed changes were a reflection of increased neuronal traffic within the reticular formation. The regulating mechanism may both receive and respond to chemical and nervous stimuli when they lie within a certain range of intensity and, despite their individuality, both types of information can evoke modifications in the rate and depth of breathing.

In experiments which were designed to detect an analogous respiratory control centre to that then considered to exist in mammals Saalfeld (1936) carried out medullary transections and showed that these inhibited respiration. Winterstein (1921) had summarized earlier, rather inconclusive, but similar results. However, the sensitivity to changes in body temperature, blood pH, and other stimuli which is comparable to that of mammals is not accurately attributable to either such a proposed medullary focus, or to the proposed panting centre in the diencephalon. In this respect some divergent and often improbable data have been reported. Orr and Watson (1913) obtained the remarkable result that an increase in the partial pressure of carbon dioxide inhibited respiration. Such a conclusion is clearly at odds with the stimulatory effects of increased carbon dioxide tensions which are observed in the case of ventilation rates amongst animals as taxonomically distant as insects and mammals. Dooley and Koppanyi (1929) explained this apparent aberration by attributing it to the irritation which is caused by high concentrations of $CO_2$ within the nasal pathways. This, they suggested, initiated reflex stimulation of inhibitory pathways. If $CO_2$ is administered through the opened humerus, or following local anaesthetization of the nasal mucosae, it stimulates respiration. Nevertheless very variable responses were obtained by Hiestand and Randall (1941). In some species a rise in the $CO_2$ partial pressure resulted in a stimulatory effect, in others it gave inhibition, and in many cases the specific form of the results appeared to be dependant upon the anaesthetics used.

If *Gallus* breathes an atmosphere of 10% $CO_2$ respiration is stimulated but the observed change is an increase in amplitude rather than ventilation rate. Following pento-barbital anaesthesia this effect is not observed, and, with other anaesthetic agents, it can totally inhibit or at least make respiration slower. It would seem that the relevant control foci are sensitive to, or affected by, anaesthetic substances as well as pH, $CO_2$ partial pressure, etc. More recently Andersen and Lövö (1964) showed that $CO_2$ has a stimulatory effect upon the respiratory mechanisms of the duck so that the respiratory minute volume approximately doubles when the $CO_2$ concentration represents 7·5% of the respired air.

Other constituent factors in respiratory control have also been elucidated. Several workers have suggested that the mechanisms which inhibit inspiration during diving result from a postural apnea which is elicited by the vestibular system (Huxley, 1913; Paton, 1931; Anderssen, 1959a,b, 1963a,b,c). It is clear that such inhibition certainly reflects trigeminal impulses which are elicited by wetting the beak and nostrils (Andersen, 1963a). Artificial pulmonary inflation inhibits inspiration and stimulates expiration according to Winterstein (1921). Complementarily artificial pulmonary evacuation can inhibit expiration and induce inspiration, although such effects were said to be abolished by vagotomy. However, Graham (1940) concluded that deflation of the lung had no effect on respiration. Furthermore, pulmonary distension did not abolish panting in *Anas* unless the preparation was under amytal anaesthesia.

Sturkie (1965) concluded that it was likely that avian respiration was under the control of a classical Hering-Breuer reflex. This would imply that there are two types of afferent fibre in the vagus, both of which have receptors in the lungs or ribs, and which are stimulated by the alternate contraction and expansion of the pulmonary region but in the presence of a non-expansive lung some other source of such stimulation would be necessary. The work of Fedde *et al.* (1961) suggested that such a Hering-Breuer type reflex situation is not of overwhelming importance in the maintenance of the normal respiratory rhythm in the chick. A sudden expansion of the pulmonary parenchyma using a unidirectional air-flow failed to initiate expiration. Sturkie (1965) found that under such conditions no reflex respiratory movements resulted when there was an adequate respiratory rate, but, if the ventilation rate fell to 200 cc/min or less, reflex movements did commence. The pathway of such neural control was not elucidated. Finally, Richards (1968, 1970) recently demonstrated that the Hering-Breuer reflex is only present in both the fowl and the pigeon at normal body temperature.

As was implied above the demonstration of such apparent Hering-Breuer type systems in birds is clearly difficult to interpret in the light of suggestions that the avian lung is relatively inexpansive. This led King *et al.* (1968a,b) to attempt to record from units in the peripheral stump of the mid-cervical vagus of 6–15-week-old chicks. They were able to demonstrate quite clearly that some units fired during inspiration but ceased doing so at its peak. A few fired at the period of the peak alone, and others which only appeared to be active during the expiratory phase were occasionally detected. Several intermediate types which fired during both inspiration and expiration were also distinguished although doing so at a slower rate in the latter case. They concluded that these units all represented the direct afferent neural activity which continuously informs the central nervous system of the state of the respiratory cycle. However, the greater variety of afferent activity in phase with eupnoea renders any direct comparison with the mammalian condition difficult. Jones (1969) also came to similar conclusions. Taking into consideration the varied data which have now been presented we may conclude that whilst peripheral stimuli such as cutaneous electric shocks together with the carbon dioxide content of the respired air exert effects on the ventilation rate, there appears to be a basic vagal drive mechanism and it was the impulses associated with this that were detected by King *et al.*

## IV. THE VESTIBULAR AFFERENT SYSTEMS

### A. Introduction

Both the vestibular and cochlear systems have been extensively studied since Retzius (1884), Krause (1906) and Gray (1908) published their contributions. The general similarity to those of diapsid reptiles was commented upon at an early stage. The innervation is via the VIIIth, or stato-acoustic nerve, which has two principal components. The static part runs to the vestibular ganglion and the acoustic to the separate cochlear ganglia. The branches to the vestibular ganglion receive three small nerves from the ampullae, the macula neglecta and the maculae. In other terms the posterior branch of the VIIIth nerve serves the posterior ampulla, the sacculus, the macula neglecta including the papilla lagenae, and the basal membrane of the cochlea. The ramus anterior normally innervates the anterior and lateral ampullae and the utricle but in *Turdus* and *Columba* it also serves the saccule.

## B. Peripheral Structures

### (i) *Gross Appearance*

The avian bony labyrinth can be excised fairly easily because of the spongy bone which surrounds it. In birds of small size the two posterior semi-circular canals may approximate to the mid-sagittal plane and sometimes the labyrinths of each side have a connection in the form of the foramen intercaniculare. The horizontal canal passes through the loop of the posterior one, and, at the point of cross-over, the bony canals but not their contents, can be fused. The positions of the canals within the head have one constant feature which is similar to that of the ribbon-like area centralis present in certain avian eyes (*q.v.*). The horizontal canal is typically directed in space so that it assumes the horizontal plane when the bird holds its head in the alert posture. A superficial inspection suggests that the anterior and posterior canals have undergone torsion by comparison with the situation in other classes. However, Portmann (1961), following Duijm (1951), drew attention to the differences which arise from different rest and alert positions in say *Struthio*, *Ardea*, *Ciconia*, *Capella* and *Phalacrocorax*, whilst Hadžiselimović and Savković (1964), whose results are outlined below, have shown other ecological and behavioural associations.

The three semi-circular canals of vertebrates generally lie in nearly orthogonal planes and serve to indicate angular motion of the head with respect to inertial space (see, for example, Young, 1969). The structure of the otoliths suggests an over-damped mass-spring-dashpot model linear accelerator. In their survey of the function of these static organs Jones and Spells (1963) directed attention to the fact that the canals themselves are accurate measuring instruments which, largely on account of their small size, can mechanically integrate the short duration angular accelerations to which they are exposed. The velocity of the endolymphatic flow relative to the canal is directly related to the actual angular acceleration of the canal. The relative fluid displacement and the consequent angular deflection of the water-tight cupula are both proportional to the change of the angular velocity of the canal itself. However, the sensitivity of the instrument requires adjustment in order to match the probable patterns of head movement which are peculiar to the size, shape and habits of the particular species. They concluded that such adjustments were largely achieved by small but suitable variations in canal bore. The avian labyrinth is of particular interest because it also has the gross directional variations of which Jones and Spells seem to have been unaware.

## TABLE 26

*The dimensions of the bony semi-circular canals in some avian species.*

| Species | Maximum diameter of vestibule (mm) | Superior Canal | | Posterior Canal | | Horizontal Canal | |
|---|---|---|---|---|---|---|---|
| | | Internal diameter (mm) | External diameter (mm) | Internal diameter (mm) | External diameter (mm) | Internal diameter (mm) | External diameter (mm) |
| *Struthio masai* | 3·5 | 7·5 | 11·0 | 5·5 | 8·5 | 4·0 | 8·5 |
| *Apteryx mantelli* | 3·0 | 4·0 | 6·0 | 3·0 | 4·0 | 2·5 | 4·0 |
| *Colymbus septentrionalis* | 4·5 | 6·0 | 7·55 | 4·0 | 6·0 | 4·25 | 6·25 |
| *Sula capensis* | 4·5 | 7·0 | 9·5 | 5·5 | 8·0 | 4·75 | 7·25 |
| *Phalacrocorax carbo* | 3·0 | 5·0 | 7·5 | 4·0 | 6·25 | 4·5 | 6·5 |
| *Nycticorax griseus* | 3·5 | 5·0 | 7·0 | 3·75 | 5·25 | 4·5 | 6·0 |
| *Anas boscas* | 4·5 | 3·0 | 4·5 | 3·5 | 5·0 | 3·5 | 5·0 |
| *Buteo vulgaris* | 4·5 | 5·25 | 7·25 | 6·0 | 9·0 | 7·0 | 9·5 |
| *Rhynchotus rufescens* | 2·5 | 3·75 | 5·25 | 3·0 | 4·5 | 3·5 | 5·25 |
| *Lagopus scoticus* | 3·5 | 4·0 | 5·5 | 3·0 | 4·25 | 3·25 | 5·0 |
| *Balearica pavonina* | 3·5 | 6·0 | 8·0 | 4·75 | 6·5 | 4·5 | 6·5 |
| *Goura coronata* | 3·5 | 5·25 | 7·5 | 4·0 | 6·5 | 5·0 | 7·0 |
| *Licmetis nasica* | 3·0 | 4·5 | 6·5 | 4·5 | 6·0 | 4·5 | 6·5 |
| *Speotyto cunicularia* | 3·5 | 6·0 | 9·5 | 4·5 | 6·0 | 5·0 | 6·75 |
| *Corvus corone* | 3·5 | 5·75 | 7·0 | 4·75 | 6·5 | 6·5 | 8·0 |
| *Turdus musicus* | 2·25 | 4·0 | 5·0 | 3·0 | 4·25 | 4·0 | 4·75 |

After Gray (1908).

The anterior canal is usually the longest, has the largest ampulla, and typically lies in the sagittal plane. The posterior, which is also vertical, lies in a plane which is at right angles to that of the anterior. Its ampulla is situated at the back of the utricle and is somewhat separated from the other two. The difference in length of these two vertical canals is 7 mm in the pigeon, the anterior measuring 14 mm in length and the posterior only 7·3 mm. The horizontal canal is the last one to develop during ontogeny, its sensory crista has a different structure, and in the order Piciformes it is the largest of the three.

The first detailed study of these structures was that of Gray (1908). He concluded that in birds the superior or anterior canal is of two distinct types—upright or drooping. The former type occurs in *Buteo*, *Goura*, *Corvus* and *Turdus*. Typically the anterior limb rises almost vertically from the ampulla and then turns rather sharply backwards in a horizontal position to finally curve downwards, forwards and re-enter the vestibule. As the result of this its long axis can be almost vertical. In Gray's second, or drooping type, the anterior limb does not rise vertically in this way but instead passes upwards and backwards before curving down and then forwards to the vestibule. In this case its long axis is therefore not vertical but instead lies in an oblique postero-dorsal plane. He suggested that this condition typified genera such as *Rhea*, *Rhynchotus*, *Columba*, *Gallus*, *Anas*, *Nycticorax* and *Licmetis*. Overall measurements of the internal and external circumferences are contained in Table 26.

The correlations which can be drawn between gross labyrinth parameters and the differing modes of life in various species of bird have now been studied by a number of authors. Ibragimova (1958) drew attention to the fact that the canals in birds are generally rather large by comparison with those of other vertebrate classes. A possible explanation for this lies in the additional observation of Lewin (1955) and Turkewitsch (1931) that the labyrinth of "good" fliers has long thin canals (e.g. *Columba* and *Larus*), whilst that of poorer fliers such as the hen, duck and goose has short thick canals. Furthermore, birds of similar ordinal status, but exposed to differing ecological pressures, also show differences. Solotuchin (1925) described the labyrinth in 74 species and more recently Hadžiselimović and Savković (1964) studied 27 species which represented some 9 orders. They concluded that, as was suggested by Lewin and Turkewitsch, there is indeed a direct correlation with the overt flying ability.

A series of labyrinthine structures is shown in Fig. 55. Hadžiselimović and Savković showed that in able fliers the long thin canals are C shaped and the upper canal particularly has a slight spiral curve, whilst the ampullary ends are pronounced. This summarized the situation in *Columba oenas*, *Asio otus*, the *Turdidae*, *Falco*, *Aquila* and *Corvus*. In what they term "exceptionally good" fliers such as *Garrulus glandarius*, *Scolopax rusticola*, *Corvus*, *Turdus viscivorus*, *Turdus merula* and *Falco*, the upper canal with its largest curve is inclined medially. In *T. philomelos* the upper canals of each side are less close. They associated these differences with differing ecological characteristics. Similarly the posterior canal is inclined laterally in good fliers so that it

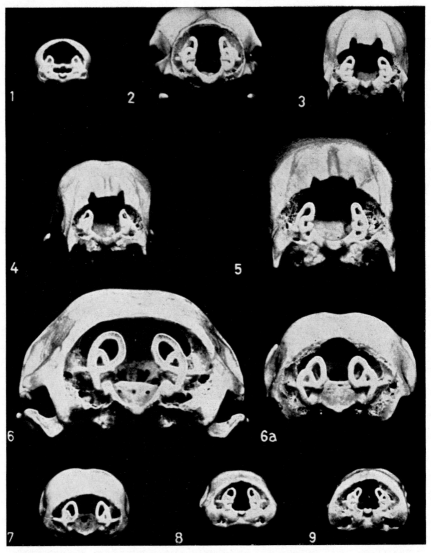

FIG. 55. The semi-circular canals of nineteen species of bird as seen following exposure from the posterior end of the skull. (Hadžiselimović and Savković, 1964).

| | | |
|---|---|---|
| 1. *Podiceps ruficollis* | 8. *Tetrastes bonasia* | 14. *Asio otus* |
| 2. *Ardea cinerea* | 9. *Phasianus colchicus* | 15. *Picus canus* |
| 3. *Anas platyrhynchos* | 10. *Coturnix coturnix* | 16. *Corvus corax* |
| 4. *Anas domestica* | 11. *Fulica atra* | 17. *Turdus viscivorus* |
| 5. *Anser domesticus* | 12. *Scolopax rusticola* | 18. *Turdus philomelos* |
| 6. *Aquila chrysaetos* | 13. *Columba oenas* | 19. *Turdus merula* |
| 7. *Falco peregrinus* | | |

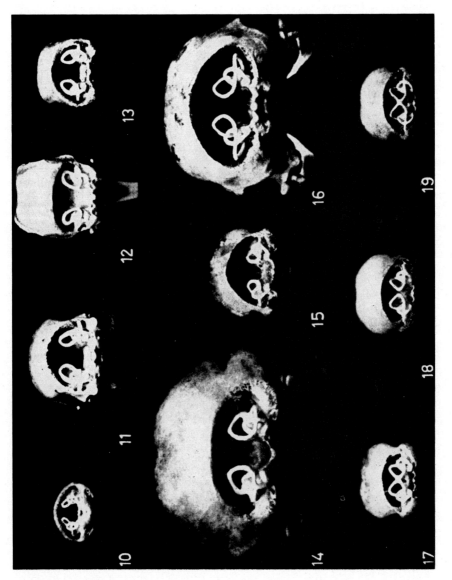

FIG. 55. (contd.)

encloses, together with the upper canal, an angle which is greater the better the specific flying ability. In good fliers there can also be a double intercrossing of the lateral canal.

## (ii) *The Utricle, Saccule and Ampullae*

Together with the utricle the semi-circular canals are often referred to as the pars superior of the inner ear. All three canals are connected to the utricle by their ampullae which contain a sensory organ or crista acoustica. In a small recess which is close to the origin of the two anterior ampullae the utricle itself has a sensory region—the macula neglecta, which corresponds to the ampullary cristae. From the utricular floor a small foramen gives access to the saccule containing the sensory units of the macula sacculi which, in *Gallus*, is inclined at 40° to the vertical. At the hind end of the saccule the sacculo-cochlear duct passes ventrally to the cochlea and lagena. It is accepted that in mammals the ampullary crests and the maculae of both utricle and saccule are the special end organs concerned with equilibratory vestibular reflexes influencing, on the one hand, the position of the eyes in relation to the movements of the head via the median longitudinal bundle and the nuclei of cranial nerves III, IV and VI, and, on the other hand, muscle tone via the vestibulo-spinal tracts. The macula lagenae at the distal end of the cochlea is situated in a vertical plane which is normal to that of the utricular macula.

These sensory regions contain two types of receptor. The cupula type occurs in the 3 ampullae and also in the utricular macula neglecta. The macula type characterizes the receptors of the main sensory areas of the utricle, saccule and lagena. Both types have a similar basic structure. The epithelium lining the labyrinthine walls is thickened by cellular differentiation, hypertrophy and an ingrowth of nerve fibres. The sensory spots are formed by a layer of basal cells overlain by a layer of supporting cells above which are the sensory hair cells. The arborizations of the nerves of the vestibular ganglion enter the organ and invest these sensory cells with a peri-cellular terminal network.

*Cupulae.* In cupulae the sensory region is composed of epithelial folds which are filled with connective tissue. The sensory hairs of this crest are enclosed in a non-crystalline and gelatinous mass. The ampullary crest of the horizontal semi-circular canal retains this simple structure. In the two other ampullae a fold, the septum cruciatum, gives the structure the appearance which is shown in Fig. 56. This fold, which Portmann (1961) erroneously stated to be peculiar to birds, is devoid of a sensory epithelium and divides the sensory zone. Various other names have been given to it. Steifensand, in de Burlet (1935), termed it the septum cruciforme; Retzius (1884) the septum transversum, whilst de Burlet himself used the terms processus cristalis and septum cruciatum. Dohlman (1961) probably influenced other workers who use the term cruciate eminence.

Usually located at the middle of the crista it protrudes perpendicularly to the long axis. Igarashi and Yoshinobu (1966) carried out a comparative study which included material from pigeons, robins, sparrows, the hawk *Falco sparverius*, the ducks *Aythya americana* and *Dendrocygna bicolor*, together with Humboldt penguins *Spheniscus humboldia*. These species were chosen as being examples of widely varying locomotory types. The hawk for its high

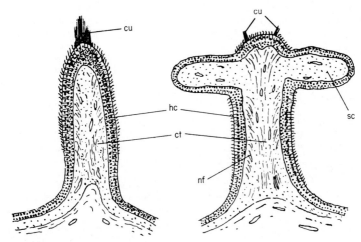

Fig. 56. Diagram of the cruciform structure in the posterior ampulla of *Chrysotis* and the static crest in the external ampulla (left), (Portmann, 1961.)

Abbreviations: ct, connective tissue; cu, hair cells connected with the covering cupula; hc, hair cells; nf, nerve fibres; sc, cruciate septum.

speed and aerial diving, the ducks for high speed but horizontal flying, and the penguins as examples of non-volant influences. All had a prominent cruciate eminence and the projection was usually slightly longer towards the canal than on the utricular side but the actual shape and size did exhibit slight variations. In the ducks it was rather shorter and wider than in pigeons and robins. However, that of *Spheniscus* was almost the same size as that of pigeons. Igarashi and Yoshinobu suggested that its widespread presence in birds and turtles may be correlated with the relatively poor development of the horizontal canal. They also suggested that when sufficiently prominent it may influence the hydro-dynamics of the endolymphatic flow to the two portions of the cristae which it serves to divide. As the canal opening to the ampulla is situated slightly to one side of the eminence the endolymphatic flow may give more direct hydrodynamic effects to the one side. Were this the case then the afferent impulses would carry different and complex physiological information. To a certain extent it would fulfil the role of two distinct cupula-endolymph systems with different orientations relative to the axis of the head. Even if the two systems had identical sensory inputs Igarashi and

Yoshinobu suggest that the resulting redundancy would be an advantage in case of injury to one. Other functions have also been suggested and Dohlman (1961) intimated that it may act as a stabilizer for the cupulae of the vertical canals during horizontal movements. The secreting properties of the dark cells which cover the eminence are not understood and Wersall (1961) and Dohlman (1964) have demonstrated that methylene blue uptake occurs in the supporting cells. In conclusion one can emphasize that the presence of a rudimentary cruciate eminence in the carnivora precludes it from having a function which is related to flight alone, but it may be related to the well-known scansorial and air-borne agility of the relevant types.

*Maculae.* In contrast to the condition in the cupulae there are crystalline components in the maculae. Each sensory area is associated with a gelatinous secretion (Werner, 1939). This is attached to the sensory patch particularly in its marginal region. Appearing at the 7th day of incubation in the chick it becomes filled with statoconia which are largely composed of $CaCO_3$, although there seems to be some calcium phosphate. These constituents never fuse to form a single compound structure, and vary in size according to their site of origin. The largest is the utricular statolith. Vinnikov (1969) has summarized the structure of the macula. The relative distribution and polarization of stereocilia and kinocilia are complex. The stereocilia are covered by a membrane which is a continuation of the plasma membrane of the cell. The diameter of the stereocilia varies from about $0.1$ to $0.2 \mu$ from apex to base. The stereocilia comprise bundles of fibrils with a diameter of 20–40 A. Kinocilia can have a diameter of $0.3 \mu$ and contain 9 pairs of peripheral and 2 central fibrils. The latter are perpendicular to the bundle of adjacent stereocilia. The kinocilium is the longest of the sensory hairs, is covered by a continuation of the cell membrane and penetrates deeply into the otolithic membrane to which it is attached at its apex. Consequently the otolith is a floating structure whose shearing movements are caused by changes of the position of the body relative to gravity but are regulated by the contraction or relaxation of the kinocilium. Usually 40–100 stereocilia form 7–9 pairs of straight rows opposite the kinocilium. There is reason to believe that the polarization of the receptor cells in birds, reptiles and amphibians differs from that in fish and mammals. Nevertheless in all classes the maculae have kinocilia facing in four directions. A single macula can respond to deviations from all horizontal axes but each cell shows maximal responses to angles around a specific axis. The Soviet workers have analysed the cytological changes that follow stimulation and consist first of all of the disappearance of RNA aggregated at the basal pole of the cell, together with nucleolar migration. Stimulation by radial acceleration in the pigeon leads to a regular cycle of cytochemical events in this scheme. These include changes in protein metabolism, synaptic vesicle appearance and mitochondrial form.

## C. The Central Vestibular Relationships

In the vestibular region of the medulla in different avian species some six different nuclear groups have been distinguished. The lateral, medial and descending nuclei of the pigeon are represented in Fig. 57. Following Kappers *et al.* (1936) and the synoptic terminology of Sanders (1929) one may cite the more general variants as follows:

(i) The small tangential vestibular nucleus which receives axons of large diameter as soon as these penetrate the medullary region. Within the nucleus their synaptic connections have been described as having a very characteristic spoon-shaped appearance.

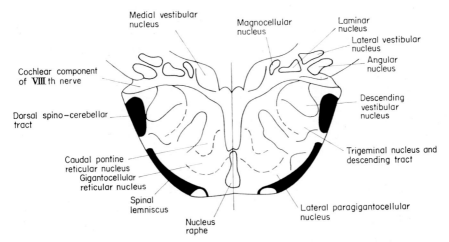

FIG. 57. Diagram showing the structure of the fore part of the avian medulla.

(ii) The ventro-lateral vestibular nucleus which has large multipolar neurons and in which similar fibres terminate. This structure is better known as Deiter's nucleus, of which it is the presumed analogue, and provides axons for the ipsilateral vestibulo-spinal tract supplying the nuchal region, trunk and limbs. Portmann and Stingelin (1961) emphasized the correlation which exists between the size of this tract, and the nuclei of the vestibular group, with the ability to fly. Some axons also decussate and run to the median longitudinal fasciculus of the opposite side, whilst ascending fibres run to the midbrain and hypothalamic region.

(iii) The descending vestibular nucleus (Fig. 57) receives fibres from the utricular region.

(iv) The dorso-lateral vestibular nucleus which has historically been sub-divided into 3 components on the basis of cyto-architectonic studies. The

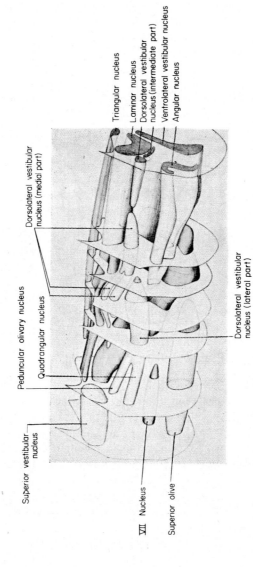

FIG. 58. The vestibular nuclei in the medulla of *Strix aluco*. (Stingelin, 1965).

neurons of the medial and intermediate regions have been described as synapsing with the cerebello-motor system, those of the lateral part with the spino-cerebellar tract.

(v) The dorso-medial vestibular nucleus which, together with the foregoing, received fibres which pass dorsally after entering the medulla (Fig. 57).

(vi) The superior vestibular nucleus which has connections with the cerebellar region. Secondary axons have also been described passing to the contralateral vestibular region alongside decussating vestibular fibres. At its area of greatest extent this nuclear grouping can be H shaped, and, situated dorsally, the upper fork surrounds a lateral cerebellar nucleus.

The actual representation of such central vestibular masses is subject to some variation from order to order (see Stingelin, 1965). Five discrete units exist in the pigeon and their positions have been expressed in the stereotaxic atlas of Karten and Hodos (1967). They are the descending, dorso-lateral, lateral, medial and superior nuclei. The medial and descending components extend further towards the hind end of the medulla, lie in a position to the side of the vagal-glossopharyngeal complex, and overlie the descending tract of the trigeminal nerve. In this region the medial is dorso-medially placed with respect to the descending nucleus, but, further forwards, it assumes a medial position as the IX-X cellular components taper off. These two vestibular units then lie at the same vertical level. At the site of the cochlear nerve it is, however, dorso-medial again when the descending nucleus assumes a position below this nerve and also below the lateral vestibular, angular and laminar nuclei. Further towards the rostral end the superior and dorso-lateral nuclei are very large and make pronounced contributions to medullary structure. They occupy a position to the side of the medial, which penetrates forward to the level of the main sensory trigeminal nucleus, and the superior replaces the dorso-lateral anteriorly. Here it lies below the brachium conjunctivum and, as one passes rostrally, above first of all the lateral vestibular mass and then the medial part of the uncinate fasciculus. The condition in *Strix aluco* is shown in Fig. 58.

These nuclear groups and more particularly the definitive ones described by Karten and Hodos (1967) are concerned with the delicate coordinations of the somatic musculature and as such they are of immense importance during swiftly moving bipedal locomotion or flight. Benjamins and Huizinga (1926, 1927) found a particularly close relationship between the macula sacculi and the rotation of the eye, which would agree with the situation in mammals referred to above. According to Portmann (1961) the semi-circular canals and their central fibres are especially involved in the control of antagonistic muscles whilst the association of the utricle is largely with the body musculature. During development the vestibular nuclei are derived from neural epithelium which is situated in areas medial to the neural lip (Harkmark, 1954a,b).

In this they contrast with the cochlear nuclear masses because these arise from the rhombic lip, although the cells do not migrate outward in a ventro-medial direction as has been described for other medullary nuclei.

## V. THE COCHLEAR AFFERENT SYSTEMS

### A. The Auditory Nerve

The peripheral systems which are involved in hearing were outlined in Chapter 5. As has already been noted, the posterior branch of the VIIIth nerve, the ramus cochlearis, contains fibres which pass to it from the macula neglecta, the sacculus and the posterior ampulla, as well as those from the cochlea itself. Furthermore, the rami cochlearis and lagenaris of an "average"

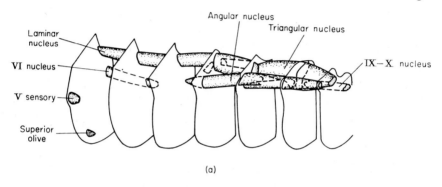

(a)

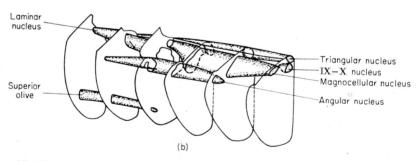

(b)

FIG. 59. The cochlear region of the medulla in (a) *Capella gallinago*. (b) *Aix galericulata*. (Stingelin, 1965).

bird weighing 100 g contain about 7500 fibres, of which slightly in excess of 10% seem to relate to the lagena. This number varies in relation to body weight with an exponential constant of 0·15, but Winter (1963) emphasized that it is not possible to compare these figures with those for mammals. Nevertheless it is clear, as was pointed out by Pumphrey (1961) and others, that weight for weight the avian auditory nerve contains only half as many elements as do those of mammals. In common with the condition of other vertebrates the nerve also includes amongst its fibres some efferent components whose activity is analogous to that of the centrifugal fibres in the avian visual system (*q.v.*). These are discussed in Chapter 7 below. Although their precise origin is not clear they appear to arise contralaterally.

At the medullary level the vestibular components separate from those which are of cochlear origin and pass to, say, the tangential vestibular nucleus. The cochlear fibres pass, on the contrary, to the magnocellular and angular nuclei. Schwartzkopf (1968) who, together with Karten (1968) and Harman and Phillips (1967), has reviewed our knowledge of the avian auditory pathway, emphasizes that these nuclei are almost certainly homologous with the mammalian cochlear nucleus. In particular it has long been assumed that the angular nucleus corresponds to the dorsal region of this mammalian structure (Brandis, 1894; Ramon y Cajal, 1908 and Winter, 1963). The experimental results of Boord and Rasmussen (1963) which are outlined in Section VD go a long way towards elucidating and substantiating the older works. They concluded that the fibres which originate in the cochlear and lagenar ganglia respectively, are distributed in an orderly manner within the secondary centres of the medulla. They also confirmed that fibres from the lagenar branch terminate in part on the vestibular and cerebellar nuclei, as well as within the reticular formation. The other components of this branch enter the magnocellular and angular nuclei and this may corroborate the suggestion of Pumphrey (1961) that the lagenar region is implicated in hearing.

## B. The Magnocellular, Angular and Laminar Nuclei

From comparative studies of the different vertebrate classes it was postulated long ago that the magnocellular nucleus is both ontogenetically and phylogenetically older than the angular nucleus. The position of both nuclei in the pigeon brain is shown in Fig. 57. In general the latter nucleus is situated along the dorso-lateral surface of the medulla and anteriorly it can penetrate to the level of the abducens nucleus (*q.v.*). Cochlear axons which enter the brain at the same level turn dorsally and enter it on the ventro-medial side. They therefore separate it from the magnocellular nucleus of the same side.

Comparative studies of the magnocellular nucleus have suggested that it can be differentiated into a number of components. In *Passer* there are, according to Sanders (1929), medio-medial, medio-lateral and ventro-lateral contributions each with similar cytological characteristics (see Section VC). The first two lie in front of the last and have a lamellar organization in places. These two fuse as one passes backwards and just behind this level the rather smaller-celled ventro-lateral moiety appears. Around the nucleus there is the often crescentic cell mass of the laminar nucleus which has a dorsal relatively micro-cellular component and a larger celled ventral part although the two can be difficult to differentiate in the anterior regions. Characteristically there are small cells with radiating dendrites. These are arranged in two rows, one above and one below the cell-bodies, and it is this organization which has resulted in the very descriptive name of the nucleus.

Despite this general tendency there is some variation in the appearance of the laminar nucleus in different genera. Such variations are particularly striking in the owls. In those strigiform species which have a well-developed aural asymmetry (see Chapter 5) the simple planar layer of neurons is replaced by an undulatory appearance. Whilst the number of neurons in the simple planar form is relatively small there are very many more in these owls. Schwartzkopf (1968) cites falconiform and strigiform genera of equivalent

## TABLE 27

*The relative cell numbers in the central auditory nuclei of some avian species.*

|  | Body weight (g) | Magnocellular nucleus | Angular nucleus | Laminar nucleus | Superior olivary nucleus |
|---|---|---|---|---|---|
| *Athene noctua* | 183 | 3,450 | 2,750 | 2,490 | 2,550 |
| *Tyto alba* | 278 | 15,760 | 9,960 | 10,020 | 11,830 |
| *Asio otus* | 293 | 10,810 | 5,060 | 8,340 | 4,460 |
| *Strix aluco* | 452 | 11,070 | 5,930 | 6,920 | 4,770 |
| *Bubo bubo* | 2,800 | 7,890 | 4,980 | 5,070 | 4,310 |
| *Pica pica* | 213 | 4,460 | 2,870 | 2,500 | 1,850 |
| *Corvus corone* | 533 | 5,160 | 4,320 | 2,540 | 2,030 |

From Winter and Schwartzkopf, 1961.

body size in which the number of neurons in the strigiform laminary nucleus is five times that in the hawk. For example, the number for *Falco tinnunculus* is 2000 whilst that for *Tyto alba* it is 10,000. Such numerical disparities also occur in the neural components of other medullary centres in these owls (Winter and Schwartzkopf, 1961). As a result of this the total number of auditory neurons in each half of the medulla varies from 28,700 in *Strix aluco* to 46,000 in *Tyto alba*. In contrast the number in *Pica* and *Corvus*, which are passeriform genera of roughly comparable size, lies between 12,000 and 14,000. That in the remaining owls which have symmetrical ears is comparable with these latter figures for other orders rather than those of their aurally asymmetrical relatives. *Athene noctua*, the little owl, has 11,700, and *Bubo bubo*, the great horned owl, has only got 18,000 although its body weight is 10 times greater than that of the barn owl. The relative numbers are as listed in Table 27.

Winter (1963) drew attention to the fact that these medullary centres show a far greater correlation with the behaviour and ecology of the species than other auditory structures do. The basilar membranes of the aurally asymmetrical strigiform genera, which hunt entirely at night and are able to catch mobile prey by acoustic location alone, are only twice as long as those of other birds. Clearly this contrasts markedly with the six- to seven-fold difference in the number of secondary and tertiary neurons in the medullary auditory pathway. It seems probable that these are information processing foci of considerable importance and that they are possibly responsible for introducing the delays and attenuations which Batteau (1967), in his consideration of mammals, has suggested would facilitate binaural and monaural sound localization and sound recognition.

## C. The Histology of the Angular and Magnocellular Nuclei

Several types of neuron have been distinguished within these two nuclei (Boord and Rasmussen, 1963; Boord, 1968). However, as Schwartzkopf (1968) has pointed out, the degree of differentiation of both the ganglion cell types and the synaptic connections in these avian, secondary, auditory nuclei, is less complex than appears to be the case in mammals. Within the medial part of the magnocellular nucleus there are large ovoidal cells with a diameter of 20–25 $\mu$ around which axonic fibres terminate in the form of peri-cellular calyces. The more lateral regions of the nucleus have a mixture of cells which vary in size from 18–24 $\mu$. Synapsing fibres again form calyces around the larger ones but terminate as peri-cellular plexi around those with smaller diameters. In contrast to both of these regions the ventro-lateral region characteristically comprises a closely packed mass of fusiform and stellate cells which are surrounded by a fibrillar network.

In the angular nucleus the cell-bodies in the dorso-lateral region can reach 34 $\mu$ in diameter. They are therefore much larger than those in the ventral area where the predominant size ranged from 8–17 $\mu$, although occasional larger cells were present at a density of one per section. The medial area of the nucleus is composed partly of cells which are similar to those of the lateral region, and, in the ventral levels, of others which compare with those in the ventro-lateral region of the magnocellular nucleus. That is, there is a mass of fusiform and stellate components. At the point where both the ventro-lateral part of the magnocellular nucleus and the ventral part of the angular nucleus are united, the narrow bridge contains cells which are similar to both regions.

Holmes (1903) considered that these actually represented a true and discrete cochlear nucleus, but Boord and Rasmussen provided evidence which suggests that they receive lagenar not cochlear projections. Their lesion studies, which are considered below, revealed point to point connections between the apical, middle and basal thirds of the cochlear nerve and the cochlear nuclei.

## D. Lagenar and Cochlear Fibre Projections

### (i) *Lagenar Fibre Projections*

If the lagenar fibres are investigated by surgical transection of the nerves from which they originate, their central terminations are found to be intimately associated with those of the apical cochlear fibres (Boord and Rasmussen, 1963). However, at their point of entry these lagenar derivatives penetrate the medulla directly in front of, and below, the apical cochlear ones. Subsequently they follow these latter to the most ventral boundary of the angular nucleus and the tract then bifurcates at this level, which is where both angular and magnocellular nuclei are connected by the cell bridge that was mentioned above. At this point a marked accumulation of the fibres has given rise to the term lagenar root.

Ascending fibres run forward and under the angular nucleus to the ventral region which is immediately under that which receives apical cochlear fibres. Some then continue in a forward direction to enter the lateral part of the superior vestibular nucleus on the one hand, and the lateral, intermediate and medial cerebellar nuclei on the other. The descending lagenar root passes posteriorly and under the micro-cellular ventro-lateral component of the magnocellular nucleus. A large number of its fibres then actually enter this structure whilst others penetrate the reticular formation. Axons which are located in a median position, and derived from both the ascending and descending components, are distributed throughout the entire length of the lateral and medial vestibular nuclei. The giant cell nuclei of Ramon y Cajal (1908), the large multipolar cells which are situated ventro-medial to the ascending root, together with cells in the dorso-medial and lateral areas of the reticular formation, all receive fibres. It is therefore clear that, although the majority of lagenar fibres are distributed in the medullary vestibular and reticular components, a sizeable number end on the ventral and ventro-lateral parts of both the angular and magnocellular nuclei.

### (ii) *Cochlear Fibre Projections*

The lesions which gave this information about the central connections of lagenar fibres also enabled Boord and Rasmussen to provide detailed evidence about the medullary sites at which fibres from the apical, middle and basal thirds of the cochlear nerve end. Lesions of one type interrupted the fibres which originate in the apical third of the cochlear ganglion and also half of the fibres within the lagenar nerve. The true cochlear fibres enter the medulla as the most anterior and ventrally situated component of the ramus cochlearis. They traverse beneath the lateral region of the angular nucleus and then bifurcate into lateral and medial branches. The first of these ends within the ventral region of the angular nucleus, whilst the second passes to the lateral part of the dorso-lateral border of the magnocellular nucleus.

Lesions of a second type resulted in fibre degeneration within the distal two-thirds of the cochlear nerve, and the total degeneration of the lagenar nerve. Fibres from the middle third of the cochlear nerve enter the medulla at a position which is intermediate between those of the apical and basal cochlear fibres. The tracts then bifurcate above and behind those associated with the apical region. The lateral branches end in the intermediate part of the angular nucleus whilst the medial branches, after traversing the dorsal border of the magnocellular nucleus, synapse with the cells that are situated within the lateral half of its macro-cellular medial component. Some fibres from the basal third of the nerve also end in this region but at positions which are more medial than those which have just been mentioned. Other, and similar fibres, pass to the most dorsal part of both the medial and lateral areas of the angular nucleus. On the basis of this evidence it is clear, as Boord and Rasmussen point out, that the distribution of afferent cochlear fibre endings shows a definite spatial arrangement at the secondary stations of the medullary auditory pathway. In view of this one may speculate that subsequent lines to higher relay stations may maintain this topographical isolation, and that this itself reflects a central tonotopic projection from the cochlea.

## E. Higher Stations on the Medullary Auditory Pathway

The foci at other levels which are related to the avian auditory pathway are referred to in various sections. A comparison of other medullary stations with those in mammals is difficult, and, without further qualifications, ill-advised. The superior olive is the principal tertiary station in mammals but the functions of the avian olivary structures are only incompletely known although, according to Winter and Schwartzkopf (1961), it is particularly well represented in the Strigiformes. Schwartzkopf (1968) intimated that its principal projections are derived from the angular nucleus as the majority of the fibres from the laminar nucleus appear to traverse it without making any synaptic connections but the situation was nevertheless somewhat controversial and some collaterals were thought to end in it.

On leaving the olive the relevant fibres decussate through the ventral stria and then continue through the lateral lemniscus to the mesencephalon. In mammals most of such decussating fibres pass ventrally and constitute the trapezoidal body. In birds the ventral stria is less prominently developed and it is the dorsal commissure which assumes greater prominence. The decussating fibres which have their site of origin within the magnocellular nucleus pass through it and enter the ventral surface of the contralateral laminar nucleus. Other fibres from magnocellular foci enter the ipsilateral laminar nucleus via the shortest route. As a result of this it is clear that the latter nucleus receives information from both the ipsilateral and contralateral ears through corresponding secondary axons. If these enter at different foci, the contralateral from below and the ipsilateral from above, there is at once an immediate topographical differentiation. This is also apparent from other information and Schwartzkopf suggested that the synaptic and fibre orientation may be involved in the processing of binaural information. We may once again remind ourselves of Batteau's suggestions that were referred to in Chapter 5. Boord (1968) concluded that the magnocellular nucleus has two projections—one to the laminar nucleus via the dorsal cochlear tracts, the other to the trapezoid body—in detail he found that selective lesions of the magnocellular nucleus showed that degenerating axons emanate from the ventro-medial border of the medial part and comprise a tract which courses medially and crosses the raphe both dorsal to, and in part through, the median longitudinal fasciculus. On the contralateral side this crossed cochlear tract turns forward below the magnocellular nucleus and forms an ascending bundle which gives off fibres that appear to terminate on the dendrites of the cells along the ventro-lateral convex border of the laminar nucleus. An occasional fibre passes through to its dorsal side. Each laminar nucleus also receives a homolateral innervation from the ipsilateral magnocellular nucleus which ascends along the dorsal side of the latter as an uncrossed dorsal cochlear tract. At the level of the laminar nucleus fibres leave the parent bundle, course both sideways and downwards, and terminate about the dendrites of the cells along the dorso-medial concave border of the laminaris. As lesions elsewhere in the cochlear nuclei give no such results Boord concluded that only the medial part of the magnocellularis projects bilaterally to the laminar nucleus. He concluded that this last is homologous with the medial superior olivary nucleus of mammals.

## F. Evoked Responses in the Medullary Auditory Nuclei

Besides his exploration of the caudal neostriatum which is referred to elsewhere (q.v.) Erulkar (1955) also inserted electrode probes into the angular, magnocellular and laminar nuclei. In each of seven experiments all three nuclei regularly exhibited activation following the presentation of click stimuli to the ipsilateral ear; never, however, following such stimulation of the contralateral ear. The latent periods which were observed when the stimulus was delivered at a distance of 3 cm from the appropriate ear were 1·8 msec for the angular nucleus and 1·9–2·0 msec for the other two. These values can be compared with latencies of 1·0 and 1·6 msec which were obtained for the first and second neural responses as measured at the round window.

Erulkar also concluded that the superior olivary nucleus, which is rather indistinct in birds, also invariably showed activation. What is more such potentials followed the presentation of click stimuli to either ear. He did, however, gain the unsubstantiated impression that responses to contralateral stimulation had a smaller amplitude than those following stimulation of the ipsilateral ear. Furthermore, the latent periods were rather longer than was the case in the three former nuclei. This was particularly interesting because the close similarity in the latent periods which he observed at laminar and magnocellular foci appeared to corroborate the long-standing contention that the laminar nucleus received direct projections from the cochlea.

More recently Stopp and Whitfield (1961) have recorded the responses to peripheral sound stimulation within 100 brain stem units most of which were in the magnocellular nucleus. They found that the threshold-response curves were essentially similar to those of mammals. The range of frequencies to which such responses were observed extended from below 100 cyc/sec to over 4000. Two-thirds of the units which they studied could be inhibited by tonal stimuli and about one-fifth were not excitable under the conditions used, and their resting activity was not inhibited. At low frequencies the responses were usually, but not always, related to the phase of the stimulating sinusoid.

In almost all cases the spontaneous firing rate was higher than that which one observes in the trapezoidal body of cats, and rates of 100/sec were predominant. There were other slower rates and these could fall to as low as 10/sec. When the stimulating tone ended almost all the observed spontaneous activity was temporarily depressed and a "silent" period ensued. In different units stimuli of equal magnitude produced increases of anything from 5–195% above these spontaneous rates. Although some such units which responded to characteristic frequencies within the range 90–4130 cyc/sec were identified, the majority showed activation at around 1800 cyc/sec (see Fig. 44). The separate application of tones to the two ears showed that the contralateral threshold was invariably 25–40 decibels above that for the ear of the same side. This agrees with the anatomical evidence which suggests that the auditory nerve terminates entirely in the ipsilateral nuclei. Although the results were not confirmed by unilateral cochlear ablation, they suggested that it would be reasonable to attribute differences of this magnitude to mechanical cross-transmission. The attenuation which would be expected to accompany such non-nervous pathways would lead to higher thresholds of approximately this order.

In general the inhibitory phenomena were quite specific. For each unit there appeared to be a particular value which exhibited maximum effects. The observed relationship between the intensity of the inhibitory tone and the band of frequencies over which it is influential follows the same general triangular pattern which is seen in the mammalian auditory pathway, and is also analogous to the excitatory threshold curves.

## VI. PRE-OTIC AFFERENT SYSTEMS

### A. The Facial Nerve

The avian facial nerve is much less significant than the trigeminal and, as was emphasized in Section IIIB (iv), is almost certainly not involved in the contributions to the taste modality. These last are all provided by the glossopharyngeal and its branches. The roots of the facial are situated rostrally to those of the stato-acoustic nerve and in a tolerably close relationship to the superior olive. The geniculate ganglion is not very large. As the majority of the taste-buds are located within the dermatome of the glossopharyngeal the disposition of the facial nerve, together with its sensory and motor nuclei, is often cited as an example of neurobiotaxis (q.v.) and the influence of the pattern of incoming sensory stimulation on the determination of medullary nuclear organization. Kappers (1920) pointed out that a nerve cell tends to remain as near as possible to its source of stimulation, and that, when there is any danger of separation resulting from the development of intermediate structures, there follows during phylogeny a migration towards the source of stimulation. This once all-important theory of neurobiotaxis is discussed further in Chapter 7 but one can note that in many works such influences of sensory input on the form of medullary organization are exemplified by the relative sizes of the facial, glossopharyngeal and vagal afferent systems. The relatively anterior position of the facial components is then due, in the theories of neurobiotaxis, to the absence of gustatory impulses which would draw them backwards. It is certainly hardly surprising that with far larger and more important sensory fields the vagal and glossopharyngeal have more prominent roots and nuclei.

### B. The Trigeminal Afferent Systems

#### (i) General Considerations

The variable form of the avian trigeminal nerve is illustrated by the conditions which pertain in *Anser anser, Cathartes burrovianus, Haliaetus albicola, Asio otus* and *Alcedo ispida*. These are shown in Figs 60–62 and are taken from the comparative studies of Barnikol (1954). Earlier contributions had previously been made by Lakjer (1926), Stresemann (1933), and Starck (1940).

Barnikol was at some pains to emphasize the considerable individual variation that one finds. Indeed there can be very marked differences between the innervation of the right and left sides of a particular individual specimen and these variations are frequently quite arbitrary.

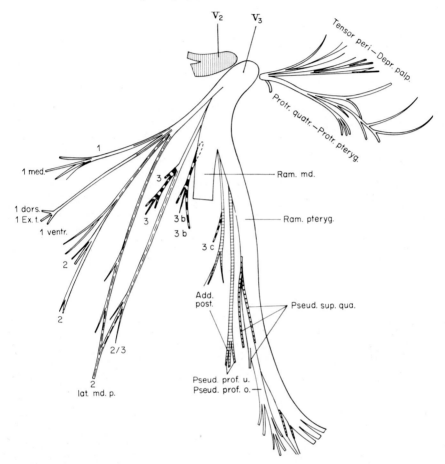

FIG. 60. The appearance of the Vth nerve in *Cathartes burrovianus*. (Barnikol, 1954). The variable form of the principal branches (Rami mandibularis and pterygoideus) can be seen by comparison with figures 61–62. $V_2$–$V_4$ = branches of Trigeminus.

The ophthalmic and maxillary nerves are principally composed of somatic sensory fibres which are associated with the corpuscles of Grandry (*q.v.*) and other cutaneous tangoceptors. As a result of this the relative development of the afferent system parallels, in general terms, the degree of development of the cranial region. In the brain it is not easy to trace separate and distinct paths

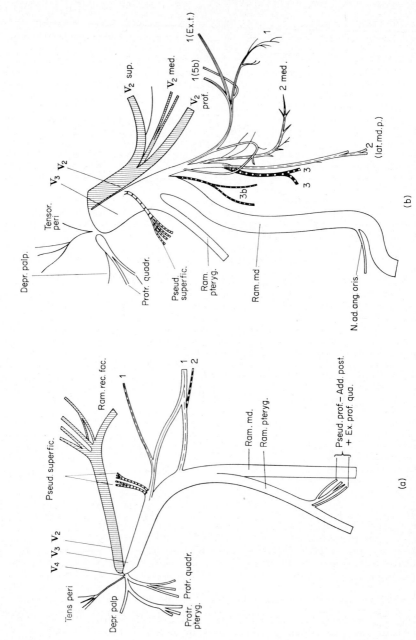

Fig. 61. The form of the Vth nerve in (a) *Alcedo ispida*. (b) *Asio otus*. (Barnikol, 1954).

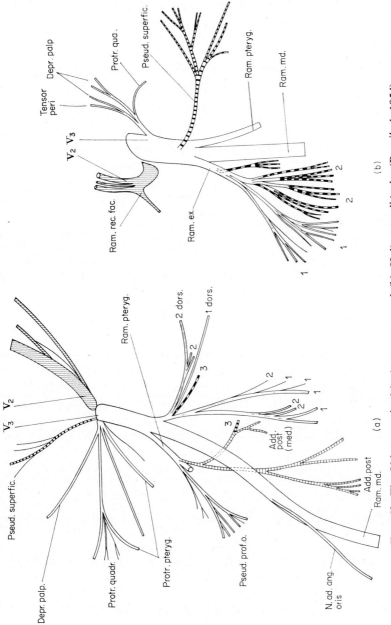

FIG. 62. The Vth nerve in (a) *Anser anser* and (b) *Haliaetus albicola*. (Barnikol, 1954).

for the maxillo-mandibular and ophthalmic branches but within the intra-medullary region the trigeminal terminations are at a well-developed main sensory nucleus and a nucleus which is associated with the descending tract. It is essential to also remind oneself of the mesencephalic trigeminal sensory nucleus which, in mammals, is associated with jaw and ocular proprioception. Similar functions also appear to characterize the avian mesencephalic component and these are considered in Chapter 10. The main sensory nucleus is situated close to the point of entry to the brain, contains large cells, and is largely isolated from the neighbouring grey matter by myelinated fibres of the descending tract. The nucleus of the descending root is often diffuse and small, comprising scattered patches of grey. As one moves towards the hind end its size increases somewhat and it is this posterior region, which also receives cutaneous fibres from the vagal inflow, which can comprise part of the so-called spinal sensory nucleus of the trigeminal. Both the main sensory nucleus and the nucleus of the descending tract are shown in Fig. 57 and 59.

A detailed analysis of the relative contributions made by the trigeminal and other medullary nuclei in various orders was provided by Stingelin (1965). This is discussed further in relation to the afferent pathways in Chapter 17; however, one can note that the volume of the medullary sensory component of V exceeded that of the angular, magnocellular and laminar nuclei in *Ibis* and *Melopsittacus*. In *Corvus corone* it was more or less equivalent to the volume of the homologous motor nucleus, which is far smaller than it is in *Ibis*, whilst in the series of species of *Capella, Ibis, Aix, Strix, Agapornis, Melopsittacus, Merops* and *Corvus* the sizes of a number of these nuclei, together with that of the reticular formation, appear to bear direct relation-ships to such parameters as body weight.

Projections from the chief sensory nucleus have long been described as passing both to the principal motor centres within the medulla and also to the cerebellum (Craigie, 1928; Sanders, 1929; Papez, 1929; Brandis, 1894; Kappers *et al.* 1936). A tract which was first described by Wallenberg (1903, 1904) also seems to arise from it, and, following a partial decussation, projects forward to the telencephalon. This, the quinto-frontal tract, is of considerable importance and will be referred to again in subsequent chapters. Other projections towards the anterior region are provided by the quinto-mesencephalic tract passing to both the lateral mesencephalic nucleus (*q.v.*) and the optic tectum. At least in the Casuariiformes crossed fibres are also said to pass from the enlarged descending root nucleus and to either end in the reticular formation or form the trigeminal mesencephalic tract.

## (ii) *The Composition of the Ophthalmic Branch*

The principal study of the individual anatomical components within the trigeminal nerve is that of Graf (1956) who used the method of nerve fibre

## TABLE 28

The total number of fibres counted, and the percentage of different calibre groups, in pigeon nerves.

| | Total number | 0-1 $\mu$ | 1-2 $\mu$ | 2-3 $\mu$ | 3-4 $\mu$ | 4-5 $\mu$ | 5-6 $\mu$ | 6-7 $\mu$ | 7-8 $\mu$ | 8-9 $\mu$ | 9-10 $\mu$ |
|---|---|---|---|---|---|---|---|---|---|---|---|
| Ophthalmic division of trigeminal sensory nerve | 1325 | 5·9 | 34·7 | 17·3 | 13·3 | 17·4 | 9·8 | 1·9 | — | — | — |
| Trochlear nerve | 935 | — | 31·7 | 27·2 | 33·9 | 6·6 | 0·6 | — | — | — | — |
| Posterior root of 22nd segment | 904 | 1·0 | 15·7 | 24·4 | 17·6 | 19·5 | 14·1 | 7·6 | 0·3 | — | — |
| Anterior root of 23 segment | 629 | — | 16·7 | 19·0 | 10·5 | 21·5 | 16·7 | 13·7 | 2·7 | — | — |
| Anterior root of 24 segment | 1069 | 0·8 | 10·2 | 17·1 | 8·8 | 16·4 | 20·2 | 20·5 | 4·9 | 1·3 | — |
| Sciatic nerve (mixed branch) | 1688 | 1·5 | 18·8 | 20·7 | 15·3 | 17·1 | 12·4 | 8·5 | 2·8 | 1·9 | 1·0 |

From Graf, 1956.

8

analysis which was developed by Haggqvist. A comparison of his results with those which he obtained from other avian nerves is contained in Table 28. In contrast to the situation in mammalian nerves those of birds are relatively rich in small sized fibres and fibres of larger diameter are scarce. In the sensory trigeminal branch which he studied the upper limits of the calibre lay between 6–7 $\mu$. In fact the actual percentage representation of fibres with different diameters resembles that of man in overall pattern but the entire avian spectrum is shifted towards the left with a peak at 1–2 $\mu$ instead of 3–4 $\mu$. Consequently in man there are fibres of up to 14 $\mu$ diameter which belong to a size group that is apparently absent in birds, or at least from the pigeon.

## (iii) *Peripheral Stimulation of the Main Sensory Trigeminal Nucleus*

The main sensory nucleus comprises a first synaptic relay station within the avian trigeminal system. Distinct from the spinal nuclear complex, and situated both in front of, and dorso-lateral to, the trigeminal motor nucleus, its fibre connections within the quinto-frontal tract undergo a partial decussation and then project forwards on both sides to the basal nucleus without any intervening synaptic relays. Portmann and Stingelin (1961) noted that the relative size of this latter nucleus is directly proportional to that of the main sensory nucleus which in turn reflects the relative development of the beak region.

Gross dissections of the trigeminal nerve show that it frequently innervates the mouth and beak, and only provides limited contributions to the orbital region because the ophthalmic division runs primarily to the upper mandible.

By isolating 130 units within the main sensory nucleus Zeigler and Witkovsky (1968) showed that 72 % were activated by mechanical stimulation of the beak. The remaining 28 % were responsive to displacement of the lower bill component from its resting position. Although a clear-cut distinction between the tactile and pressure units was by no means straightforward, it was at least tolerably easy to discriminate between those which are activated by a light touch as opposed to those responding to tapping. Erulkar (1955) had previously explored the nucleus and recorded electrical responses which were readily evoked by light tactile stimulation of the beak region, or by displacing a few feathers on the head and neck. In all his experiments there was typically a latency of 1·8 msec which is comparable with the condition in mammals. He concluded that the small size of the nuclear mass, coupled with the large size of the tip of his electrode and the difficulty of delivering a precisely and finely delimited mechanical stimulus, prevented him from obtaining any information about such topological representation of the various somatic regions which may exist within the nucleus.

Zeigler and Witkovsky in their subsequent investigations found that the receptive fields which supplied their units were generally restricted to the mouth and beak. The mouth and palate were very well represented, as was the horny substance of both upper and lower beak components and also the beak tips. They did not isolate any units which showed activation following stimulation of the tongue. The actual somatic fields were usually small, involved only a few square millimetres, and were generally restricted to one or other of the upper and lower beak components. However, a few cells did respond to stimulation of both, and others to tactile stimuli applied to the nictitating membrane or the skin in the orbital region. The adaptive characteristics of all these units were similar to those previously observed in mammals. Both fast adapting and slow adapting units existed but it was the latter which predominated.

When fixed in the stereotaxic head holder the bird's mouth is slightly open. Under these conditions 31 units underwent a steady discharge and a decrease in the rate of firing which appeared to be directly proportional to the rate of jaw displacement. When it was held in a new fixed position a new and characteristic non-adapting discharge rate was established.

Although no signs of somatotopic representation were observed in either the medio-lateral or anterio-posterior planes, a clear sequence of unit organization was apparent in the dorso-ventral plane. Here there was a progressive tendency for units in different topographical positions to respond to mandibular, maxillary and ophthalmic stimulation. There was also a successive representation of distal, and then more proximal foci, as the electrode penetrated downwards in a vertical plane. Although the series of mandibulo-maxillary sites sometimes underwent a reversal, units which responded to orbital stimuli were always situated ventrally. There was no evidence for actual modality segregation and along a given line of the nucleus the electrode probes indicated that jaw-displacement responses were intermingled between areas that were activated by touch and pressure. Furthermore there were *no* exclusively contralaterally stimulated units and almost all the loci were ipsilateral although a few large bilateral fields clearly existed.

## (iv) *The Corpuscles of Grandry*

The corpuscles of Grandry, or Grandry-Merkel (Merkel, 1875), which are illustrated in Fig. 63, are present in the skin of the bill region amongst Anseriformes. Measuring some 50 $\mu$ in length they are also widely distributed on the tongue and palate of many species (Botezat, 1906; Boeke, 1926a,b; Dijkstra, 1933; Klein, 1931–32; Portmann, 1950, 1961). They are typically composed of two or more cells which are associated with the terminal fibrillum of a nerve fibre. Large corpuscles may contain up to five secondary cells, and four or so corresponding corpuscles may be related to one nerve fibre.

The flattened disc which is formed by the two apposed cells is surrounded by satellite cells which are themselves within a connective tissue capsule. In the past the origin of the principal cells has been the subject of some controversy. Izquierdo (1879) thought that they were of epithelial origin but Szymonowicz (1897) that they were derived from connective tissue. They

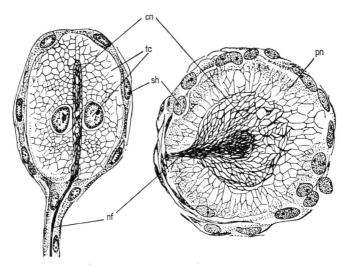

FIG. 63. Corpuscles of Grandry from the duck. (Portmann, 1961).
Abbreviations: Longitudinal section on left, horizontal section on right, cn, central neuro-fibrillar plexus; nf, nerve fibre; pn, peripheral neuro-fibrillar plexus; sh, sheath cells; tc,

possess a fibrillar structure and the apposed surfaces are excavated, forming the disc shape, with the nuclei situated at the inner borders. Some authors considered that the terminal nerve fibre distributed fibrils that penetrated the cells and both Dogiel (1904) and Botezat (1906) also reported a fine unmyelinated fibre which formed a peri-cellular plexus within the capsule. Boeke (1926b) suggested the presence of a similar peri-terminal plexus which joined the neuro-fibrillar disc of the primary fibre with the cytoplasm of the secondary cells.

## References

Akerman, B., Anderson, B., Fabricius, E. and Svensson, L. (1960). Observations on central regulation of body temperature, and food and water intake in the pigeon (*Columba livia*). *Acta Physiol. Scand.* **50**, 328–336.

Anderssen, H. T. (1959a). A note on the composition of alveolar air in the diving duck. *Acta Physiol. Scand.* **46**, 240–243.

Anderssen, H. T. (1959b). Depression of metabolism in the duck during experimental diving. *Acta Physiol. Scand.* **46**, 234–239.

Anderssen, H. T. (1963a). Reflex nature of physiological adjustments to diving and their afferent pathways. *Acta Physiol. Scand.* **58**, 263–273.

Anderssen, H. T. (1963b). Factors determining the circulatory adjustments to diving. I. Water immersion. *Acta Physiol. Scand.* **58**, 173–185.

Anderssen, H. T. (1963c). Factors determining the circulatory adjustments to diving. II. Asphyxia. *Acta Physiol. Scand.* **58**, 186–200.

Anderssen, H. T. and Lövö, A. (1964). Effect of $CO_2$ on respiration of avian divers (ducks). *Comp. Biochem. Physiol.* **12**, 451–456.

Baer, M. (1896). Beiträge zur Kenntnis der Anatomie und Physiologie der Atemwerkzeuge bei den Vögeln. *Z. Zool.* **61**, 421–498.

Barnikol, A. (1954). Zur Morphologie des Nervus Trigeminus der Vögel unter besonderer Berucksichtigung der Accipitres, Cathartidae, Striges, und Passeriformes. *Z. wiss. Zool.* **157**, 285–332.

Bath, W. (1906). "Die Geschmachsorgane der Vögel." Dissertation. Berlin University, p. 46.

Batteau, D. W. (1967). The role of the pinna in human sound localization. *Proc. Roy. Soc. B.* **168**, 158–180.

Benjamins, C. E. and Huizinga, E. (1926). Die Raddrehung wird bei den Tauben von den Sacculolithen ausgelost. *Z. Hals-nasen Ohrenheilk.* **15**. Part II.

Benjamins, C. E. and Huizinga, E. (1927). Untersuchungen uber die Funktion des Vestibularapparates der Taube. *Pfluger's Arch.* **217**, 105.

Bert, P. (1870). "Leçons sur la Physiologie Comparée de la Respiration." Paris.

Bing, H. I. (1959). On the physiology of cutaneous sensory modalities. *Acta Physiol. Scand.* **46**, 88–96.

Boeke, J. von (1913). Die döppelte (motorische und sympathische) efferente Innervation der quergestreiften Muskelfasern *Anat. Anz.* **44**, 343–356.

Boeke, J. (1926a). Die Beziehungen der Nervenfasern zu den Bindegewebselementen und Tastzellen. *Z. mikros. anat. Forschung.* **4**, 448–509.

Boeke, J. (1926b). Noch einmal das periterminale Netzwerk, die Struktur der motorischen Endplatte und die Bedeutung der Neurofibrillae. *Z. mikros. anat. Forschung.* **7**, 95–120.

Boerhaave, A. K. (1795). "Impetum faciens dictum Hippocrati per corpus consentiens phylologia et physiologia illustratum." Luchtmans and Son.

Boord, R. L. (1968). Ascending projections of the primary cochlear nuclei and nucleus laminaris in the pigeon. *J. Comp. Neurol.* **133**, 523–542.

Boord, R. L. and Rasmussen, G. L. (1963). Projection of the cochlear and lagenar nerves on the cochlear nuclei of the pigeon. *J. Comp. Neurol.* **120**, 463–475.

Bordoni, L. (1888). Sull'apnoea sperimentale. *Sperimentale* **61**, 113–132.

Botezat, E. (1904). Geschmacksorgane and andere nervose Endapparate im Schnabel der Vögel. *Biol. Centralbl.* **24**.

Botezat, E. (1906). Die Nervenendapparate in den Mundteilen der Vögel. *Z. wiss. Zool.* **84**, 205–260.

Botezat, E. (1910). Morphologie, Physiologie und phylogenetische Bedeutung der Geschmacksorgane der Vögel. *Anat. Anz.* **36**, 428–461.

Bourgeois, N. (1896). Notes sur l'innervation respiratoire chez les oiseaux. *Arch. Biol.* **14**, 343–350.

Brandis, F. (1894). Untersuchungen uber das Gehirn der Vögel. II. Das Kleinhirn. *Arch. mikr. Anat.* **43**, 787–813.

Brown-Séquard, F. (1860). Récherches expérimentales sur la physiologie de la moelle allongée. *J. Physiol. Homme et Animaux.* **3**, 151–157.

Bubien-Waluszewska, A. (1968). Le groupe caudal des nerfs craniens de la Poule domestique. *Acta Anat.* **69**, 445–457.

Burlet, H. M. de (1935). Die Ungleichwertigkeit der Bogengange. *Z. Anat. Entwicklungsges.* **104**, 79–102.

Cordier, R. (1954). Le système nerveux-centrale et les nerfs cerebrospinaux. *Traité de Zoologie.* **12**, 202–232.

Couvreur, E. (1892). Sur le pneumogastrique des oiseaux. *Ann. Univ. Lyon.* **2**, H 3.

Craigie, E. H. (1928). Observations on the brain of the humming bird (*Chrysolampis mosquitus* Linn., *Chlorostilbon caribaeus* Lawr.) *J. Comp. Neurol.* **48**, 377–481.

Dijk, J. A. van (1930). The effect of stimulation of the cervical sympathetic cord upon the function of cross-striated muscle in the pigeon. *Arch. neerl. Physiol.* **15**, 126–137.

Dijkstra, C. (1933). De- und Regeneration der sensiblen Endkorperchen des Entenschnabels. *Z. mikros. anat. Forsch.* **34**, 75–158.

Dogiel, A. S. (1904). Ueber die Nervenendigung in der Grandoyschen und Herbstschen Korperchen. *Anat. Anz.* **25**, 558.

Dohlman, G. F. (1961). On the case for repeal of Ewald's second law. *Acta otolaryngol.* Suppl. 159; 15–24.

Dohlman, G. F. (1964). Secretion and absorption of endolymph. *Ann. Otolaryng.* **73**, 708–723.

Dooley, M. S. and Koppanyi, J. (1929). The control of respiration in the domestic duck, (*Anas boscas*). *J. Pharm. exp. Therap.* **36**, 507.

Doyle, W. L. and Watterson, R. L. (1949). The accumulation of glycogen in the "glycogen body" of the nerve cord of the developing chick. *J. Morph.* **85**, 391–404.

Duijm, M. (1951). On the head posture in birds and its relations to some anatomical features. *Proc. Kon. Ned. Akad. Wet. C.* **54**, 202–211; 260–271.

Duncan, C. J. (1960). The sense of taste in birds. *Ann. appl. Biol.* **48**, 409–414.

Duncan, C. J. (1962). Salt preference in birds and mammals. *Physiol. Zool.* **35**, 120–132.

Duncan, C. J. (1963). The response of the feral pigeon when offered the active ingredients of commercial repellants in solution. *Ann. appl. Biol.* **51**, 127–134.

Duncan, C. J. (1964). Sense of taste in the feral pigeon. *Anim. Behaviour.* **72**, 77–83.

Engelmann, C. (1937). Weitere Versuche uber den Geschmacksinn des Huhns. *Z. vergleich. Physiol.* **24**, 451–462.

Engelmann, C. (1960). Weitere versuche uber die Fütterwahl des Wasserflugels. Uber die Schmeckempfindlichkeit der Ganse. *Arch. Gefluchelzucht, Kleintierkunde.* **3**, 91–102.

Erulkar, D. S. (1955). Tactile and auditory areas of the brain of the pigeon. An experimental study by means of evoked potentials. *J. Comp. Neurol.* **703**, 420–458.

Fano, G. and Masini, G. (1894). Sur les rapports functionnels entre l'appareil auditifs et le centre respiratoire. *Arch. ital. Biol., Pisa* **27**, 309–312.

Fedde, M., Burger, R. E. and Kitchell, R. (1961). On the influence of the vagus nerve on respiration. *Poultry Sci. Abs.* p. 34.

Flourens, P. (1824). Récherches expérimentales sur les propriétés et les fonctions du système nerveux dans les animaux vertebrés." Crecot, Paris.

Foa, C. (1911). Récherches sur l'apnée des oiseaux. *Arch. ital. Biol., Pisa* **55**, 412–422.

Fowle, A. S. E. and Weinstein, S. (1966). Effect of cutaneous electric shock on the ventilation response of birds. *Amer. J. Physiol.* **210**, 293–298.

Francois-Frank, C. A. (1906). Études de mecanique respiratoire comparée. Analyses graphiques des mouvements du sternum et des cotes de l'abdomen chez les oiseaux. *C.R. Soc. Biol., Paris.* **2**, 370–372.

Fredericq, L. (1883). Expériences sur l'innervation respiratoire. *Arch. Physiol.* (1883). 51–68.

Friede, R. L. (1966). "Topographic Brain Chemistry," pp. 543. Academic Press, New York and London.

Friedlander, A. (1898). Untersuchungen uber das Ruckenmark und das Kleinhirn der Vögel. Neurol. Centralbl. 17, 397–409.

Gage, S. H. (1917). Glycogen in the nervous system of vertebrates. J. Comp. Neurol. 27, 451–465.

Graf, W. (1956). Caliber spectra of nerve fibres in the pigeon. (Columba domestica). J. Comp. Neurol. 105, 355–360.

Graham, J. D. (1940). Respiratory reflexes in the fowl. J. Physiol. 97, 525.

Gray, A. A. (1908). "The Labyrinth of Animals." 2 vols. Churchill, London.

Gray, Sir James (1968). "Animal Locomotion", pp. 479. Weidenfeld and Nicholson, London.

Grober, J. A. (1899). Uber die Atmungsinnervation der Vögel, Pflugers Arch. 76, 427–469.

Grunwald, J. (1904). Plethysmorgraphische Untersuchungen uber die Atmung der Vögel. Arch. Physiol. Suppl. B., 182–192.

Hadžiselimović, H. and Savković, L. (1964). Appearance of semi-circular canals in birds in relation to mode of life. Acta Anat. 57, 306–315.

Halpern, B. P. (1961). Gustatory nerve responses in the chicken. Amer. J. Physiol. 203, 541–544.

Hamilton, H. L. (1952). "Lillie's Development of the Chick." Henry Holt, New York.

Harkmark, W. (1954a). Cell migrations from the rhombic lip to the inferior olive, the nucleus raphe and the pons. J. Comp. Neurol. 100, 115–210.

Harkmark, W. (1954b). The rhombic lip and its derivatives in relation to the theory of neurobiotaxis. In "Aspects of Cerebellar Anatomy" (Jansen and Brodal, eds), pp. 264–284. Johan Grundt, Oslo.

Harman, A. L. and Phillips, R. E. (1967). Responses in the avian midbrain, thalamus and forebrain evoked by click stimuli. Exptl. Neurol. 18, 276–286.

Head, H., Rivers, W. H. R. and Sherren, J. (1905). The afferent nervous system from a new aspect. Brain 28, 99.

Herbst (1848). Quoted by Ramon y Cajal, 1911.

Hiestand, W. A. and Randall, W. C. (1941). Species differentiation in the respiration of birds following carbon dioxide administration, and the location of inhibitory receptors in the upper respiratory tract. J. Cell. Comp. Physiol. 17, 333–340.

Hiestand, W. A. and Randall, W. C. (1942). Influence of the proprioceptive vagal afferents on panting and accessory panting movements in mammals and birds. Amer. J. Physiol. 138, 12–15.

Heistand, W. A., Stemler, F. W. and Wiebers, J. E. (1953). Gasping patterns of the isolated respiratory centres of birds and mammals. Physiol. Zool. 126, 167–173.

Hinde, R. A. (1966). "Animal Behaviour; a Synthesis of Ethology and Comparative Psychology", pp. 534. McGraw-Hill.

Holmes, G. (1903). On the comparative anatomy of the nervus acusticus. Trans. Roy. Irish Acad. 13, 32–101.

Hunter, J. I. (1934). The influence of the sympathetic nervous system in the genesis of the rigidity of striated muscle in spastic paralysis. Surg. etc. 39, 721–743.

Huxley, F. M. (1913). On the reflex nature of apnoea in the duck in diving. II. The reflex postural apnoea. Quart. J. Exp. Physiol. 6, 159–182.

Ibragimova, Z. I. (1958). Sravniteljna anatomija kostanoga labirinta. VI. Kongr. anat. histol. embriol. U.S.S.R. Zborn. referata Kongr. 214–215.

Igarashi, M. and Yoshinobu, T. (1966). Comparative observations of the eminentia cruciata in birds and mammals. Anat. Rec. 155, 269–278.

Ingvar, S. (1918). Zur Phylo-und Ontogenese des Kleinhirns. *Folio neurobiol.* **11**, 205–495.

Izquierdo, V. (1879). Beiträge zur Kenntnis der Endigung der sensiblen Nerven. *Inaugural dissertation. Strasburg.*

Jones, D. R. (1969). Afferent vagal activity related to respiratory and cardiac cycles. *Comp. Biochem. Physiol.* **28**, 961–965.

Jones, G. M. and Spells, K. E. (1963). A theoretical and comparative study of the functional dependance of the semicircular canal upon its physical dimensions. *Proc. Roy. Soc. B.* **157**, 403–419.

Kappers, C. U. A. (1920–21). "Vergliechende Anatomie des Nervensystems". Bohn, Haarlem.

Kappers, C. U. A., Huber, G. C. and Crosby, E. C. (1936). "The Comparative Anatomy of the Nervous System of Vertebrates including Man." Macmillan, London. (1960 reprint, Hafner and Co. 3 vols.)

Kare, M. R. and Pick, H. Jr. (1960). The influence of the sense of taste on feed and fluid consumption. *Poultry Sci.* **39**, 697–706.

Karten, H. J. (1963). Ascending pathways from the spinal cord in the pigeon (Columba livia). *Proc. Int. Congr. Zool.* **2**, 23.

Karten, H. J. (1968). The ascending auditory pathway in the pigeon (*Columba livia*). II. Telencephalic projections of the nucleus ovoidalis thalami. *Brain Res.* **11**, 134–153.

Karten, H. J. and Hodos, W. (1967). "A Stereotaxic Atlas of the Brain of the Pigeon (*Columba livia*)", pp. 193. Johns Hopkins University Press.

Kilmer, W. L., McCulloch, W. S. and Blum, J. (1969). Embodiment of a plastic concept of the reticular formation. *In* "Biocybernetics of the Nervous System" (Proctor, L. D., ed.), pp. 213–260. Little, Brown & Co., Boston.

King, A. S., Molony, V., McLelland, J., Bowsher, D. and Mortimer, M. F. (1968a). Afferent respiratory pathways in the avian vagus. *Experientia.* **24**, 1017.

King, A. S., Molony, V., McLelland, J., Bowsher, D., Mortimer, M. F. and White, E. S. (1968b). On vagal afferent control of avian breathing. *J. Anat.* **104**, 182.

Kitchell, R. L., Strom, L. and Zotterman, Y. (1959). Electrophysiological studies of thermal and taste reception in chickens and pigeons. *Acta Physiol. Scand.* **46**, 133–151.

Klein, M. (1931–32). Sur la différentiation des élements tactiles dans le neurone d'amputation des nerfs du bec de Canard. *Arch. Anat. Hist. Embr. Strasburg.* **14**, 267–300.

Knoll, P. (1880). Uber Myocarditis und die ubrigen Folgen der Vagussektion bei Tauben. *Prag. Z. Heilk.* **1**, 180–254.

Krause, R. (1906). *Gehörorgan. Handb. Entw. lehre wirbeltiere.* II. Part I.

Kuhn, A. and Trendelenburg, W. (1911). Die exogenen und endogenen Bahnen des Ruckenmarks der Taube. *Arch. Anat. Physiol. Leipzig.* 35–48.

Lakjer, T. (1926). Studien uber die trigeminusuersogte Kaumuskulatur der Sauropsiden Copenhagen.

Lamettrie (1751). Oevres philosophiques. Cited by Ten Cate (1936).

Langelaan, J. W. (1915). On muscle tonus. *Brain.* **38**, 235–380.

Larsell, O. (1948). The development and subdivisions of the avian cerebellum. *J. Comp. Neurol.* **89**, 123–190.

Legallois, C. (1812). Expériences sur le principe de la kie. Paris.

Lervold, A. M. and Szepsonwold, J. (1963). Glycogenolysis in aliquots of glycogen bodies of the chick. *Nature, Lond.* **200**, 81.

Lewin, M. A. (1955). Zavisimost anatomiceskogo stroenia kostnogo labirinta ptić ot obraza i žižni. *Zool. Zh. Akad. Nauk. U.S.S.R.* **34**, 3.

Malinovsky, L. (1962). Contribution to the anatomy of the vegetative nervous system in the neck and thorax of the domestic pigeon. *Acta Anat.* **50,** 326–347.

Masai, H. (1961). Comparative neurobiological studies on the glycogen distribution in the central nervous system of sub-mammals. *Yokohama Med. Bull.* **12,** 239–260; 261–264.

Merckel, F. (1875). Tastzellen und Tastkörperchen bei den Haustieren und beim Menschen. *Arch. mikros. Anat.* **11,** 636.

Meyer, O. (1884). Uber den Glycogenhalt embryonaler und jugendlicher Organe. Dissertation. Breslau University.

Moore, C. A. and Elliott, R. (1946). Numerical and regional distribution of taste buds on the tongue of the bird. *J. Comp. Neurol.* **84,** 119–131.

Nagel, W. A. (1901). Uber kunstliche Atmung mit kontinuierlichen Luftström bei Vögeln. *Zbl. Physiol.* **14,** 553–555.

Nikolaides, R. (1914). Untersuchungen uber die Regulierung der Atembewegungen der Vögel. *Arch. Physiol.* 553–564.

Noback, C. R. and Shriver, J. E. (1966). Phylogenetic and ontogenetic aspects of the lemniscal systems and the pyramidal system. *In* "Evolution of the Forebrain" (Hassler, R. and Stephan, H., eds), pp. 316–325. Georg Thieme, Stuttgart.

Olivo, D. M. (1925). Do alcuni caratteri fisici e chimici del corpe glicogenico del midollo lombo-sacrale degli ucelli. *Boll. Soc. Ital. Biol. Sper.* **1,** 81–84.

Orr, J. B. and Watson, A. (1913). Study of the respiratory mechanism in the duck. *J. Physiol.* **46,** 337–348.

Oscarsson, O., Rosen, I. and Uddenberg, N. (1963). Organisation of the ascending tracts in the spinal cord of the duck. *Acta Physiol. Scand.* **59,** 143–153.

Pannese, E. (1968). Developmental changes of the endoplasmic reticulum and ribosomes in nerve cells of the spinal ganglia of the domestic fowl. *J. Comp. Neurol.* **132,** 331–364.

Papez, J. W. (1929). "Comparative Neurology." Crowell and Co., New York.

Paton, D. M. (1913). The relative influence of the labyrinthine and cervical elements in the production of postural apnoea in the duck. *Quart. J. Exp. Physiol.* **6,** 197–207.

Paton, D. M. (1927). I. Submergence and postural apnoea in the swan. II. Reflex postural adjustments of balance in the duck. *Proc. Roy. Soc., Edinburgh* **47,** 283–293.

Portmann, A. (1950). Oiseaux. *Traité de Zoologie.* Vol. 15. Masson, Paris.

Portmann, A. (1961). Sensory organs, Part I. *In* "Biology and Comparative Physiology of Birds", Vol. 2. (Marshall, A. J. ed.), pp. 37–54. Academic Press, London and New York.

Portmann, A. and Stingelin, W. (1961). The central nervous system. *In* "Biology and Comparative Physiology of Birds". (Marshall, A. J. ed.), vol. 2, pp. 1–36. Academic Press, London and New York.

Pumphrey, R. J. (1961). Sensory organs, Part 2. *In* "Biology and Comparative Physiology of Birds". (Marshall, A. J., ed.), vol. 2, pp. 55–86. Academic Press, London and New York.

Ramon y Cajal, S. (1908). Los ganglios centrales del cerebelo de las aves. *Trab. lab. biol. Madrid.* **6,** 195.

Ramon y Cajal, S. (1909–11). "Histologie du Système Nerveux de l'Homme et des Vertebrés." Maloine, Paris.

Retzius, O. (1884). "Das Gehörorgan der Reptilien, Vögel und Saugetiere." Stockholm.

Richards, S. A. (1968). Vagal control of thermal panting in mammals and birds. *J. Physiol.* **199,** 89–101.

Richards, S. A. (1970). The biology and comparative physiology of thermal panting. *Biol. Rev.* **45,** 223–264.

Rogers, K. T. (1957). Ocular muscle proprioceptive neurons in the developing chick. *J. Comp. Neurol.* **107**, 427–435.

Rogers, K. T. (1965). Development of the XIth or spinal accessory nerve in the chick. *J. Comp. Neurol.* **125**, 273–286.

Saalfeld, F. E. von (1936). Untersuchungen uber das Hacheln bei Tauben. *Z. vergleich. Physiol.* **23**, 727.

Salmoiraghi, G. C. and Burns, B. D. (1960). Notes on the mechanism of rhythmic respiration. *J. Neurophysiol.* **23**, 14–26.

Sanders, E. V. (1929). A consideration of certain bulbar, midbrain and cerebellar centres and fibre tracts in birds. *J. Comp. Neurol.* **49**, 155–221.

Schartan, O. (1938). Die periphere Innervation der Vögelhaut. *Zoologica.* **95**, 1–17.

Schwartzkopf, J. (1955). Schallsinnesorgane, ihre Funktion und biologische Bedeutung bei Vögeln. *Acta. 11th Congr. Intern. Ornithol.* 1954, pp. 189–208.

Schwartzkopf, J. (1968). Hearing in birds. CIBA symposium on "Hearing in Vertebrates", pp. 41–59.

Sherrington, C. S. (1906). "The Integrative Action of the Nervous System." Scribner Sons, New York.

Shimazono, J. (1912). Das Kleinhirn der Vögel. *Arch. mikrok. Anat.* **80**, 397–449.

Siefert, E. (1896). Uber die Atmung der Reptilien und Vögel. *Pflugers Arch.* **64**, 321–506.

Sina, M. P. (1958). Vagal control of respiration as studied in the pigeon. *Helv. Physiol. Pharm. Acta.* **16**, 58–72.

Sinclair, D. A. (1955). Cutaneous sensations and the doctrine of specific energy. *Brain.* **78**, 584–614.

Solotuchin, A. A. (1925). Vergleich Anatomie der Vögelgleichgewichtsorgan. *Proc. Z. Congr. Zool. Anat. Histol. U.S.S.R.* 219–220.

Starck, D. (1940). Beobachtungen an der Trigeminusmuskulatur der Nashornvögel nebst Bemerkungen uber einige Besonderheiten des Vögelschadels. *Morph. Jahrb.* **84**, 585–623.

Steding, G. (1962). Experimente zur Morphogenese des Ruckenmarks. *Acta Anat.* **49**, 199–231.

Stehlik, V. (1927). Regulation der Atmung bei Vögeln. *Zit. nach. Physiol. Abs.* **11**, 217.

Stingelin, W. (1965). Qualitative und quantitative Untersuchungen an Kerngebieten der Medulla oblongata bei Vögeln. *Bibliotheca Anatomica.* **6**, 1–116.

Stopp, P. E. and Whitfield, I. C. (1961). Unit responses from the brain stem nuclei in the pigeon. *J. Physiol.* **158**, 165–177.

Stresemann, E. (1933). Aves. *In* "Handbuch der Zoologie", vol. 7.

Stübel, H. (1910). Beiträge zur Kenntnis der Physiologie der Blutkreislaufes bei verschiedenen Vögelarten. *Pfluger's Arch.* **135**, 249–365.

Sturkie, P. D. (1965). "Avian Physiology", pp. 766. Ballière, Tindall and Cox, London.

Szymonowicz, L. (1897). Uber den Bau und die Entwicklung der Nervenendigung im Entenschnabel. *Arch. mikr. Anat.* **48**, 329.

Ten Cate, J. (1936). Physiologie des Zentralnerven-systems der Vögel. *Ergebnisse der Biologie.* **13**, 93–173.

Terni, T. (1924). Richerche sulla considetta sostanza gelatinosa (corpo glicogenico) del midollo lombo-sacrale degli ucelli. *Arch. Ital. Anat. Embriol.* **21**, 55–86.

Terni, T. (1926). Sui nuclei marginali del midollo spinale dei Sauropsidi. *Arch. Ital. Anat. Embriol.* **23**, 610–628.

Tiegs, O. W. (1931). Note on the posture of the birds wing and its supposed control by sympathetic nerves. *Amer. J. Physiol.* **98**, 547–550.

Tossetti, Urb. (1755). Lettera secunda al Signor Dottore G. Valdambrini, medico primario in Cortona. Rome, 9 April (cited by Ten Cate).

Trèves, Z. and Maiocco, F. (1905). Osservazioni sull'apnoea degli ucelli. *Arch. Fisiol.* **2**, 185–206.

Turkewitsch, B. G. (1931). Beitrage zur Frage uber die Abhangigkeit der Anatomischen Struktur des Knochenlabyrinthes von der Lage des Körpers und Grad der Beweglichkeit der Vögel. *Russ. J. Physiol.* **14**, 71–76.

Viet, F. (1911). Der Einfluss der Apnoe auf die Errigbarkeit der Nervenzentren. *Sitz. Abh. naturf. Ges. Rostock.* N.F. **3**, 299–314.

Vinnikov, Y. A. (1969). The ultrastructural and cytochemical bases of the mechanism of function of the sense organ receptors. *In* "The Structure and Function of Nervous Tissue" (Bourne, G. H. ed.), vol. 2, pp. 265–392. Academic Press, London and New York.

Wallenberg, A. (1903). Die Ursprung des Tractus isthmo-striatus der Taube. *Neurol. centralbl.* **22**, 98–101.

Wallenberg, A. (1904). Neue Untersuchungen uber den Hirnstamm der Taube. *Anat. Anz.* **24**, 357–369.

Watterson, R. L. (1949). Development of the glycogen body of the chick spinal cord. I. Normal morphogenesis, vasculogenesis and anatomical relationships. *J. Morph.* **85**, 337–390.

Watterson, R. L. (1952a). Development of the glycogen body of the chick spinal cord. III. The paired primordia as revealed by glycogen-specific stains. *Anat. Rec.* **113**, 29–52.

Watterson, R. L. (1952b). The glycogen body as a derivative of the avian neural tube. *Biol. Bull.* **103**, 389–390.

Wechsler, W. and Kleihues, P. (1968). Protein metabolism and cytodifferentiation in the nervous system. *In* "Macromolecules and the Function of the Neuron" (Lodin, Z. and Rose, S. P. R., eds), *Excerpta Medica*, pp. 73–90.

Werner, C. F. (1939). Die otolithen im Labyrinth der Vögel, besonders beim star und der Taube. *J. Ornithol.* **87**, 10–23.

Wersall, J. (1961). Discussion to H. Engstrom. "The innervation of the vestibular sensory cells." *Acta oto-laryngol.* Suppl. **163**, 30–41.

Whitlock, D. G. (1952). A neuro-histological and neurophysiological study of the afferent fibre tracts and receptive areas of the avian cerebellum. *J. Comp. Neurol.* **97**, 567–635.

Winkelman, R. K. and Myers, T. T. III (1961). The histochemistry and morphology of the cutaneous sensory end organs of the chicken. *J. Comp. Neurol.* **117**, 27–31.

Winter, P. (1963). Vergleichende qualitative und quantitative Untersuchungen an der Hörbahr von Vögeln. *Z. Morph. Okol. Tiere.* **52**, 365–400.

Winter, P. and Schwartzkopf, J. (1961). Form und Zellzahl des akustischen Nervenzentren in der Medulla oblongata von Eulen (Strigen). *Experientia* **17**, 515–516.

Winterstein, H. (1921). "Handbuch der Vergleichenden Physiologie", vol. I, part II. Fischer, Jena.

Young, (1969). Biocybernetics of the vestibular system. *In* "Biocybernetics of the Nervous System" (Procter, L. D. ed.), pp. 79–117. Little, Brown and Co., Boston.

Zeigler, H. P. and Witkovsky, P. (1968). The main sensory trigeminal nucleus in the pigeon. A single unit analysis. *J. Comp. Neurol.* **134**, 255–264.

# 7 The Medullary Efferent Systems, The Olive, The Pons and Allied Structures

## I. INTRODUCTION

Since Kappers first enunciated his theory of neurobiotaxis the medullary motor nuclei have been intimately associated with general concepts of brain organization. Some of these are based on definitive factual information but others are the results of often stimulating, but speculative hypotheses. In the case of the theory of neurobiotaxis, although more recent embryological work has relegated certain of its tenets to the realm which is the prerogative of historians of science, it has had a far-reaching influence on neurological theories and in certain respects it remains a useful and descriptive phylogenetic generalization. Kappers first put forward his theory in 1907 and then elaborated it further in the following year (1908a,b). It was largely summarized in his book (1920, 1921), in his paper (1917), and in the chapter which he contributed to Penfield's book (1932). From that period onwards it is implicit in many of the statements of his later work, notably Kappers *et al.* (1936) and Kappers (1947). Some brief reference to it at this point is essential if the currents of thought which underlie many of the works published in the 1920s and 1930s are to be understood. In general there are two principal parts to the theory. The first is concerned with the apparent shifts which have occurred in the situation of the motor nuclei during phylogeny; the second relates to the factors which have contributed to the union of simultaneously excited fibres—fasciculation—which tend to have a comparable anatomical path.

In the first place Kappers suggested that if the amount of stimulation which is reaching a motor nucleus from a certain source increases during phylogeny, then that nucleus will change its location in the brain and undergo a migration towards the source. It is as well to emphasize that, as this refers to long-term phylogenetic changes, the term "migration" is being used in a figurative sense. We merely find that the nucleus in a given taxon is situated in a different position from that which it occupies in the supposedly ancestral forms. It was, however, originally envisaged that actual topographical migrations always occurred during ontogeny. In view of the very considerable emphasis that was at one time placed on the theory of ontogeny recapitulating phylogeny, and the relatively slow pace at which suggestions spread out from their discipline of origin to other disciplines, this is a clearly comprehensible situation. Some such cell migrations certainly do occur. For example, the nucleus of cranial nerve IV (*q.v.*) is frequently reported as occupying in the embryo an initial position behind that of nerve III. In most vertebrates it subsequently undergoes a change of position and reaches forward towards the latter, and in a similar way the abducens nucleus also undergoes an apparent forward movement during embryological differentiation and development. Besides the very large number of such data which were adduced by Kappers, others have been added by, for example, Addens (1933), and the overall tenets are widely cited by anatomists (see Grays, 1958).

As far as relevant embryological events of the avian brain are concerned the principal consideration which has been undertaken in the last 20 years is that of Harkmark (1954a,b, 1956). He concluded that the neurobiotactic theory certainly did not explain the very marked cellular migrations which take place during ontogeny from the region of the rhombic lip. He pointed out that the migrating cell strands which he observed consisted of young and undifferentiated cells. They migrate from a site in which they have no paths on their dorsal side, in a ventro-medial direction and therefore towards those loci from which they ultimately receive impulses. He considered that according to the strict tenets of Kappers theory the migration of such cells at this stage should be stimulofugal not stimulopetal. However, it would be foolish to permit these very specific supposed inadequacies to obscure the tremendous value that the neurobiotactic theory has had over the last 60 years in providing an intuitive conceptual framework for comparative neurological studies. A summary of the various nuclei is provided by Stingelin (1965) and Jungherr (1969).

## II. THE HYPOGLOSSAL COMPLEX

The distribution of the vertebrate cranial nerves, and their segmental origin, which is most easily interpreted by a study of the conditions in the Pteraspidomorpha, Cephalaspidomorpha and aphetohyoid fishes, are summarized in Fig. 52. They were surveyed by Portmann (1950), Cordier (1954) and Portmann and Stingelin (1961). As in the case of the afferent systems the efferent nuclei and pathways which are associated with the avian medullary region present some variations from those which one is accustomed to, in, say, mammals. As with other such avian peculiarities these have in the past been cited as representative of an organization which is intermediate between that of reptiles on the one hand and mammals on the other.

Typically the muscle buds in the more anterior of the post-otic somites of vertebrates are found, not in the dorsal region, but in the ventral region and the hypoglossal nerve serves them. From Fig. 52 it can be seen that this represents the ventral root components of the more posterior segments of the vagus–accessorius series and its general origin from the medulla floor indicates that it is a ventral root (Young, 1950). In birds this hypoglossal component has been particularly controversial owing to its site of origin being involved in the somewhat confused occipital/anterior cervical region of the embryo. The muscle buds with which it is associated contribute to the tongue and syrinx and during development some 5 ventral roots have been described in association with them. In the adult there are only 2–3 and during maturation the processes of assimilation are not always easy to follow.

Bubien-Waluszewska (1968) has summarized the peripheral hypoglossal components as comprising two principal axonal bundles. These join to form a single trunk close to the atlas vertebra. Of the two it is the dorsal one which predominates and has an anastomosis with the first cervical root. A descending branch from a further anastomosis with the second cervical root joins the fourth cervical. This resembles the *ansa cervicalis* of some mammals from which, in man, branches pass to the sternohyoid, sternothyroid and the inferior belly of the omohyoid muscle.

In the pigeon the position of the nuclear structures which are associated with this nerve complex have been accurately mapped in terms of stereotaxic coordinates by Karten and Hodos (1967). They occur in association with the central medullary nucleus and are situated laterally in relation to the median longitudinal fasciculus against which the hypoglossal nucleus is closely apposed (Fig. 50). Towards the hind end of its longitudinal extent its size increases. In this region it is in close proximity to the ventral funiculus below, and is overlain by the very reduced posterior prolongation of the ventral part of the central medullary nucleus. Such a relatively straightforward description belies the very considerable discussions that have taken place in the past about central hypoglossal representation. Earlier workers contrasted the condition which they observed in birds and reptiles. The latter have a discrete nucleus for the homologous nerve and this is situated within the somato-motor region of the medullary bulb. The suggestion that in birds part of this has become separated from the main mass and forms a separate unit, the intermediate nucleus, interposed in the direction of the vagal-glossopharyngeal complex has been the basis of much discussion. Amongst the older papers those of Kosaka and Yagita (1903), Kappers (1911, 1912, 1947), Beccari (1923), Black (1922), Groebbels (1922), Sanders (1929), Addens (1933) and Kappers *et al.* (1936) all identified and argued about various elements which they attributed to a hypoglossal complex and other medullary efferent structures. Kappers (1912) first suggested that the somatic efferent column of the spinal cord was continued into the medulla and represented there by both a ventral and dorsal column. Beccari and Addens were responsible for demonstrating that the ventral component was divisible anatomically into a smaller, and more dorsal, hypoglossal nucleus, together with a larger and more ventral prolongation of the ventral horn grey.

Kappers (1912) also concluded that in many genera there exists a close relationship with the vagal region. Kappers et al. (1936) thought that this region of proximity was somewhat posteriorly situated in the Anseriformes and Columbiformes, and that it was minimal in the chick. Such close topographical relationships have also been described as even more intimate with cell bodies of vagal origin attributed to the intermediate nucleus in, for example, the Sphenisciformes. Nevertheless correlations which have been shown to exist between the relative size of this last-named nucleus and the relative importance of the musculature which is served by the hypoglossal, have been taken as suggesting an involvement with this nerve. It is a conspicuous medullary component in psittaciform species such as the budgerigar but less prominent in, for example, the sparrow. Specific details of the relationships in *Coturnix* are provided by Fitzgerald (1969).

## III. THE SPINAL ACCESSORY NERVE

For a very long period in the history of avian brain studies the existence of a discrete spinal accessory nerve was strenuously denied and it was written off as being represented by a component of the vagus. In mammals it is formed by the union of cranial and spinal roots and represents the separated caudal rootlets of the vagus. The separation is, even here, only partial, because after a very short common course the cranial components join the vagus and are distributed with its branches. Addens (1933) considered that birds closely resembled reptiles in the structure of their spinal accessory and that in most genera its nucleus was a posterior continuation of that of the XIIth nerve. In the fowl he thought that it had become free, although remaining attached to the XIIth nucleus during embryonic development. Lying at the level of the 2nd and 3rd cervical roots he described the fibres leaving it as running upwards to emerge in the angle which is formed by the lateral and dorsal aspects of the spinal cord. Once they had emerged he considered that they passed forwards to join the vagus so that the most posterior of them traversed the second cervical ganglion. It was left to Rogers (1965) to demonstrate unequivocally that a definite spinal accessory nerve exists in birds.

From a study of the embryological development of the posterior cranial and cervical regions Rogers dispelled any doubts about the existence of an avian spinal accessory. Developmentally nerve fibres emerge at stage 19 in the chick, and do so simultaneously from both the posterior occipital and upper two cervical levels. By stage 21 they unite to form a trunk for the XIth nerve and their growth at this time is towards the vagal root. During the later stages of ontogeny the caudo-rostral increase in the fibre contribution to the trunk of XI is closely correlated with the number of fibres which are added by emerging rootlets at the upper cervical and lower occipital levels. Rogers interpreted this as arguing against any substantial contribution to the nerve from cells which are located in the root ganglion of the vagus. He further suggested that the motor fibres of the spinal accessory can probably be considered as the anatomical equivalents of the fibres of von Lenhossek at the 2nd to 4th cervical levels. These had figured large in doctrinal discussions of avian neurology during the years after their initial description.

Following this description the principal difference that one can observe between the avian and mammalian spinal accessory nerves is the more posterior origin of the spinal portion in mammals. In the chick it merges with the ganglion of the second cervical, and, during the period that the first cervical ganglion is present, passes straight through it. The observations of Kaupp (1918) and Watanabe (1961) which suggest that it innervates the anterior part of the *m. cutaneous colli* lateralis, also imply that in birds the XIth nerve is a true spinal accessory. Indeed, Watanabe suggested that this muscle is the avian homologue of the mammalian trapezius on the basis of this innervation. However, the cervical origin of XI is less extensive in birds than in mammals (see Lubosch, 1899 for the condition in the Strigiformes) so that it can obviously only represent a relatively small part of the components which are contributed to its mammalian homologue.

# IV. THE VAGAL AND GLOSSOPHARYNGEAL NERVES

## A. Central Relationships

The position of the dorsal motor nucleus of the vagus is shown in Fig. 50. Merging with the glossopharyngeal nucleus in its more anterior region it is also closely apposed to the intercalated nucleus within the same general area. Further back the vagal nucleus itself leaves the rather latero-dorsal position which all these three occupy in front, and lies above the intermediate nucleus and very close to the small central canal. In this region the mass of the median longitudinal fasciculus is aggregated ventrally and the vagal nucleus disappears whilst the hypoglossal nucleus, inserted in the angle of the fasciculus, continues to increase in transverse diameter. In view of their continuity Fig. 50 also indicates the position which is assumed by the glossopharyngeal further forward. At its most anterior extent this last is both overlain by, and in close apposition to, the principal median component of the vestibular nuclear complex.

The older anatomists concluded that as in the diapsid reptiles the caudal efferent column of the avian central nervous system has a relatively lateral position. As a direct result of this they suggested that the vagal efferents first of all pass to the lateral region of the median longitudinal fasciculus (Fig. 50)

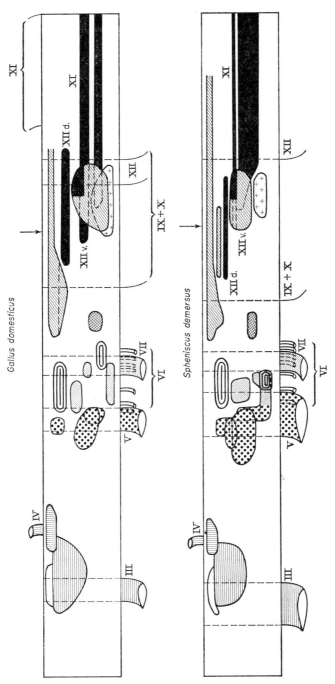

FIG. 64. The distribution of medullary nuclear masses according to Addens (1933).

and then run laterally to give a caudally directed loop. The nerve can actually leave the cranium at the same point as the glossopharyngeal (Portmann, 1950) but Bubien-Waluszewska (1968) has described independent foramina at least in the chick (see Chapter 6, Section III A).

In *Anas, Gallus, Columba* and *Passer*, Sanders (1929) described three distinct groups of cells within the efferent vagal nucleus. The most anterior of these had large multipolar cells, the ventro-medial had rather smaller ones which were comparable with those of the dorsal glossopharyngeal nucleus, whilst the ventro-lateral had an even more micro-cellular composition. He concluded that it was from this last component that the bulk of the vagal fibres originate. In contrast he only found one cell type in the vagal nuclear complex of the parakeet. Craigie (1928) emphasized that in both of the trochiliform genera *Chlorostilbon* and *Chrysolampis*, the dorsal glossopharyngeal and vagal motor nuclei could only be identified with difficulty, and Addens (1933) considered that the dorsal glossopharyngeal nucleus was simply a vagal component. He identified a ventral glossopharyngeal nucleus and this suggestion was tentatively confirmed by Craigie (1928) for the Trochiliformes. It is obvious that Addens subscribed to different interpretations of vagal-glossopharyngeal relationships from those which are generally accepted. He did, however, conclude that the IXth nucleus was single. A summary of the distribution of such nuclei in the longitudinal direction is provided by Fig. 64.

The vagus is a mixed sensori-motor nerve in all vertebrates. Its known functions in the control of pulmonary ventilation were outlined in Chapter 6, Section III C. These are limited to afferent impulses concerned with a vagal drive mechanism. The motor control of cardiac and alimentary canal function is outlined below. One may, however, note at this point that Dahl *et al.* (1964) demonstrated the diverse form of its constituent fibres. An analysis of the action potentials which can be detected within it suggested two main fibre populations. They assumed that these were non-myelinated C, and small, myelinated B types. The first elevation which one can record is generated by the B fibres, the second by the non-myelinated ones.

## B. The Vagal Control of the Heart Rate

It was emphasized in Chapter 6, Section III C that the principal vagal branches go to the heart, lungs and gut. As in the case of presumed pulmonary ventilation rate control many older workers considered the problem of the vagal control of cardiac activity and the best known contributions from an historical standpoint are those of Claude Bernard (1858), Einbrodt (1859), Knoll (1880), Couvreur (1892), Jurgens (1909), Stübel (1910) and Paton (1912). In nature there are distinct seasonal circadian variations in avian blood pressure (Weiss *et al.* 1961) which suggest the involvement of diencephalic and other factors. As far as the vagus is conerned Johansen and Reite (1963), who worked on both the gull *Larus argentatus*, and also ducks, showed that bilateral vagotomy caused both tachycardia and a rise in the systemic blood pressure. Butler (1967) using chickens, found that bilateral cervical vagotomy and adrenergic receptor blockade gave *no* tachycardia. Complementarily, Johansen and Reite demonstrated that direct electrical stimulation of the peripheral ends of the cut vagus induces bradycardia. Again it is worth noting that bradycardia occurs naturally during hypothermia in the Caprimulgiformes and was seen in the giant humming bird by Lasiewski *et al.* (1967). In the homeothermal condition the heart rate in the latter species, *Patagona gigas*, ranges from 300–1020/min. It is thus comparable with the rates recorded from *Estrilda troglodytes* (500–1020) and *Melospiza melodira* (450–1020). During the torpid condition which was induced by keeping it in the dark with no food for 12–24 hours at 20°C, the rate fell to 60/min. Similar changes have also been recorded in the Inca dove, *Scardafella inca*, during nocturnal hypothermia in poor nutritional states (MacMillen and Trost, 1967). The opposite condition occurs in hyperthermia. Whittow *et al.* (1964) observed an increase in cardiac output when chickens were exposed to ex- perimentally induced rises in body temperature. These increases were as- sociated with, but occurred at a higher rectal temperature than, rises in the respiratory rate and they reflected an increase in both the stroke volume and heart rate. That afferent impulses of vagal origin are involved in such changes was shown by Jones (1969). These compare with those involved in the Hering- Breuer reflex.

The relative contributions which are made by the vagal nerves and the cardiac sympathetic system to the experimentally induced conditioning of heart rate after light/shock presentations were thought by Cohen and Pitts (1968) to be principally on the side of the sympathetics. The shortest response latency, was, nevertheless, a reflection of vagal control and either system is capable of mediating the differentiation following light/shock pairing and the non-reinforced presentation of light of a different colour.

Variations in heart rate that accompany the phases of the respiratory cycle have been known since the last century. McCrady *et al.* (1966) have surveyed the situation. Cardiac acceleration accompanies inspiration, cardiac slowing occurs during expiration (see Koppányi and Dooley, 1928). It has been suggested that this is the result of a wave of neural excitation which spreads from the inspiratory to the cardiac accelerating foci, or from expiratory to cardiac inhibitory centres. The first would stimulate or the second depress heart rate. McCrady *et al.* concluded that such phenomena involve vagal afferent pathways carrying impulses to the "cardiac regulatory centres"; that central radiation of impulses from respiratory to cardiac centres occurs; and that a sympathetic outflow to the heart is synchronized with respiration. Many investigators have noted that, whilst unilateral vagal transection does not abolish the respiratory heart rate, bilateral section does.

Butler and Jones (1968) were unable to obtain any evidence for a postural reflex cardiac control in ducks and they were of the opinion that neither the position of the head, nor the temperature of the water, affected the cardiac responses which occur during diving. The level of the heart rate appeared to be closely related to the respiratory frequency and no bradycardia occurred if the bird was submerged but permitted to breathe through a tracheal cannula. When apnea and bradycardia did occur during submersion the first inspiration which occurred after surfacing was some 2–3 times larger than normal, and was accompanied by an instantaneous rise in heart rate. Atropine abolished the diving bradycardia, as did vagal cold block. Only one vagal trunk seemed to be involved in this chronotropic control at any one time and at that time it also appeared to be an important factor in the control of respiratory frequency. Butler and Jones concluded that, as $\beta$-adrenergic receptor blockade had no effect on either diving bradycardia or the post-dive tachycardia, the cardiac response was solely a result of parasympathetic vagal activity.

Following the foregoing considerations Cohen et al. (1970) and Cohen and Schnall (1970) have demonstrated quite clearly that the vagus has a cardio-inhibitory function in the pigeon. Retrograde degeneration studies following cervical vagotomy indicated an incompletely inverted topographical representation of the vagus in the dorsal motor nucleus. No evidence for a nucleus ambiguus emerged. The vagal cardio-inhibitory fibres appear to be represented throughout the rostral half of the nucleus but are particularly concentrated in the ventral portion approximately 0·75 mm in front of the obex. Electrical stimulation of the dorsal medullary regions gives bradycardia. This is particularly true of the lateral dorsal motor nucleus, the solitary tract, and the commissural nucleus of Cajal. When elicited from regions other than the dorsal motor nucleus such bradycardia is frequently accompanied by apnea. Cohen and Schnall therefore concluded that the efferent cardio-inhibitory fibres are restricted to the lateral dorsal motor nucleus and that the bradycardia resulting from stimulation elsewhere reflects activation of afferent units.

## C. The Vagal Control of Alimentary Canal Motility

An alternative name for the vagus nerve is the pneumogastric and as this name implies the nerve supplies branches to the gut as well as to the heart and lungs. That it is certainly implicated in the control of movements within the alimentary canal was shown by Nolf (1927) who reported that bilateral vagal transection results in a permanent lowering of gastric motility, and that a similar, but temporary, reduction of this nature follows the application of hemilateral ligatures. Groebbels (1932) also reported evidence for the presence of both excitatory and inhibitory fibres to the gizzard. Nevertheless Nolf (1929, 1934) could find no incontrovertible evidence for changes in the pattern of rhythmic intestinal movements following vagal transection, although there were indications of a slight slowing of aboral waves. He did, however, find that vagal stimulation increased motility.

## V. THE FACIAL AND TRIGEMINAL NUCLEAR COMPLEXES

As a working generalization one can say today that the various components of the facial and trigeminal motor nuclei all lie laterally, or dorso-laterally in respect to the gigantocellular part of the caudal pontine reticular nucleus. Alternatively they can be described as falling on a convex arc that runs from the lateral pontine nucleus (q.v.) ventrally, to the median longitudinal fasciculus above. Posterior to the gigantocellular pontine constituent the facial moieties are lateral to the paramedian nucleus from which they are separated by fibres of nerve VI. Complementarily, further forward the trigeminal motor masses overlie the ventral sub-coerulean nucleus with fibres of the occipito-mesencephalic tract passing above them (see Fig. 140).

As was the case with the vagal and glossopharyngeal nuclei the central representation of the facial and trigeminal nerves has been the subject of both discussion and controversy during the last 70 or so years. Kappers et al. (1936) placed great emphasis on their contention that the avian facial nuclear complex differs from that of other vertebrates in lying in front of the transverse plane at which the nerve arises from the brain, not behind it. According to their scheme the situation in the large, cursorial, ratite genera such as *Casuarius* is different from that in most neognaths and resembles that in *Alligator*. In any case there is some variation in the degree of independence which is exhibited by its component parts, and their close relationship with the trigeminal motor nucleus. Addens (1933) considered that there are only two nuclear groupings associated with the facial in at least some *Sphenisciformes* and *Ciconiiformes*, but three in the genera *Gallus* and *Cacatua* together with other *Psittaciformes*.

As the result of this supposedly rather anterior position of the facial motor nucleus in birds it has long been cited as an example of neurobiotactic influences within the medulla. One of the major discussions of this topic is that of Addens (1933) who had rather different ideas on nuclei to those of earlier workers. In the chick and cockatoo he considered that there were large dorsal and ventral nuclei and a smaller intermediate one. In view of the results of the degeneration studies of Kosaka and Hiraiwa (1905) he concluded that the dorsal nucleus was concerned with the innervation of the *m. depressor mandibulae*, the intermediate with the *posterior mylohyoideus*, and the ventral with the *sub-cutaneous colli*. He suggested that in *Cacatua roseicapilla* the dorsal nucleus contributes fibres from its anterior half to a bundle which joins the motor root of the trigeminal. Leaning on Kappers's theory of neuro-biotaxis he explained such a relationship in terms of fasciculation. In opening the beak the *depressor mandibulae* which is represented in the facial nucleus collaborates with the *orbito-quadratus* which is a trigeminal muscle. As the depressor mandibulae pulls down the lower jaw the upper is lifted to a varying extent by the orbito-quadratus. The greater degree of mobility which occurs in the upper jaw in the kinetic Psittaciformes led, said Addens, to the union of the relevant parts of VII and V. The overall distribution of such components is shown in Fig. 64.

The compound nuclear complex from which the trigeminal outflow origin-ates was also said by the older anatomists to have a more reptilian form in the Casuariiformes. This comprised two efferent foci, the larger of which occupied a relatively dorsal position close to the ventricular floor. The smaller lies in a more ventro-lateral position. They are, however, again widely separated from the dorso-lateral position of the main sensory nucleus, which overlies fibres of the occipito-mesencephalic tract, and the nucleus and descending tract of the nerve.

As with the case of the sensory nucleus the condition in *Capella, Ibis, Aix, Strix, Agapornis, Melopsittacus, Merops* and *Corvus* was described by Stingelin (1965). Approximately equal in volume to the sensory nucleus in *Corvus* and *Merops*, it was about half the size of this latter structure in *Strix* and *Melop-sittacus*, and far smaller in the other genera. In *Ibis* whilst the sensory nucleus had a volume of 1236 cubic units, that of the motor nucleus was 81.

# VI. THE ABDUCENS NERVE

Although in both ontogeny and phylogeny the abducens nerve is situated between the facial and the trigeminal these last were considered together be-cause of the close inter-relationships which exist between them in birds. The abducens, together with the trochlear and oculomotor, is of course concerned with the ocular muscles but in spite of this its nucleus is separated from theirs. Whilst that of the abducens lies within the medulla and close to those of the facial and trigeminal, the trochlear and oculomotor nuclei are situated within the mesencephalic region and are dealt with in Chapter 10. According to the older anatomists the actual position of the abducent nucleus is somewhat further back in *Casuarius* and comparable with the position which it occupies in the earlier phases of development in other genera. As such it also compares with the position of some of the components of the root complex in *Passer*. Passing out ventrally the fibres supply the lateral recti and the bulbar retractor muscles. Broadly speaking these fibres originate in a large and more or less uniform cellular mass which is situated to the side of the median longitudinal fasciculus. Above the dorsal component of the facial nuclear complex, it extends forwards from the transverse plane at which the vestibular root emerges (see Karten and Hodos, 1967). From this well-defined position at the side of the median longitudinal fasciculus the fibres pass down through the caudal pontine reticular nucleus and the trapezoidal body. A number of authors have described accessory nuclei in the species with which they worked. These include Craigie (1928), Gehuchten (1893), Sanders (1929) and Terni (1922a,b) who featured such components in the neighbourhood of their superior olivary nuclei. Addens (1933) thought that others lay at the ventro-lateral medullary angle, near to the periphery and widely separated from the principal nucleus. Sagittally both nuclei lay at about the same level but the accessory VI cells were close to the medial side of the descending root of V, to which they sent dendrites.

Historically many such authors have described the relationships which the abducens nucleus has with other structures. For a long time it has been considered that it received more or less equal fibre projections from both the homolateral and contralateral cerebellar and vestibular regions. These would conform with the suspected function of vestibular organs, and more especially the ampullary crests, saccular and utricular maculae, in the control of ocular movements in relation to those of the head.

## VII. THE EFFERENT COCHLEAR BUNDLE

One of the interesting developments in sensory physiology during the latter part of the 1950s was the realization of the very important part which is played by centrifugal systems in the control of receptor organ sensitivity. One such system is the efferent cochlear bundle. In the pigeon this decussates immediately in front of the abducent nucleus and then pursues a ventral course from the raphe in order to intersect the facial genu. The efferent cochlear fibres only accompany the facial nerve for a short distance along its course before they become dispersed amongst the widely scattered vestibular root fibres and leave the brain stem as small fascicles (Boord, 1961). After this they pass through the superior vestibular ganglion and proceed downwards and backwards through the inferior vestibular ganglion, the saccular nerve and the vestibulo-cochlear anastomosis, to join the cochlear nerve.

The demonstrated function of the bundle is apparently directly comparable with that which it exercises in carnivores. According to Boord (1961) if stimulated electrically that of the pigeon elicits a dual response. First of all it potentiates the cochlear microphonic potential which is possibly generated at the apex of the hair cells (see Chapter 5, Section VIII). Secondly it simultaneously reduces the voltage of the auditory nerve's response to sound. This is generally considered to result from an inhibitory effect, and produces a marked reduction in the acoustic input to the ascending auditory pathway as far as this is reflected by direct recordings of the potentials which are evoked following stimuli. Both of these effects, the peripheral on the microphonics and the central at the first auditory relay station within the brain stem, are eliminated by the administration of strychnine or brucine but they are not affected by other convulsants. Picrotoxin has no effect. This being the case then the neuro-pharmacological properties of the avian cochlear efferents seem to resemble those of the cat amongst mammals. Nevertheless certain differences do exist and Desmedt and Delwaide (1963) reported that the liminal convulsant dose of strychnine is much higher in the pigeon than in the cat. As noted above this alkaloid exerts a relatively specific antagonism to the class of inhibition of which the efferent nerves are an example and it is a useful measure of the degree of similarity in different species. Desmedt and Delwaide (1965) also speculated upon the reasons for microphonic potentiation. If, as has been suggested, the microphonics are, or are related to, the generator potential of nerve excitation, then the enhancement may be a measure of the increase which occurs in this potential when current flow through the synapse is prevented. There are some differences in the dynamics of such negative feedback control in birds and mammals. It is also worth noting that alongside the cycle of cytological events reported in the utricular maculae (Vinnikov, 1969) there may be efferent fibre involvement. The origin of the efferent fibres and endings in the vestibular part of the labyrinth is not known but some investigators consider that the cholinesterase in the macula is entirely related to the efferent innervation.

## VIII. THE INFERIOR OLIVE

### A. Anatomy and Histology

Following the initial description of an olivary nucleus in birds, which was the work of Kreis, this brain component has been quite widely studied. Prominent amongst the names which are associated with further descriptions are those of Turner (1891), Brandis (1894), Friedlander (1898), Frenkel (1909), Williams (1909), Yoshimura (1909), Sinn (1913) and Craigie (1928, 1930). Williams in his account featured it as a more or less round accumulation of cells except in the flamingo, *Phoenicopterus ruber*, in which two olivary components were confluent and formed a loop which was open below. According to Yoshimura this form is applicable to the olive of other birds, indeed including all those species which Williams had studied.

In his comparison of olivary structure within the birds and mammals Kooy (1915, 1917) suggested that the condition in the echidnas (*Tachyglossus*) provided a bridging link. This enabled him to put forward tentative suggestions about the homologies of the respective lamellae. These are summarized in Table 29.

## TABLE 29

*The homologies of the avian inferior olive and mammalian structures.*

| Avian structures | Mammalian structures |
|---|---|
| Ventral lamella | Principal olive |
| Dorsal lamella | Accessory olives |
| Medial part | Medial accessory olive |
| Lateral part | Dorsal accessory olive |

From Kooy (1917).

Much more recently Vogt-Nilsen (1954) investigated the olivary structure of the following species:

| | |
|---|---|
| *Gallus gallus* | *Picus viridis* |
| *Columba livia* | *Pica pica* |
| *Anas platyrhynchos* | *Garrulus glandarius* |
| *Capella gallinago* | *Passer domesticus* |
| *Larus argentatus* | *Muscicapa striata* |
| *Alca torda* | *Turdus merula* |
| *Strix aluco* | *Erithacus rubecula* |
| *Melopsittacus undulatus* | *Hirundo rustica* |

From these comparative studies he concluded that the avian olivary structures are always situated near to the ventral border of the medulla at the level of the calamus scriptorius. As a number of individual cell groups are certainly distinguishable he followed the terminology of Kooy when describing them and his general conclusions can be summarized as follows. The dorsal lamella is much larger than the ventral and has both medial and lateral components. The longest axis of the median accessory olive is more or less parallel to that of the raphe, whilst that of the dorsal accessory olive parallels the ventral medullary border. Both structures are themselves compound. The median accessory olive has clearly distinguishable dorsal, intermediate and ventral parts; and the dorsal accessory olive, the largest component of the dorsal lamella, has lateral and medial parts. In the case of the ventral lamella, which Kooy had suggested as the homologue of the principal olive in mammals, there are both lateral and median nuclear moieties. It is the former of these which is the most prominent and best developed, and topographically it extends further back than its partner does. The overall distribution of all these elements relative to each other is shown in Fig. 65.

In his embryological studies of the events which take place at the rhombic lip (see Section I) Harkmark (1954a, b) demonstrated that, in the chick, the

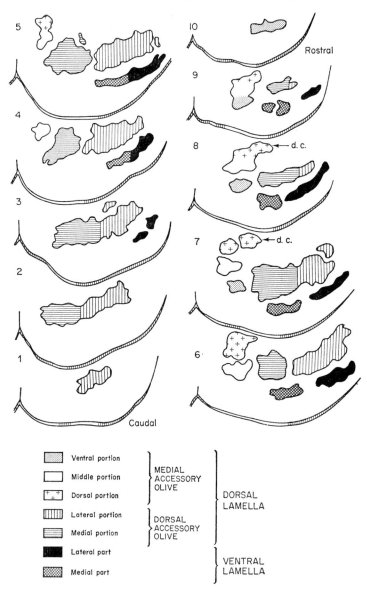

FIG. 65. The structure of the avian inferior olive. (Vogt-Nilsen, 1954).

dorsal lamella originates in the deep cell strand which arises as a result of cell migration away from the lip. In contrast the ventral lamella has an origin which is analogous to that of the pontine nuclei (*q.v.*) described by Brodal *et al.* (1950) and considered in Section IX. Such data corroborate the suggested homologies between the avian and mammalian conditions which Kooy had put forward nearly 40 years earlier.

Histological investigations which have been carried out on thionin stained sections reveal that the constituent perikarya are predominantly polygonal, although both round and spindle-shaped cells can also be present. An apparent correlation between overall olivary dimensions and the relative size of the constituent cells is suggested by the fact that it is the larger olives which often contain the largest cells and the smaller ones have a rather more micro-cellular composition. Furthermore in individual specimens the cell size is more or less homogeneous although this is not an invariable rule. In some birds, where such variations as this implies do occur, a roof of larger cells occurs within the dorsal cap of the median accessory olive whilst others can be found within the most lateral part of the ventral lamella.

Vogt-Nilsen (1954) concluded that, with the sole exception of the pigeon, in all the species which he studied there was a similar organization. It was only the dorsal cap of the median accessory olive which exhibited any marked differences from one species to the next. This structure is particularly well-defined in *Pica pica, Muscicapa striata* and *Picus viridis*. In species other than the magpie, flycatcher and woodpecker it is smaller and its actual presence in the blackbird, *Turdus merula*, was doubtful. Kooy (1917) had stated that in the penguin *Spheniscus demersus* the median accessory olive possesses a ventral knot which is synonymous with the ventral portion which Vogt-Nilsen described. He had also concluded that within a given order the structure retained a more or less constant appearance. In particular he cited the conditions amongst the Cuculiformes. Vogt-Nilsen (1954) followed this up by drawing attention to the fact that in the seven passeriform species included in his own list the ventral lamella extended to the posterior olivary pole more often than was the case in species from other orders. Taking account of this one may conclude that, although the overall pattern of the inferior olivary complex shows similar characteristics in all birds, both specific, and indeed individual, differences do occur.

## B. Olivo-cerebellar Connections

Experimental and histological indications of olivo-cerebellar projections emerge from the work of Williams (1909), Yoshimura (1909), Shimazono (1912), Sinn (1913), Craigie (1928) and Sanders (1929). Similar information was also provided by Whitlock (1952) whose work is considered in Chapter 8.

Vogt-Nilsen (1954) went to some lengths to discover whether there was any direct correlation between the overt morphology of the olive and the relative degree of differentiation undergone by the cerebellum. The factors which contribute to such variations in avian cerebellar organization are also discussed in Chapter 8 and one may merely note at this juncture that overall body size appears to be very influential. The only such relationship which Vogt-Nilsen could detect was between the size of the ventral olivary lamella on the one hand, and that of the cerebellar auricle on the other. In several paleognathous species the ventral lamella is rather small to judge from the records contained in earlier papers. Vogt-Nilsen concluded that this is also true of the auricular component in *Rhea*, *Struthio* and *Dromiceius*, although that of *Casuarius* is rather larger. However, it is only fair to emphasize that comparable conditions occur in some neognathous genera. Vogt-Nilsen concluded that the floccular part of the cerebellum receives projections from the contra-lateral, medial accessory olive.

Returning to embryological considerations we can note that in his account of ontogenetic cell movements and the theory of neurobiotaxis (see Section I above), Harkmark (1956) also discussed the influences which the cerebellum may exert on the development of both the pontine and olivary regions. He observed that following experimentally induced lesions of the cerebellar plates defects occurred in the contralateral inferior olive and in the pontine area of both sides. Extensive cerebellar lesions produced widespread defects in the contralateral olive together with even more massive changes in the pons. Furthermore, bilateral cerebellar ablation led to bilateral olivary disturbance and there certainly seemed to be a definite, and probably causal, relationship between the two series of events. Preoccupied with a consideration of generalized neurobiotactic theory he suggested that, although it is essential for axons to establish connections with their normal terminal field, the area of axonal proliferation has no influence on the distribution, migration, mutual organization or initial differentiation of nerve cells. One can certainly agree with the latter component of his list.

# IX. THE PONTINE REGION

## A. Historical Introduction

To students of mammalian anatomy the pontine region is a well-defined area. In Primates it comprises a clear component of the forepart of the hindbrain; is situated in front of the cerebellum; from its upper part there emerge the cerebral peduncles, whilst behind and below it is continuous with the medulla from which it is separated by a transverse furrow in which the abducens, facial and stato-acoustic nerves appear. Although Papez (1929),

Craigie (1930) and Kappers *et al.* (1936) had all suggested that a primordial pontine grey area existed in the avian brain the definitive pontine structures were at that time generally thought to be peculiar to mammals. Kappers (1947) drew attention to the possibility that the grey which is present in the basal medullary region, between the levels at which nerves V and VIII emerge, might be part of a tecto-pontine-cerebellar system. He also pointed out that, if this was subsequently shown to be the case, then it might be a *forerunner* of the mammalian cortico-pontine-cerebellar interaction. Once again one observes the direction of such interest even in Kappers, much of whose personal authority can be ascribed to the way in which he investigated individual neurological components within the central nervous systems of different phyla and classes at least partially for their own inherent interest.

Following Wallenberg's (1903, 1904) description of conspicuous fibre bundles which pass along the ventral surface of the forepart of the medulla, Shimazono (1912) and Craigie (1930) traced them into the cerebellum. It was these same bundles that Papez (1929) homologized with the mammalian ponto-cerebellar fibres. He considered that they originated in a small group of cells which are enclosed within them and had been reported more or less incidentally, by Brouwer (1913) and Sinn (1913), and were to be described again by Scholten (1946) amongst others (see Brodal *et al.* 1950).

A further and all-important stage in the recognition of these cells and axonal bundles was reached when Brodal *et al.* (1950) undertook partial extirpation of the cerebellar region in 3–8-day-old chicks. Using thermocautery techniques to induce lesions they observed that retrograde changes subsequently occurred in these cellular masses and that these changes were comparable to those seen under similar circumstances in the pontine nuclei of cats and rabbits. From that time it has been universally accepted that the nuclei do represent pontine analogues and their more specific characteristics are outlined below.

## B. Normal Anatomy and Histology

The avian pontine nuclei comprise two fairly well defined cellular groupings which lie on each side of the mid-ventral line. When seen in transverse sections they appear as dorso-ventrally flattened cell-groups which are situated inside the ventral surface of the rhombencephalon. From the ventral aspect they are "covered" by the axons of the so-called Wallenberg's commissure. Certain of the component commissural fibres penetrate the nuclear masses and, in particular, they isolate the medial component from the neighbouring reticular formation. Similar fibres, which actually pass through the nuclei, can be seen to have their origins amongst the component cells.

Although both nuclear groups of a given side are of approximately similar size the actual lines of demarcation of the medial mass are clearer than those of the lateral. Furthermore the two differ in their longitudinal positions. The lateral projects further forward than the medial one which, on the other hand, is more extensive posteriorly. In both cases their most posterior limits are, however, at the level of the vestibular nerve where this is found to the side of the brain stem. They are therefore situated somewhat behind the front end of the group of cells which form the intermediate nucleus of Sanders. Anteriorly they extend to just in front of the dorsal end of the motor complex of V and to the level of its most anterior root fibres. It is only in the very rostral areas that one can detect any coalescence between the two. Although also isolated from the reticular substance in general, the medial nucleus can be connected to it by way of cellular strands.

The component cells of these pontine regions are of two principal types. The majority are of medium size, slightly smaller than those of neighbouring motor nuclei and very much smaller than those of the reticular formation. In Nissl-stained preparations they appear as multipolar, and sometimes spindle-shaped or pyriform units. The other cytological constituents are less numerous and their size is only about half of the foregoing. The distribution through the nuclei of these two cell types is more or less homogeneous and both appear intermingled in sections from varying levels. This relatively uniform composition facilitates the differentiation of the nuclear moieties from the more dorsally situated reticular formation which has cells of a generally smaller size. Such a difference is, nevertheless, not so clear in the more posterior regions where the two have cells of a more comparable appearance, but, even at this level, they are well separated by the cell-free space.

The anterior part of the lateral nucleus is sometimes only imprecisely defined as the percentage representation of the medium sized cells diminishes and the cells themselves are more scattered. Again in the hindmost region the lateral in particular is somewhat loosely structured, and is less sharply demarcated than the medial one. Indeed although they usually retain their individuality Brodal et al. (1950) did observe them to actually coalesce behind in some specimens. At more intermediate antero-posterior levels the fibres of the abducens nerve leave the brain from a position which lies between them and a fine bundle occasionally penetrates the most lateral part of the medial grouping.

## C. Ponto-cerebellar Connections

When outlining the data which refer to olivo-cerebellar projections it was noted that unilateral lesions of the embryonic cerebellar plates produced defects within the pons of both side (see Section VIII B). According to Harkmark (1956) such pontine changes were most conspicuous in that part of the contralateral medial nucleus which is closest to the mid-line, and also in the lateral part of the ipsilateral one. Furthermore small, experimentally induced lesions in the cerebellum produced discrete localized defects within the pons. In their earlier, and now classic paper, Brodal *et al.* (1950) had also provided analogous information which was derived from their electro-cautery studies in young chicks. Following extensive hemilateral cerebellar ablation or lesions, practically all of the cells contained within the contralateral medial pontine nucleus, together with the majority of those in the ipsilateral lateral one, underwent retrograde changes and eventually disappeared. After more precisely defined and restricted lesions these pontine degenerations are less severe. By

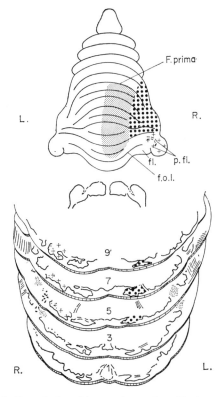

FIG. 66. Ponto-cerebellar relationships as shown by ablation experiments. (Brodal *et al.*, 1950).

comparison of the results which they obtained by making such lesions in a variety of cerebellar foci, Brodal and his co-workers concluded that the bulk of the pontine fibres pass to the lateral regions of folia VI–VII (see Chapter 8), as well as to other non-foliated cortex which lies to the side of these. Although pontine projections involving the medial parts of the cerebellum are less numerous the para-flocculus is very much involved. Fewer fibres were also observed connecting the pons to the anterior lobe. In keeping with the rather lateral representation in folia VI–VII they also found that within Va and b, it was the more lateral areas in close proximity to non-foliated cortex which had the richest representation. Connections with more medial areas were far fewer in number.

Although the distribution of the retrograde changes within the pontine nuclei is relatively diffuse, some degree of direct topographical representation of the cerebellar regions does seem to exist. The most clearly differentiated relationship is that which exists between the contralateral medial nucleus and the lateral parts of the cerebellum. The source of the parafloccular connections is rather less clear cut. Fibres to this region seem to have a widespread origin although the area of maximum representation lies in the rostral half of the ipsilateral lateral nucleus, particularly within its medial area. Also Brodal *et al.* concluded that the fibre connections with those medial foliar regions which are situated behind the "primary fissure" (see Chapter 4), take their origin from the lateral part of the lateral pontine nucleus of the same side. There are therefore some relatively point-to-point relationships and the overall pattern is represented in Fig. 66. Over and above these there are also a varying number of connections between a given area of the cerebellum and the entirety of the pontine regions. From such results Brodal *et al.* concluded that the lateral parts of cerebellar folia VI–VIII (the presumed declive-folium-tuber-pyramis complex in Larsell's terms) together with the lateral region of folium V (the culmen) may well be the avian equivalent of the mammalian cerebellar hemispheres.

## X. THE NUCLEUS RAPHE

Studies on the afferent fibre connections which exist between the ascending spinal tracts and medullary structures, show that in this latter region the nucleus raphe and the hind part of the inferior olive receive such projections. This was confirmed by Karten (1963) using Nauta-Gygax impregnation after cervical hemisection at the level of roots 12–14. The raphe itself begins caudally at the level of the posterior horizontal ventral olivary lamella. Its relative position is shown in Figs 50 and 57. In transverse sections it is visible as a cell column on either side of the mid-line. In front of the olive the nuclei from the two sides converge ventrally and together occupy a more or less triangular field within the ventral part of the raphe and the adjacent grey. In the caudal part of the pons they can assume the form of thin stripes and gradually disappear (Harkmark, 1954a,b). Sinn (1913) thought that the dorsal part was a reticular tegmental nucleus. In his embryological studies Harkmark showed that there was no foundation for this idea. Using preparations with experimentally produced lesions together with normal embryos he also demonstrated that the nucleus has an origin which is comparable to that of both the pontine nuclei and the olive. All receive cells from the superficial cell strands and in the case of the raphe Harkmark concluded that these too are derived from the contralateral rhombic lip.

## XI. MEDULLARY HISTOCHEMISTRY

The overall pattern of avian brain biochemistry, as far as it is known, was outlined in Chapter 3. We can, however, note a few of the salient points that refer to the medullary region. The acetylcholine, 5-hydroxytryptamine,

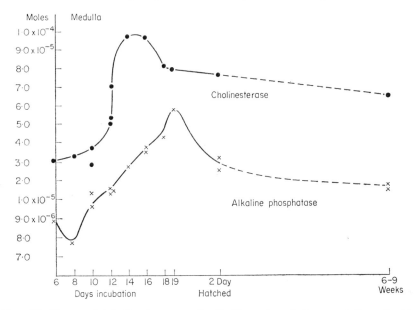

Fig. 67. The acetylcholinesterase and alkaline phosphatase activity of chick medulla. (Rogers *et al.*, 1960). Molecules of substrate split per hour for each milligram of tissue nitrogen plotted against developmental age.

dopamine and nor-adrenalin levels in m$\mu$moles/g were established for the pigeon and are contained in Table 15. The highest nor-adrenalin values anywhere in the brain were obtained from the medullary preparations (5·67 m$\mu$moles) and this was also true for 5-hydroxytryptamine. The acetylcholine and dopamine levels were, however, from 1–6 times lower than those which were recorded from other regions (18·1 and 1·70 m$\mu$moles respectively). Nevertheless the former value exceeds that of the cerebellum and telencephalon and the latter the levels in the cerebellum and diencephalon plus optic lobes.

9

The results of Lowry type estimations of the concentrations of acetyl cholinesterase, using acetylcholine bromide as a substrate, enabled Rogers *et al.* (1960) to present comparative date for all stages of embryonic development between the 6th day of incubation and 2 days after hatching, with a further point at 6–9 weeks. These data are summarized in Fig. 67. It can be seen from this figure that a marked and steep increase in the levels took place at the 12th day of incubation and that this led to a peak which persisted to the 14th day. Subsequent lower levels which were achieved on day 16 were not dissimilar to those of the post-hatching period.

Using a modified Gomori technique Rogers and his collaborators also monitored the alkaline phosphatase concentrations during the same period. These are also depicted in the graphs of Fig. 67. Unlike the acetyl cholinesterase values these did not undergo a rapid increase at the 12th day but climbed relatively steadily to a maximum at 18 days when the cholinesterase maximum had been passed. There was also a significant drop during the early period in which data were obtained, and this is not reflected in the cholinesterase curve. At 12 days of incubation the pattern in all other regions of the brain was different from that in the medulla. In these other locations the phosphatase peaks preceded those of esterase and declined during the period in which there were maximal levels of the latter. Rogers and his co-workers explained this by suggesting that the early appearance of medullary electrical activity preceded the total and final differentiation of its component areas. This would certainly comply with the observed fact that Gomori-type staining leaves the large motor neurons of the motor areas virtually untouched whilst all the way from the ventricle to the ventral medullary margin fairly intense staining occurs between them. Although this staining appears to be particularly marked on the neuroglial components it also occurs on the axons. To what extent it represents Schwann cell and other glial activity in such sites is not clear.

# References

Addens, J. L. (1933). The motor nuclei and roots of the cranial and first spinal nerves of vertebrates. *Zeitschr. Anat. Entwickl. Abt. Zeitschr. ges. Anat.* **101,** 307–410.

Beccari, N. (1923). Intorno al primo differenziamento dei nuclei motori dei nervi cranici. *Monit. Zool. Ital.* **34,** 161–166.

Bernard, C. (1858). "Leçons sur la physiologie et la pathologie du système nerveux", p. 394.

Black, D. (1922). The motor nuclei of the cerebral nerves in phylogeny: IV Aves. *J. comp. Neurol.* **34,** 233–275.

Boord, R. L. (1961). The efferent cochlear bundle in the caiman and pigeon. *Exp. Neurol.* **3,** 225–239.

Brandis, F. (1894). Untersuchungen uber das Gehirn der Vögel, II das Kleinhirn. *Arch. f. Mikr. Anat.* **43,** 787–813.

Brodal, A., Kristiansen, K. and Jansen, J. (1950). Experimental demonstration of a pontine homologue in birds. *J. comp. Neurol.* **92**, 23–69.

Brouwer, B. (1913). Uber das Kleinhirn der Vögel. Nebst bemerkungen uber das Lokalisation problem im Kleinhirn. *Folia Neurobiol.* **7**, 349–377.

Bubien-Waluszewska, A. (1968). Le groupe caudal des nerfs craniens de la Poule. *Acta Anat.* **69**, 445–457.

Butler, P. J. (1967). The effect of progressive hypoxia on the respiratory and cardio-vascular systems of the chicken. *J. Physiol.* **191**, 309–324.

Butler, P. H. and Jones, D. R. (1968). Onset of, and recovery from, diving brady-cardia in ducks. *J. Physiol.* **196**, 255–272.

Cohen, D. H. and Pitts, L. H. (1968). Vagal and sympathetic components of con-ditioned cardio-acceleration in the pigeon. *Brain Res.* **9**, 15–31.

Cohen, D. H., Schnall, A. M., MacDonald, R. L., and Pitts, L. H. (1970). Medullary cells of origin of cardio-inhibitory fibres in the pigeon. *J. Comp. Neurol.* **140**, 299–320.

Cohen, D. H. and Schnall, A. M. (1970). Medullary cells of origin of cardio-inhibitory fibres in the pigeon. II. Electrical stimulation of the dorsal motor nucleus. *J. Comp. Neurol.* **140**, 321–342.

Cordier, R. (1954). Le système nerveux centrale et les nerfs cerebrospinaux. "Traité de Zoologie", vol. 12, pp. 202–332. Masson, Paris.

Couvreur, E. (1892). Sur le pneumogastrique des oiseaux. *Ann. Univ. Lyon.* **2**, H3.

Craigie, E. H. (1928). Observations on the brain of the humming bird (*Chrysolampis mosquitus* Linn. *Chlorostilbon Caribaeus* Lawr). *J. comp. Neurol.* **48**, 377–481.

Craigie, E. H. (1930). Studies on the brain of the Kiwi (*Apertyx australis*). *J. comp. Neurol* **49**, 223–357.

Dahl, N. A. Samson, F. E. Jnr. and Balfour, W. M. (1964.) Adenosine triphosphate and electrical activity in chicken Vagus. *Amer. J. Physiol.* **206**, 818–822.

Desmedt, J. E. and Delwaide, P. J. (1963). Neural inhibition in a bird; effect of strychnine and picrotoxin. *Nature, Lond.* **200**, 583–585.

Desmedt, J. E. and Delwaide, P. J. (1965). Functional properties of the efferent cochlear bundle revealed by stereotaxic stimulation. *Expl Neurol.* **11**, 1–26.

Einbrodt, A. (1859). Uber den Einfluss der Nervi vagi auf die Herzbewegung der Vögel. *Arch. Anat. Physiol. u. wiss Med.* 439–459.

Fitzgerald, T. C. (1969). "The Coturnix Quail", p. 306. Iowa State University Press.

Frenkel, B. (1909). Das Kleinhirnbahnen der Taube. *Bull intern. Acad. Sci. Cracow.* **2**, 123–147.

Friedlander, A. (1898). Untersuchungen uber das Ruckenmark und das Kleinhirn der Vögel. *Neurol. Centralbl.* **17**, 397–409.

Gehuchten, A. van (1893). Anatomie du système nerveux de l'homme, Leuven, Van, In.

"Gray's Anatomy: Descriptive and Applied", pp. 1604. Longmans Green and Co., London.

Groebbels, F. (1922). Der Hypoglossus der Vögel. *Zool. Jahrb. (Abt. Anat.)* **43**, 465–484.

Groebbels, F. (1932). "Der Vögel. I. Atmungswelt und Nahrungswelt." Gebruder Borntraeger, Berlin.

Harkmark, W. (1954a). Cell migrations from the rhombic lip to the inferior olive, the nucleus raphe and pons. *J. comp. Neurol.* **100**, 115–210.

Harkmark, W. (1954b). The rhombic lip and its derivatives in relation to the theory of neurobiotaxis. *In* "Aspects of Cerebellar Anatomy" (Jansen, J. and Brodal, A., eds), pp. 264–284. Johan Grundt, Tanum Forlag, Oslo.

Harkmark, W. (1956). The influence of the cerebellum on development and maintenance of the inferior olive and pons. *J. exp. Zool.* **131**, 333–355.

Johansen, K. and Reite, O. B. (1963). Cardiovascular responses to vagal stimulation and cardio-accelerator nerve blockade in birds. *Comp. Biochem. Physiol.* **12**, 479–487.

Jones, D. R. (1969). Afferent vagal activity related to respiratory and cardiac cycles *Comp. Biochem. Physiol.* **28**, 961–965.

Jungherr, E. (1969). "The Neuroanatomy of the Domestic Fowl," pp. 126. *Amer. Ass. Avian Pathol.*

Jurgens, H. (1909). Uber die Wirkung des Nervus vagus auf das Harz der Vögel. *Pfluger's Arch.* **129**, 506–524.

Kappers, C. U. A. (1907). Phylogenetische Verlagerungen der motorischen Oblongata kerne ihre Ursache und Bedeutung. *Neurol. Zeitbl.* **26**, 834–840.

Kappers, C. U. A. (1908a). Weitere Mitteilungen bezuglich der phylogenetischen Verlagerung der motorischen Hirnnervenkerne. *Folia Neuro-biol.* **1**, 157–172.

Kappers, C. U. A. (1908b). Weitere Mitteilungen uber Neurobiotaxis. *Folic Neuro-biol.* **7**, 507–534.

Kappers, C. U. A. (1911). Weitere Mitteilungen uber Neurobiotaxis VI. The migrations of the motor root-cells of the vagus group and the phylogenetic differentiation of the hypoglossus nucleus from the spino-occipital region. *Pschiat. Neurol. Bi. Amsterdam.* **15**, 408.

Kappers, C. U. A. (1912). idem. VII. Die phylogenetische Entwicklung der motorischen Wurzelkerne in oblongata und Mittelhirn. *Folia Neuro-biol.* **6**, 1–42.

Kappers, C. U. A. (1917). idem. IX. An attempt to compare the phenomena of neurobiotaxis with other phenomena of taxis and tropisms. *J. comp. Neurol.* **27**, 261–298.

Kappers, C. U. A. (1920–21). "Vergleichende Anatomie des Nervensystems." Bohn, Haarlem.

Kappers, C. U. A. (1932). Principles of the development of the nervous system. *In* "Cytology and Cellular Pathology of the Nervous System" (Penfield, W., ed.), pp. 45–89. Hoeber, New York.

Kappers, C. U. A. (1947). "Anatomie Comparée du Système Nerveux", pp. 754. Masson, Paris.

Kappers, C. U. A., Huber, G. C. and Crosby, E. C. (1936). "The Comparative Anatomy of the Nervous System of Vertebrates Including Man." 3 volumes. Macmillan (1960 reprint). Hafner Press.

Karten, H. J. (1963). Ascending pathways from the spinal cord in the pigeon, (*Columba livia*). *Proc. Int. Cong. Zool.* **2**, 23.

Karten, H. J. and Hodos, W. (1967). "A Stereotaxic Atlas of the Brain of the Pigeon" (*Columba livia*), pp. 193. Johns Hopkins University Press.

Kaupp, B. F. (1918). "The Anatomy of the Domestic Fowl." Saunders.

Knoll, P. (1880). Uber Myocarditis und die ubrigen Folgen der Vagussektion bei Tauben. *Prag. Z. Heilk.* **1**, 180–254.

Kooy, F. H. (1915). De phylogenese van de oliva inferior. *Ned. Tijdss. v. Geneesk.* **51**, 2533–2536.

Kooy, F. H. (1917). The inferior olive in vertebrates. *Folia Neuro-biol.* **10**, 205–369.

Koppányi, T. and Dooley, M. S. (1928). The cause of cardiac slowing accompanying postural aponea in the duck. *Amer. J. Physiol.* **85**, 311–323.

Kosaka, K. and Hiraiwa, K. (1905). Uber die Facialiskerne beim. Huhn. *Jahrb. Psychiatr.* **25**, 57–69.

Kosaka, K. and Yagita, K. (1903). Experimentelle Untersuchungen über den Ursprung des Nerven Hypoglossus und seines absteigenden Astes. *Jahrb. Psych. Neurol.* **24**, 150–189.

Lasiewski, R. C., Weathers, W. W. and Bernstein, M. H. (1967). Physiological responses of the giant humming bird (*Patagona gigas*). *J. Comp. Biochem. Physiol.* **23**, 797–813.

Lubosch, W. (1899). Vergleichend anatomisch Untersuchungen uber den Ursprung und die Phylogenese des N. accessorius Willisii. *Arch. Mikr. Anat.* **54**, 514–602.

MacMillen, R. E. and Trost, C. H. (1967). Nocturnal hypothermia in the Inca dove *Scardafella inca*. *Comp. Biochem. Physiol.* **23**, 243–253.

McCrady, J. D., Vallbona, C. and Hoff, H. E. (1966). Neural origin of the respiratory heart rate response. *Amer. J. Physiol.* **211**, 323–328.

Nolf, P. (1927). Du role des nerfs vague et sympathétique dans l'innervation motrice de l'estomac de l'oiseaux. *Arch. Int. Physiol.* **28**, 309–428.

Nolf, P. (1929). Le système nerveux enterique. *Arch. Int. Physiol.* **30**, 317.

Nolf, P. (1934). Les nerfs extrinsique de l'intestin chez les oiseaux. *Arch. Int. Physiol.* **39**, 165.

Papez, J. W. (1929). "Comparative Neurology." Crowell and Co., New York.

Paton, D. N. (1912). On the extrinsic nerves to the heart of the bird. *J. Physiol.* **45**, 106–114.

Portmann, A. (1950). Système nerveux. *In* "Traité de Zoologie", vol. **15**, pp. 185–203.

Portmann, A. and Stingelin, W. (1961). The central nervous system. *In* "Biology of Birds" (Marshall, A. J. ed.), Vol. 2, pp. 1–36. Academic Press, New York and London.

Rogers, K. T. (1965). Development of the XIth or spinal accessory nerve in the chick. *J. comp. Neurol.* **125**, 273–286.

Rogers, K. T., DeVries, L., Kepler, J. A., Kepler, C. R. and Speidel, E. R. (1960). Studies on the chick brain of biochemical differentiation related to morphological differentation. II. Alkaline phosphatase and cholinesterase levels and the onset of function. *J. exp. Zool.* **145**, 49–60.

Sanders, E. B. (1929). A consideration of certain bulbar, midbrain and cerebellar centres and fibre tracts in birds. *J. comp. Neurol.* **49**, 155–221.

Scholten, J. M. (1946). "De plaats van der paraflocculus in het geheel der cerebellaire correlaties". Acad. Proefschr., Amsterdam.

Shimazono, J. (1912). Das Kleinhirn der Vögel. *Arch. Mikr. Anat.* **80**, 397–449.

Sinn, R. (1913). Beitrag zur Kenntnis der Medulla oblongata der Vögel. *Monatschr. Psychiatr. Neurol.* **33**, 1–39.

Stingelin, W. (1965). Qualitative und quantitative Untersuchungen an der Kerngebieten der Medulla oblongata bei Vögeln. *Bibliotheca Anatomica.* **6**, 1–116.

Stübel, H. (1910). Beitrage zur Kenntnis der Physiologie der Blutkreislaufes bei verschiedenen Vögelarten. *Pfluger's Arch.* **135**, 249–365.

Terni, T. (1922a). Ricerche sul nervo abducente e in special modo intorno al significato del suo nucleo accessorio d'origine. *Folia Neuro-biol.* **12**, 277–327.

Terni, T. (1922b). Il sostrato anatomico del riflesso di chiusura della membrana nittitante ne Sauropsida. *Arch. Fisiol.* **20**, 305–311.

Turner, C. H. (1891). The morphology of the avian brain, *J. Comp. Neurol.* **1**, 39–92.

Vinnikov, Y. A. (1969). The ultrastructural and cytochemical bases of the mechanism of function of the sense organ receptors. *In* "The Structure and Function of Nervous Tissue" (Bourne, G. H. ed.), Vol. 2, pp. 265–392. Academic Press, New York and London.

Vogt-Nilsen, L. (1954). The inferior olive in birds. *J. Comp. Neurol.* **101,** 447–481.

Wallenberg, A. (1903). Die Ursprung des Tractus isthmo-striatus der Taube. *Neurol. Centralbl.* **22,** 98–101.

Wallenberg, A. (1904). Neue Untersuchungen uber den Hirnstamm der Taube. *Anat. Anz.* **24,** 357–369.

Watanabe, T. (1961). Comparative and topographical anatomy of the fowl. VIII. On the distribution of the nerves in the neck of the fowl. *Jap. J. vet. Sci.* **23,** 85–94.

Weiss, H. S., Fisher, H. and Giminger, P. (1961). Seasonal changes in avian blood pressure related to age, sex, diet, confinement and breed. *Amer. J. Physiol.* **201,** 655–659.

Whitlock, D. G. (1952). A neurohistological and neuro-physiological study of the afferent fibre tracts and receptive areas of the avian cerebellum. *J. comp. Neurol.* **97,** 567–635.

Whittow, G. C., Sturkie, P. D. and Stein, G. Jnr. (1964). Cardiovascular changes associated with thermal polynea in the chicken. *Amer. J. Physiol.* **207,** 1349–1353.

Williams, E. M. (1909). Vergleichend anatomische Studien uber den Bau und die Bedeutung der Oliva inferior die Saugetiere und Vögel. *Arb. Neur. Inst. Weiner. Univ.* **17,** 118–149.

Yoshimura, K. (1909). Experimentelle und Vergleichend anatomische Untersuchungen uber die untere Olive der Vögel. *Arb. Neur. Inst. Weiner. Univ.* **18,** 46–59.

Young, J. Z. (1950). "The Life of the Vertebrates", pp. 767. Oxford University Press.

# 8 The Cerebellum

## I. GENERAL INTRODUCTION

As the cerebellar region is intimately involved in the control and maintenance of muscle tonus and therefore equilibrium, it is particularly important in the more active vertebrate genera. This is especially true in the case of bipedal cursorial or flying forms. The mechanisms which are involved in mammalian cerebellar control and inhibitory activity, as far as these are known, were reviewed by Eccles *et al.* (1967) and Marchiafava (1968). Comparable recent data for birds are relatively scarce. However, in both classes dysfunction generally leads to dyskinesia, atonia or dystonia. In view of the clinical importance of such conditions it is not surprising that there were a fair number of studies on the avian cerebellum prior to that of Larsell (1948), which established the principal anatomical homologies, and that of Goodman *et al.* (1965) which has provided an elegant picture of the intra-cerebellar functional organization.

The ontogenetic development has already been outlined in Chapter 4, Section VII but one may note the historical importance of investigations carried out in the two first decades of the present century by Mesdag (1909), who established the existence of a cerebellar *anlage* in 4·5-day-old chick embryos, and Ingvar (1919) whose work is a basis for all more recent discussions. Indeed many of these have featured extensive and discursive considerations of the synonymies and homologies of those fissures which he termed x, y and z. In the early stages of embryonic differentiation the lateral eminences which subsequently fuse to form the dorsally situated cerebellum are independent and unattached to one another, except for their connection via the membranous ventricular roof. Their surfaces are also smooth and traces of fissures only appear later.

The elaborate adult anatomy has been referred to in greater or less detail by some 17 authors of importance. From an historical point of view one can cite Ramon y Cajal (1908), Shimazono (1912), Brouwer (1913), Hoevell (1916), Kuenzi (1918), Kappers (1920), Groebbels (1927), Craigie (1928), Housman (1929), Sanders (1929) and of course the reviews of Kappers (1947), Kappers *et al.* (1936), Portmann (1950), Portmann and Stingelin (1961) and Cordier (1954). Even the earlier workers established that in birds the organ system is subject to a fair amount of gross anatomical variation. Usually somewhat rounded in outline, it appears to be compressed backwards in *Turdus*, projected forward in *Haliaetus* and backwards in *Cygnus*. Kuenzi (1918) drew attention to a number of such differences. He concluded that it is relatively broad in genera such as *Alca, Ibis, Charadrius, Podiceps, Phoenicopterus* and *Tadorna*; relatively narrow in others such as the gruid genus *Anthropoides*, together with *Micropus (Apus* auctt) and *Fringilla.* In dorsal view it appears parallel-sided in *Micropus*, converges anteriorly in *Sula, Corvus* and *Lorius*, and posteriorly in *Dromiceius, Alca, Ardea, Ciconia, Ibis, Columba, Larus, Charadrius* and *Tadorna.* He also considered that its sides could be arched as in *Anthropoides, Pelecanus*, the rallid genus *Porphyrio*, the honey-buzzard genus *Pernis* and the *Turdidae.*

Brandis (1894) was the first person to draw attention to the apparent increase in the number of folia which accompanies an increase in body size from one species to another, and he was followed by Kappers in 1921. They pointed out that a comparison between the cerebellum of small birds such as the genera *Nucifraga, Regulus, Parus* or the pycnonotid *Otocompsa*, with those of large birds such as *Struthio, Oligyps* or *Sula bassana*, suggested that in these the more posterior regions were overlain by the anterior. The generalized appearance of an avian cerebellum is shown in Fig. 68.

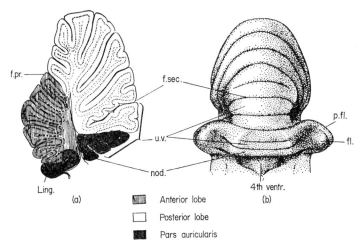

FIG. 68. The general organization of the avian cerebellum. (Portmann and Stingelin, 1961). (a) mid-sagittal plane, (b) posterior view.
Abbreviations: f.pr., primary fissure of Larsell; f.sec., secondary fissure; fl., flocculus; p.fl., paraflocculus; Ling., Lingula; nod, nodulus; uv., uvula; 4th vent, 4th ventricle.

Turning to functional studies these, like those of the medulla, date from very early periods in the history of neuro-physiology. Contributions of varying merit and validity were made in the last century by Magendie (1825), Longet (1842), Flourens (1824), Wagner (1858–60), Brown-Sequard (1860), Dalton (1861), Renzi (1863), Weir-Mitchell (1869), Luciani (1893) and

Friedlander (1898). These were followed in the early decades of the present century by the works of Reisinger (1916), Bremer (1924) and Groebbels (1928a–e). Historical reviews which refer to these works are those of Ten Cate (1936, 1965), von Buddenbrock (1953) and Karamyan (1956). Generally speaking it was widely considered by such workers that the most important afferent connections were tracts which arose within the vestibular system and mesencephalon, but the question of direct links between the vestibular apparatus and the cerebellum continued to figure in controversies up to the time of publication of Whitlock's study (1952). Frenkel (1909), Shimazono (1912) and Ingvar (1918) denied that they existed. Wallenberg (1907) and Groebbels (1928b) disagreed with this and claimed that they did. The data of Wallenberg, Shimazono and others also suggested that tecto-cerebellar fibres run from the deep layer of the optic tectum to the cerebellum. Kappers et al. (1936) thought that these were less prominent than the connections with both the medulla and the spinal cord. Karamyan (1956) made the slightly obscure suggestion that, by comparison with the condition in both Amphibia and Reptilia, the basic tracts have been transferred from the frontal to the caudal part of the brain as the result of the development of locomotor organs which require precise coordination of their musculature during walking, flying and swimming.

## II. CEREBELLAR FOLIAR TOPOGRAPHY

The work of Larsell (1948), Larsell and Whitlock (1952) and Saetersdal (1956a,b, 1959a) forms the basis for our understanding of cerebellar foliation. To these have been added the more recent and extensive comparative study of adult structure which was undertaken by Senglaub (1963), and the posthumous review of Larsell's work on vertebrate cerebella by Jansen (Larsell, 1967). Prior to Larsell (1948) it had been assumed that the bulk of the avian cerebellum corresponded to the mammalian vermis. The auricle, a lateral extension from the basal caudal region, was puzzling and variously called flocculus, floccular node, auricle and paraflocculus. Brouwer (1913) had given the designation "fissure x" to a furrow which was well represented in the 25 species which he studied. He considered that the posterior "vermis" was divided into a median and posterior part by another furrow—"fissure y". Ingvar followed this terminology and added two further widely represented fissures "z" and "un". He stated in a footnote that x, y and z were the equivalents of the primary and prepyramidal sulci and the secondary fissure in mammals. Kappers et al. (1936) understandably concurred with such decisions. A summary of Larsell's conclusions, together with those of Saetersdal was given in Chapter 4, Section VII B. Larsell's conclusions are also tabulated in Table 30, and Saetersdal's comments are referred to below. It is, however,

## TABLE 30

*A comparison of the cerebellar nomenclature of Ingvar and Larsell according to Larsell (1967).*

| Nomenclature of Ingvar (1918) | Nomenclature of Larsell (1967) |
|---|---|
| un | postero-lateral fissure |
| x | primary fissure |
| y | secondary fissure |
| z | first uvular sulcus |

worth recalling that there is a basic difference between the cerebellar development of birds and mammals. In mammals it is the fissura prima, as its name implies, that first divides the median part of the corpus cerebelli. In birds it is the fissura secunda, or, in the case of *Chlorostilbon aureoventris* and *Colibri serrirostris* the intraculminary.

The now classical paper of Larsell (1948) was based upon a study of the cerebellar development in *Anas* and *Gallus* alongside a comparison of the adult anatomy in both these genera and *Columba*, the sooty grouse *Dendragapus*, the humming bird *Lampornis*, and the bald eagle *Haliaetus*. Later, 1952, working with Whitlock he added studies on the dusky horned owl, *Bubo virginiatus saturatus*, which is the Pacific coast variant of the great horned owl; the American short eared owl, *Asio flammeus*; the barn owl, *Tyto alba pratincola*; and Brandt's cormorant *Phalacrocorax penicillatus*. He concluded that the superficial appearance of both folia and fissures is very similar in all of them, that they are difficult to distinguish in most avian species, and that the cerebellum can, on the basis of primary afferent connections, be divided into a flocculo-nodular node and corpus cerebelli.

More specifically he considered that the fissures result in 10 primary folia. These are numbered, using Roman numerals I to X, from the rostral region backwards. The corresponding and resultant medullary rays he numbered 1–10. Some primary folia clearly developed synchronously with other secondary ones and variations occur between the relative speeds of development in different genera. He homologized the avian and mammalian conditions in the manner that is shown in Table 31. In this scheme folium I is the avian equivalent of the lingula and he concluded that this is relatively large in birds. Folia II and III correspond to the mammalian central lobule which in all his avian material comprised two folia. This was also true for the culmen (IV and V). Folium VI he considered to be the avian equivalent of the declive; VII the folium vermis and tuber vermis, VIII the pyramis and IX the uvula. In his duck and chick material the margin to which the posterior medullary velum is attached was drawn forward and redoubled on itself. As this corresponded to the manner in which the nodulus develops in man, Larsell designated it folium X or the nodulus.

## TABLE 31

*The homologies of avian folia with mammalian cerebellar structures according to Larsell (1948, 1967).*

| Avian folium | Mammalian structure |
|---|---|
| I | Lingula |
| II and III | Central lobule |
| IV and V | The culmen |
| VI | The declive |
| VII | Folium and tuber vermis |
| VIII | The pyramis |
| IX | The uvula |
| X | The nodulus |

In general Larsell (1948) agreed with the earlier conclusions of Ingvar although he directed attention to the latter's incorrect identification of fissure y in the adult. In particular he confirmed the synonymy of the avian fissure x and the mammalian primary fissure (Table 30). In contrast Saetersdal (1959a) concluded from his study of the representatives of 9 orders that Larsell's intraculminary fissure is the true homologue of the mammalian primary fissure. His descriptions, which were referred to in Chapter 4, suggested that this landmark always appears very early on in ontogeny and, indeed, more or less at the same time as the prepyramidal fissure. It is therefore one of the three primary fissures that dissect the corpus cerebelli. The first to appear in the frontal and middle cerebellar regions, it is also the deepest and most prominent fissure amongst Passeriformes. Saetersdal concluded that fissure x of Brouwer and Ingvar does not appear until a much later stage of ontogeny, and usually at the same time as the secondary and tertiary sub-folia of the posterior region. He attributed his points of difference from the story of Ingvar and Larsell to their inadequate embryological material. Fissure formation in the posterior region would therefore proceed at a speed which is considerably in excess of that occurring in the front half of the disc and contrasts with cerebellar differentiation in mammals which tends to be most intensive anteriorly. On this scheme avian fissure formation can be localized into two distinct regions, the one frontal, the other posterior. Dissection of the disc commences with the establishment of the fissura secunda in the hind region and then new lobes are divided off successively in both rostral and caudal directions. Anteriorly the appearance of the primary fissure marks the beginning of the frontal phase and subsequently fissures are again established both in front of, and behind this point.

The result is that, if Saetersdal's suggestions are correct, then there are four primary lobes in the anterior region of Larsell's system. Saetersdal was therefore forced to the reluctant conclusion that it is not possible to present a complete homologization of the avian and mammalian cerebellar lobes on the basis of embryological development. In contrast to the scheme outlined in Table 31, he suggested that within the avian anterior lobe folium I represents the lingula, II the central lobe, with III and IV representing the culmen. In the posterior lobe folium V will be included in the declive. He considered that this is far larger in many Galliformes and Anseriformes than it would be in Larsell's scheme although he suggested that such size variations are less in the Passeriformes. As another possible alternative he intimated that V may correspond to the whole declive and both VI and VII to the folium and tuber vermis.

## III. VARIATIONS IN SUB-FOLIAR SIZE AND NUMBER

Following his suggestions about the possible homologies of avian folia with mammalian cerebellar areas (Table 31), Larsell suggested that the relative degree of development of the various folia and sub-folia in the different species which he had at his disposal might reflect the differential development of the

### TABLE 32

*The possible anatomical involvement of various folia in the avian cerebellum as deduced from their relative size in different species.*

| Folium | Area of the body with which it is concerned |
|---|---|
| I | "Caudal" musculature |
| II and III | Hind limbs |
| IV, V, VI | Anti-gravity muscles and flight |
| VII | Auditory region |
| VIIIa and VIIIb | Somatic sensory projections |
| IX Big in all | Co-ordination of spino-cerebellar and vestibular |
| Larsell's specimens | impulses involved in equilibrium. |

From Larsell (1948).

parts of the body with which they were principally involved. In view of the results of Whitlock (1952), who monitored the effects of peripheral stimuli at various foci (Table 34), and Goodman *et al.*, who observed the motor activities which follow cerebellar stimulation, it seems unlikely that such

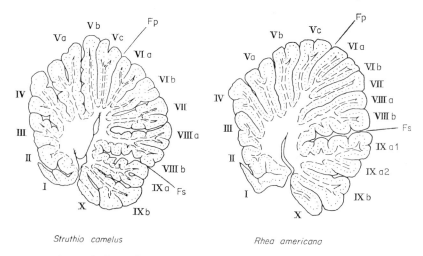

*Struthio camelus*                *Rhea americana*

FIG. 69. The cerebellum of *Rhea* and of *Struthio*. (Senglaub, 1963).

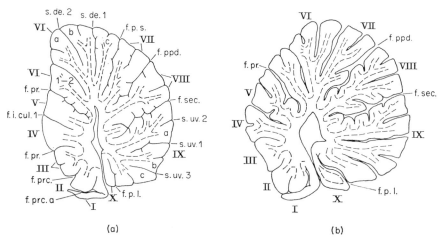

(a)                                        (b)

FIG. 70. The cerebellum of (a) the rockhopper penguin, *Eudyptes crestatus*. (b) the emperor penguin, *Aptenodytes forsteri*. (Larsell, 1967).

Abbreviations: f.i.cul., Intraculminary fissures; fprc, Precentral fissures; fpl, postero-lateral fissure; fppd, prepyramidal fissure; fpr, fissura prima; fps, posterio-superior fissure; fsec, fissura secunda; med. obl., medulla; nod, nodulus; s. uv. l., first uvular sulcus; v. m. a., anterior medullary velum; v. m. p., posterior medullary velum.

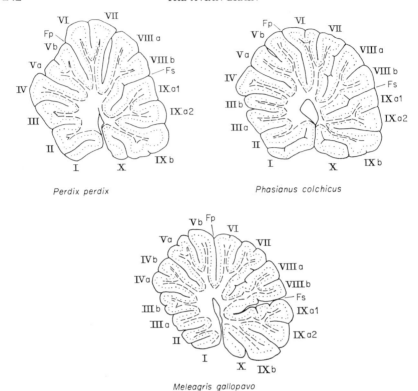

FIG. 71. The cerebellar structure of three genera of Galliformes. (Senglaub, 1963).

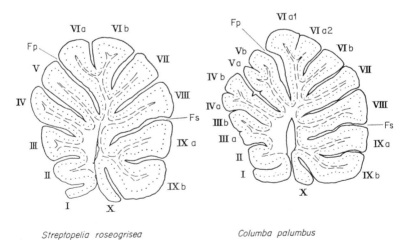

FIG. 72. The cerebellar structure of two species of Columbiformes. (Senglaub, 1963).

foliar involvement is a reflection of topographical efferent representation. However, a comparison of the data in Tables 32 and 34 will show that some of Larsell's suggestions do conform with Whitlocks experimental data. This is most obviously true in the case of folia IV, V and VI which, incidentally, are more accessible for experimental investigation than are say I and X.

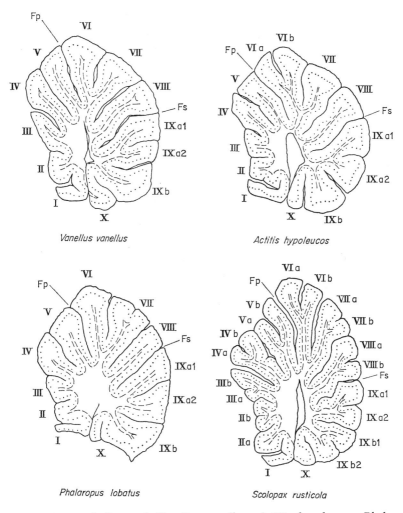

FIG. 73. The cerebellum of *Vanellus vanellus*, *Actitis hypoleucos*, *Phalaropus lobatus*, and *Scolopax rusticola*. (Senglaub, 1963).

Larsell's conclusions have influenced all more recent synoptic works on avian cerebellar structure. As such they are the basis for the short considerations presented in the Traité de Zoologie and also the major study of adult foliation and sub-foliation made by Senglaub (1963). Together with observations on a number of varieties of domestic hen, Senglaub produced details of the gross form of the cerebellum in some 70 species which were representative of 14 avian orders. Alongside the information on the Sphenisciformes which

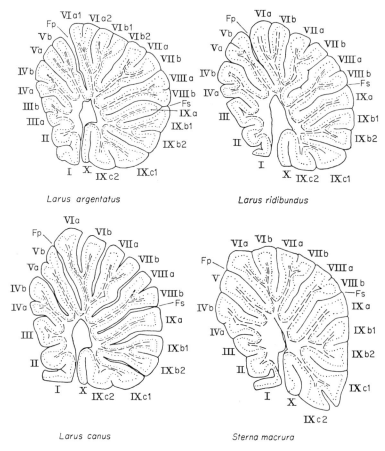

FIG. 74. The structure of the cerebellum in three species of *Larus* and in *Sterna macrura*. (Senglaub, 1963).

is contained in Larsell (1967) this enables us to obtain a very clear picture of gross cerebellar variation in birds. A comparison of many genera is provided by Figs 69–78.

The differences between the cerebella of small genera on the one hand, and the large cursorial ratites on the other was commented upon by Brandis (1894) and mentioned in Section I. As can be seen from Fig. 69, in both *Struthio* and *Rhea* there is a similar overall appearance. In *Struthio* there are well-marked

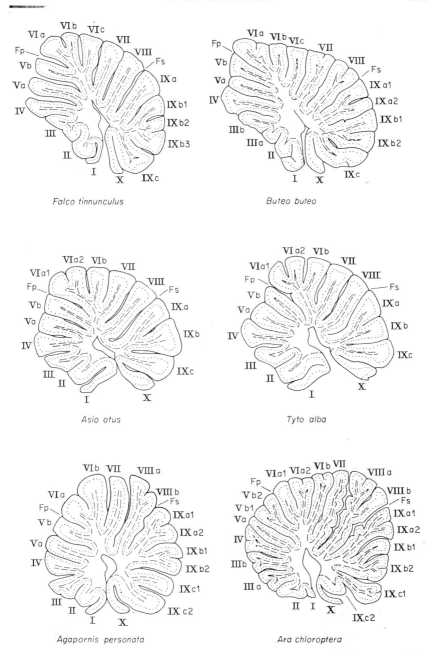

FIG. 75. The cerebellar structure of some Falconiformes, Strigiformes and Psittaciformes. (Senglaub, 1963).

fissures and almost all of the primary folia are sub-divided into secondary and
tertiary sub-folia. Folium V in the well-developed anterior lobe has three very
well-marked sub-folia. A comparison of the region of folia IX and X with its
appearance in other orders intimates, however, that it is somewhat under-
developed. Folium IX is not particularly large and the nodulus or folium
X has a relatively simple appearance.

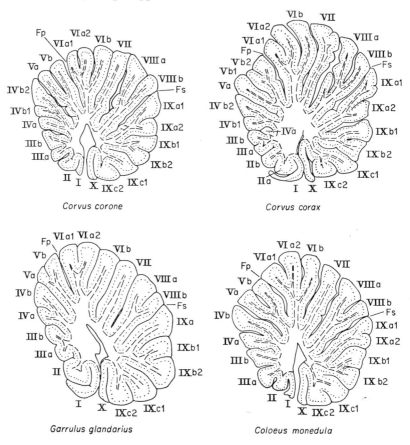

FIG. 76. The cerebellum in four species of Corvidae. (Senglaub, 1963).

Senglaub did not deal with the Impennae but Larsell (1967) provided data
for three species of penguin, *Eudyptes crestatus*, *Pygosceles adelinae* and
*Aptenodytes forsteri*. These are, respectively, the rock-hopper, adelie and
emperor penguins. As can be seen from Fig. 70 folia VI–IX, comprising
Larsell's posterior lobe, are relatively large. Folium VI of *Eudyptes* is divided
into three sub-folia apparently corresponding to VIa, VIb and VIc in owls,
eagles and pigeons. There is also an elongated fold between V and VIa. A

similar fold occurs in the species of *Aptenodytes* where V is also very deeply divided and it is also present in *Pygosceles*. In all three species folium VII is divided into two superficial folia by a relatively shallow furrow, and the walls of folium VII are somewhat indented. Since Whitlock (1952) found that in

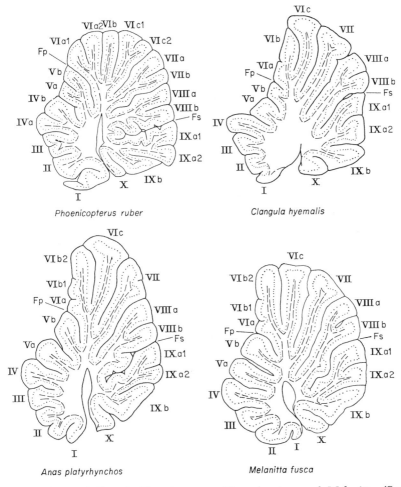

Fig. 77. The cerebellum in *Phoenicopterus*, *Clangula*, *Anas* and *Melanitta*. (Seng-laub, 1963).

pigeons, owls and ducks VII is predominantly activated by the retina (Table 34 below) a rather large folium would not be unexpected in the "large eyed" penguins. Folium VIII is also large and sub-foliated in all three. Two, and sometimes three, sub-folia reach to the surface in *Aptenodytes forsteri*, and when only two do so they are more deeply separated from one another than is the case in *Pygosceles* and *Eudyptes*. The hind wall of VIII is also very folded to produce some four or five small sub-folia. It was this folium which Whitlock found to be predominantly associated with stimuli of auditory

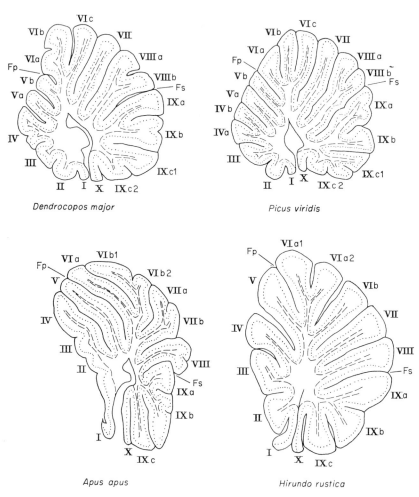

FIG. 78. The cerebellum in *Dendrocopus major*, *Picus viridis*, *Apus apus* and *Hirundo rustica*. (Senglaub, 1963).

origin. The large size of folium IX, which is common to all birds, is nevertheless accompanied by rather more sub-foliation than is usually the case.

In view of the ease with which chicks and ducks can be reared in the laboratory it is not surprising that many authors have studied their cerebellar regions. However, it will be remembered that Saetersdal (1959a) drew attention to the apparent differences which existed between the results of Brouwer, Ingvar, Larsell and himself. He concluded that in the case of embryonic material these reflected variations in the incubating conditions, but that there were definite anatomical variations in the case of adult material. This is irrespective of the differing interpretations of foliar structure which exist in the various accounts. Our knowledge of such variations has been amplified and corroborated by Senglaub's study of different varieties of poultry. It is clear that both foliar and overall differences do exist and that these fit the general pattern which will emerge from the following paragraphs.

Considering species other than the domestic fowl it is clear that amongst the Galliformes a number of differences exist between the different species which have been studied. The drawings of Fig. 71 enable a comparison to be made between the conditions in the common or Hungarian partridge, *Perdix perdix*, the pheasant, *Phasianus colchicus* and the turkey, *Meleagris gallopavo*. In the species of *Perdix* the anterior lobe has relatively simple folia and only V is further sub-divided into two sub-folia. Behind this region both VI and VII are relatively weakly developed, whilst VIII is sub-divided into two, and IX into three sub-folia. The situation in the larger species *Phasianus colchicus* is somewhat different. One may note that the overall weight of this last-named bird is some four times greater than that of the partridge, the weights being about 1200 and 340 g respectively. In the cerebellum folium III is represented by two sub-folia and folium VI has a small secondary fissure. This tendency towards a greater degree of overt anatomical complexity is carried even further in the still larger *Meleagris gallopavo* where its effects are greatest in the anterior lobe. These apparent relationships between body size and the degree of cerebellar foliation, which are analogous on the one hand to the inter-specific differences between the penguins (which were discussed above) and on the other to members of the orders Columbiformes, Charadriiformes, Lariformes, Falconiformes, Strigiformes, Psittaciformes and Passeriformes (represented in Figs 72–78 and discussed below) led Senglaub (1963) to suspect a direct and causal interaction. In this he was clearly reverting to conclusions which are similar to those which were reached by Brandis (1894) some 70 years earlier.

A comparison of cerebellar folia III–VI in the dove *Streptopelia roseogrisea* and the analogous region of the wood-pigeon *Columba palumbus*, both of which are featured in Fig. 72, will again demonstrate a greater degree of differentiation within the anterior lobe of the latter heavier bird. By comparison with the situation in Columbiformes and the galliform species of Fig. 71, the anterior regions of the cerebellum are less prominent in the Charadriiformes and Lariformes illustrated in Figs 73 and 74. Alongside this the posterior region is very strongly developed and folium IX has three massive sub-folia, especially amongst the Laridae.

An increased complexity, which is comparable with that outlined for the galliform species also parallels species size in, for example, the Scolopacidae. From Fig. 73 it can be seen that *Scolopax rusticola*, with a body weight of 300 g, has many more sub-folia than *Actitis hypoleucos* which only weighs 50 g. Additional and corroborative data of this nature are also provided by the parrots *Agapornis personata* and *Ara chloroptera* (Fig. 75) and the jay, jackdaw and carrion crow (Fig. 76). *Agapornis*, with a total weight of 35–45 g, has a less sub-divided and more diffusely organized cerebellum than *Ara*, which weighs about 1000 g; and a comparable increase in complexity occurs in the series *Garrulus glandarius, Coloeus (Corvus) monedula* and *Corvus corone*, whose body weights are 140, 220 and 450 g respectively. In general, Senglaub (1963) concluded that, although similarities between both species and genera can extend to the details of fissure pattern, differences are clearly introduced by contrasting body size.

In view of all the foregoing examples it seems reasonable to agree with Senglaub. One may say that cerebellar cortical divisions vary in association with overall body size amongst the members of a given taxonomic grouping. In genera and species of large body weight the fissures tend to penetrate deeper into the organ than is the case in other related genera and species of smaller size. As a result of this the sub-foliation is more extensive amongst the former. There also tends to be a greater degree of differentiation of the "auricular" region in large bodied genera. This overall increase in the extent of the foliation clearly implies a direct linear relationship between the area of the cerebellar surface and body size as portrayed by total weight. However, species from different orders whose weights are nevertheless comparable can have very different patterns of foliation. This is exemplified by a comparison of the charadriiform species *Phalaropus lobatus* and the psittaciform species *Agapornis personata* (Figs 73 and 75) both of which weigh approximately 40 g. Alternatively one can cite the pheasant, *Phasianus colchicus*, and the herring gull, *Larus argentatus*, amongst the heavier forms. Specimens of both these species weigh about 1100–1200 g. Complementary and corroborative conclusions are also provided by a comparison of the degree of foliation in the species pairs *Perdix perdix* and *Garrulus glandarius*, or *Streptopelia roseogrisea* and *Agapornis personata*.

In contrast to these differing foliar patterns in species pairs of comparable weight, Senglaub also drew attention to some *similarities* which can occur between species which are widely separated taxonomically but whose overall morphological organization has certain characters in common. Although his first two examples may be considered by some to beg the question, others are both striking and unquestionable. He emphasized the similarity which exists between the cerebellar regions of the Struthioniformes and Rheiformes, and between various Anseriformes and the still somewhat controversial flamingoes

(Fig. 77) whose exact taxonomic relationships are disputed. More striking are the similarities between the anterior regions of *Micropus* (*Apus* auctt), various Trochiliformes and the passeriform family Hirundinidae. In the swift which appears in Fig. 78, the front part of the cerebellum is very poorly developed. Indeed Senglaub suggested that the lack of development which is shown by folia II and III may be an expression of the relatively reduced use of the hind limbs in this genus and its relatives. The considerable development of IV, V and VI he attributed to the great versatility which they show during flight. These ideas are identical to those put forward 15 years earlier by Larsell to explain the comparable situation in the trochiliform genera *Cyanolaemus* and *Lampornis*. The humming birds are of course placed by many ornithologists and zoologists within a separate sub-order of the order containing the swifts, but such a close taxonomic relationship does not exist between the swifts and swallows as the latter are members of a passeriform family. It is therefore of considerable potential interest that Senglaub considered folium VI to be somewhat comparable in the two groups although an inspection of Fig. 78 reveals that the resemblance is not particularly marked. The weakest development of VI amongst all these species occurs in *Phalaropus lobatus* where it contrasts with the prominence of its homologue in swifts. In conclusion, it is essential to recall the warning of Saetersdal (1959a), who said that complete homologies between the avian cerebellar lobes are difficult and any specific suggestions are best treated with at least some scepticism, no matter from where or from whom they originate.

## IV. THE CEREBELLAR NUCLEI

In the pigeon three principal nuclear masses are associated with the cerebellum and it was these that Groebbels (1929) found were associated with vestibular activities. He concluded that they exerted a reflex inhibitory control over the vestibular centres in the medulla which are involved in the turning and raising of the neck. Injury to the cerebellar nuclear masses of the right side frees the right medullary vestibular centres from this inhibition and as a result the tone of the muscles which turn the neck to the left is increased. The lateral nuclei also appeared to be of importance for spreading the tail as following bilateral extirpation the ability to fan the tail is lost on both sides.

The three nuclei of the pigeon are generally referred to as the internal, intermediate and lateral. The latter lies to the side of the basal cerebellar region and in its more posterior part is above the closely juxtaposed angular, laminar and lateral vestibular nuclei (*q.v.*). Further forward, as the laminar nucleus takes up a more median position, the lateral cerebellar cell mass is embraced by the dorsal component of the dorso-lateral vestibular nucleus. In the anterior and mid-longitudinal extent of its range it is situated below and to the side of the intermediate cerebellar nucleus. This projects forward as a discrete unit above the lateral cerebello-vestibular process and to the lateral side of the internal cerebellar nucleus. In the more anterior regions the internal and intermediate nuclei are closely apposed and lie above the ventral cerebellar commissure, which, as it assumes its maximum distribution, first of all displaces the lateral cerebello-vestibular process to the side and then finally occludes it. Abutting the ventricular cavity for much of its length, the internal nucleus tapers away from this in the most anterior area where it projects forward in front of the eclipsed intermediate cerebellar nucleus.

Historically the component cell groupings have been the subject of some controversy. As is so often the case a certain amount of this reflects nomenclatural confusions and in particular this is true of the disagreement which has occurred over whether the individual nuclei are discrete entities. Doty (1946) showed that those of *Passer domesticus* are continuous with each other through "Cell-bridges", and he was therefore of the opinion that they all comprised a single and greatly folded nuclear mass. More recently Yamamoto *et al.* (1957) came to similar conclusions. Nevertheless this does not seem to negate the usefulness of anatomical terms indicating the principal unitary components. More confusing discrepancies occur in the earlier literature. Brandis (1894), Shimazono (1912), Kappers (1921) and Bartels (1925) only described two nuclei on each side which they generally called the medial and lateral. Turner (1891) had similar opinions and considered that a large, irregular cell mass situated close to the ventricle was homologous with the mammalian dentate. The cell groups were, however, often sub-divided further, and in *Alauda* Bartels described up to seven smaller components within the lateral nucleus. The works of Ramon y Cajal (1908) carry these observations to a rather different conclusion. He thought that the deep cerebellar grey in the avian brain could be attributed to five nuclei—the superior lateral, inferior lateral, intermediate, intercalate and internal, and he homologized the two lateral masses with the dentate of mammals. Further confusion is added by the fact that followers of Koelliker (1896) homologized the entire cerebellar grey with the fastigial nucleus of man and other mammals, whilst Yamamoto, Ohkawa and Lee use the somewhat different terms medial, interpostal and lateral.

# V. CEREBELLAR HISTOLOGY

Throughout all vertebrates the organization of the cells and axons within the cerebellum has a similar and distinctive appearance. The grey matter is found in two situations, as the surface forming the cortex and as independent nuclear masses, the cerebellar nuclei, in the interior. The gross anatomical relations of these latter nuclei were considered in Section IV. Externally the cortex is covered by the pia mater, internally it overlies the white matter.

The *molecular layer* consists of superimposed superficial, intermediate and deep strata of nerve cells and a large number of non-myelinated fibres. In general terms the cells of the superficial stratum comprise small pyramidal structures which send their dendrites in all directions. The axons terminate by synaptic contact with the Purkinje cells which are located in the deepest layer. Between these two levels are the large pyramidal basket cells whose dendrites and axons project in a sagittal direction. They not only end by arborizing around a Purkinje cell, but also give rise to collaterals which behave in a similar way so that one basket cell can come into contact with a number of Purkinje cells. The deep layers consist of Purkinje cells and Golgi cells. The former are peculiar to the cerebellum, and form a single layer of large flask-shaped cells at the junction of molecular and granular layers and rest on the last-named layer. One or more dendrites then pass in to the molecular layer where their sub-divisions produce a rich arborescence. Like the Purkinje cell itself this arborization is flattened in a plane at right angles to the long axis of the folium so that in sections across the folium both are broad and expanded, whilst in sections which originate parallel to the long axis of the folium both are limited to a narrow area. From the bottom of this flask-shaped cell an axon passes through the granular layer to the deep nuclei and gives off collaterals as it does so. The cells of Golgi which lie alongside the Purkinje cells have short axons that end by synapsing with the short dendrites of the granular cells. Such cells provide a potential mechanism for re-stimulation of the granular cells and thereby the Purkinje cells.

The *granular layer* is composed of numerous small nerve cells together with nerve fibres. The granular cells themselves are small, round, and equipped with relatively large nuclei. Each cell is provided with a number of dendrites which radiate out and terminate close to the cell-body in fine terminal tufts. Many fibres enter the granular layer from the white substance. Some merely traverse it and pass to the molecular layer whilst others divide into a number of branches and go no further. These bear rather moss-like appendages, are termed moss fibres, were first described in birds, and according to Ramon y Cajal they originate from the spino-cerebellar and olive-cerebellar projections. As each moss-fibre discharges into several granular cells the number of Purkinje cells, which comprise the primary neurones of the efferent pathway, stimulated by one moss-fibre is increased, and the whole organization provides a mechanism for the wide dissemination of a response to an afferent impulse.

The specific works on avian cerebellar histology are largely rather old— Stieda (1864), Ramon y Cajal (1888, 1904, 1908, 1909–11), Falcone (1893), Dogiel (1896), Craigie (1926), Hirako (1935, 1940) and Larsell (1967). The small number of published contributions in more recent decades reflects the overall similarity between avian and mammalian cerebella. This was first emphasized by Ramon y Cajal and has resulted in the avian structure being neglected. There are, however, some differences and amongst these one can cite the relatively reduced arborization of the Purkinje cell dendrites. Two or three processes were described by both Dogiel (1896) and Hirako (1935) but these multiple units are less common than cells with a single dendritic process and seem to occur beneath the deep parts of the fissures. In all cases the primary processes divide at some distance from the cell body and the resulting secondary processes diverge away from each other at a wide angle. Tertiary terminal dendrites arising from these secondary ones are largely distributed in a plane which is perpendicular to the cortical surface but some, particularly the lower ones, course horizontally or even towards the granular layer. Golgi preparations suggest that they are less numerous in the young chick than in the adult pigeon, and in both cases their number is less than that which is customarily seen in mammals. Hirako described their slender punctilinear appearance following Weigert-Pal staining and it looks as though the actual number of "spines" with which they are studded is smaller than in mammals and that the interspine distances are correspondingly greater.

Usually the Purkinje cell axons have only got two collaterals in contrast to the three or four which are present in mammals, and these are recurved so that they enter the zone immediately above the cell body. Ramon y Cajal concluded that the axons reach to the cerebellar nuclei and experimental evidence for this was presented by Shimazono (1912). In one experiment he actually obtained data which suggested that some also project as far as Deiter's nucleus. Such fibres would parallel the long cortico-fugal units that have been described in mammals as reaching the vestibular system from the entire vermal cortex.

A plexus of peri-cellular fibres surrounds the Purkinje cells and is analogous to the peri-cellular plexus of mammals. If Shimazono's conclusions are to be believed the components of this plexus are derived from afferent systems within the optic lobe, spinal cord and medulla since lesions to these areas are followed by the appearance of degenerating fibrils. The granular layer includes both granule cells and star or Golgi type II neurons, but again, according to Ramon y Cajal (1909–11), the avian granules have fewer and shorter terminal dendritic branches than do those of mammals. These processes end in cerebellar glomeruli which are analogous to those first described in mammals by Denissenko (1877). The axons ascend and bifurcate within the molecular layer in the typical fashion.

Basket cells are numerous, usually either triangular or spindle-shaped, and two or three dendritic processes arise from opposite poles. At first of large size they subsequently branch and the resulting long, slender, varicose fibres produce many finer branches. In some cases this occurs at various levels of the molecular layer, but in others just below the foliar surface. In the outer-most zone of the molecular layer small and superficial cortical cells occur in large numbers, many lying close to the pia. According to Dogiel they are of two major types, the one with ascending axons and the other with axons run-ning at right angles to the principal foliar axis. Cells with an appearance which is between that of the cortical cells and that of the basket cells are distributed

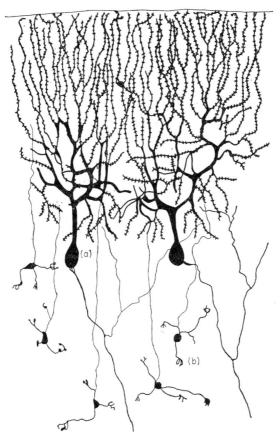

FIG. 79. Purkinje cells (a) and Granular cells (b) from the cerebellum of the pigeon. (Ramon y Cajal, 1909–11).

within the zone separating the deep and superficial layers of the molecular layer.

Mossy fibres are of a similar overall appearance in both birds and mammals although there are slight variations. Craigie (1926) found the ones with the simplest appearance in the blackbird which was the strongest flier of the five species which he studied. Synarmotic cells which connect the granular layer on one side of a medullary ray with that on the other side, were described in birds by Löwenberg (1938, 1939) following Landau's earlier description in mammals. They are largely localized within the medullary substance and from them arise associative fibres that connect different parts of the same folium. The star cells have also had a similar associative function ascribed to them and their number is said to be greater in birds than in mammals.

## VI. CEREBELLAR HISTOCHEMISTRY

There are a number of histochemical studies of avian cerebellar material in the literature, and one may consider them here alongside the biochemical determinations which have already been mentioned in Chapter 4 and other unpublished observations. Singh (1966) reported that phosphorylase activity could be detected in the cortex, Purkinje cells and throughout the white matter. The fibre tracts exhibited positive reactions in the axis cylinder but the ependymal cells appeared to give a negative result. He also gave a description of periodic acid Schiff staining which suggests that the bulk of the positive material is intercellular. The granular layer was unstained except in the proximity of blood vessels and the ramifications of the Purkinje cells also gave negative results (Singh, 1967).

Data on phosphatase activity during embryological development, which are comparable with those of Fig. 67 for the medullary region, are contained in Fig. 80. Rogers (1960) concluded that there is at first relatively little positive activity, and that that which is detectable is associated with glial activity rather than axons. It occurs in two places. On the one hand one can see it just external to the Purkinje cells, and, on the other, within the tracts that make up the arbor vitae—the numerous fibres which are interdigitated between the deep-lying cerebellar nuclei and enter or leave by the cerebellar peduncle. By the 19th day of incubation very intense staining occurs in both these areas and it is particularly heavy on the nerve fibres. Only the deeply lying tracts which make connection with the brain stem retain this staining ability in the young adult although a new phenomenon appears at that time. Throughout the cerebellum stained bundles approach or leave the Purkinje cell surfaces on the granular layer side, but it was by no means clear that these were Purkinje cell axons. As was the case in the medulla the large cell bodies of neurons never exhibit activity and these therefore stand out as pale spots.

No evidence for 5-hydroxytryptamine was obtained by Eiduson (1966) when using cerebella of chick embryos younger than the 20th day of incubation. From then on he was able to detect approximately $0.2 \, \mu g/g$ wet weight of tissue. Aprison and Takahashi (1965) found that of the four principal brain parts which they studied (the medulla, cerebellum, diencephalon plus optic lobes, and telencephalon) only the cerebellum had values for acetylcholine,

5-hydroxytryptamine, dopamine and nor-adrenalin which were of comparable magnitude. These were 1·60, 1·30, 0·85 and 1·4 mμmoles/g respectively (Table 15). Far greater variations of all four neuro-humors occurred between the four regions and the "mid-brain" values for acetylcholine were some 18 times greater than those of the cerebellum.

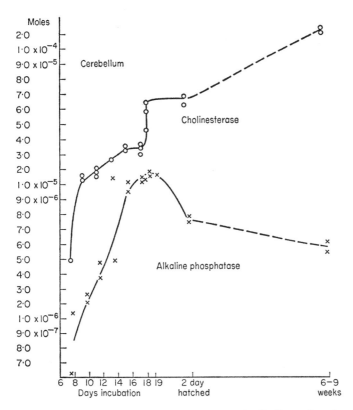

FIG. 80. The embryonic developmental pattern of alkaline phosphatase and cholinesterase activity in the chick cerebellum expressed as moles of substrate split per milligram of tissue nitrogen for the different developmental periods. (Rogers *et al.*, 1960).

Rogers *et al.* (1960) observed a sharp increase in cerebellar cholinesterase on the 19th day of incubation (Fig. 80). As in both the forebrain hemispheres and retina the level had by this time reached $5 \times 10^{-5}$ moles of substrate split per mg of tissue per hour. They assumed that the sharp increase was associated with the functional changes that take place prior to hatching. The differential distribution of such activity within the cerebellum of pigeons, canaries and parakeets was considered by Friede and Fleming (1964), and I have obtained comparable results in other anseriform, falconiform, psittaciform and passeriform species. Their results are contained in Table 33. It is worth noting that besides these esterases the use of other differential inhibitors in the concentrations which are recommended by Pearse (1960) suggests the presence of some A, B and C esterase activity.

## TABLE 33

*The distribution of cholinesterase activity within the cerebellum and the effects of various inhibitors on it.*

| | Acetylcholinesterase | | | | | Butyrylcholine esterase | |
| | Molecular layer | Granular layer | Eserine | DFP | BW | Molecular layer | Granular layer |
| --- | --- | --- | --- | --- | --- | --- | --- |
| Pigeon | ++ | —— | total | total | — | + | — |
| Canary | ++ | —— | total | total | — | + | — |
| Parakeet | ++ | —— | total | total | — | + | — |

From Friede and Fleming (1964).

These data broadly support the suggestion of Feldberg and Vogts that chains of cholinergic and non-cholinergic neurons exist. They also suggest that these are not necessarily arranged in an alternate sequence and that the type of transmission seems to be somewhat haphazard. The organization within the cerebellar cortex permits one to study two subsequent synaptic regions in the same section. These are provided by the contacts between afferent mossy fibres and the granular cells (cerebellar glomeruli) and between the granule cells and Purkinje cells. The anatomical stability of the architecture, is, as has already been emphasized, common to, and stable throughout, all the vertebrates. There are no obvious morphological differences which would offer themselves as a potential explanation for the variation which is sometimes observed in the acetylcholinesterase patterns. As a result of this it has been suggested that either cholinergic or non-cholinergic transmissions can be used at a particular synaptic type and that the results are very much species dependant.

## VII. CEREBELLAR ABLATION STUDIES

In view of the delicate nature of the postural control which is exerted by the cerebellum of mammals, and the disorders such as intention tremors which accompany its malfunction, it is not surprising that gross extirpation leads to massive disorders in birds. Similarly, if the results of Goodman *et al.* (1965) are of general application (see Section VI), it is equally understandable that the results are not always easy to analyse. Certainly it is fair to say that the most interesting and informative data have been obtained by stimulation and activation experiments rather than the production of preparations with extensive cerebellar lesions.

The first experimental investigation of avian cerebellar function seems to be that of Flourens (1824). He reported that the ablation of the upper layer in *Columba* was not accompanied by any marked changes although extirpation of the deeper layers resulted in both postural and locomotory disturbances. Weir-Mitchell (1869) also observed such locomotor disabilities after cerebellar extirpation, noted that such preparations became fatigued quickly and concluded that it had an important role in the maintenance of energy. Lange (1891) described analogous disorders after extensive but incomplete extirpation and suggested that they were the expression of a sharp increase in muscle tonus at the extremities. These last went into a state of spastic extension if the preparation attempted to move. A few days after the operation these symptoms showed a slight remission and at that time the preparations could undertake a reeling and disordered pecking. After a few months they could again undertake both eating and drinking although a *slight* tremor persisted. Under similar circumstances Bremer and Ley (1927), and Martin and Rich (1918) reported extensor rigidity, head-tossing and locomotor disorders.

Groebbels (1928a, b, c, d, e, 1929) attempted to investigate the possible interrelationships between the cerebellum and the vestibular apparatus. He concluded that impulses from those vestibular centres which are involved in turning and lifting the neck included reflex mediation by the cerebellum. He envisaged that the latter exerted a regulatory or inhibitory function. Unilateral destruction of the cerebellar nuclei or "restiform body" led to a contralateral bending of the neck and ablation of the vestibular centres had the same result. After ablation of both the vestibular area and the lateral cerebellar nucleus of the same side, head-turning appeared a few days following the operation. This was at first ipsilateral and subsequently contralateral. If this extirpation was carried out bilaterally there was no neck twisting and as a result he described a regulatory pathway which involved the cerebellum, fastigial nucleus, uncinate fasciculus and vestibular centres. Whilst bilateral extirpation of the connecting commissure or extirpation of the nuclei gave opisthotonus, hemilateral section or extirpation did not, and such results differed markedly from those which he obtained by destroying the frontal ampullae.

More recent works on gross cerebellar ablation are closely associated with Karamyan (1956) and his collaborators. They undertook extirpation from behind forwards, removing the anterior region "layer by layer", and being very careful to avoid any extensive haemorrhage from the occipital region. Using cockerels, ducks and doves they found that following total or extensive ablation the prognosis in the last-named was often poor. In the immediate post-operative period there were several locomotor disturbances and some idea of the extent of the operation could be obtained by observing these. They found, by subsequent histological investigations, that preparations in which the head, tail or wings were held in an unnatural position, and in which there was distortion of the neck, had had incomplete extirpation and the material had been removed asymmetrically. Alternatively such symptoms indicated that other, and non-cerebellar, foci had been damaged. Those preparations in which total symmetrical ablation was achieved and in which there were no accessory lesions did not show such postural asymmetry. However, the head swayed backwards, forwards and sideways, and there was a marked nystagmus. Karamyan also described a condition which he called severe motor excitation. The preparations "tried to stand up and run", turning over from side to side and sharply extending their wings. During such seizures of uncoordinated motor activity they exhibited opisthotonus, and, as a result, the head was thrown back so that it almost touched the neck in the vicinity of the wings. The feet and tail were also extended, with the feet stretched in the direction of the tail which was itself in an almost vertical position. This phase of extensor hypertonus of the neck, feet and tail muscles was not a constant one and the manifestations increased when the preparations "attempted to walk". The "motor irritability" was so excessive that any internal or external stimulus increased the overt extensor tonus. It decreased, however, during "Absolute rest".

This sharp increase in extensor muscle tonus, coupled with the prevalent state of hypotonus at rest, suggested to Karamyan that in cases of cerebellar extirpation both atonia and hypertonia reflected the general or central excitatory state of the nervous system (see Sherrington, 1906). This conforms with the findings of Moruzzi and also with those of von Holst and Saint Paul (1963) which were referred to in Chapter 1, Section I. Moruzzi undertook focal stimulation of the cerebellum and reported that increases or decreases in muscle tonus were dependent upon what he referred to as "the initial level of irritability".

Such muscular symptoms as those which have been described above were also accompanied by severe alimentary disorders which the Soviet workers ascribed to autonomic dysfunction. These could include diarrhoea together with peripheral disturbances such as loss of feathers. Ulceration of the extremities followed when the preparations survived for longer than 10 days but most specimens became severely emaciated within 2–3 days of the operation.

The results of similar cerebellar ablation in *Gallus* and *Anas* were somewhat different. Whilst the doves exhibited excessive motor disorders and postural dystonia, Karamyan found such symptoms were secondary or absent in the rooster. During the immediate post-operative period motor activity was reduced or entirely absent. The preparation lay quietly with limbs extended and he concluded that unnatural rest comprises an important component of the cerebellar ablation syndrome. There were, however, marked changes in muscle tonus of the neck and extremities. Hypertonia persisted and was subsequently replaced by atonia. As in the case of the doves such de-cerebellate rooster preparations showed severe alimentary dysfunction although it was not possible to determine satisfactorily the possible causal relationship between alimentary muscle atonia and muscular atrophy. According to the Soviet workers it is not difficult to induce conditioned reflexes in such de-cerebellate preparations. Fifteen days after the operation, using either light or sound stimuli, they were able to elicit such reflexes as easily as with normal birds.

As a result of the difficulties which are attendant upon such total ablation experiments Karamyan (1956) also undertook partial extirpations. Although Ten Cate (1936, 1965) and also Bremer and Ley (1927) described the results of extirpating the anterior lobe alone as identical to those of total ablation Karamyan found that in his preparations the symptoms of partial extirpation were different. Many of the 27 preparations persisted for a considerable post-operative period. In the first few days their movements were disordered and ended at once in prostration with wings and tail extended. Each sporadic seizure of this kind was associated with hypertonus of both the caudal and appendicular musculature. However, after 5–7 days they could stand erect although any attempt at movement was inevitably accompanied by severe swaying. The limbs were extended to the front and side of the mid-line, and there was a marked hyperaesthesia. As a result of this any weak external tactile stimuli gave rise to violent reactions, and attempts at both preening and scratching were very noticeable. With the passage of time all these motor disorders decreased, as did many of the dystrophic phenomena, although some traces of dystonia, astasia and ataxia always remained.

## VIII. CEREBELLAR ACTIVATION BY PERIPHERAL STIMULI

It has already been noted that Larsell (1948) suggested the possible topographical involvement of different cerebellar folia on the basis of their relative degree of development in genera of differing habits and build (Table 32). After these suggestions had been made Whitlock (1952) carried out electrophysiological experiments to determine the afferent pathways which lead to cerebellar activation. Using specimens of *Columba livia*, *Bubo virginiatus* and

*Anas platyrhynchos* held under barbiturate anaesthesia, he utilized both natural physiological stimuli and also single, thyratron controlled, electrical condenser discharges on peripheral nerves. These studies made a tremendous contribution to our knowledge of the distribution of various receptor zones within the cerebellar cortex. The first unequivocal demonstration of spontaneous electrical activity within isolated avian cerebellar tissue not exposed to such peripheral activation was provided by Cunningham and Rylander (1961). This compares with the similar activity which has been shown to occur in cats by Snider *et al.* (1967) but its precise significance in life is not yet clear since much cerebellar function is consequent upon primary sensory activity at other and peripheral foci.

Whitlock used an electro-magnetically controlled stylus to move single feathers and provide tactile stimulation. The tip of the stylus moved through 3 mm and was an adaptation of the technique which Snider and Stowell (1944) developed. Stimulation of the visual receptors was achieved by a flash from a Strobotron placed at a distance of 10 cm from the dark adapted eye and triggered by a thyratron controlled condenser. For the auditory system Whitlock used the customary click stimuli emitted from an earphone located 5 cm away from the homolateral ear and again controlled by a thyratron condenser combination but these were of unknown pitch and volume. Cerebellar activation was only recorded as positive if the electrical discharges occurred when the feather was actually being moved, the light stimulus falling on the eye or the click stimulus being emitted. As a result they give a measure of the latency within the direct afferent pathways, but do not indicate those of indirect pathways which may lead to subsequent and more delayed activation of identical or different cerebellar loci.

The results which were obtained from such experiments are summarized in Table 34. A comparison of this Table with Table 32 shows that there is some agreement between Larsell's earlier deductions and inductions, and the experimentally verified representation of sensory projections. This is true in the case of folia IV, V and VI. Direct stimulation of the peripheral nerves at various points of the body gave activation within the ipsilateral regions of folia III, IV, VIa, VIb and VIc. Whitlock concluded that there is a definite somatotopic representation with the caudal part of the body projecting most rostrally and the leg, wing and facial areas following in sequence behind. An analogous topographical organization was also indicated by the activation which accompanied tactile stimulation of the feathers. After such stimuli activation potentials occurred within the cerebellar folia following a latent period of 22–30 msec in the case of the leg feathers, 16–22 msec in the case of the wing feathers and 10–16 msec for facially evoked responses. The leg responses were recorded ipsilaterally in folia IV, V and in one case VIa. Tail responses occurred ipsilaterally in folia III and IV, wing responses in IV, V

## TABLE 34

*Summary of the cerebellar folia in which activation potentials were recorded following discrete peripheral nerve and receptor stimulation.*

| Animal and structure of stimulation | Regions stimulated and locations of responses | | | | | | |
|---|---|---|---|---|---|---|---|
| | Tail | Leg | Wing | Face | Cochlea | Retina | Optic tectum |
| *Columba* | | | | | | | |
| Nerve | IV | IV, V | IV, V, VIa | VIa, b | | | |
| Receptor | III, IV, Va | IV, V, Va | V, VIa | VIa, b, c | VII, VIII | VII, VIII | |
| *Bubo* | | | | | | | |
| Nerve (tectum directly) | III | III, IV, Va | IV, Va, b, VIa | | | | VII, VIII |
| Receptor | III, IV | IV, Va | IV, Va, b | VIa, b, c | VIc, VII, VIII | VII, VIII | |
| *Anas* | | | | | | | |
| Nerve | IV | Va, b, VI | Vb, VI | VIb | | | |
| Receptor | | | | | VIIa, VIIIa, b | VIc, VIIb | |

From Whitlock (1952).

and VIa, and the facial responses in folium VI. These areas were, as far as could be detected, co-extensive with those which exhibited activation after direct stimulation of the peripheral nerves.

Activation by auditory click stimuli was seen at 16 foci within folia VIc, VII and VIII. The latent period which intervened between the stimulus presentation and the appearance of cerebellar potentials was rather less than that following tactile stimulation of the facial region and lasted from 4–10 msec. As one would expect in view of the predominance of decussating fibres within the optic tracts retinal stimuli contrasted with the foregoing and produced only contralateral activation. The responses were located in folia VIc, VII and VIII and had considerably longer latent periods than those which followed auditory stimulation ranging from 24–26 msec. However, the areas involved in the case of both types of peripheral phenomena appeared to be broadly coextensive and as a result Whitlock suggested that folium VIc, folium VII and folium VIII should be designated the audio-visual area. In mammals, Snider and Stowell (1944) had come to a similar conclusion for the lobulus simplex, the folium and tuber vermis, and the pyramis complex, a fact that could be cited as corroboration of Larsell's (1948, 1967) homologizations which are contained in Table 31.

# IX. THE EFFECTS OF DIRECT CEREBELLAR STIMULATION

Direct electrical stimulation of the cerebellar cortex has been carried out on a number of occasions and has recently resulted in the very elegant hypotheses of Goodman and his collaborators. Bremer and Ley (1927) used weak electrical currents on the anterior lobe of pigeons which had extensor spasm following forebrain extirpations. They observed an inhibition of the extensor tone in both neck and leg muscles and concluded that this reflected an inhibitory mechanism which is concerned with the maintenance of posture. These experiments were extended by Raymond (1958) who decided that the responses which he evoked could be broadly categorized as those obviously related to the act of balancing and others, in which such a relationship was less apparent.

Raymond used implanted electrodes in chickens coupled with direct cerebellar stimulation in anaesthetized pigeons. As would be expected both from *a priori* consideration and the widely cited results of von Holst and Saint Paul (1963), he found that the movements which he elicited by the stimulation of unrestrained birds varied with the position of the electrode, the strength and duration of the stimulating current, the prevailing posture of the preparation, the acts being performed at the moment of stimulation and, finally, the interval between stimuli. Three general phases could be observed which, in terms of the time that elapsed between the application of the stimulus and their appearance, he termed stimulus, rebound and long after effect. All his preparations responded by either swaying, rotating or squatting with flexion of the anti-gravity muscles. The precise form of the appendicular responses accompanying partial or complete squatting varied and appeared to be directly related to the stimulus intensity. With a strong current they were executed quickly, with a low current rather slower, and in the first case they comprised the first overt action which the preparation made.

Head-turning always occurred if the stimuli were applied at foci which were situated away from the mid-line. With the exception of sites in the culmen and centralis (*sensu* Larsell) this involved an ipsilateral motion. Stimulation at these foci, but away from the mid-line, gave mixed head-turning responses. Neck responses were also universal but the pattern varied from folium to folium. Extension of the neck muscles occurred when the stimuli were applied to folia III, IV, VIa, VIb, VIc, IXa and IXb. Retraction followed from folia Vb, and mixed extension/retraction from folia Va, VIIIa and VIIIb. Movements of the beak, tongue and nictitating membranes resulted from stimulation of III, VIc, VII, VIIIa and VIIIb. All these data are summarized in Table 35.

As far as the wings are concerned the most consistent source of motor responses was folium VIII, but Va, VIa and VIb, VII, VIIIa and VIIIb also produced effects. Stimulation of III, IV, IXa and IXb gave definitive stimulatory responses in the tail, with Vb, VIa, VIb, VII and VIII giving other mixed results. It is also worth noting that although the application of electrical stimuli to the regions behind the secondary fissure was associated with a hyper-active after effect, primary movements which were elicited from elsewhere were usually followed by a period of quiescence that could sometimes be broken by brief periods of further activity.

After Raymond's work Goodman *et al.* (1963, 1965) also undertook an investigation of the possible functional localization within the cerebellum. The pattern which is suggested by their results is rather different from the discrete focal organization suggested by the peripheral afferent activation studies. Using 125 $\mu$ diameter, stainless steel implanted electrodes which were insulated with polyethylene they found that the elicited motor responses were slow tonic movements of the head, neck, legs and occasionally the wings. All the components of a postural pattern which was obtainable at a particular locus did not necessarily occur with each individual stimulation. However, on the occasions when the complete pattern was not observed the movements which were evoked comprised a recognizable fraction of it. In the unanaesthetized duck they demonstrated that the corpus cerebelli is divided functionally into three longitudinal cortical zones, while the flocculus represents a fourth zone. This suggests that the avian cerebellum conforms to the cortico-nuclear theory of cerebellar function which was proposed by Chambers and Sprague (1955a,b) and Goodman and Simpson (1960, 1961) for the *Alligator,* cat and monkey. This states that bilateral and reciprocal control of posture by the cortical regions of the cerebellar hemispheres is phylogenetically an old organization of cerebellar function. The regulation of ipsilateral movements is a more recent specialization.

## TABLE 35

*A summary of the motor responses which Raymond* (1958) *elicited by electrical stimulation of the cerebellar folia.*

| Folium | Face | Neck | Wing | Tail |
|--------|------|------|------|------|
| III | Nictitating membrane | Extension | | Down |
| IV | | Extension | | Erect |
| Va | | Extension-retraction | Present | |
| Vb | | Retraction | | Mixed |
| VIa and b | Eyelids | Extension | Present | Mixed |
| VIc | Beak | Extension | | |
| VII | Beak | Retraction | Present | Mixed |
| VIIIa | Beak and nicitating membrane | Extension-retraction | Present | Mixed |
| VIIIb | Beak and nictitating membrane | Extension-retraction | Present | Mixed-down |
| IXa | | Extension | | Down |
| IXb | | Extension | | Down |

At 15 electrode placements within the medial third of the corpus cerebelli, and situated approximately 2–5 mm on either side of the mid-line from folium Vb to IXb, the results of electrical stimulation were ipsilateral hind limb flexion and protraction, coupled with contralateral hind limb extension and retraction. The head and neck rotated to the ipsilateral side. If the animal was currently floating on water these effects gave rise to a circling movement in the ipsilateral direction with the contralateral leg providing the propulsion. Opposite and complementary results were obtained from a 2 mm strip of cortex situated to the side of the foregoing. Electrical stimulation at 8 sites distributed between folia Va and IXb and within this second strip resulted in the extension and retraction of the ipsilateral leg with that of the other side undergoing flexion and protraction. In this case the head and neck turned towards the contralateral side. To these two contrasting combinations of flexion, protraction, extension and retraction Goodman *et al.* gave the terms vermal and para-vermal zone patterns. The sites from which they can be elicited are represented in Fig. 81 which shows a schematic dorsal view of the cerebellum.

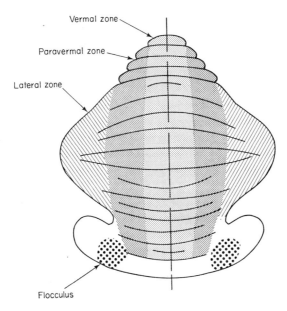

FIG. 81. The distribution of the bands evoking gross movements in the avian cerebellum (Goodman, *et al.*, 1965). The longitudinal zones have been extended to include those folia which were not reached by an electrode.

Stimulation of foci which were situated within the lateral third of the cerebellar cortex and therefore in the most lateral foliated cortex together with those non-foliated regions associated with this, gave a pattern which was comparable with that of the vermal zone and comprised ipsilateral hind limb flexion and contralateral extension. Owing to their relatively inaccessible position no focal stimulations were carried out upon anterior folia I–IV nor upon XIc, X or the paraflocculus. The flocculus gave a para-vermal zone pattern and this suggested that it was closely associated functionally with the corpus cerebelli. However, in this region, there was a superimposed contralateral rotation of the whole animal about its longitudinal axis. When freely swimming preparations were used this was reflected in a complete 360° turn. Such a direct rotational response was not observed during the stimulation of any other area.

In regions which are transitional between the definitive vermal and para-vermal zones stimulation sometimes caused one response type at first and then this reversed into the other. What is more, the electrode placements which were situated close to the border area between two zones suggested that this intermediate cortex was effective during bilaterally symmetrical responses. As a result of this, although bilateral reciprocal postures were the rule, there were occasionally bilaterally symmetrical effects, and these were evoked from many widely scattered foci. Goodman *et al.* suggested that the relatively in-frequent occurrence of forelimb movements might be a consequence of the young age of their experimental ducklings and the lesser importance of the wings in locomotion prior to adulthood. When they were involved wing move-ments only took the form of extension from the normal flexed position. Although such extension occasionally occurred simultaneously on both sides it was predominantly unilateral and elicited from sites within all three zones. Furthermore these wing movements were generally interpretable as related to the other postural characters of the region with contralateral extension evoked from the vermal and lateral zones and ipsilateral extension from the inter-mediate para-vermal zone.

Conjugate eye-movements were never observed but both eyelid closure and retraction of the nictitating membrane consistently occurred following stimulation of three grouped foci in the vermal zone of folia VIb and VIc. These actions were predominantly associated with the contralateral eye although bilateral effects occasionally resulted from one electrode in VIb. This was the only indication of any somatotopic organization within the efferent pathways. In view of the observed termination of trigemino-cerebellar tract fibres in folia Vb to VIc and sometimes VII, together with the termina-tion of afferents from the facial region in folium VIb (Table 34), both afferent and efferent studies concur.

At the end of the final testing period six of the preparations were marked by thermo-coagulation. After this, as would be expected, the threshold neces-sary to evoke somatic responses was at least doubled. There were also some changes in the observed postural patterns by comparison with the situation prior to coagulation. For example, the results of pre-coagulation stimulation of one electrode located at the vermal/para-vermal transition were pre-dominantly bilateral reciprocal movements with, nevertheless, frequent sym-metrical movements. Following thermo-coagulation only symmetrical re-sponses occurred, and then only at supra-threshold stimulus levels.

Post-stimulatory rebound was usually observed at sites within all three zones. At relatively high stimulus intensities it was then followed by a sequence of more generalized movements, which began with an immediate movement from the rebound posture into a slow and coordinated stepping sequence. This gradually increased in speed whilst the head and neck were retracted. Such generalized activity rarely lasted longer than a minute after the cessation of stimulation. The orderly series of tonic movements which Raymond (1958) reported as constituting a long after-effect in the chick never occurred in either the duckling or gosling. A further analysis of the earlier work of Raymond showed that both his results, and those of Chiarugi and Pompeiano (1956) who used a thalamic pigeon preparation, could be fitted to the general vermal/para-vermal zone pattern. Raymond used older birds than Goodman and his collaborators and he also tested them as they stood on a perch, walked or flew. As has been made clear in the foregoing paragraphs Goodman *et al.* stimulated their ducklings whilst they were either swimming freely, or, alternatively, suspended in a sling. During these tests they frequently observed that volitional movements obscured and interrupted the postural patterns that resulted from cerebellar stimulation. As they themselves point out such factors could well explain those differences which exist between the two sets of results. Raymond, as will have been seen above, did not report any medial/lateral differences in the postural responses. However, he did direct attention to a cortical area, some 2–3 mm to the side of the midline, where the necessary threshold levels rose sharply. On either side of this area responses could be elicited at far lower current strengths and the lowest thresholds which he recorded were in the more lateral regions. In contrast, Goodman *et al.* only noted significant differences in the necessary currents in the case of electrodes that actually penetrated the cerebellum and those simply impinging upon its surface. Movements which Raymond considered as overtly related to the maintenance of posture were generally in agreement with those evoked by stimulating the vermal and lateral zones of *Anas*. Goodman *et al.* suggest that his medial and lateral low threshold regions may therefore represent the vermal and lateral cortical zones, the intermediate high threshold area being the para-vermal zone. It is of interest that Chiarugi and Pompeiano (1950) found that direct cerebellar stimulation of their thalamic pigeon preparations gave ipsilateral hind limb flexure, and contralateral extension. As it would appear that the sites of their stimulating electrodes lay between 2–2·5 mm on each side of the mid-line their observations appear to tally very closely with those from the vermal zone in the duckling.

## X. SPONTANEOUS ELECTRICAL ACTIVITY IN CEREBELLAR EXPLANTS

The first detailed description of isolated foci capable of spontaneous potential production was provided by Cunningham and Rylander (1961). The period during which they made their most extensive observations lasted from 25·5 hours to 97·5 hours after explantation but the material continued to show electrical activity until 120 hours. Two and a half hours after explantation potentials were just visible as very small wave forms which were scarcely distinguishable from the background noise. By 18 hours in culture this activity was decidedly more marked and comprised a five-fold repetition of a complex series of potentials. Each repetition lasted for 5 sec and the entire sequence took 50 sec and achieved a potential of 10 $\mu$V.

After 25·5 hours a simpler repetitive signal of lower magnitude occurred and this showed some variability between sequences. It appeared to comprise the central part of the former complex with the same 5 sec interval separating the potential maxima. If such a complex failed to appear a further 5 sec period of silence followed after which activity appeared at the time when the subsequent one would have been expected. Towards the end of 750 consecutive peaks, lasting a total of 70 min, the interval separating the individual maxima was slightly increased. By 30–31 hours in culture the intervals between the impulses became more irregular and two distinct types of potential emerged. The first involved impulses which were aggregated into regularly spaced groups. The second consisted of smaller, more or less irregular impulses, which gave a continuous background. The intervals which separated the larger variety gradually increased and, whilst they lasted about 20 sec after 31 hours of culture, they reached 140 sec at 41 hours. A change also occurred in the pulse maxima between 69–72 hours, by which time they were markedly weaker, and the intervals between them more irregular. Cunningham and Rylander concluded that the progressive re-grouping of the pulses was actually a natural phenomenon and not a cultural artifact which reflected progressive cellular damage, although they ascribed the increasing length of the intervals between each burst to actual deterioration of the active sites. Two principal foci were included in their suggestions, one responsible for the major spikes and the other for the lower and rather more regular potentials.

# XI. CEREBELLAR FIBRE TRACTS

## A. Historical Introduction

The connections which appear to exist between the cerebellum on the one hand and the inferior olive and pons on the other were outlined in Chapter 7, Sections VIII B and IX C. Historically, many workers have studied both the afferent paths leading to the cerebellum and the efferent paths which leave it. A summary of the main afferent paths described in the literature would include the dorsal and ventral spino-cerebellar systems which were referred to by Friedlander (1898), Ingvar (1919) and Sanders (1929); the cochleo-cerebellar fibres which were reported by Bok (1915); direct vestibular root fibres that were described by Groebbels (1929, etc.), Ramon y Cajal (1908), Sanders (1929) and Shimazono (1912); direct trigeminal root fibres together with both homolateral and contralateral trigemino-cerebellar tracts which were cited by Biondi (1913), Craigie (1928) and Sanders (1929); a tecto-cerebellar tract that figured in the descriptions of Munzer and Weiner (1898), Wallenberg (1900), Frenkel (1909) and Shimazono (1912); and a semilunar–cerebellar tract which was described by both Craigie (1928) and Sanders (1929). All these tracts feature in Kappers *et al.* (1936), some are also cited in Kappers (1947) and represented in Fig. 140.

Complementarily, the principal efferent connections are with the reticular nuclei, vestibular system, tegmentum, and possibly other (diencephalic) regions. These include the cerebellar-spinal tracts which were referred to by Ramon y Cajal (1908), Frenkel (1909) and Shimazono (1912); both the inferior and superior cerebello-reticular tracts of Sanders (1929); the cerebellar motor system of Frenkel, Wallenberg and Sanders (*loc. cit*), the cerebello-vestibular fibres of Sanders; and the brachium conjunctivum or cerebello-tegmental tract, and the cerebello-diencephalic tract of Muskens (1930). Many of the afferent pathways were subsequently confirmed by Whitlock (1952).

## B. The Cerebellar Commissures

The commissura cerebelli traverse the mid-line within the rostral part of the cerebellar region. As such they underlie the ventricle in the posterior part of their longitudinal extent and both the internal and intermediate cerebellar nuclei further forward, where the latter tend to be closely juxtaposed and in a medial position. Following Whitlock (1952) one can conclude that there is a trigeminal bundle and a ventral spino-cerebellar tract. The trigeminal projection comes, in part, directly from the trigeminal root with other components originating in the mesencephalic trigeminal sensory nucleus. These axons pass into the area where the cerebellum is attached to the anterior medullary velum and at that point both trigeminal components and also those of spino-cerebellar origin intermingle and cross the mid-line. In the adult of *Columba* the decussation occurs within that medullary material which is situated just behind folia II and III. (Larsell's central lobule.)

According to both Larsell (1948) and Whitlock (1952) the lateral commissure in embryonic duck or chick material consists of fibres that decussate near the caudal margin of the cerebellum. They thought that these fibres arose in part from the vestibular nuclei although a small percentage pass directly from the vestibular root of the VIIIth nerve to the posterior cerebellar region. Both these direct and secondary projections can be followed into the floccular region and also towards the mid-line. Their overall path tends to follow the postero-lateral fissure which is such an important landmark in the earlier stages of ontogeny. As the mid-line is approached the number of fibres decreases and they seem to terminate in those regions close to the fissure itself.

## C. The Spino-cerebellar Tracts

As was mentioned in Chapter 6, Section II A, Karten found degenerating fibres within the cerebellar region following cervical hemisection. Within the cord the spino-cerebellar tracts are easily visible in the superficial portion of the lateral white funiculus and in the embryo they can be followed forward along the lateral surface of the medulla. The dorsal spino-cerebellar fibres (see Figs 50 and 57) traverse the rootlets of the VIIIth nerve and then swing abruptly upwards to the cerebellum. The ventral spino-cerebellar tract runs forward to the level of the roots of the Vth nerve and then passes upwards below the medullary surface to enter the cerebellar peduncle and cerebellar commissure. Similar overall distributions seem to occur in such widely different genera as *Columba, Gallus, Cyanolaemus, Lampornis, Bubo* and *Asio* which, together with the duck, represent the orders Galliformes, Anseriformes, Columbiformes, Trochiliformes and Strigiformes.

Whitlock (1952) found that in the pigeon lesions of the spino-cerebellar tract result in fibre degeneration within the superficial zone of the posterior part of the lateral funiculus. Anteriorly the fibres run through the medulla to the cerebellar peduncle and then into the cerebellum, with the dorsal spino-cerebellar tract fibres passing slightly behind them. As such his results agree with, and therefore corroborate, those of earlier workers. In the cerebellum itself the degenerating axons occur in folia II, III, IV, V, VIa and VIb. A smaller, but nevertheless quite definite, concentration also occurs within the uvula (folium IX) and scattered bundles pass to VIII. Although Sanders (1929) described them as running to the deeper nuclei Whitlock was unable to trace them to this level.

## D. The Vestibulo-cerebellar Tracts

Vestibulo-cerebellar pathways were found in both the embryonic and adult stages by Whitlock (1952) who therefore corroborated the general conclusions of Ramon y Cajal, Shimazono, Groebbels and Sanders (*loc. cit.*). Both direct and secondary projections exist and most of them seem to end within the cortex around the postero-lateral fissure, although at least in embryonic stages some pass to the deep cerebellar nuclei. In the adult the exact pathways are obscured by allochthonous fibres which also traverse the vestibular region. Following lesions of the VIIIth nerve and of the vestibular nuclei degenerating fibres could be traced into folia IX and X. Secondary fibres also appeared to pass to the so-called auricular region which is a lateral extension of these self-same folia. It is of considerable interest that Whitlock was not able to detect any cochlear projections comparable with those which were referred to in Section XI A. Furthermore lesions of the cerebellar folia never resulted in any chromatolysis within the cells of the laminar and magnocellular components of the medullary auditory complex. This is particularly surprising in view of the very distinct activation potentials which occur in at least folia VII and VIII following the presentation of click stimuli to the ipsilateral ear (Table 34).

## E. The Trigemino-cerebellar Tracts

Ablation of the superior (main) sensory nucleus of the Vth nerve resulted in other unavoidable damage to the overlying cerebellar fibre systems and this necessarily gave a confused picture of trigemino-cerebellar relations. There were, however, in all lesioned preparations, degenerating fibres which passed to folia V, VI and occasionally VII. These fibres from the main sensory nucleus correspond to the dorsal trigemino-cerebellar tract of Sanders (1929) and conform with the activation potentials which were elicited in folium VI (Table 34). They appear as a direct connection between the Vth root and the cerebellar commissures.

## F. Tecto-cerebellar Tracts

Tecto-cerebellar tracts can be seen within the anterior medullary velum of both *Gallus* and *Anas*. Fibres of such bundles run to the velar region from both the medial and superior regions of the optic tectum. Following experimental ablation of the medial tectal grey, especially around the ventricle, and also of the anterior medullary velum itself, there is some damage to other tracts. However, using the Marchi method it is possible to trace projections to folia VI, VII, VIII and IX. On the other hand cortical ablation of folia VI and VII gives rise to chromatolytic symptoms of degeneration within both the nucleus isthmi and tectal complexes (*q.v.*). In view of the activation potentials which can be monitored within three of these folia after presentation of visual stimuli the fibre studies again conform with the electro-physiological data, and one may justifiably assume a close functional relationship between both the tectal and cerebellar regions.

## References

Aprison, M. H. and Takahashi, R. (1965). 5-hydroxytryptamine, acetylcholine, 3,4,dihydroxyphenylethylamine and nor-epinephrine in several discrete areas of the pigeon brain. *J. Neurochem.* **12,** 221–230.

Bartels, M. (1925). Uber die Gegend des Deiters und Bechterewskernes der Vögeln. *Zeit. f. ges. Anat* **77,** 726–784.

Biondi, G. (1913). Il nuclei d'origine e terminali del nervo trigemino nel pollo. *Riv. ital neuropat. psichiat ed elettroterapia.* **6,** 50.117.

Bok, S. T. (1915). Die Entwicklung der Hirnnerven und ihre zentralen Bahnen. *Folia neuro-biol.* **9,** 475–565.

Brandis, F. (1894). Untersuchungen uber das Gehirn der Vögel. *Arch. f. mikr. Anat.* **43**, 787–813.

Bremer, F. (1924). Récherches sur la physiologie du cervelet chez le pigeon. *C.R. Soc. Biol. Paris* **90**, 381–384.

Bremer, F. and Ley R. (1927). Récherches sur la physiologie du cervelet chez la pigeon. *Arch. internat. Physiol.* **28**, 58–95.

Brouwer, B. (1913), Uber das Kleinhirn der Vögel nebst bemerkungen uber das Lokalisationproblem im Kleinhirn. *Folia neuro-biol.* **7**, 349–377.

Brown-Sequard, E. (1860). Récherches experimentales sur la physiologie de la moelle allongée (1859). *J. Physiol. Homme et Animaux.* **3**, 151–157.

Buddenbrock, W. von. (1953). "Vergleichende Physiologie II. Nervenphysiologie", pp. 396. Birkhauser, Basle.

Cate, J. Ten (1936). Physiologie des Zentralnervensystem der Vögel. *Ergebn. Biol.* **13**, 93–173.

Cate, J. Ten (1965). The nervous system of birds. *In* "Avian Physiologie" (Sturkie, P. D., ed.), pp. 697–751. Baillière, Tindall and Cassell, London.

Chambers, W. W. and Sprague, J. M. (1955a). Functional localization in the cerebellum. I. Organization in longitudinal corticonuclear zones and their contribution to the control of posture, both extrapyramidal and pyramidal. *J. comp. Neurol.* **103**, 105–129.

Chambers, W. W. and Sprague, J. M. (1955b). Functional localization in the cerebellum II. Somatotopic organization in cortex and nuclei. *A.M.A. Arch. Neurol. Psychiat.* **74**, 653–680.

Chiarugi, E. and Pompeiano, O. (1956). Effeti della stimolazione electtrica localizzata della cortecia e die nuclei del cervelletto nel piccione talamico. *Arch. di. Sci. biol.* **40**, 25–37.

Cordier, R. (1954). Lè système nerveux centrale et les nerfs cerebrauxspinaux. "Traité de Zoologie". Vol. 12, pp. 202–332. Masson, Paris.

Craigie, E. H. (1926). Notes on the morphology of the mossy fibres in some birds and mammals. *Trav. Lab. Rech. biol. Madrid.* **24**, 319–331.

Craigie, E. H. (1928). Observations on the brain of the humming bird. (*Chrysolampis mosquitus* Linn., *Chlorostilbon caribaeus* Lawr.) *J. Comp. Neurol.* **48**, 377–481.

Cunningham, A. W. B. and Rylander, B. J. (1961). Behaviour of spontaneous potentials from chick cerebellar explants during 120 hours in culture *J. Neuro-phys.* **24**, 141–149.

Dalton, C. (1861). On the cerebellum as the centre of coordination of the voluntary movements. *Amer. J. med. Sci.* **41**, 83–98.

Denissenko, G. (1877). Zur Frage uber den Bau der Kleinhirnrinde bei verschiedenen Klassen von Wirbelthieren. *Arch. mikr. Anat.* **14**, 203–242.

Dogiel, A. S. (1896). Die Nervenelemente in Kleinhirne der Vögel und Saugethiere. *Arch. mikr. Anat.* **47**, 707–719.

Doty, E. J. (1946). The cerebellar nuclear grey in the sparrow (*Passer domesticus*). *J. comp. Neurol.* **48**, 471–499.

Eccles, J. C., Ito, M. and Szentágothai, J. (1967). "The Cerebellum as a Neuronal Machine." Springer, Heidelburg, Berlin, Göttingen, New York.

Eiduson, E. (1966). 5-hydroxytryptamine in the developing chick brain. *J. Neurochem.* **13**, 923–932.

Falcone, C. (1893). "La Corticcia del Cerveletto." Naples.

Flourens, P. (1824). "Récherches Expérimentales sur la Propriétés et les Fonctions du Système Nerveux dans les Animaux Vertebrés." Crecot, Paris.

Frenkel, B. (1909). Das Kleinhirnbahnen der Taube. *Bull. internat. Acad. Sci. Cracowie.* (math. nat.) **2**, 123–147.

Friede, R. L. and Fleming, L. M. (1964). Comparison of choline esterase distribution in the cerebellum of several species. *J. Neurochem.* **11**, 1–7.

Friedlander, A. (1898). Untersuchungen uber das Rückenmark und das Kleinhirn der Vögel. *Neurol. Centralbl.* **17**, 397–409.

Goodman, D. C. and Simpson, J. T. Jr. (1960). Cerebellar stimulation in the unrestrained and unanaesthesized. *Alligator. J. Comp. Neurol.* **114**, 127–136.

Goodman, D. C. and Simpson, J. T. Jr. (1961). Functional localization in the cerebellum of the albino rat. *Exptl Neurol.* **3**, 174–188.

Goodman, D. C., Horel, J. A. and Freemon, F. R. (1963). Functional localization in the cerebellum of the bird (*Anas domesticus*). *Anat. Rec.* **145**, 233–234.

Goodman, D. C., Horel, J. A. and Freemon, F. R. (1965). Functional localization in the cerebellum of the bird and its bearing on cerebellar function. *J. comp. Neurol.* **123**, 45–53.

Groebbels, F. (1927). Die Lage- und Bewegungsreflexe der Vögel. V. Die physiologische Gruppierung der Lage- und Bewegungsreflexe der Haustaube und ihrer weitere Analyse durch Labyrinthentfernung und galvanische Reizung nach Entfernung des Labyrinths und seiner Teile. *Pfluger's Arch.* **217**, 631–654.

Groebbels, F. (1928a). Die Lage und Bewegungsreflexe der Vögel. VI. Degenerationsbefunde im Zentralnervensystem der Taube nach entfernung des Labyrinth und seiner Teile. *Pfluger's Arch.* **218**, 89–97.

Groebbels, F. (1928b). Die Lage und Bewegungsreflexe der Vögel. IX. Die Wirkung von Kleinhirnlasionen und ihre anatomisch-physiologische Analyse. *Pfluger's Arch.* **221**, 15–40.

Groebbels, F. (1928c). Die Lage und Bewegungsreflexe der Vögel. X. Die Analyse der Beziehungen zwischen Labyrinth und Kleinhirn. *Pfluger's Arch.* **221**, 41–49.

Groebbels, F. (1928d). Die Lage und Bewegungsreflexe der Vögel. XI. Die Analyse der Stutzreaktion. *Pfluger's Arch.* **221**, 50–65.

Groebbels, F. (1928e). Anatomisch-physiologische Untersuchungen uber die Beziehungen zwischen Labyrinth und Kleinhirn. *Dtsch. Z. Nervenheilk.* **107**, 154–160.

Groebbels, F. (1929). Die Wirkung von Kleinhirnläsionen und die anatomisch-physiologish Analyse. *Pflugers. Arch.* **221**, 15.

Hirako, G. (1935). Beitrage zur wissenschaftlichen Anatomie des Nervensystems. Demonstration der Purkinjeschen Zellen des Kleinhirns, die durch Weigert-Palisch Markscheidenfarbung dargestellt sind. *Folia anat. jap.* **13**, 561–566.

Hirako, G. (1940). Differenzierung, Entwicklung und Formveranderung der Purkinjeschen Zellen beim Huhner Kleinhirn. *Jap. J. med. Sci.* **8**, 97–110.

Hoevell, J. J. L. D., van (1916). The phylogenetic development of the cerebellar nuclei. *Proc. kon. ned. Akad. Wet. Sect. Sci.* **18**, 1421–1434.

Holst, E., von and Saint Paul, U., von (1963). On the functional organization of drives. *Anim. Behav.* **11**, 1–20.

Ingvar, S. (1918). Zur Phylo- und Ontogenese des Kleinhirns. *Folia neuro-biol.* **11**, 205–495.

Ingvar, S. (1919). "Zur Phylo- und Ontogenese des Kleinhirns nebst ein Versuch zu einheitlicher Erklarung der Zerebellaren Funktion und Lokalization. Lunds.

Hausman, L. (1929). The comparative morphology of the cerebellar vermis, the cerebellar nuclei and the vestibular mass. *Rcs. Publ. Ass. Res. Nerve. Ment. Dis.* Williams and Wilkins, Baltimore.

Kappers, C. U. A. (1920–21). "Vergleichende Anatomie des Nervensystems." Bohn. Haarlem.

Kappers, C. U. A. (1947). "Anatomie Comparée du Système Nerveux", pp. 754. Masson, Paris.

Kappers, C. U. A., Huber, G. C. and Crosby, E. C. (1936). "The Comparative Anatomy of the Nervous System of Vertebrates Including Man." 3 vols. Macmillan (1960, reprint). Hafner, New York.

Karamyan, A. I. (1956). Evolution of the function of the cerebellum and cerebral hemispheres. *Gors. izd. med. lit. Leningrad.* Trans. Israel programme 1962, pp. 160.

Koelliker, A. (1896). "Handbuch der Gewebelehre des Menschen." 2. 6. Aufl, Leipzig.

Kuenzi, W. (1918). Versuch einer systematischen Morphologie des Gehirns der Vögel. *Rev. Suisse. Zool.* **26,** 17–112.

Lange, B. (1891). Inwieweit sind die symptome, welche nach Zerstorung des Kleinhirns beobachtet wirden, auf Verletzungen des Acusticus zuruckzufuhren. *Pfluger's Arch.* **50,** 615–625.

Larsell, O. (1948). The development and subdivisions of the cerebellum of birds. *J. comp. Neurol.* **89,** 123–190.

Larsell, O. (1967). "The Comparative Anatomy and Histology of the Cerebellum from Myxinoids Through Birds", pp. 291. University of Minnesota.

Larsell, O. and Whitlock, D. G. (1952). Further observations on the cerebellum of birds. *J. comp. Neurol.* **97,** 545–566.

Longet, F. A. (1842). "Anatomie et Physiologie du Système Nerveux de l'Homme et des Animaux Vertébrés." Paris.

Löwenberg, H. (1938). The presence of the synarmotical cell in the cerebellum of birds. *Bio-Morphosis.* **7,** 273–280.

Löwenberg, H. (1939). Études sur les cellules de Golgi, les cellules interstitielles et les voies d'association intracerebelleuses chez les mammifères et les oiseaux. *Arch. int. Med. exp.* **14,** 51–102.

Luciani, L. (1893). "Das Kleinhirn." Deutsche ausgabe, Leipzig.

Magendie, M. (1825). "Précis Élémentaire de Physiologie." (5th ed.), pp. 261. Bruxelles.

Marchiafava, P. L. (1968). Activities of the central nervous system: motor. *Ann Rev. Physiol.* **30,** 359–400.

Martin, E. G. and Rich, W. H. (1918). The activities of decerebrate and decerebellate chicks. *Amer. J. Physiol* **46,** 396–411.

Mesdag, T. M. (1909). *Bijdrag tot de ontwikkelingsgeschiedenis van de structuur der hersenen bij het kip.* Dissertation, Groningen.

Munzer, E. and Wiener, H. (1898). Beitrage zur Anatomie und Physiologie des Zentralnervensystems der Taube. *Monatschr. Psychiatr.* **3,** 379–406.

Muskens, L. J. J. (1930). On tracts and centers involved in the upward and downward associated movements of the eyes after experiments in birds. *J. comp. Neurol.* **50,** 289–331.

Pearse, A. G. E. (1960). "Histochemistry; Theoretical and Applied." Churchill, London.

Portmann, A. (1950). Les oiseaux. "Traité de Zoologie", vol. 15. Masson, Paris.

Portmann, A. and Stingelin, W. (1961). The central nervous system. *In* "The Biology and Comparative Physiology of Birds" (Marshall, A. J., ed.), vol. 2, pp. 1–36. Academic Press, New York and London.

Ramon y Cajal, S. (1888). Estructura de los centrios nerviosus de las aves. I Cerebelo. *Rev. Trim. Histol. norm. patol.* 1. (cited by Ramon y Cajal, 1909–11).

Ramon y Cajal, S. (1904). "Textura del Sistema Nerviosa del Hombre y de los Vertebrados." Moya, Madrid.

Ramon y Cajal, S. (1908). Los ganglios centrales del cerebello de las aves. *Trav. lab. biol., Madrid* **6**, 177–194.

Ramon y Cajal, S. (1909–11). "Histologie du système nerveux de l'homme et des vertébrés." Maloine, Paris.

Raymond, A. (1958). Responses to electrical stimulation of the cerebellum in unanaesthetised birds. *J. comp. Neurol.* **110**, 299–320.

Reisinger, L. (1916). Das Kleinhirn der Hausvögel. *Zool. Anz.* **47**, 189–198.

Renzi, P. (1863–64). Saggio di fisiologia sperim sui centri nervosi della vita psic. *Schmidts Jb.* **123**, 151–161.

Rogers, K. T. (1960). Studies on the chick brain of biochemical differentiation related to morphological differentiation. III. Histochemical localization of alkaline phosphatase. *J. Exp. Zool.* **145**, 49–60.

Rogers, K. T., DeVries, L., Kepler, J. A., Kepler, C. R. and Speidel, E. R. (1960). Studies on the chick brain of biochemical differentiation related to morphological differentiation. II. Alkaline phosphatase and cholinesterase levels and onset of function. *J. Exp. Zool.* **145**, 49–60.

Saetersdal, T. A. S. (1956a). On the ontogenesis of the avian cerebellum. I. Studies on the formation of fissures. *Univ. Bergen Arb. Nat.* no. 2, 1–15.

Saetersdal, T. A. S. (1956b). On the ontogenesis of the avian cerebellum. II. Measurements of the cortical layers. *Univ. Bergen Arb. Nat.* no. 3, 1–53.

Saetersdal, T. A. S. (1959a). On the ontogenesis of the avian cerebellum. III. Formation of fissures with a discussion of fissure homologies between the avian and mammalian cerebellum. *Univ. Bergen Arb. Nat.* no. 3, 1–44.

Saetersdal, T. A. S. (1959b). On the ontogenesis of the avian cerebellum. IV. Mitotic activity in the external granular layer. *Univ. Bergen Arb. Nat.* no. 4, 1–39.

Sanders, E. B. (1929). A consideration of certain bulbar, midbrain and cerebellar centres and fibre tracts in birds. *J. comp. Neurol.* **49**, 155–221.

Senglaub, K. (1963). Das Kleinhirn der Vögel in Beziehung zu phylogenetischer Stellung, Lebensweise und Körpergrosse. *Z. wiss. Zool., Leipzig*, **169**, 1–63.

Sherrington, C. S. (1906). "The Integrative Action of the Nervous System." Scribner Sons, New York.

Shimazono, J. (1912). Das Kleinhirn der Vögel. *Arch. mikr. Anat.* **80**, 397–449.

Singh, R. (1966). Some observations on the histochemical localization of phosphorylase in the nervous tissue of the parrot (*Psittacula krameri*). *Acta Anat.* **63**, 281–288.

Singh, R. (1967). Some observations on the histochemistry of the neuropile tissue of the brain of the parrot (*Psittacula krameri*). *Acta anat.* **68**, 567–576.

Snider, R. S. and Stowell, A (1944). Receiving areas for the tactile, auditory and visual systems in the cerebellum. *J. Neurophysiol.* **7**, 331–358.

Snider, R. S., Teramoto, S. and Ban, J. T. (1967). Activity of the Purkinje and basket cells in chronically isolated chick folia. *Exptl Neurol.* **19**, 443–454.

Stieda, L. (1864). Zur vergleichenden Anatomie und Histologie des Cerebellum. *Arch. Anat. Physiol.* 407–433.

Stingelin, W. (1962). Ergebnisse der Vögelgehirn forschung. *Verhandl. Naturf. Ges. Basle.* **73**, 300–317.

Turner, C. H. (1891). The morphology of the avian brain. *J. comp. Neurol.* **1**, 39–92.

Wagner, R. (1858–60). Kritische und experimentelle Untersuchungen uber die Funktionen des Gehirns. *Nachr. Ges. wiss. Gottingen. Math. Phys.*

11

Wallenberg, A. (1900). Uber centrale Endstatten des Nervus octavus der Taube. *Anat. Anz.* **17,** 102–108.

Wallenberg, A. (1907). Beitrage zur kenntnis des Gehirns der Teleostier und Selachier. *Anat. Anz.* **31,** 369–399.

Weir-Mitchell, S. (1869). Researches on the physiology of the cerebellum. *Amer. J. med. Sci.* **57,** 320–338.

Whitlock, D. G. (1952). A neurohistological and neurophysiological study of the afferent fibre tracts and receptive areas of the avian cerebellum. *J. comp. Neurol.* **97,** 567–635.

Yamamoto, S., Ohkawa, K. and Lee, L. (1957). On the cerebellar nuclei of birds. *Arch. hist. jap.* **13,** 129–139.

# 9  The Avian Eye and Vision

## I. INTRODUCTION

The structure of the avian eye has been surveyed on a number of occasions. Outstanding amongst such general reviews are those of Franz (1934), Rochon-Duvigneaud (1943, 1950, 1954), Walls (1942) and Kare (1965), whilst Pumphrey provided a detailed discussion of certain more specific aspects. The close phylogenetic affinities which exist between birds and diapsid reptiles are reflected in the fact that the eyes of the former contain no features of any importance which are absent from those of the latter. Nevertheless vision is of immense importance to most birds, the principal exceptions being certain Strigiformes and Caprimulgiformes, and this has been summed up in the aphorism that "birds are eyes powered by wings", or "wings guided by eyes". The great size of the avian eyeball is often ignored by the casual observer as only the rather small cornea shows in the circular lid-opening. However, it is only amongst the very small species of humming-birds, finches and warblers that the size of the eyes is comparable with that of most amphibians and reptiles. The weight of the eyes often exceeds that of the brain and those of hawks and owls are as large as man's, despite the huge disparity in body size.

This is not the place for a detailed discussion of homolysis in visual pigments which gives free radicals whose physical properties may make them the fundamental agents in photo-sensitivity. The basic attributes are common to all vertebrate rhodopsins and the free radical mechanism almost certainly accounts for the high quantum efficiency of visual receptors. Instead we can largely confine ourselves to those features which are peculiar to, and characteristic of, avian eyes. These have been the subject of a number of papers which are of historical importance and besides the reviews noted above one can cite Beer (1892), Dogiel (1895), Nussbaum (1901), Collin (1903), Grynfelt (1905), Hess (1912), Blochmann and Husen (1911), Leplat (1912), Slonaker (1918), Walls and Judd (1933), Stresemann (1933), Verrier (1936) and Eck (1939).

Whilst there are more diurnal than nocturnal species of bird, none are as exclusively diurnal in their ocular characteristics as are some reptilian genera (Rochon-Duvigneaud, 1943). The observation that some galliform and columbiform species return to their roosts earlier than other species has led to the suggestion that they may require greater light intensities for vision. Most other species have a more or less well-developed crepuscular vision and, leaving aside *Falco vespertinus*, some Falconiformes for example will continue to hunt after sunset. Complementarily, the circumpolar strigiform genera

279

*Surnia* and *Nyctea* habitually hunt by day whereas the majority of owls do not. Guhl (1953) studied the effect of limited light intensities on bird behaviour, and found that chicks would start *feeding* when the light was 1-foot candle and would begin to *peck one another* at 2-foot candles. They were seen by Benner (1938) to use shadows as a means of determining depth and in recent years they have been shown to respond to a number of standard optical illusions. They are said to respond to photographs which have peas with appropriate shading but fail to respond to pictures in which the shadows are in all directions. Hess (1956) reported a bimodal colour preference in the case of chicks with one peak in the orange and another in the blue but found a narrower preference range for Peking ducklings. These had a sharp peak within the green to yellow-green region. Donner (1953), using a micro-electrode technique, concluded that the photochemical processes in the pigeon are similar to those of other vertebrates, but Adler and Dalland (1959), using an operant conditioning technique, found that the dark adaptation curve in the starling was rather different from that of man.

Under normal conditions the eyes of the pelecaniform *Phalacrocorax* do not seem to completely fill the orbits but those of most birds do. This is especially so in the case of the large eyes of the Falconiformes, and even more so in the Strigiformes despite their use of hearing for directional location (*q.v.*). As a result, most avian eyes have only a limited freedom of movement although many species are capable of a forward convergence towards the beak tip. This is best known in the bittern, *Botaurus stellaris*, which can direct its eyes almost vertically downwards in the morphological sense and scrutinize an intruder with both of them whilst it maintains its characteristic cryptic posture with the beak pointing upwards. Nevertheless other species are also able to move their eyes and those of the Psittaciformes appear to have some mobility as do those of the Gruidae, Laridae and Pelecaniformes. In the Rhamphastidae and Bucerotidae, in which the eyes are separated from each other by the relatively large beak, they seem to have a very remarkable mobility.

In relation to the great and important part which vision plays in the life of birds Lorenz (1952) and Pumphrey (1961) have commented upon the fact that corvines, Strigiformes, Alciformes and Galliformes appear to rarely attack eyes. Both these authors concluded that this "deeply rooted inhibition" originates in the selective value of mutual eye protection amongst nestlings and that it may be a contributory factor to the protective value of eyespots in insect coloration (Blest, 1957; Pumphrey, 1961). However, it is generally assumed, from the incidence of damage to them, that such eyespot patterns function by drawing attention away from more vulnerable areas of the body and this would conflict directly with such suggestions of avoidance.

Pumphrey summarized the predominance of the avian eye by comparing

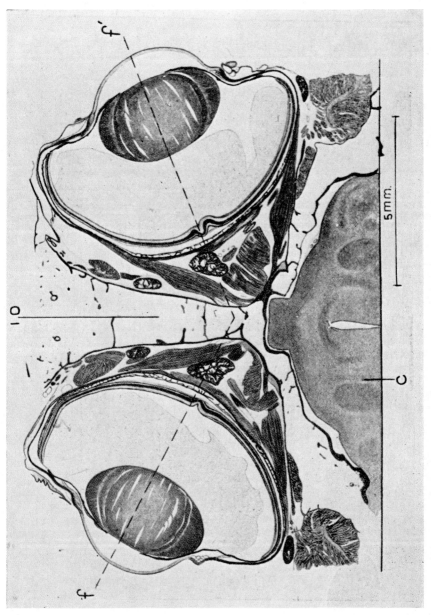

FIG. 82. Horizontal section of the eyes of the bullfinch. The foveae make an angle of 130°. (Rochon-Duvigneaud, 1943).

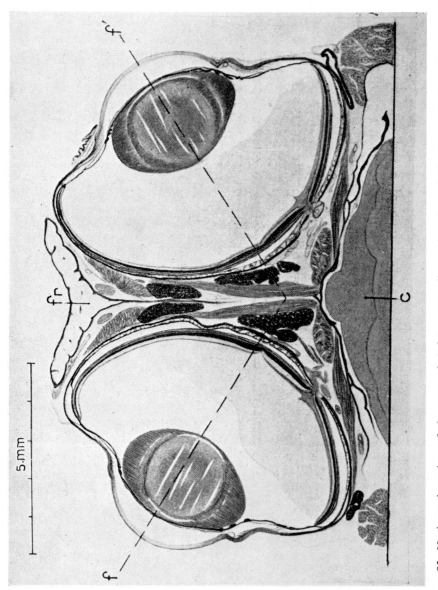

Fig. 83. Horizontal section of the eyes of *Sylvia atricapilla* passing through the fovea. (Rochon-Duvigneaud, 1943). fr, frontal bone; c, brain. The foveae form an angle of *ca.* 120°.

the conditions in birds and man. *Struthio*, standing as high as us but with a lighter body weight, has an eye which is five times larger. In smaller species the disproportion is even greater. In the starling, *Sturnus vulgaris*, the head is, as in man, one-tenth of the total body. However, our eyes are less than 1 % of the weight of the head whereas in the starling they are approximately 15%. He concludes that birds have eyes which are as heavy as is aerodynamically practicable. Their actual position varies, and it is most easily defined in terms of the angle between the two optic axes. Genera with relatively narrow heads such as *Erolia*, *Eurynorhynchus*, *Tringa* and *Columba* have their eyes placed laterally and this angle is large, in the latter genus 145°. In genera with more rounded heads the angle is less. In *Pyrrhula* (Fig. 82) it is 130°, and in *Sylvia atricapilla* (Fig. 83) 120°, whilst in *Falco tinnunculus* it is only about 90°.

## II. THE MORPHOLOGY OF THE EYES

### A. Gross Appearance

The near spherical form of the mammalian eyeball renders it mechanically strong and also facilitates its rotation within the orbit. In birds this is not the case and, although the retina usually follows a segment of a spherical surface which is centered on the posterior nodal point, the overall shape is flat, globose or tubular. There is, broadly speaking, a small anterior component, the cornea, which is a segment of a sphere, and a larger posterior component which is more or less globular. These two are united by the intermediate region, or scleral ring, which is composed of some 12–15 ossicles. What would otherwise be a line of mechanical weakness then acts as an origin for the striated, and rapidly acting, muscles of accommodation. This ring has a small anterior opening for the cornea and a larger posterior one which is continuous with part of the sclerotic. The relative size of these two openings, coupled with the particular angle of inclination of the scleral walls which make it longer or shorter, provides the basis for the eye outline.

In Galliformes, Columbiformes and other narrow-headed birds with a short anterio-posterior optic axis, the scleral ring is almost disc-shaped. In genera such as *Passer* and *Corvus* the posterior segment of the eye is more globular and the scleral ring more conical. In this respect the appearance of the whole eye is related to the size of the head and varies with this in, say, the round-headed *Falconiformes* and in *Fringilla*, *Sylvia* and *Hirundo*. In the Strigiformes the bony cone tends to be elongated into a tube so that the retinal surface is reduced. Some idea of these variations can be gained by comparing Figs 82–85. In yet other species, such as *Charadrius hiaticola* and *Burhinus oedicnemus*, there is a large cornea and lens which recalls a nocturnal condition.

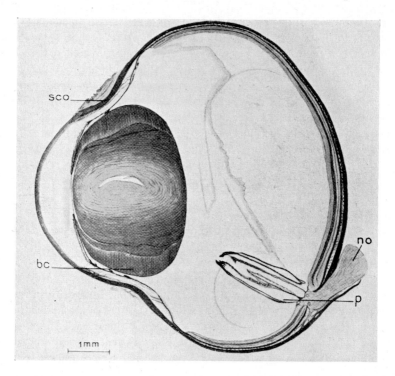

FIG. 84. Vertical section of the eye of *Hirundo* passing through the pecten and optic nerve. (Rochon-Duvigneaud, 1943). bc, periphery of lens; sco, ossified scleral ring; no, optic nerve; p, pecten.

## TABLE 36

*A comparison of the retinal surface of two diurnal and two nocturnal species.*

| Species | Retinal surface in terms of the basal area of the cornea |
|---------|------------------------------------------------|
| *Falco tinnunculus* | 11·5 |
| *Columba livia* | 10·0 |
| *Strix aluco* | 4·0 |
| *Tyto alba* | 3·5 |

From Rochon-Duvigneaud (1943).

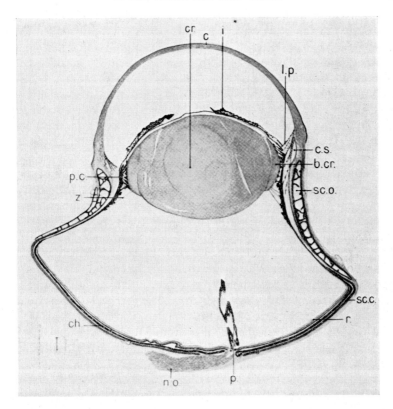

FIG. 85. Horizontal section of the eye of *Strix aluco*. (Rochon-Duvigneaud, 1943).
Abbreviations: C, Cornea; ch, Choroid; cr, lens; cs, canal of Schlemm; Lp, pectinate ligament; No, optic nerve; P, pecten; Pc, Ciliary process; r, retina; Sco, ossified component of sclerotic; Z, the Zonule.

The result of this is that the avian eye is always asymmetrical, and, furthermore, the upper part is better developed. This usually results in an increased ability to form an image of the ground during flight, and a lesser developed ability to form an image of the sky. There is, however, some variation in ocular structure, particularly as between habitually diurnal and nocturnal birds. The data of Table 36 portray this by expressing the retinal surface area in terms of the basal area of the cornea for two diurnal and two nocturnal species.

## B. The Sclerotic and the Choroid Membrane

The gradual development of sclera and choroid in the chick was described by O'Rahilly (1962). A periodic acid Schiff positive membrane is found externally on the optic cup by stage 16. At stage 24 a mesodermal condensation external to this represents the primordial sclera and by stage 28 cartilaginous plates can be detected. At stage 36 the choroid membrane comprised vascular chorio-capillary and basal laminae. The definitive scleral ossicles are visible by the succeeding stage 37. In the adult the sclerotic coat is a fibrous membrane which is reinforced on the one hand by the scleral plates and elsewhere by a layer of cartilage. This cartilage is very often ossified in the region of the optic nerve and such ossifications occur in the Corvidae, *Turdus merula*, *Sitta europaea*, *Motacilla alba*, the Fringillinae, *Phoenicurus phoenicurus*, the Hirundinidae, *Upupa*, *Sula* and the Piciformes. They are apparently absent from *Caprimulgus*, *Columba*, the Burhinidae, *Larus ridibundus* and at least some Charadriidae, the Falconiformes, Strigiformes and *Phalacrocorax*. When it is present this posterior ossification arises from cartilaginous precursors and it therefore differs from the scleral ossicles which are membrane bones.

In *Phalacrocorax* the anterior fibrous part bulges out in the neighbourhood of the cilio-scleral space as a ring surrounding the cornea. The actual thickness of the structure varies from species to species and can be 0·17 mm in eyes of medium size or 0·7 mm in larger ones. There are also intra-ocular size variations in different parts of the eye of a single species. Thus close to the scleral ring in *Spizaetus bellicosus* it measures 0·36 mm in thickness but is 0·085 mm at the equator. In *Caprimulgus*, which has fairly thin-walled but large eyes, it is 0·070 mm, whilst amongst members of the *Sittidae* and Hirundinidae, together with *Hydrobates pelagica*, it can be as thin as 0·035–0·040 mm. In all cases the very thin lamina fusca is closely adherent to the cartilaginous coat so that this is in close contact with the choroid vessels.

The choroid itself is basically composed of the two standard layers but it has fewer pigment cells and collagen fibrils than is customarily the case in mammals. The considerable development of the venous system makes it rather thick and this vascularization is particularly marked amongst many divers. A layer of medium sized vessels resemble those of the pecten. An external vascular layer contains large thin-walled venous lacunae and if empty these can give the impression that the choroid is composed of two pigmented layers separated by a clear space.

## C. The Cornea

According to Rochon-Duvigneaud (*loc. cit.*) the corneal curvature is absolutely regular so that the avian eye does not suffer from corneal astigmatism. Its thickness varies, and is generally greater in larger birds but there is no direct linear relationship with gross body size. Rochon-Duvigneaud cites values of 270 $\mu$ for *Corvus corone* and 215 $\mu$ in *Asio otus*. Amongst species of

smaller size, such as those included in the Paridae and Fringillidae, it is only
130–150 $\mu$, whilst in some crepuscular and nocturnal genera with voluminous
but thin-walled eyes, as is the case in *Caprimulgus* and certain Micropodiformes, it can measure as little as 110 $\mu$. Complementarily, within the particularly thick walls of *Sula bassana* it can attain 810 $\mu$ whilst in *Aquila chry-
saetos* it is 612 $\mu$ at the centre and 936 $\mu$ at the periphery.

The histological structure is fairly similar in all cases. Bowman's membrane,
which, together with the epithelium, substantia propria, Descemet's membrane and the mesenchymal membrane, is a constant component of the
mammalian structure, is not always differentiated (Walls, 1942). The matrix
is denser in the superficial regions where it frequently has a lamellar appearance. The deeper layers are fibrillar. In *Bubo*, Rochon-Duvigneaud found
that the laminar component represented the anterior one-third and the
remaining two-thirds were fibrillar. The proportions were reversed in *Spizaetus bellicosus* and the cornea of *Strix aluco* is entirely composed of very fine
lamellae.

The strongly curved cornea, coupled with the rather flattened anterior
surface of the lens, results in a comparatively deep anterior chamber to the
eye. Within this there is abundant aqueous humour. Together with the
vitreous humour this helps to keep the eyeball distended and changes in the
pressure which they exert can cause visual disturbances. Rochon-Duvigneaud
was able to extract 2 cc in *Bubo* and suggested that it facilitated iris movements.

## D. The Iris

The predominant colour of the avian iris is a deep brown verging to black,
but it can be grey (as in *Balearica*, *Garrulus* and *Coloeus*), pale blue (as in
*Sula bassana*), green (as in *Phalacrocorax*), clear yellow as in many Falconiformes, and a yellowy orange in *Asio otus* and *Bubo*. In some species there is
also a sexual dimorphism. Clear yellow in the male golden pheasant, brown
in the female, red in the male condor, and again brown in the female. In the
first case the colour matches that of the nearby feathers.

The pupil is always round when contracted, but can form a horizontal oval
when it is dilated as is the case in the Burhinidae. Its degree of dilation or
contraction is under the control of the nucleus isthmi complex (*q.v*). In the
majority of species the muscle layer of the iris is equally thick throughout
although this actual thickness varies from species to species. In *Larus ridibundus* the principal component is a ring in the median area and there are
only a few fibres peripherally. The well-developed condition of *Phalacrocorax*,
and more especially *Puffinus puffinus* and *Podiceps ruficollis*, contrasts markedly with the reduced musculature of many Caprimulgiformes and Strigiformes

in which it can be reduced to a few circular bundles in the internal zone. Scattered between the muscular elements there are blood vessels, nerves, collagen fibrils and the processes of the melanophores which contribute the typical deep brown colour. The yellow to red colorations of other species seem to be attributable to an, often thick, layer of pigment-containing vesicular cells, and irises of these colours are frequently thicker than the dark brown or black ones. The iris dilator muscles lie behind the sphincter and usually include two sets of radial fibres which are of differing size and inserted on the ciliary body. These can, however, be reduced in some species to a single set of intermingled fibres.

## E. The Ciliary Zone and Lens

Light entering the eye is refracted by the cornea, aqueous humour, the lens and the vitreous humour. The distance which separates the lens from the point of focus is its focal length. For a lens of given refractive index this focal length can be varied by changing the curvature of the lens. The refractory power is traditionally expressed as the reciprocal of the focal length in meters and a lens of one diopter (1 D) has a focal length of one metre. The distance over which such a lens can maintain a focus at the retina is known as the accommodation range and in mammals this range results from changing the anterior and posterior curvature of the lens by means of the ciliary muscles. Contraction of Brucke's muscle compresses the lens and shortens its focal length. Contraction of Crampton's muscle alters the corneal shape. In nocturnal birds these muscles seem to be reduced and the focal length relatively fixed. If the image of a distant object is focused correctly on the eye the condition is known as emmetropia. If the point of focus is in front of the retina it is known as myopia or near-sightedness and if beyond the retina it is hypermetropia or far-sightedness. With the exception of the apparently myopic kiwi most birds are either emmetropic or slightly hypermetropic.

In all birds, and particularly *Phalacrocorax*, the lens is very soft although it has a harder core. It therefore retains considerable powers of accommodation throughout life. The gross appearance shows more variation than in other vertebrate classes. In many Psittaciformes, in *Hydrobates* and in *Upupa* there is a flat anterior and a very convex posterior surface. In certain Anseriformes, Strigiformes and *Caprimulgus* it is strongly curved at both front and back whilst in a number of Falconiformes and oscine passerines the posterior curvature exceeds that of the anterior.

One result of the large size of the anterior chamber and the anterior region in general is the large size of the ciliary region. This can be drawn out into long processes of which 130 occur in *Spizaetus* and *Milvus migrans*. The fibrous lamina is both thin and elastic. Where it is continuous with the choroid it is contiguous with the sclerotic and elsewhere it leaves a free cilioscleral space as a peripheral continuation of the anterior chamber. There are three ciliary muscle zones and a large canal of Schlemm which is either attached to the sclera or partially separated from it by Crampton's muscle. Its free border is formed by a lamina of thin reticular tissue and the lumen is divided into a number of secondary canals. Forming a ring around the entire anterior chamber in the eyes of *Bubo, Alectorix* and *Phasianus*, it is less complete in *Sterna, Falco, Burhinus, Caprimulgus* and *Pyrrhula*.

Pumphrey (1961) discussed the actual process of accommodation in birds. As stated above very little occurs in nocturnal genera. Crampton's muscle is strong and Brucke's weak in many Strigiformes, and in *Caprimulgus*, where it is again weak, Crampton's muscle is lacking from the inner side. In diurnal birds the lateral compression of the lens which shortens its focal length does not result from a simple slackening of the tension in the zonal fibres together with the residual elasticity of the lens capsule as is the case in man. The lens margin is firmly in contact with the ciliary body and this is actually driven towards the axis of the eyes by Brucke's muscle. The position and appearance of Crampton's muscle suggests that when contracting it pulls the margin of the cornea down towards the fundus so that the central corneal curvature is increased. The relative contributions which are made by changes in both the lens radius and that of the corneal surfaces are not known. Cursory observations demonstrate the extreme speed with which focal changes occur and the act of accommodation is also accompanied by a rapid, strong, and sometimes transient, pupillary contraction. This does not normally shorten the focus of the eye but results in a considerable increase in its depth of focus. The muscles which are involved in accommodation therefore appear to be functionally synergic with the *sphincter iridis* which contracts the pupil. Pumphrey suggested that this helps to explain the anomalous condition in the cormorant. When underwater, the cornea no longer performs an appreciable percentage of the refraction in the eye and the sudden need for an additional 20 D of refraction above the normal requirements is met by compressing the anterior part of the lens by means of the iris sphincter. The relatively reduced condition of Crampton's muscle reported from diving birds also provides another argument in favour of its altering corneal curvature.

## III. THE PECTEN

The avian eye, capable of high resolution over a large part of the visual field, is necessarily completely free of the vascular component of the mammalian retina. It is fairly widely agreed that nutrients reach the retina by diffusion through the vitreous humour and that their source may lie in the pecten, whatever its other functions. This structure is situated at the blind spot caused by the entry of the optic nerve, has its apex directed towards the lens, and in the majority of species its basal length equals half the horizontal diameter of the eye. Unlike the analogous structure in reptiles, which is a simple cone or peg, the pecten of birds is folded and has a much greater surface area. In view of the thicker and more complex retinal structure in birds this increased complexity is not in itself surprising. Conical and reminiscent of the reptilian condition in *Apteryx*, it usually comprises membranous plates. A series of 20 such plates in *Struthio* is carried on an axial structure of

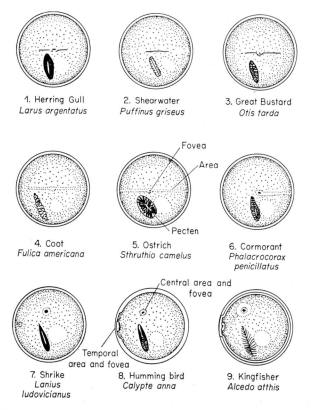

FIG. 86. The pecten, area and fovea of the avian eye as seen through an ophthalmoscope. (Pumphrey, 1961).

similar composition. In *Rhea* it is more complicated and in *Casuarius* there is an overall conical outline with four large and two small folds. Twenty-four folds occur in the Tinamiformes. In general the pecten is similar amongst the carinate orders and comprises a folded lamina with a variety of specific characteristics. In the Strigiformes, Caprimulgiformes and the parrot *Strigops habrotilus* there are 4–8 folds. In *Micropus* there are 11, and amongst the Anseriformes usually 10–16. However, amongst the majority of diurnal avian species there are 20 or more and the eye of *Bubo* is therefore equipped with a pecten which is proportionately much smaller than those of passerines such as the Paridae and Fringillidae.

Electron-microscopic studies of pecten fine structure were undertaken by Fischlschweiger and O'Rahilly (1966), Francois *et al.* (1963), Seamen and Storm (1963), Semba (1962) and Tanaka (1960). The blood vessels have several rather unusual features such as peri-endothelial cells and numerous microvilli which project into the lumen. In contrast to the other workers Fischlschweiger and O'Rahilly found that the endothelial lining comprised cells with clear cell membranes and numerous terminal bars. The basement membranes are extensively thickened by additional ground material containing reticular micro-fibrils. There is considerable variation in the arrangement of the pigment cells and there may or may not be spaces into which their processes can penetrate. Some of the available evidence suggests that the covering membrane is continuous with the internal limiting membrane of the retina but the role and origin of the peri-pectinate cells remains obscure. Fischlschweiger and O'Rahilly were at some pains to emphasize the apparent differences which can arise from simple technological differences and to urge caution in the interpretation of conflicting data.

Raviola and Raviola (1967) described a comparable situation in the pigeon pecten. Arterioles with a thin and unspecialized endothelium give rise to a dense capillary network. The endothelial cells are provided with thin lamellar processes on both their luminal and basal surfaces which are constricted at their point of origin from the cell body. The luminal lamellae are both long and erect and have a predominantly longitudinal orientation. The basal ones are shorter, more compressed and also somewhat irregularly orientated. Besides the usual complement of sub-cellular organelles there are a moderate number of coated vesicles and uncoated invaginations of the plasmalemma. Also, a labyrinthine system of interstitial cavities surrounds the stalks of the basal lamellae. These contain a material whose electron density is comparable with that of the basement lamina, and similar material occurs on the outer surface of the lamellae where it gradually merges with the matrix of the adventitial layer. This also contains many collaginous fibrils. The whole capillary network is drained by venules whose walls have a comparable structure to those of the capillaries.

The pigmented cells, which tend to be sandwiched between the two basement laminae and to separate the blood vessels from the vitreous body, contain melanin granules, large mitochondria and numerous bundles of filaments. The nature of such cells has been discussed on a number of occasions. Mann (1924a, b) assumed that they were of ectodermal origin; Blochmann and von Husen (1911), and also Kauth and Somner (1953) that they were glial derivatives and Bachsich and Gellert (1935) that they were connective tissue cells. Raviola and Raviola (1967) were clear that they certainly differed from the glial cells of the central nervous system.

The presence of massive vascularization together with "microvillar" structures would clearly agree with the nutritive or secretory functions which have been assigned to the pecten by Denissenko (1881), Slonaker (1918), Mann (1924a, b) and Walls (1942). This would also be true of the unusually high carbonic anhydrase activity which has been reported by Leiner (1940) and Kauth and Somner (1953). Many workers have compared the pecten with other highly vascular structures which can be found in vertebrate eyes. These

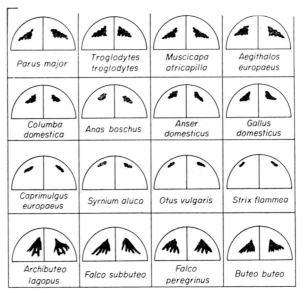

Fig. 87. The shadow of the pecten on the retina. (From Pumphrey, 1961 after Menner, 1938).

include, besides the papillary cone of lizards (Duke-Elder, 1958; Walls, 1942), the choroid rete (Prince, 1956; Francois and Neetens, 1962) and the falciform process of fishes. However, Raviola and Raviola focused attention on to the fact that a counter-current system of capillaries with an unspecialized lining, such as is found in the choroid rete, is lacking in the pecten (Tanaka, 1938). Collateral capillaries shunt the blood from the afferent to the efferent vessels over the entire extent of the eye.

Menner (1938), having carried out an experiment with an artificial eye, consisting of a camera with a ground glass screen into which a cardboard model of the pecten could be inserted, pointed out another, and previously unsuspected, relationship between the pecten and the species habits. He claimed that if his apparatus was pointed at swifts wheeling and turning high in the sky, their barely detectable images became more prominent after the cardboard was introduced. Although Crozier and Wolff (1943a,b) adduced further evidence Pumphrey (1961) was unable to completely convince himself of either the validity of the experiment or its conclusions. Certainly, if one plots the outline of a variety of avian pectens as they are outlined on the retina in life (as shown in Fig. 87) then it is clear that the extent of the intrusion of their shadow on the retina is not simply proportional to the number of visual cells requiring nutrition (Menner, 1938). It does, therefore, seem possible that besides its trophic function the pecten may also serve an optical function and enhance the eyes sensitivity to the movement of objects within the visual field.

## IV. THE RETINA

### A. Retinal Structure and Colour Vision

The sensitive units of the eye are the rod and cone cells and, if more complex, the essential organization of the avian retina is still similar to that of other vertebrates. The innermost layer of the eye, it arises during embryological development as a protrusion on each side of the brain. The light passing through the other ocular components passes through most of the retina and forms an image at the pigmented epithelium. When observed through an opthalmoscope the eye appears grey or yellowish in *Phalacrocorax, Sagittarius, Cacatua, Cuculus, Saxicola, Passer* and *Turdus* since the retinal pigment masks the reddish choroid. Elsewhere this is not the case and in species with crepuscular or nocturnal habits such as *Apteryx mantelli,* or in *Spheniscus demersus, Fratercula arctica, Oedicnemus scolopax, Cancroma cochlearia, Strigops habroptilus, Caprimulgus europaeus* and *Strix* the retinal pigment layer is less opaque and the red colour of the vascular membrane shows through.

Schultze (1866) together with Walls and Judd (1933) established the presence of small coloured droplets within the avian retina. Green in *Milvus milvus,* they are generally yellow or red. The brighter ones occur in species with diurnal habits and in nocturnal birds they tend to be either pale yellow or colourless. A number of such colourless ones do however also occur in small acromyodine birds such as *Phylloscopus, Troglodytes, Parus* and *Sylvia,* together with *Garrulus* and *Phalacrocorax.* Coloured oil-droplets only are found in *Falco tinnunculus, Picus viridis, Tringa subarcuata, Charadrius hiaticola, Larus ridibunda, Hydrobates pelagicus, Fringilla* and *Coloeus.* In general the red and orange ones are larger but 3–6 times less numerous than the yellow and colourless together (Rochon-Duvigneaud, 1943). It is possible that the possession of these coloured droplets enables birds to exceed man's ability to discriminate mixed or pigmentary colours. It is certainly easy to make up different mixtures of spectral colours which appear identical to our unaided vision but are at once distinguishable if one uses extra-ocular filters. Pumphrey (1961) therefore conceded the possibility that the oil-drop-lets may act as *intra-ocular* filters thereby endowing birds with a colour vision which man can only achieve by using external aids. At an earlier period such a suggestion had been challenged by Walls (1942) and the various arguments for both points of view are summarized by Alpern (1968). Pedler (1969) felt that in view of our ignorance of their function oil-droplets are irrelevant to any understanding of visual function. Walls suggested that the yellow filters reduce the effects of chromatic aberration, minimize glare and dazzle, and also enhance contrast. They are analogous to the macular pigment of man and in the avian foveal cones there are only such yellow ones. Complementarily, he thought that the red filters reduce the effects of Ray-leigh scattering in the atmosphere, a function which would be of particular importance during sunrise. As Alpern emphasized, it is very difficult to imagine an easy method of testing such hypotheses experimentally. Indeed, whether or not all avian cones even have the same visual pigment is still not known although Fujimoto *et al.* (1957) attempted to investigate this using a micro-spectrophotometer. Watson (1915) and Lashley (1916) showed that the chicks spectral limits are 700–715 nm at one end of the spectrum, and be-tween 395–405 nm at the other. Kare (1965) thought that a shift of peak sensitivity during the period of maturation of chicks to adult hens, with a value of 560 nm in the former and 580 nm in the latter (Honigmann, 1921), might reflect an increase in the density of oil droplets with age. Hawks, woodpeckers and owls with few or no droplets would, according to him, see blues and violets as man does (see also Sluckin, 1970).

Despite the lack of widespread data on cone pigments the oil-droplets are another matter. Fujimoto *et al.* investigated the situation in hens, ducks and doves, and Strother (1963) added to the list by looking at the American bronze turkey. He found that irrespective of either oil-droplet size or the species studied the absorption peak was relatively flat at 80% for *all colours* studied. All the colours also exhibited a rather steep slope at "cut off" which was 570 nm for the red, 540–530 nm for the orange, 510 nm for the yellow, 440 nm for the green and 390 nm for the colourless ones. The values for all frequencies were consistently higher according to Fujimoto *et al.* but they also concluded that there were only trivial differences from one species to the next (see also Wallman, 1970).

Turning to gross retinal parameters the relative thickness of the retina varies and is by no means proportional to the overall size of the eye. That of *Parus* is almost as thick as an eagle's, and thicker than that of *Alectorix rufa, Strix aluco* or *Bubo*. The retinal thickness in a number of species is tabulated in Table 37. From these data it is clear that the thicker conditions are generally found in the passerines and falconiforms. The excessively thick condition in *Circaetus gallicus* is accompanied by a very small cell size so that the total number of cells is immense. In any case the large size of an eagle's eye will result in a larger image and it is not surprising that its visual acuity exceeds that of small birds (Rochon-Duvigneaud, 1943).

## TABLE 37

*The thickness of some avian retinas according to Rochon-Duvigneaud (1943).*

| Species | Retinal thickness |
|---|---|
| *Alectorix rufa* | 290 $\mu$ |
| *Asio otus* | 252 $\mu$ |
| *Strix aluco* | 306 $\mu$ |
| *Corvus frugilegus* | 324 $\mu$ |
| *Sula bassana* | 342 $\mu$ |
| *Bubo bubo* | 360 $\mu$ |
| *Accipiter nisus* | 400 $\mu$ |
| *Sterna minuta* | 400 $\mu$ |
| *Cotyle riparia* | 400 $\mu$ |
| *Aquila chrysaetos* | 420 $\mu$ |
| *Buteo buteo* | 480 $\mu$ |
| *Falco tinnunculus* | 450 $\mu$ |
| *Falco sub-buteo* | 420 $\mu$ |
| *Lanius excubitor* | 460 $\mu$ |
| *Circaetus gallicus* | 630 $\mu$ |

## B. Areae and Foveae

There are usually one or more regions of the retina where the concentration of cones exceeds that elsewhere. In such an *area* the number of cells within the nuclear layers is also higher and there is a 1:1 ratio between cones and ganglion cells. The retina is consequently thick in such regions. Within the *area* there is usually a steep-sided depression, or fovea, at the bottom of which the cones attain their closest packing (1,000,000 per square millimeter in the larger members of the Falconiformes). In many such birds, together with the Hirundinidae, Sternidae and Alcedinidae, etc., there is a second smaller fovea at the posterior or temporal border of the retina. This is separated from the central fovea by some 6 mm. The Strigiformes have a single fovea which is placed excentrically and in the region occupied elsewhere by this lateral fovea. In many cases one or both foveae are located on a horizontal or oblique retinal stria.

According to Rochon-Duvigneaud the foveae of the Passerinae, Corvinae, Piciformes and Falconiformes are more clearly differentiated than those of the Galliformes or Columbiformes. In the latter orders they can be shallower and lack the histological complexity seen amongst the four former groups. An area is a place of maximum optical resolution. The function of the central fovea is less obvious but Pumphrey (1961) suggested that refraction from its sloping side is a device which enables the eye to be locked to a given object and increases the eye's sensitivity to movements of that object. The investigations of Duijm (1958) and Wood (1917) show that three principal types of area-fovea arrangements exist. In many graminivorous species there is a single area, which may or may not be foveate, lying close to the optic axis. This is known as the *area centralis*. In many water birds and in some species which inhabit open plains the area is extended into a horizontal band within which the fovea can assume the form of a trough. Thirdly, in the Falconiformes, Hirundinidae, Alcedinidae and Trochiliformes there can be two areas both of which are foveate.

In these cases the central fovea is close to the optic axis and the lateral fovea is so placed that the image of an object ahead can, with a slight degree of convergence, be formed on the temporal foveae of both eyes simultaneously. They are, therefore, presumably used in binocular, stereoscopic vision during the hunting of prey in free flight. As Pumphrey pointed out, it is interesting in this respect that such temporal foveae are more open and less steep sided than the central fovea of the same eye. They are closer structurally to the primate fovea which is certainly implicated in stereoscopic vision. It may be that for this purpose the advantages of the steep-sided fovea are outweighed by the disadvantages. On the other hand the extended area of water birds suggest that they are a device for accurately fixating the horizon as a datum to which other objects in the visual field can be referred.

Following his work on the definitive head posture of birds (Duijm, 1951) this author was particularly interested in Pumphrey's earlier suggestion that the ribbon-like areas were uniformly orientated in a horizontal position. Each species of bird has a specific head posture, which is particularly noticeable when the specimen is alert, and is characterized by occupying a fixed position in relation to gravity. Such postures are very striking when, as in the case of the oyster-catcher, the bill of the bird is pointing distinctly downwards and Hofer (see Duijm) introduced the term clinorhynchous to describe the skulls of the species.

Duijm (1958) compared the position of the central area of the retina in various species exhibiting different degrees of clinorhynchy. He chose *Haematopus ostralegus*, *Vanellus vanellus*, *Fulica atra* and *Larus argentatus*. The position of the bill in the normal head posture showed considerable, and highly significant, specific differences. The positions of both the lateral

## TABLE 38

*The position of the bill, lateral semi-circular canal and ribbon-like* area centralis *relative to the horizon in the normal head posture.*

| Species | Angles included between the horizon and | | |
| --- | --- | --- | --- |
| | Bill | lateral semi-circular canal | ribbon-like *area centralis* |
| *Haematopus ostralegus* | 28° | 5° | −9° |
| *Vanellus vanellus* | 36° | 2° | 2° |
| *Fulica atra* | 25° | 1° | 6° |
| *Larus argentatus* | 7° | 3° | 4° |

From Duijm (1958).

## TABLE 39

*The species in which a ribbon-like area has been reported by one or more of* Chievitz (1891), Slonaker (1897), Wood (1917), Rochon-Duvigneaud (1943), Lockie (1952) *and* Duijm (1958).

Struthionidae
*Struthio camelus* L.

Podicipitidae
*Podiceps cristatus* (L)

Procellariidae
*Puffinus griseus* (Gm)
*Puffinus puffinus* (Brunn)
*Fulmarus glacialis* (L)

Sulidae
*Sula bassana* (L)

Phalacrocoracidae
*Phalacrocorax penicillatus* (Brandt)

Phoenicopteridae
*Phoenicopterus ruber* L.

Anatidae
*Anas platyrhynchos* L. (domesticated)
*Melanitta deglandi* (Bonap)
*Fuligula glacialis*
*Anser anser* (L) (domesticated)
*Anser hyperboreus* Pall.

Rallidae

*continued overleaf*

Recurvirostridae
*Recurvirostra avocetta* L.

Laridae
*Larus argentatus* Pontopp
*Larus canus* L.
*Larus ridibundus* L.
*Sterna hirundo* L.
*Sterna paradisaea* Pontopp
*Sterna albifrons* Pallas
*Sterna sandvicensis* Lath.
*Sterna dougallii* Mont.

Alcidae
*Alca torda* L.
*Uria aalge* (Pontopp)
*Fratercula arctica* (L)
*Cepphus columba* Pall.

Micropodidae
*Apus apus* (L)

Hirundinidae
*Hirundo rustica* L.
*Riparia riparia* (L)
*Tachycineta bicolor* (Vieill)

*continued overleaf*

Table 39 (*contd.*)

| | |
|---|---|
| *Fulica atra* L. | Motacillidae |
| *Fulica americana* Gm. | *Motacilla alba* L. |
| | *Motacilla flava* L. |
| Otidae | |
| *Otis tarda* L. | Turdidae |
| | *Saxicola rubetra* (L.) |
| Haematopodidae | *Oenanthe oenanthe* (L) |
| *Haematopus ostralegus* L. | |
| | Emberizidae |
| Charadriidae | *Emberiza calandra* L. |
| *Vanellus vanellus* (L) | |
| *Charadrius hiaticula* L. | Fringillidae |
| *Pluvialis apricaria* (L) | *Carduelis cannabina* (L) |
| *Charadrius semipalmata* Bonaparte | |
| | Corvidae |
| *Squatarola squatarola* (L) | *Corvus brachyrhynchos* Brehm. |
| Scolopacidae | |
| *Ereunetes pusillus* (L) | |
| *Calidris alpina* (L) | |
| *Calidris canutus* (L) | |
| *Arenaria interpres* (L) | |
| *Tringa glareola* (L) | |
| *Tringa hypoleucos* L. | |
| *Totanus melanoleucos* (Gm) | |
| *Limosa lapponica* (L) | |
| *Numenius arquata* (L) | |
| *Numenius hudsonius* Lath | |

semicircular canal and the ribbon-like central area relative to the horizon when the bill is held in this position are contained in Table 38. Duijm concluded from these rather limited data that where a ribbon-like central area exists the species has a tendency to hold the eye so that the long axis of the area is close to the horizontal. Such a position also characterizes the lateral semi-circular canal.

A list of species in which such a ribbon-like area exists is contained in Table 39. A complementary list of the species in which it has been shown to be absent is provided by Table 40. Duijm concluded that as a relatively similar condition very often seems to occur throughout large taxa and even families, it may well represent a relatively old adaptation of general importance. As implied above it has not been found in either birds of prey or inhabitants of forests but rather in those species which live in open spaces. As birds receive a generally detailed picture of the environment, without moving their head this can include 300°, such a plane of preferential sensitivity to movements in relation to the horizontal position is apparently favoured as a reference position for the fixed coordinates. Duijm emphasized that it must be of

## TABLE 40

*Species in which a positive search has failed to reveal a ribbon-like area.[a]*

Tinamidae
*Rhynchotus rufescens* (Temminck)

Ardeidae
*Ardea cinerea* L.
*Nycticorax nycticorax hoactli* (Gm)
*Botaurus lentiginosus* (Mont)

Accipitridae
*Buteo borealis* (Gm)
*Buteo buteo* (L)

Falconidae
*Falco sparverius* L.

Tetraonidae
*Bonasa umbellus* (L)

Phasianidae
*Colinus virginianus* (L)
*Lophortyx californicus vallicola* Ridway
*Perdix cinerea* (L)
*Phasianus colchicus* L.
*Meleagris gallopavo* L.
*Gallus gallus* L. (domesticated)
*Guttera pucherani* (Hartlaub)

Scolopacidae
*Gallinago media* (Lath)
*Gallinago gallinago* (L)

Columbidae
*Columba palumbus* L.
*Columba livia* Gm. (domesticated)
*Leucosarcia melanoleuca* (Lath)

Cuculidae
*Coccyzus americanus* L.

Psittaciformes
*Kakatoe galerita* (Lath)

Strigidae
*Athene noctua* (Scop)
*Asio otus* (L)

*continued overleaf*

Tyrannidae
*Tyrannus tyrannus* (L)

Alaudidae
*Alauda arvensis* (L)

Hirundinidae
*Delichon urbica* (L)

Laniidae
*Lanius ludovicianus gambeli* Ridgway

Troglodytidae
*Troglodytes troglodytes* (L)

Mimidae
*Mimus polyglottis leucopterus* (Vigors)

Vireonidae
*Vireo flavifrons* Vieill

Prunellidae
*Prunella modularis* (L)

Parulidae
*Dendroica palmarum* (GM)
*Seiurus aurocapillus* (L)
*Opornis philadelphia* (Wils)
*Wilsonia citrina* (Bodd)

Turdidae
*Turdus migratoria* L.
*Sialia sialis* (L)

Sylviidae
*Hypolais icterina* (Vieill)
*Acrocephalus schoenobaenus* (L)
*Sylvia borin* (Bodd)
*Sylvia communis* Lath
*Regulus regulus* (L)
*Regulus satrapa* Lichtenst

Paridae
*Parus major* L.
*Parus coeruleus* L.

*continued overleaf*

Table 40 (*contd.*)

---

*Tyto alba* (Scop)
*Strix nebulosa* Forster
*Strix aluco* L.
*Otus asio* (L)

Trochilidae
*Calypte anna* (Lesson)

Alcedinidae
*Alcedo atthis* (L)
*Ceryle alcyon* (L)

Upupidae
*Upupa epops* L.

Picidae
*Dendrocopus major* (L)
*Colaptes auratus* (L)
*Colaptes cafer mexicanus* Swainson
*Melanerpes erythrocephala* (L)

Ploceidae
*Passer domesticus* (L)
*Passer montanus* (L)

Icteridae
*Agelaius phoeniceus* L.

Sturnidae
*Sturnus vulgaris* L.

Corvidae
*Corvus corone cornix* L.
*Corvus frugilegus* L.
*Pica pica* (L)
*Garrulus glandarius* (L)
*Cyanocitta cristata* (L)
*Cyanocitta stelleri* (Gm)

*Parus atricapillus* L.

Certhiidae
*Certhia familiaris* L.

Emberizidae
*Emberiza citrinella* L.

Fringillidae
*Fringilla coelebs* L.
*Serinus canaria* (L)
*Spinus tristis* (L)
*Pooecaetes gramineus* (Gm)
*Spizella pusilla* (Wilson)
*Junco hyemalis* (L)
*Melospiza melodia* (Wilson)
*Passerina cyanea* (L)
*Passerella iliaca* (Merv)

---

[a]Authors as for Table 39.

the utmost importance that such sensory systems as the eye and labyrinth are adjusted to the same reference system. Whilst this is clearly not necessarily a reason for them both to have the same zero datum, merely to have a fixed relationship, it is certainly more straightforward if they do. Such an organization of the labyrinth and eye can be viewed as the primary condition enabling a close interaction of labyrinthine and eye reflexes, and also the integration of gravitational and visual information.

## C. The Visual Cells

### (i) *The Rod Cells*

The avian retina contains both rods and cones. The cones are most numerous in diurnal species and the rods in nocturnal species, where they tend to be long and thin. However, the actual number of rods in diurnal species varies and they tend to be both larger and more numerous towards the periphery of the retina. The fine structure of these photosensitive cells has been studied in the pigeon by Cohen (1963a, b), and in the chick by Morris and Shorey (1967), Matsusaka (1967a, b), Meller and Glees (1965), Mountford (1964) and Pedler (1969). Cohen concluded that the outer segment of both types consists of flattened saccules enclosed within a membrane. The continuity of saccule and cell membrane occurred along the entire length of the outer segment in cones but only at its base in rods. The characteristics of both rods and cones are summarized in Table 41 and Figs 88 and 89.

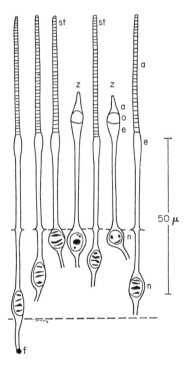

FIG. 88. The rods and cones in the eye of *Strix aluco* as seen by light microscopy. (Oehme, 1961).

Abbreviations: e., ellipsoid; f., basal region; n., nucleus; o., oil-droplet; st., rods; z., cones.

12*

## TABLE 41

*The dimensions of various receptor cells and their components.*

| Receptor cell type | Length of inner segment | Length of outer segment | Distance from outer limiting membrane to vitreal surface of synaptic body |
|---|---|---|---|
| Rods | 25 $\mu$ | 16 $\mu$ | 18 $\mu$ |
| Principal cones | 25 $\mu$ | 16 $\mu$ | 18 $\mu$ |
| Accessory cones | 22 $\mu$ | 19 $\mu$ | 18 $\mu$ |
| Type I Single cones | 22 $\mu$ | 19 $\mu$ | 22 $\mu$ |
| Type II Single cones | 22 $\mu$ | 19 $\mu$ | 22 $\mu$ |

From Morris and Shorey (1967).

The external segment of the rod cells in the chick is broader than those of cones. The mitochondria comprising an ellipsoid are closely packed, elongated, and have densely packed cristae. A paraboloid is situated vitreal to the mitochondria in the inner segment and a long cytoplasmic cylinder joins the inner segment and the nuclear region. The cytoplasm close to the outer limiting membrane has prominent vesicles. At this level the rod is separated from other receptor cells by extensions of the supporting Muller cells. The nucleus is located in close proximity to a convex synaptic body in which the vesicles range from 400–900 A in diameter. They are more densely packed and include larger specimens than do those of cones. The base of the body is associated with a number of indenting neurites. Paired proximal ones, which resemble those of cones, were only found occasionally, and distal ones were also rare. The actual appearance of such synaptic bodies also contrasts with those of cones. They are longer and not necessarily provided with proximal neurites. Several such long structures may be disposed in parallel to one another or arranged radially as in a fan.

In the other cell types as well as the rods microtubules run longitudinally along the inner segment. They are more common in the vitreal part and especially prominent in the rods where they follow a sinuous course, curving from one side to the other. Sometimes they lie alongside, and even appear to emerge from, the peripheral filaments of the inner segment. However, the fine structure of the two organelle types is different. The microtubules have a tri-laminar organization with two electron-dense components separated by

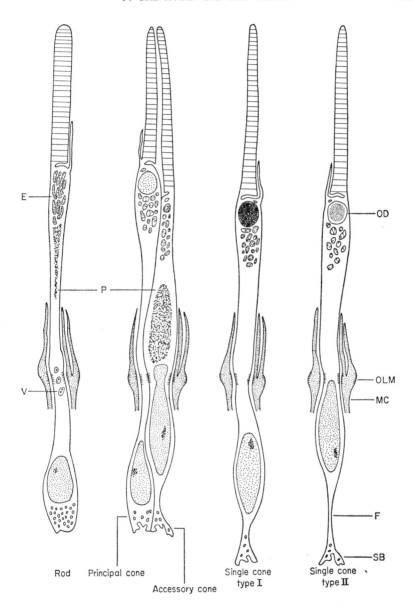

Rod    Principal cone    Accessory cone    Single cone type I    Single cone type II

FIG. 89. The structure of the rods and cones in the chick eye. (Morris and Shorey, 1967).

Abbreviations: E, ellipsoid; F, receptor fibre; MC, Muller cell; OD, oil-droplet; OLM, outer limiting membrane; P, paraboloid; SB, Synaptic body; V, prominent vesicle.

an electron transparent one. All three layers are of approximately similar thickness.

The paraboloids which were mentioned above occur in both the rods and the accessory cones. The granules within those of rods can either be grouped into a compact mass or distributed in a longer column down the inner segment. As they are particularly obvious following lead staining Morris and Shorey understandably suggest that they are glycogen bodies. Unlike those of the accessory cones those of the rods are not interspersed amongst smooth endoplasmic reticulum. However, between the ellipsoid and nucleus all receptor cell types contain ribonucleoprotein particles arranged singly or in clusters, spirals or flattened rings. There can also be a considerable amount of endoplasmic reticulum some of which is rough. In both rods, and single cones of Type I variety, these two components occupy most of the region between the nucleus and the outer limiting membrane, a region that by virtue of the more vitreal location of the nucleus, is larger in these than other receptors.

## (ii) *Principal and Accessory Cones*

As shown in Fig. 89 the principal cone is closely related to the accessory one so that the two together form a double cone structure. The internal segment projects sclerally beyond that of the accessory cone and ends in a pale staining oil-droplet. In the accessory cone a granular vesicle lies at the tapered scleral end of the internal segment and as in the rod cells there is a paraboloid in the enlarged vitreal region. In this case it is a collection of granules, probably glycogen, lying within cytoplasmic channels which are enclosed by smooth endoplasmic reticulum. Reticulum material can often also form a border to the whole paraboloid.

The nucleus of the accessory cone is more scleral in position than that of the principal one, and is connected to the synaptic body by a narrow fibre. That of the principal cone is elongated and abuts against the synaptic body. Both bodies are at the same level as was the case in the rods, and that of the principal cone is larger than those of other cell types with a broad connection to the nuclear region at the level at which they are apposed. Sections cut in different planes suggest that the synaptic body of the accessory cone is the smaller of the two and partly surrounded by the other. In any event both bodies form a common hood which envelopes a cluster of neurites in the outer plexiform layer and that of the principal cone is indented by a number of larger neurites some of which appear to be enclosed within it. Those just inside the base have few contents and are often arranged in pairs with a synaptic lamella between them. Other neurites extend towards the apical region of the body and are sometimes flattened and bent over. Referred to by Morris and Shorey as distal neurites, some contain vesicles whilst others are empty like the proximal ones.

Bordering the basal region of the synaptic body in the principal cone there are small neurites which are sometimes organized into a palisade. At points of contact between the membranes of synaptic body and neurite both can be dense and such areas may represent desmosomes. In the case of the principal cone the basal region of the body itself can extend laterally within the outer plexiform layer. The smaller synaptic body of the accessory cone tapers distally to the fine fibre which makes the connection with the nuclear zone. The edge next to the principal cone forms a long process which projects obliquely downwards into the outer plexiform layer and always seems to traverse underneath the basal part of the principal cone. It was traced for distances in excess of $6\,\mu$. Near to its point of origin one or more small branches penetrate the neuropile underlying the synaptic body of the principal cell and some such branches terminate on this body. Alongside the main process there is a long neurite which enters the accessory cell synaptic body. In the photographs of Morris and Shorey its terminal region appeared to form one of the proximal neurites described for the principal cone and tangential sections suggest that both the synaptic body process and its long associated neurite may be multiple. The accessory cones also have paired proximal neurites, which are comparable with those described for the principal cones, together with associated synaptic lamellae. Some of these paired neurites can fuse to form a large, compound structure which extends beneath the base of the synaptic body. Indeed the neurites in such locations are larger than those of the principal cone.

The two cone cells are not separated from each other at the level of the outer limiting membrane as is the case with the other sensory cells and they are jointly separated from other units at this level by Muller cells. The membrane discs in the outer segments are connected to the plasma membrane on one side whilst on the other lies the connecting cilium which joins the inner to the outer segments. The most basally situated discs are often connected to the membrane on the vitreal surface where it overlies the inner segment. In principal cones the connections on the side of the outer segment face towards the analogous structures in the accessory cone.

The connecting cilium arises from a pair of centrioles on one side of the inner segment. In the principal and accessory cones, it will be apparent from the connections and relations of the basal discs that it lies diametrically opposite to the region of apposition of the two cells. In these and all other receptor cell types inner segment processes arise from the scleral end of the inner segment and invest the base of the outer segment. They contain fine longitudinal filaments which on occasions traverse a short distance into the inner segment. Similar structures partially line the sides of the inner segment down to the level of the other limiting membrane and oblique sections show that these are grouped into separate bundles in the peripheral cytoplasm. In both the inner segment and its processes they are 40–50 A in diameter and the centres of adjacent filaments are 70–100 A apart.

### (iii) *Single Cone Cells and their Connections*

In the pigeon Cohen (*loc. cit*) described varieties of cones with up to three independant columns of saccules. Pedler (1965) included this species amongst those vertebrates which have incompletely filled and disordered outer segments to the cones although in invertebrates such orderliness in photoreceptor units has been shown to be very labile and dependant upon the presence of incident light (Curtis, 1970). Morris and Shorey (1967) distinguished between two distinct types of single cone cells in both of which the internal segment is shorter than those of principal cones so that the terminal oil-droplet is situated in a more vitreal position. At the level of the outer limiting membrane each is separated from its neighbours by intervening Muller cells and, as in the accessory cones, the nuclear region is connected to the synaptic body by a narrow fibre. The synaptic body itself is roughly triangular in outline and the fibre connecting it to the receptor is directed vertically so that it joins the unit which is situated directly above it. However, occasional oblique fibres were detected running through the outer nuclear layer and making contact with a nucleus two or three receptors away. The synaptic bodies of two independant cones are sometimes closely associated within the outer plexiform layer and only separated by a thin layer of Muller cell cytoplasm. The base of the body can extend both deeply into, and laterally, within the outer plexiform layer, and it is indented by distal neurites. Paired proximal neurites which are associated with the synaptic lamella parallel those of the principal cones. The synaptic vesicles within the synaptic bodies are smaller than those of rod cells and range from 400–600 A. Occasionally an irregular mass of dark staining, short branched tubules was seen. These masses may represent fused vesicles.

Cones of Type I are distinguished from those of Type II by the presence of electron dense oil-droplets in the apical region of the inner segment, and by the greater number of mitochondrial cristae which are detectable. Type II cones have a lighter staining oil-droplet and the number of cristae is less. A further differentiating feature is the level at which the nucleus lies, this is more vitreal in Type I cones. In a total of 286 receptor cells a count showed that 15% were rods, 36% were double cones, 25% were single cones of the Type I category and 11% were of the Type II category. The remaining 13% were not accurately ascribable to such divisions. More recently Morris (1970) found that the mean percentages of these receptors in the central retina were rods, 14%; double cones, 32% and single cones 54%. At the periphery the respective values were 33%, 30% and 37%. A statistical analysis showed that each type of cell was evenly distributed in the receptor mosaic but single cones of the type I variety tended to occur in pairs. The overall pattern was a hexagonal lattice and this reflected the hexagonal distribution of receptor precursor cells during embryology.

## D. Pigment and Muller Cells

The long processes of the cells making up the pigment epithelium invest the outer segments of both rods and cones, and may also extend as much as two-thirds of the way along the inner segments. In the case of single cones the melanin granules are generally restricted to the $3-4\,\mu$ region which intervenes between the level of the terminal bars and that of the oil-droplets, and they are arranged end-to-end down the pigment cell processes with their long axes parallel to that of the outer segments. Very often they are accompanied within the processes by longitudinal filaments which have a diameter of 40 A, and are separated from neighbouring filaments by about 30 A.

Although the projections of the pigment epithelium are not as long as those of Amphibia they penetrate further up between the rods and cones than is the case in mammals. The avian eye therefore resembles that of lower vertebrates in the pigment movements that result from changes in the amount of incident light. A decrease in light intensity results in retraction of the pigment within the epithelium so that it lies at the extreme tip of the rods and cones. In nocturnal species, and particularly *Caprimulgus europaeus*, the degree of pigmentation and the length of the interbacillary processes can both be reduced.

The glial cells of Muller pass between the synaptic and nuclear regions of the receptor cells, expand into club-shaped structures at the level of the outer limiting membrane, and finally terminate in microvilli which invest the base of the inner segments. Fine filaments can sometimes be distinguished within these microvilli. At the outer limiting membrane, where the Muller cells are expanded, the receptor cells are correspondingly constricted and the junctions lined by terminal bars. Within the Muller cell cytoplasm near to the outer limiting membrane there are mitochondria and occasionally a centriole. The cytoplasm itself stains heavily with lead. The main tracts coming from the bipolar layer can be full of microtubules and the lateral ramifications amongst the synaptic bodies with ribonucleoprotein particles. The swellings of the various Muller cells are aligned one with another, although the reciprocal constrictions on the receptors occur at differing levels which are dependant upon the particular rod or cone type. Morris and Shorey (1967) suggested that the presence and level of these swellings may therefore be critical.

Strengthened by the terminal bars they comprise a fenestrated surface beyond which the outer and inner segments project towards the pigment epithelium. The function of this surface may therefore be to maintain the receptor cells at a sufficient distance from the pigment epithelium so that the component segments can project radially. Such a radial orientation appears, not unexpectedly, to be essential for accurate visual function and amongst invertebrates, the

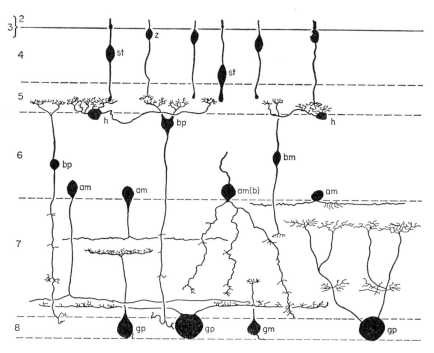

Fig. 90. The inter-relationships of the various cells in the retina of *Strix aluco*. (Oehme, 1961).

Abbreviations: am., Amacrine cell; am(b)., Centrifugal bipolar; bm., monosynaptic bipolar; gm., monosynaptic ganglion cell; gp., polysynaptic ganglion cell; st., rods; z., cones.

Phalangida, for example, deviations from it are found in species with degenerate eyes. It may also result from the slight turgidity of the inner segments together with the action of various skeletal structures such as the peripheral fibres of the inner segment. By analogy the inner segment processes may serve to align the inner and outer segments in a region where the only other connection between them is through the narrow connecting filament. Should the orientation be achieved in this way then the pigment cell processes would merely occupy the residual spaces between the various outer segments and that this may certainly be the case is suggested by the ease with which the pigment epithelium can be detached from the receptor cells. The absence of any Muller cells between the partners of double cones is rather difficult to explain satisfactorily. Morris and Shorey suggested that it might reflect the derivation during development of both principal and accessory cones from a single cell which has already been invested by its sheathing Muller cell.

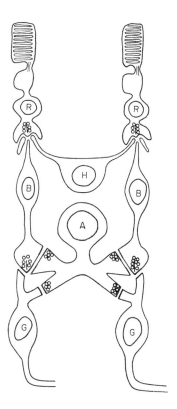

FIG. 91. The synaptic contacts in the retina. (Dowling, 1968). R, receptors; H, horizontal cell; B, bipolar cells; A, amacrine cells; G, ganglion cells.

## E. Synaptic Organization of the Retina

A review of the synaptic organization within the pigeon retina was provided by Dowling (1968) and the relative numbers of various cell types at different stations by Oehme (1961). Figure 91 shows a simplified representation of the features which are common to all vertebrate retinas. No attempt has been made to indicate any spatial array of processes or of synaptic contacts, but only the probable synapses and synaptic pathways which exist between the various cell types. For simplicity the amacrine-amacrine synapses have been omitted. In the outer plexiform layer processes from the bipolar and horizontal cells penetrate into invaginations in the receptor terminals. In the inner plexiform layer the bipolar terminals contain synaptic ribbons and make dyad contacts. At least one of the post-synaptic units involved is an amacrine process. The amacrine cell processes themselves extend laterally within the inner plexiform layer and thereby connect adjacent bipolar terminals in the reciprocal synaptic relationship. They also make feed-forward contacts directly onto ganglion cells. Amacrine cells thus both fulfil a feed-back function on to bipolar terminals and provide a pathway for lateral and reciprocal interactions between bipolar terminals and ganglion cell dendrites.

Amongst the vertebrates which were studied by Dowling and his collaborators pigeons and frogs seemed to have the most complex retinas. Although they have only relatively few contacts between bipolar cells and ganglion cells in the inner plexiform layer, there are abundant amacrine-amacrine and amacrine-ganglion cell synapses. Clearly Dowling's conclusion that in man and monkeys, where the cerebral cortex is best developed, the retina is simplest, whilst in frogs and pigeons, in which the cortex is least developed, the retina is more complex, begs a few questions. However, in terms of such phylogenetic comparisons it is certainly interesting that in the pigeon few if any of the the ganglion cell receptor fields seem to be of a simple concentric nature and a variety of stimuli are required if they are to be stimulated maximally. Simple spots of light work poorly and Dowling suggested that the cells in these retinas may be described as inhomogeneous, or non-uniform, in the sense that individual cells are responsive to quite specific features of the visual image. Examples of such stimulatory effects are considered in Section IV F.

Pedler (1969) found that in some areas of the pigeon retina the number of bipolar cells equalled that of the receptors but concluded that the individual receptor processes are so numerous and the dendritic fields of the bipolar cells, horizontal cells and dendritic filaments so broad that there is considerable overlapping. As a result he concluded that both convergent and divergent channels probably exist within the same matrix of connections in the outer plexiform layer. In an area where the number of bipolars equalled the number of receptors the average width of the bipolar dendritic tree covered an area equal to the width of five adjacent polysynaptic receptor pedicels. Polysynaptic pedicels predominate in animals such as birds which have a high visual dependance, in terms of their behaviour patterns. For example, the avian retina has to provide information about spatial orientation, form, colour, and also relative rate of movement, rapidly, accurately and more or less simultaneously.

## F. Directional Movement and Horizontal Edge Detectors

The activity of single retinal cells from cut and intact optic nerves in curarized pigeons was measured by Maturana and Frenk (1963) using metal-filled micro-pipettes. In this way they were able to study six classes of ganglion cells which differed in the visual configuration to which they responded. In general the size of the response, in terms of the number of spikes and their frequency, depended on the direction of contrast, the intensity of the contrast and the speed of the movement. They considered that the directional sensitivity is not the result of any asymmetry within the receptive field, but that directional movement detectors comprise about 30% of the accessible cell population and have five fundamental characteristics. These are:

1. They have small receptive fields when these are defined as the area from which a response can be elicited, and these vary from 55–110 $\mu$ on the retina.

2. They have an optimal or an exclusive response to the movement of an edge in one direction but not the other. The sharpness of the required edge depends on the size of the receptive field and the smaller the field the sharper the edge that is required.

3. They show no response to phasic changes of the ambient light.

4. The directional mode of response is independant of both the intensity of the ambient light and the direction of contrast across the moving edge so that the mode of response is the same for moving objects which are either lighter or darker than the background. It is also independant of both colour and the part of the receptive field in which the object moves.

5. There is usually a uniformly *on-off* receptive field and if there are exclusive *on* or *off* spots then these show no special relationship to the direction of the optimal response.

Horizontal edge detectors form about 5% of the accessible cell population and these have four functional characteristics which may similarly be summarized as follows:

1. If the head of the bird is in its "normal" position relative to the direction of gravity the detectors respond maximally to a horizontal edge moving vertically up or down across the receptive field. However, they do not respond to a small object which is moved within the receptive field. To be effective the horizontal edge has to extend across retinal fields, which appeared to measure from 40 to 80 $\mu$, and into their surroundings.

2. They again do not respond to phasic changes of the ambient light, nor to the switching on and off of a spot light.

3. They vary with respect to the deviation from the horizontal which they will tolerate. Some will respond to the moving edge even when this is inclined at 45° to the horizontal, but most do not respond to the edge when such inclination exceeds 20–30°.

4. As in the case of the directional movement detectors their mode of response is independent of the intensity of the ambient light, the direction of contrast across the edge and the colour of either the objects or the background.

In none of these cases was it possible to provide an easy explanation of the responses using classical schemes of interaction between excitatory and inhibitory areas. However, Maturana and Frenk thought that two things were of considerable significance. Each ganglion cell needs to "take account" of what is happening both within its receptive field and the surrounding area. Since each one is connected to numerous bipolar cells it receives afferent impulses other than those which are due to any specific visual stimulus, and some degrees of integration must occur at the level of the ganglion cells themselves if they only respond to the specific stimulus. In the second place the overlapping of the dendritic arbors suggests that many ganglion cells which perform different and/or similar functions are connected to the same bipolar cells (Maturana and Frenk, 1965). It therefore follows that any cell which shows a selective response to a particular configuration of afferent influences must probably be doing so independantly of the specific pathways which are peculiar to this particular pattern. A discussion of the part which is played by centrifugal fibres in the control of retinal function can be delayed for consideration alongside the isthmo-optic nucleus in the mesencephalon (Chapter 10, Section VI B).

## G. Acetylcholinesterase in the Retina

The temporal sequence of the appearance of cholinesterase in the embryonic retina was studied by Rogers et al. (1960) along with the various brain parts. There was a continuous increase during the period between the 8th and 19th day of incubation during which the levels rose from $1·0–6·0 \times 10^{-5}$ moles. A sharp rise in the alkaline phosphatase levels between the 12th and 16th day, at which they reached their maximum, precedes the final achievement of the cholinesterase maximum which occurred around the time of hatching (Fig. 92). The appearance and cellular localization of such acetylcholinesterase activity was also observed by Shen et al. (1956). They found that significant concentrations occur in the ganglion cells at the 4th day and that these are followed at the 6th day by activity in the amacrine cells and, finally, in the inner plexiform layer at the 8th day. Some slight but quite definite activity was also detectable both in the horizontal cells and the *myoids* of the visual elements.

Within the inner plexiform layer the enzyme is not distributed uniformly but is instead quite distinctly stratified. The number, pattern, location and sequence of appearance of these active strata during histogenesis are identical with those of the synaptic terminals of the ganglion, amacrine and bipolar cells. In both avian and mammalian retinas it had earlier been assumed that the primary neurones, the rods and cones, and the tertiary neurones or ganglion cells were non-cholinergic (Feldberg and Vogt, 1948; Feldberg *et al.* (1951). The absence of acetylcholine esterase from the synapses of the outer plexiform layer indicated that it was absent from the axonal terminals of the non-cholinergic neurones as well as the dendritic terminals of the cholinergic ones. It also appeared to be absent from the cell body of the cholinergic bipolar cells.

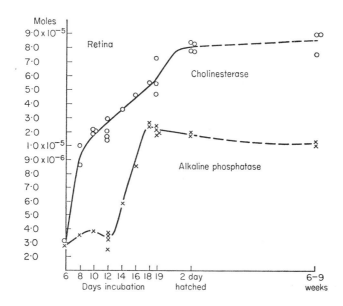

FIG. 92. The gradual maturation of the cholinesterase activity in the retina and the earlier alkaline phosphatase maximum. Data expressed as moles of substrate split per hour for each milligram of tissue nitrogen.

# V. A COMPARISON OF VISION IN MAN AND BIRDS

The high density of cones in the fundus of *Motacilla*, where they reach 120,000 mm², is associated with a cell density in the ganglion layer of 110,000 mm² (Franz, 1934). Almost every visual cell can therefore be individually represented in the optic nerve. This contrasts with the condition in man where the cone density falls to 10,000 mm² at a distance of 3 mm from the fovea, and the ganglion cells are thinly scattered not packed together in a multilaminar mass. Pumphrey (1961) summarized the differences in these two situations. In doing so he emphasized that the two eyes must be used in very different ways and that glib comparisons of their efficiency, using tests appropriate to man, are misleading.

For example it is often assumed that avian visual acuity greatly exceeds that of man. The Falconiformes and insectivorous genera see objects which would, it is asserted, be invisible to us. If, in the case of insect eaters, man retained the power of accommodation which he has at the age of 2, and could therefore get his eye as close to his prey as a bird habitually does, he would have no difficulty in seeing it. He also concluded that the visual ability of soaring birds is less remarkable than it might appear. Human acuity is reckoned to achieve about 0·5 min of arc. A man should therefore be able to see a large ungulate or its shadow, to determine whether it is stationary or moving, and differentiate it from neighbouring objects, when he is looking down on it from a height of 10,000 feet, provided his attention is directed at it. What he could not do would be to keep an adequate watch on all the antelopes in, say, 40 square miles of territory, because his good acuity is limited to the small angle of 2·5° subtended by his fovea (see also Fite, 1968).

As the result of these considerations Pumphrey concluded that, whilst the visual acuity of men and birds is of a comparable order, the rate of assimilation of the details in the visual field is much higher in birds. Avian vision is therefore no sharper but a good deal faster than that of man. Donner (1951) demonstrated that the visual acuity of small passerines is about 1·5 min of arc. This is some three times less than man's. He also showed that the acuity was in close agreement with that computed from measurements of the cone density and the posterior nodal distance. Pumphrey therefore concluded that large birds such as members of the Falconiformes may have a visual acuity up to, but not exceeding, three times that of man. A bird may absorb in one second a picture which man would have to laboriously construct by scanning the visual field piece by piece with the most accurate part of his retina. However, on the debit side, the bird's picture is flat and man's laborious binocular search provides him with information about distances which is not easily available to birds, but has to be constructed from a series of glances made in different places. Birds make a very obvious use of parallax, and the rapid lateral or vertical movements of the head in owls confronted by a strange object presumably enable them to obtain a stereoscopic view with eyes that are incapable of convergence. Furthermore for migrating birds to navigate as accurately as they do they must probably be capable of high resolution over a visual angle of 70–80° in order to ascertain the instantaneous position of, for example, the sun, and its direction of movement in the sky.

# References

Adler, H. E. and Dalland, J. (1959). Spectral thresholds in the starling (*Sturnus vulgaris*). *J. Comp. Physiol. Psychol.* **52**, 438.

Alpern, M. (1968). Distal mechanisms of vertebrate colour vision. *Ann. Rev. Physiol.* **30**, 279–318.

Bachsich, P. and Gellert, A. (1935). Beitrage zur Kenntnis der Struktur und Funktion des Pecten im Vögelauge. *Graefes Archiv. Ophthalmol.* **133**, 448–460.

Beer, T. (1892). Accommodation des Vögelauges. *Arch. Gesammte Physiol.* **53**.

Benner, J. (1938). Untersuchungen uber die Raumwahrnehmung der Huhner, *Z. Wissensch. Zool.* **151**, 382.

Blest, A. D. (1957). The function of the eyespot patterns in the Lepidoptera. *Behaviour* **11**, 210–254.

Blochmann, F. and von Husen, E. (1911). Ist der Pecten des Vögelauges ein Sinnesorgan? *Biol. Zbl.* **31**, 150–156.

Chievitz, J. H. (1891). Ueber das Vorkommen der Area centralis retinae in den vier hoheren Wierbelthierklassen. *Arch. Anat. Entwickl. Gesch.* 311–334.

Cohen, A. L. (1963a). Fine structure of the visual receptors of the pigeon. *Exp. Eye Res.* **2**, 88–97.

Cohen, A. L. (1963b). Vertebrate retinal cells and their organization. *Biol. Rev.* **38**, 427–459.

Collin (1903). Premiers stades du développement du *m. sphincter* de l'iris chez les oiseaux. *C.R. Soc. Biol.* **55**, 1055.

Crozier, W. J. and Wolff, E. (1943a). Modifications of the flicker response contour and the significance of the avian pecten. *J. Gen. Physiol.* **27**, 287–313.

Crozier, W. J. and Wolff, E. (1943b). Flicker response contours for the sparrow and the theory of the avian pecten. *J. Gen. Physiol.* **27**, 315–324.

Curtis, D. J. (1970). Comparative aspects of the fine structure of the eyes of the Phalangida and certain correlations with habitat. *J. Zool.* **160**, 231–265.

Denissenko, G. (1881). Ueber den Bau und die Funktion des Kammes (Pecten) im Auge der Vögel. *Arch. mikros. Anat.* **19**, 733–741.

Dogiel, A. S. (1895). Die Retina der Vögel. *Arch. f. Mikr. Anat.* **44**.

Donner, K. O. (1953). The spectral sensitivity of the pigeon's retinal element. *J. Physiol.* **122**, 524.

Dowling, J. E. (1968). Synaptic organization of the frog retina. *Proc. Roy. Soc. B.* **170**, 205–228.

Duijm, M. (1951). On the head posture in birds and its relation to some anatomical features. *Proc. Kon. Ned. Akad. Wet. C.* **54**, 202–211; 260–271.

Duijm, M. (1958). On the position of a ribbon like central area in the eyes of some birds. *Arch. neerl. Zool.* **13**, 128–145.

Duke-Elder, S. (1958). The eye in evolution. In "System of Ophthalmology" (Duke-Elder, S., ed.), vol. 1. Kimpton, London.

Eck, G. J. van (1939). Farbensehen und Zapfenfunktion bei dem Singdrossel (*Turdus ericetorum*). *Arch. neerl. Zool.* **3**, 450–499.

Feldberg, W., Harris, G. W. and Lin, R. C. Y. (1951). Observations on the presence of cholinergic and non-cholinergic neurons in the central nervous system. *J. Physiol.* **112**, 400–404.

Feldberg, W. and Vogt, M. (1948). Acetylcholine synthesis in different regions of the central nervous system. *J. Physiol.* **107**, 372–381.

Fischlschweiger, W. and O'Rahilly, R. (1966). The ultrastructure of the pecten oculi in the chick. *Acta Anat.* **65**, 561–578.

Fite, K. V. (1968). Two types of optomotor response in the domestic pigeon. *J. comp. physiol. Psychol.* **66**, 308–314.

Francois, J. and Neetens, A. (1962). Comparative anatomy of the vascular supply of the eye in vertebrates. *In* "The Eye", (Davson, H., ed.), vol. 1, pp. 369–416. Academic Press, London and New York.

Francois, J., Rabaey, M. and Lagasse, A. (1963). Electron microscopic observations on the choroid, pigment epithelium and pecten of the developing chick in relation to melanin synthesis. *Ophthalmologica* **146**, 415–431.

Franz, V. (1934). Hohere sinnesorgane (Auge). *In* "Handb. vergleich. Anat. Wirbelt" (Bolk, L., Goppert, E., Kallius, E. and Lubosch, W., eds), vol., 2, part 2, pp. 989–1292. Urban and Schwarzenberg, Berlin.

Fujimoto, K., Yanase, T. and Hanoaka, T. (1957). Spectral transmittance of retinal coloured oil globules re-examined with a microspectrophotometer. *Jap. J. Physiol.* **7**, 339–346.

Grynfelt, J. (1905). Epithelium posterior de l'iris de quelques oiseaux. *Ass. Anat. Geneva.*

Guhl, A. M. (1953). The social behaviour of the domestic fowl. *Kansas State College Agric. Exp. Stat. Tech. Bull.* **73**, 48.

Hess, E. H. (1956). Natural preferences of chicks and ducklings for objects of different colours. *Psych. Reports* **2**, 477.

Hess, C. (1912). *Vergleichende Physiologie des Gesichtsinnes.* Gustav Fischer, Jena.

Honigmann, H. (1921). Untersuchungen uber Lichtempfindlichkeit und Adaptierung des Vögelauges. *Pfluger's Arch.* **189**, 1.

Kare, M. R. (1965). The special senses. *In* "Avian Physiology" (Sturkie, P. D., ed.), pp. 406–446. Baillière, Tindall and Cassell.

Kauth, H. and Somner, H. (1953). Das Ferment Kohlensaurenanhydratase im Tierkorper. IV. Uber die Funktion des Pekten im Vögelauge. *Biol. Zbl.* **72**, 196–209.

Lashley, K. S. (1916). The colour vision of birds. I. The spectrum of the domestic fowl. *Anim. Behav.* **6**, 1–126.

Leiner, M. (1940). Das Atmungsferment Kohlensaurenanhydrase im Tierkörper. *Naturwiss.* **28**, 165–171.

Leplat, G. (1912). Développement et structure de la membrane vasculaire de l'oeil des oiseaux. *Arch. Biol.* **27**, 403–524.

Lockie, J. D. (1952). A comparison of some aspects of the retinae of the Manx shearwater, fulmar petrel and house sparrow. *Quart. J. micros. Soc.* **93**, 347–356.

Lorenz, K. Z. (1952). "King Solomon's Ring." Methuen, London.

Mann, I. C. (1924a). The pecten of *Gallus domesticus. Quart. J. micros. Soc.* **68**, 413–442.

Mann, I. C. (1924b). The function of the pecten. *Brit. J. Ophthalm.* **8**, 209–226.

Matsusaka, T. (1967a). The fine structure of the basal zone of the pigment epithelial cells of the chick retina. *Exp. Eye Res.* **6**, 38–41.

Matsusaka, T. (1967b). Lamellar bodies in the synaptic cytoplasm of the accessory cone in the chick retina. *J. Ultrastructure Res.* **18**, 55–70 (see also *Folia ophthalmol. Japan,* **17**, 320; *Jap. J. ophthalmol.* **10**, 266, *Z. Zellforsch.* **81**, 100).

Maturana, H. R. and Frenk, S. (1963). Directional movement and horizontal edge detectors in the pigeon retina. *Science* **142**, 977–979.

Maturana, H. R. and Frenk, S. (1965). Synaptic connections of the centrifugal fibres in the pigeon retina. *Science* **150**, 359–361.

Meller, K. and Glees, P. (1965). The differentiation of neuroglia—Muller cells in the retina of the chick. *Zeit. zellfersch.* **66,** 321–332.

Menner, E. (1938). Die Bedeutung des Pecten im Auge des Vögels fur die Wahrnehmung von Bewegungen. *Zool. Jahrb. Abt. Allgemein Zool. Physiol Tiere* **58,** 481–538.

Morris, V. B. (1970). Symmetry in a receptor mosaic demonstrated in the chick from the frequencies, spacing and arrangement of the types of retinal receptor. *J. comp. Neurol.* **140,** 35–398.

Morris, V. B. and Shorey, C. D. (1967). An electron microscope study of types of receptor in the chick retina. *J. comp. Neurol.* **129,** 313–340.

Mountford, S. (1964). Filamentous organelles in the receptor-bipolar synapses of the retina. *J. Ultrastructure Res.* **10,** 207–216.

Nussbaum, L. (1901). Die pars ciliaris retinae des Vögelauges. *Arch. mikros. Anat.* **57,** 346–353.

Oehme, H. (1961). Vergleichend histologisch Untersuchungen an der Retina van Eulen. *Zool. Jahrb.* **79,** 439–478.

O'Rahilly, R. (1962). The development of the sclera and the choroid in staged chick embryos. *Acta anat.* **48,** 335–346.

Pedler, C. (1965). Rods and cones, a fresh approach. *In* "Colour Vision" (Reuck, A. V. S. de and Knight, J., eds), pp. 52–82. CIBA.

Pedler, C. (1969). Rods and cones, a new approach. *Int. Rev. Gen. Exp. Zool.* **4,** 219–274.

Prince, J. H. (1956). "Comparative Anatomy of the Eye." Thomas, Springfield.

Pumphrey, R. J. (1961). Sensory organs, vision. *In* "Biology and Comparative Physiology of Birds" (Marshall, A. J., ed.), vol. 2, pp. 55–68. Academic Press, London and New York.

Raviola, E. and Raviola, G. (1967). A light and electron microscopic study of the pecten of the pigeon eye. *Amer. J. Anat.* **120,** 427–462.

Rochon-Duvigneaud, A. (1943). "Les yeux et la Vision des Vertébrés." Masson. Paris.

Rochon-Duvigneaud, A. (1950). Les yeux et la vision. *In* "Vertébrés Generalités", *Traité de Zoologie*, vol. 12.

Rochon-Duvigneaud, A. (1954). Les yeux et la vision. *In* "Oiseaux", *Traité de Zoologie*, vol. 15, pp. 221–242.

Schultze, M. (1866). Zur Anatomie und Physiologie der Retina. *Arch. mikros. Anat.* **2,** 175.

Seamen, A. R. and Storm, H. (1963). A correlated light and electron microscopic study on the pecten oculi of the domestic fowl. *Exp. Eye Res.* **2,** 163–172.

Semba, T. (1962). The fine structure of the pecten studied with the electron microscope. *Kyushu. J. med. Sci.* **13,** 217–232.

Shen, S-C., Greenfield, P. and Boell, E. J. (1956). Localization of acetylcholinesterase in chick retinas during histogenesis. *J. comp. Neurol.* **106,** 433–462.

Slonaker, J. R. (1897). A comparative study of the area of acute vision in vertebrates. *J. Morphol.* **13,** 445–494.

Slonaker, J. R. (1918). A physiological study of the anatomy of the eye and its accessory parts in the English sparrow. *J. Morph.* **31,** 351–459.

Sluckin, W. (1970). "Early Learning in Man and Animals", pp. 123. Allen and Unwin, London.

Stresemann, E. (1927–33). Aves. "Handb. der. Zoologie", vol. 7, pp. 900. Kukenthal, Krumbach, Berlin and Leipzig.

Strother, G. K. (1963). Absorption spectra of retinal oil globules in turkey, turtle and pigeon. *Exptl. Cell. Res.* **29,** 349–355.

Tanaka, A. (1960). Electron microscopic study of the avian pecten. *Dobutsugaku Zasshi.* **69,** 314–317.

Tanaka, H. (1938). The blood vessel distribution and pigmentation in pectens of Japanese birds. *Jap. J. med. Sci.* **7,** 133–151.

Verrier, L. (1936). Récherches sur la vision des oiseaux. *Bull. Biol. France-Belg.* **70,** 197–232.

Wallman, J. (1970). The role of oil-droplets in the colour vision of the Japanese quail. *Amer. Zool.* **10,** 506–7.

Walls, G. L. (1942). "The Vertebrate Eye." Cranbrook Inst. Sci. Michigan.

Walls, G. L. and Judd, H. D. (1933). The intra-ocular colour filters of vertebrates. *Brit. J. Ophthalmol.* **17,** 641–675, 705–725.

Watson, J. B. (1915). Studies on the spectral sensitivity of birds. *Pap Dep. Marine Biol. Carnegie Inst. Washington.* **7,** 87–104.

Wood, C. A. (1917). "The fundus Oculi of Birds." Lakeside Press, Chicago, Illinois.

# 10 The Mesencephalon

## I. INTRODUCTION

The mid-brain is derived from the intermediate member of the three primary brain vesicles. By comparison with the rather complex cell migrations which take place from the rhombencephalic lip it retains a relatively simple appearance throughout ontogeny. Associated with a number of important auditory and visual foci, its roof gives rise to the optic tectum. As in the hindbrain and spinal medulla the sulcus limitans differentiates basal and alar laminae during development. In the adult brain the efferent column of the spinal cord has its mesencephalic analogues in the nuclear masses of the third and fourth cranial nerves. In these respects the adult mesencephalon still shows traces of the ancestral metamerism which is suggested by the cranial regions of early craniates such as the Cephalaspidomorpha, Pteraspidomorpha and Acanthodii.

During embryological development the progressive increase in the size of the cerebellar and forebrain hemispheres leads to important topographical displacements within both the mesencephalic and diencephalic regions which were outlined in Chapter 4. Similar changes which occurred during phylogeny are suggested by the median position of the tectal lobes in *Archaeopteryx* and their lateral position in Cretaceous genera (Edinger, 1926, 1951; see Chapter 1, Section IV). During ontogeny the combined mass of the telencephalon and cerebellum, within the confined volume of the cranial cavity, forces the intervening mesencephalic structures generally, and the optic tectum in particular, first of all into a lateral, and then into a relatively ventral position. As a direct result of this many mesencephalic structures assume a close topographical relationship to those of the diencephalon. Kappers (*loc. cit*, Chapter 7, Section I) argued that neurobiotactic influences played a large part in this process but it is more easily explained (Cobb, 1959) in terms of the size, shape and position of the eyes (Fig. 93). In charadriiform genera such as *Capella* the telencephalon is forced so far back by the large eyeballs that its ventral surface can be seen from above. This is due to an upward and backward rotation in relation to the axis of the bill. Amongst pelecaniform genera such as *Phalacrocorax* the relatively small eyes do not have this effect and as a result the axes of both brain and bill are nearly parallel (Portmann and Stingelin, 1961). In the emu the angle between cerebral and bill axes is 27° (Cobb and Edinger, 1962). This is somewhat smaller than those of the gull *Larus argentatus*, and the grouse *Bonasus umbellus* which, are 34° and 36°

respectively. The angle of 15° for *Phalacrocorax* emphasizes that it has the straightest and most extended type of skull plus an exceptionally small brain-bill angle (see also Starck, 1955). With these qualifications the avian diencephalon and mesencephalon are comparable with those of other vertebrates.

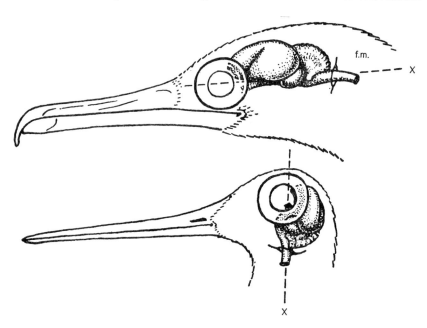

FIG. 93. The orientation of the head and brain in *Phalacrocorax* and *Capella*. (Portmann and Stingelin, 1961). X, axis of brain stem; f.m., foramen magnum of skull.

In the lower vertebrates the mid-brain is often considered to serve as a focus for the correlation and integration of impulses arising from optic stimulation on the one hand, and those representing general sensibility on the other. The efferent discharge passes via the tegmental motor nucleus (? = nucleus ruber) to the primary motor centres. In both fishes and amphibians various behavioural patterns have therefore been ascribed to mesencephalic function. In the reptiles, birds and mammals much of this overt integrative activity is assumed or regulated by the forebrain, but nevertheless the avian mesencephalon remains a primary sensory focus and retinal fibres terminate in the optic tectum. In view of the supreme importance of vision amongst the sensory modalities of many birds it is not very surprising that this tectum comprises two lateral eminences which are proportionately larger than those in many other vertebrate taxa. Furthermore the combined sensorimotor functions of preparations which have undergone telencephalic extirpation seem to be particularly centred around these and other related foci (Portmann and Stingelin, 1961).

The tectal effector systems are generally recorded in the literature as passing to the diencephalic nucleus rotundus, to the geniculate complex, and, by way of the tecto-bulbar tract, to the medullary bulb. Some other ascending pathways besides those from the retina reach, and terminate in, the mesencephalon. The dorsal part of the lateral mesencephalic nucleus (= torus semicircularis or Wallenberg's lateral ganglion) is buried within the mid-brain (Figs 94 and 95). It receives fibres from the trigeminal sensory nucleus and,

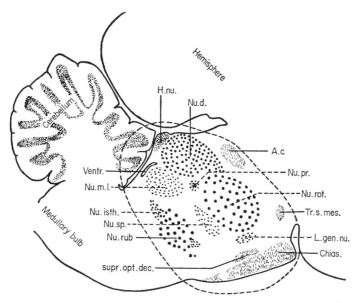

Fig. 94. Longitudinal section of the mesencephalic and diencephalic regions. (Portmann and Stingelin, 1961). The more laterally situated nuclei are indicated by stippled reference lines.

Abbreviations: a.c., anterior commissure; Chias, Chiasma; H.nu., habenular nucleus; L.gen.nu., lateral geniculate nucleus; Nu.d., dorsal nucleus; Nu. isth., nucleus isthmi; Nu.ml., lateral mesencephalic nucleus; Nu. pr., pretectal nucleus; Nu. rot., rotund nucleus; Nu.rub., red nucleus; Nu.sp., spiriform nucleus; supr. opt. dec., supra-optic decussation; Tr. s.mes., Septo-mesencephalic tract; Ventr., Ventricle.

from the vestibulo-cochlear zone. Karten (1963) confirmed other sensory relationships. After cervical hemisection at $C_{12}$–$C_{14}$ Nauta tracing of the degenerating axons showed that dense endings exist within the lateral reticular nuclei (Fig. 95) up to the level of the interpeduncular nucleus; and a diffuse spray of such degenerating axons can be seen throughout both the reticular formation and the central grey of the mesencephalon. Dorso-laterally arching fibres also run to the dorsal part of the lateral mesencephalic nucleus, although not to its central core which Karten homologizes with the semicircular torus. He suggested that the region of termination was possibly the avian homologue of the stratum profundum in the mammalian superior colliculus. Other fibres also ran to Craigie's lateral nucleus, to the tectum, and to both pretectal and subpretectal nuclei (*q.v.*).

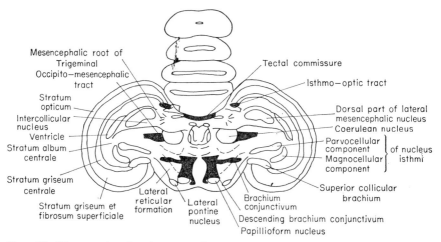

FIG. 95. Diagram showing the generalized organization of the mesencephalon.

As far as studies on mesencephalic function and physiology are concerned the first author to refer to the avian midbrain was Flourens (1824) who found that the destruction of the optic lobe led to blindness in the contralateral eye, and that pupillary movement was arrested if the lesion penetrated to deeper layers. His results were later corroborated by Longet (1842) and Renzi (1863–64). Ferrier (1879) reported that electrical stimulation of the lobes in the pigeon resulted in contralateral pupillary dilation coupled with a backward movement of the head, which was also twisted to one side, together with extension of the contralateral wing and leg. Stefani (1881) found that bilateral extirpation of the tectum gave a degree of blindness which was comparable with that thought at that time to follow forebrain ablations. Similar results were subsequently described by Bechterev (1883, 1909) who found total blindness, Schrader (1892), Boyce and Warrington (1898) and Munzer and Wiener (1898). The question of light induced mesencephalic control of pupillary dilation was later studed by Noll (1915).

Kschischkowski (1911) first used topical applications of strychnine and observed, as did Martino (1926) and Popa and Popa (1933), that it elicited spasm of the wings and legs together with rotation and lowering of the head. Using direct electrical stimulation Popa and Popa concluded that there existed within the optic lobe a definitive motor area which was lacking from the forebrain. Subsequently Vorgas-Pena (1932) tried to determine the relationship between cutaneous sensitivity and such tectal functions. He found a reflex generating zone for raising the wing on the dorsal surface, and another for the adductors on the ventral surface. He therefore concluded that the tectal activity can be influenced by reflexes from the skin. Nikolaides (1914) demonstrated a mesencephalic effect on the supposed medullary respiratory centre which prolonged inspiration. He suggested that impulses from the midbrain maintained the regularity of the medullary centre. Such papers have been reviewed by von Buddenbrock (1953) and Ten Cate (1936, 1965).

## II. THE MESENCEPHALIC NUCLEI OF THE CRANIAL NERVES

### A. The Trochlear Nucleus

The nuclear groupings which are associated with cranial nerves III and IV are complex and contribute a large part of the tegmental region. This complexity is partly the result of the differentiation of the oculomotor nucleus and in part due to the degree of overlap of the two nuclei. Their generalized neuro-anatomy was fairly well established by the older workers although the classical description of the oculomotor pathways is that of Muskens (1930) who understandably concluded that many neuro-anatomical pathways were involved in ocular movement. The best known of the earlier workers are Brandis (1893), Jelgersma (1896), Mesdag (1909), Ramon y Cajal (1909), Carpenter (1911), Kappers (1912), Brouwer (1918) and Sanders (1929). These have all been reviewed in varying detail by subsequent workers especially Kappers (1947), Kappers et al. (1936), Cordier (1954), Portmann and Stingelin (1961), Karten and Hodos (1967) and Jungherr (1969).

The trochlear nuclei are separated by the medial part of the oculomotor nuclei which extend between them. During the early phases of ontogeny this overlap does not exist and the two nuclear masses are both distinct and sharply separated. In the chick it is not until the 9th day of embryonic development that overlap occurs and can then lead to a fusion which reduces the trochlear nucleus of each side to a posterior prolongation of the dorso-lateral oculomotor nucleus. The trochlear nerve emerges from near to the hind end of this trochlear mass which is situated on the dorso-lateral side of the median longitudinal fasciculus, in close association with the central substantia grisea, and below the trochlear decussation.

In all living vertebrates the trochlear component has this position within the ventral part of the central grey and this location immediately suggests an analogy with the more posterior nuclei of the abducens and hypoglossal nerves. Its outflow passes laterally and caudally, decussates, and then emerges through the dorsal aspect of the medullary velum. This path has been widely discussed. The "migration" of the trochlear nucleus resulting in its fusion with that of the oculomotor has been cited as direct evidence that the rhombencephalic isthmus is, or was in the past, a relatively unstable topographical region but this does not really clarify the situation very much. The final result, with the nerve emerging dorsally, has also been accepted on pragmatic grounds as a reasonable path for a nerve which supplies the superior oblique muscle of the eyeball. This dorsal point of emergence has also been implicated in trochlear involvement with the pineal eye of ancestral craniates. Frazer, who is followed by the contributors to *Gray's Anatomy*, suggested that the isthmus region is partly telescoped into the mesencephalic region and that during this process the basal lamina extends laterally and over the alar lamina. In doing so it carried with it the exit of the trochlear. He has also suggested, by analogy with the ventral spinal roots which possess a large ipsilateral and a small contralateral axonal contribution, that, in the case of the trochlear nerve, it is the contralateral component which predominates. This gives rise to the decussation.

The fibre contribution of an avian trochlear nerve was analysed by Graf (1956) and is presented along with other comparative data in Table 28. Counts of the number of cells within the nucleus of normal chicks and those of preparations from which the optic vesicle had been extirpated, showed a very marked difference between the two. In normal specimens the number of cells fell from 1400 to 700 between the 9th and 17th day of incubation (Dunnebacke, 1953; Cowan and Wenger, 1967). Following vesicular excision this fall took place earlier and only 1000 persisted on the 11th day. Those that remained were assumed to relate to the 10–20% of the superior oblique muscle which had not been extirpated.

## B. The Oculomotor Nucleus

Even after fusion with, or close apposition to, the oculomotor nuclear complex, the trochlear nucleus retains a discrete and unified form. In contrast the oculomotor complex has four well-defined components which are situated in greater or lesser proximity to each other. These are:

(a) A ventro-medial component lying between the median longitudinal fasciculus of each side.

(b) A dorso-medial component which can be separated from the former but is clearly a prolongation of it.

(c) A dorso-lateral component situated, as its name implies, above and to the side of the fasciculus.

(d) The smaller mass known as the Edinger–Westphal nucleus which lies above, and sometimes lateral to, the dorso-lateral nucleus, or the dorsal nucleus. Its relations on all other sides are with the central grey. Together with the two dorsal parts it is topographically an anterior homologue of the cells of the trochlear nucleus.

The original suggestion of an homology with the mammalian Edinger–Westphal nucleus relied upon observations of its degeneration following the complete atrophy of the eye in sparrows, and the fact that fine fibres from the ciliary ganglion could be traced to it (see Section V D). This, since in mammals the nucleus contributed pre-ganglionic fibres to the intrinsic eye muscles via the ganglion. Alongside these fine fibres run others of larger diameter many of which decussate. The origin of this decussation has also evoked considerable interest in the past and authors such as Kappers et al. (1936) considered that it involved a cellular migration which affected in particular the cells of the ventro-medial nuclear component. They concluded that such movements occurred during the 6th to 8th day of incubation in the chick.

In chick embryos which have been incubating for 5–17 days typical ganglion cells occur on the peripheral course of cranial nerves IV and VI, and proximal to the ciliary ganglion of III. The number of cells is greatest on VI, but some of those on III may well be intermixed with ciliary ganglion cells (Rogers, 1957), and therefore not identifiable. A proprioceptive function has been assumed for these cells. Rogers suggested that neural crest cells are probably secondarily included in the brain to form the mesencephalic nucleus of V and certain of such cells, destined to become proprioceptors for III, IV and VI, then migrate towards their respective nerves. Some reach their respective nuclei and, of these, some migrate even further out along the nerves. The overall function of the oculomotor, as its name suggests, is in eyeball movements, pupillary contraction and accommodation.

## C. The Mesencephalic Trigeminal Nucleus

The separation of the mesencephalic sensory nucleus of the trigeminal nerve from both the main sensory nucleus and the motor nucleus is more or less extreme in the Trochiliformes (Craigie, 1928). In general scattered neurons which relate to it can be found at a number of sites in the region of the avian optic tectum. According to Valkenburg (1911), Weinberg (1928), Kappers et al. (1936) and Kappers (1947) there are also two more definitive groupings. The precise location of the component cells in the pigeon, together with those of the trochlear and oculomotor nuclei are plotted on stereotaxic coordinates by Karten and Hodos (1967). In this species the structure lies in close association with the tectal commissures and near to the mid-line above the oculo-motor cells. The relative prominence of the medial and lateral cell groups in other species varies somewhat and was described by earlier authors. In the sparrow both have approximately the same number of cells. The apparent presence of a single moiety on each side in the pigeon reflects the pre-dominance of the median group whose cells are some three times as numerous as those of the lateral one. This predominance is somewhat less in *Anas*, and there appears to be very little difference between the two in *Gallus*. However, it does seem that the lateral grouping is frequently the larger, and in some Ardeiformes no medial cells are reported.

As in other vertebrates the mesencephalic root leaves to supply the sense organs of the mandibular muscles. Weinberg (1928) suggested that some fibres pass to the maxillo-mandibulary branch and Wallenberg (1904) long ago suggested that collateral fibres run to the cerebellar region, the trigeminal motor nucleus and the reticular formation. The long standing assumption has been that the nucleus is at least partially concerned with proprioceptive im-pulses. Extracellular recordings of the electrical activity of mesencephalic tri-geminal cells enabled Manni *et al.* (1965) to investigate the effects of jaw movements. The input of these cells is from those proprioceptors which are associated with the superficial external mandibular adductor and the retractor anguli oris muscles. About 35% of the units exhibited spontaneous activity and had discharge frequencies varying between 42–70/sec. The remaining 65% were silent prior to jaw movements or the mere stretching of the isolated muscles. Then they fired with an amplitude of 150–200 $\mu$V.

About 85% of the total units were in fact activated by jaw movements at some time or other. Normally a discharge rate of 260 pulses/sec occurred as soon as the jaw was opened. The rate then fell to 130/sec with some slow adaptation. When the jaw movements ceased the activated units first of all stopped discharging, and then, after a few msec, recommenced at the resting rate. In a later study Azzena and Palmieri (1967) showed that these cells were all part of monosynaptic reflex arcs in which the receptors were muscle spindles within the muscles themselves. The afferent pathway is represented by the mesencephalic trigeminal neurons in the posterior commissure. The processes of these unipolar cells send a peripheral branch to the muscle spindles of the masticatory muscles. The central components make synaptic contact with the neurons of the ipsilateral motor nucleus. It is axons within the mandibular nerve and arising from this last-named nucleus that represent the efferent limb of the reflex arc and the conduction velocity along the peri-pheral afferent axons was 100 m/sec therefore implicating fibres of type 1A. On the basis of these data Azzena and Palmieri concluded that the avian masticatory muscles act as anti-gravity muscles and that the stretch reflex plays a fundamental role in the jaw closing mechanism.

# III. THE OPTIC TECTUM

## A. Neuro-anatomy and Histochemistry

The avian optic nerves reflect the importance of vision as a sensory modality. Their principal characteristics were described long ago and Ramon y Cajal dealt with their supposed complete decussation at the optic chiasma which results in the right nerve reaching the left tectum and the left nerve the right tectum. There is no superficial grey matter and the nerves spread over the entire tectal surface as a fibrous envelope. Much more recently Knowlton (1964) using preparations with experimentally induced anophthalmia showed that at least some uncrossed fibres probably exist. Further a large band of fibres which enter the inner surface from the so-called geniculate complex led Papez (1929) and others to suggest that the optic tectum has a double optic innervation. The middle grey layers have been described as a sort of "optic cortex" from which the tecto-spinal tract originates. The constituent fibres cross over below the cerebellum as the tectal commissure, pass through the mesencephalon medial to the oculomotor nerves, and then run caudally as the tectospinal tracts on each side. Many of these fibres penetrate the oculo-motor nuclei and all of them contribute to a system of synaptic connections that has been thought for a long time to mediate movements of the limbs, head and ocular region in response to visual stimuli. The passage of such tracts is shown diagrammatically in Fig. 140, they have also been envisaged as a direct reflex pathway from the tectal region to the motor nuclei.

In all vertebrates the optic tectum appears during embryonic life as a thick nervous mass lying above the mesencephalic ventricle. It later becomes divided into two components, the corpora bigemina, and, as has been emphasized above, in birds the development of both forebrain and cerebellar regions is accompanied by the separation of these lobes and their marked lateral displacement. Their final ventro-lateral position was, however, ascribed by Ingvar (1923), at least in part, to the neurobiotactic influences which are exerted by the optic tract. Although very voluminous in all birds they are particularly large relative to the total brain volume amongst the diurnal birds of prey and some Corvidae (see Tables 78–79 below). Portmann and Stingelin (1961) concluded that the Struthioniformes, Casuariiformes and Phoenicop-teridae have the least developed tectum in comparison with the standard unit which comprises the condition in galliform birds of comparable size. This is discussed further in Chapter 16.

Despite differences of opinion about the actual number of cytoarchi-tectonic layers which can be differentiated within it, the principal features of the avian tectum were established during the closing decade of the nineteenth century and the early decades of the twentieth. The major contributions were those of Bellonci (1888), Ramon y Cajal (1891, 1911), Gehuchten (1893),

Kolliker (1896), Edinger and Wallenberg (1899), Ris (1899), Schupbach (1905) and Craigie and Brickner (1927). Within the tectum these authors described from 7 to 15 superimposed cellular and fibrillar layers. After the work of Ramon y Cajal it became clear that, as in the Reptilia, 15 could be detected under optimum conditions. These are usually grouped into six strata but may be briefly summarized as follows:

1. An inner epithelial layer whose ependymal elements have processes which radiate towards the tectal surface.

2. A molecular layer which is largely composed of dendritic processes whose cell bodies lie in layers 3 and 5. There is also a small number of medullated axons and scattered nerve cells.

3. A cellular layer with two or three rows of cells. These are of three quite distinct types.

    (i) A pyriform group has basilar dendritic processes penetrating layers 2 and 4, apical dendrites projecting towards the peripheral region and a neuraxis that, after giving off collaterals, enters the peripheral fibre zone.

    (ii) A group whose cell body is similar to (i) but of which the neuraxis is arched. Originating from a dendrite at the level of layer 9 it passes towards the ventricle until reaching and entering layer 6.

    (iii) A group in which the neuraxes arise at the point of bifurcation of the apical dendrite in layer 7. From this point it passes outwards and returns to end in a number of terminal branches in layer 7.

4. A molecular layer with some myelinated fibres and a dendritic plexus.

5. A cellular layer with 3–5 levels of cells each separated one from another by bands of fibres. As in layer 3 there are the three principal cell types.

6. Another molecular layer largely composed of myelinated fibres. Amongst the relatively small number of neurons there are some giant ganglion cells with widely diverging dendrites which pass towards the periphery. Besides these there are plumose cells with elongated cylindrical cell bodies which give off a large number of dendrites some of which course horizontally. The fibre bundles are sagittally arranged at the inside of the layer, oblique in its intermediate region and tangential in the outer and lower regions. Distinctly delimited along its inner boundary the layer gives off broad fascicles from its peripheral border which project out at least to the region of arborization of the optic nerve fibres. Layer 6 sends projections to the posterior commissure which, in its dorsal pathways, links the tectum with that of the other side and also the pretectal and spiriform nuclei. Some fibres from layer 6 may connect with the so-called inferior colliculus.

7. A neuronal layer of three distinct regions:

    (i) A deep region with many variable cells including plumose and pyriform types comparable with those of layer 6.

(ii) An intermediate region of irregular cell groups. Many of these cells are of the pyriform type and the groups send a well-developed dendritic plexus towards the periphery. It is from these dendrites that the cell type noted under 3(iii) above arise.

(iii) A peripheral region with a variety of scattered cells including both pyriform and fusiform types, with neuraxes passing towards layer 6, and the dendrites contributing to the peripheral plexi.

8. A cellular layer with 2 or 3 irregular cell rows which are more clearly differentiated in the lateral proximal region. The peripherally directed dendrites of these cells penetrate the peripheral plexi whilst the centrally directed dendrites pass to layer 7 and branches pass to layer 6.

9. A further molecular layer comprising a rich dendritic plexus within which the most deeply penetrating of the optic fibres arborize.

10. A slim and irregular layer with two cell types. One consists of small cells with a centrally directed axon ending in fine branches at the level of layer 7, the other of cells with horizontally arranged dendrites and cell bodies.

11. A molecular layer containing a massive terminal arborization of the optic fibres and with cells which are broadly analogous to those of layer 10 although one additional type projects both peripherally and centrally.

12. A layer in which an irregular row of cells with small cell bodies are interposed in a plexus of myelinated fibres which are continuous with the optic tract. The cells are either analogous to the last cited for 10 and lie horizontally, or have no definite orientation.

13. A molecular layer with a few cells and comprising the most peripheral region of termination of optic tract fibres. The giant ganglion cells of layer 6 also penetrate and end in this layer.

14. A layer receiving a large number of projections from deeper tectal layers and also the optic tract fibres. These pass obliquely to the underlying layers in which they terminate.

All these were summarized by Huber and Crosby (1933, 1934) and Kappers *et al.* (1936) and grouped into the following six strata which have been used by most subsequent workers. Once again these are directly analogous and homologous to those of reptiles (Fig. 95).

1. The periventricular fibrous layer which lies between (2) below and the ventricular lining.
2. The various layers comprising the periventricular grey which swathes the mesencephalic trigeminal nucleus.
3. The efferent tectal layer.
4. A more external layer of grey matter from which the efferent fibres originate.
5. A field of alternating fibre bundles and afferent/efferent synaptic regions.
6. The external or optic layer.

These are usually referred to as the stratum fibrosum periventriculare; the stratum griseum periventriculare; the stratum album centrale; the stratum griseum centrale; the stratum fibrosum et griseum superficiale and finally the stratum opticum. Their development during ontogeny was studied by Fujita (1964) using auto-radiography and tritiated thymidine as a marker. He found that in the chick the central and periventricular grey layers appear as distinct entities during the 5th and 6th day of incubation. Then throughout the 6th, 7th and 8th days the external half of (5) above becomes progressively differentiated from the matrix. During the self-same period the internal half is also maturing but this is not complete until the 9th day. Furthermore neuroblast development within the 4 layers of the dorso-medial part of the tectum is delayed.

Rogers (1960) reported that the alkaline phosphatase activity of the tectum exceeded that in the forebrain but was less than that in the brain stem. The first positive traces of such activity appeared at the 14th day of incubation and the level remained fairly high in the adult brain. By the 18th day of incubation layers with light staining characteristics alternated with others which had a more marked reaction. The appearance of the lightly stained layers was in part due to the large number of cells which were unstained apart from their nuclei. A comparison of the situation at this time and that of the adult is provided by Table 42. The values which were obtained for both cholinesterase

## TABLE 42

*The distribution of alkaline phosphatase activity in the optic tectum of adult and embryonic chicks.*

|  | 18-day embryo | Adult |
|---|---|---|
| Stratum opticum | Moderately dark | Moderately dark with the layer next to the stratum griseum heavily stained |
| Stratum griseum | Pale | Pale |
| Stratum fibrosum | Moderately dark | Moderately dark |
| Stratum griseum centrale | Pale | Moderately dark |
| Stratum album centrale | Moderately dark | Not so dark |
| Stratum griseum periventriculare | Very heavy intercellular stain | Very heavy intercellular strain |

From Rogers (1960).

and phosphatase activity by Rogers *et al.* (1960) are shown in Fig. 96. The pattern was similar to that observed elsewhere in the embryonic brain and only differs from the condition in the medulla. A period of increasing phosphatase activity accompanies neuronal differentiation. The cholinesterase levels then undergo a sharp rise during the period from the 14th to the 19th days of incubation and electrical activity increases from its very low initial

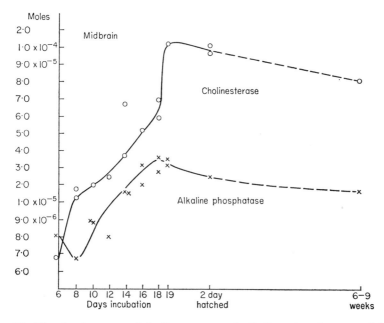

FIG. 96. The development of cholinesterase and alkaline phosphatase activity in the mesencephalon. Data as for Fig. 92. (Rogers *et al.*, 1960).

values at days 17 and 18. A similar rise in the 5-hydroxytryptamine concentration occurs at the time that the cholinesterase activity increases. The concentration fluctuates around $0 \cdot 1$ $\mu g/g$ wet weight during the 14th–17th days and then rises sharply. Until the immediate post-hatching period the values then remained at $0 \cdot 4$–$0 \cdot 5$ $\mu g/g$ wet weight.

## B. Retino-tectal Topographical Projections

Following the surgical ablation of known retinal quadrants in 3–5-day-old chick embryos de Long and Coulombre (1965) were able to observe the resulting pattern of optic fibre distribution on the tectum at 12 days. In this way they established that there is a definite topographical representation of the various retinal areas on the tectum. This topographical specificity is established between the 3rd and 4th day of incubation and retinal ablation prior to this period is followed by the development of normal tectal connections. The pattern of representation is illustrated in Fig. 97.

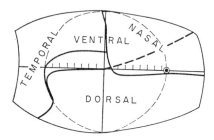

FIG. 97. The topographical representation of the retinal areas on the optic tectum. (de Long and Coulombre, 1965).

Extirpation of the nasal quadrant of the retina resulted in the postero-dorsal quarter of the tectum being devoid of fibres. Ventral retinal ablation produced bare tectal areas which overlapped with the nasal and temporal projection areas. The tectal maps which followed dorsal retinal ablations established that the entire ventro-lateral and postero-ventral tectal surfaces normally receive projections from the upper areas. The line which separates these projection areas for the dorsal and ventral retinal components was well-demarcated on the anterior tectum but not so clearly defined posteriorly. De Long and Coulombre suggested that this can probably be explained in terms of the relatively delayed maturation of these hindermost regions. The area may either itself be incompletely developed at 12 days of incubation, or, alternatively, its fibre growth may not have finished.

The pattern of Fig. 97 is also in close agreement with the results of Hamdi and Whitteridge (1954) who worked on pigeons. It is clear that the general relationships are established by the time that the optic fibres completely cover the tectal region, and some 6 days prior to the time at which electrical activity is first detectable within the system. It is also evident that by this time both the retina and the tectum have got fixed mosaic maps which may be independantly determined.

## C. Tectal Fibre Connections

### (i) Recent Anatomical Work

The projections of the tectum were studied by Karten (1965) following surgical lesions. These showed that in the absence of any injury to either the underlying reticular formation or to the nucleus isthmi, axonal degeneration could be traced to a number of important foci by using the Nauta-Gygax technique. Ipsilateral degenerating fibres could be traced to the lateral pontine nucleus, the lateral mesencephalic reticular formation, the cuneiform nucleus, the intercollicular nucleus, the isthmo-optic nucleus and the external cellular layer of the sub-thalamic tegmentum. Contralateral projections ran to the opposite tectal cortex via the tectal commissure, and to both the papilioform nucleus and the predorsal medullary cell groups via the decussation of the tecto-bulbar tract. Efferent tectal fibres passing via the superior collicular brachium ended in the sub-pretectal nucleus, and others continued forward to the ipsilateral rotund nucleus and the adjacent intercalated nucleus. Passing via the ventral supraoptic decussation some could be followed to the posteroventral nucleus. Further and significant numbers followed an extra-brachial pathway to the dorso-lateral posterior thalamic nucleus of the same side, to the principal pretectal nucleus and other cell groups in close proximity to it. Clearly these results corroborate those of the older workers who had concluded that the tectum maintained widespread connections.

## (ii) *The Optic Nerves*

The most important afferent fibres to the tectal region are of course those of the optic nerves themselves. As long ago as 1898 Perlia concluded that they undergo a complete decussation in *Cuculus*, *Gallus* and *Passer* to give rise to the medial and lateral tracts of each side. This last tract was described as passing to the tectum as well as to the lateral geniculate nucleus, the external nucleus and the superficial synencephalic nucleus. Knowlton (1964) has now shown that some uncrossed fibres almost certainly exist (see above).

Although his suggestions are now largely of antiquarian interest it is worth noting that Ramon y Cajal gave considerable thought to the possible reasons for the presence of an optic chiasma. He decided that amongst the lower vertebrates with their relatively lateral eyes the two visual fields were totally independent and only juxtaposed at the mid-line. The right eye sees objects which are to the right, the left eye those on the left. When transmitted to higher centres the two retinal images would hence give a somewhat dissected panoramic picture. The optical properties of the lens would result in the discordant images on the left of Fig. 98. He suggested that the central processing of such visual information would be greatly simplified if there was a total decussation of the primary input lines as shown in Fig. 98. Although there are some undoubted fallacies in this argument nevertheless it had a considerable appeal as a neat theoretical explanation. It is only relatively recently that experiments with inverted spectacles have demonstrated the extensive compensations which the central nervous system can undertake.

13*

Also, at the turn of the century, the telegraphic systems, which have provided analogies for much cybernetic research, were simple. It is therefore not unduly surprising that in his memoirs Ramon y Cajal said that, although his scientific contributions were of unequal value, he thought that his explanation of the optic chiasma was fundamental.

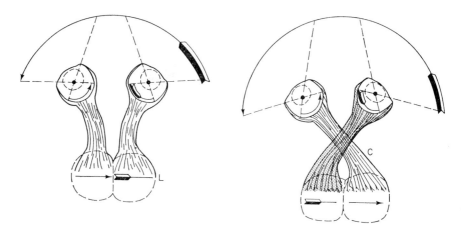

FIG. 98. Diagrams illustrating Ramon y Cajal's theory of the chiasma. With a total, or near total cross over the panoramic field is not discordant.

## (iii) *Other Connections of the Tectum*

The recent work of Karten which was outlined in (i) shows that tectal connections, apart from optic tract fibres, link it to a number of other areas. A brief survey of the literature reveals that these include the external, ecto-mammillary, oculomotor, sub-pretectal, superficial magnocellular and pre-commissural nuclei, together with other diencephalic cell masses. Effector fibres have long been considered to run to the rotund nucleus (*q.v.*), the dorsal part of the lateral mesencephalic nucleus, and, via the tecto-bulbar tract, to the medullary bulb. Although the total number of tracts and fibre bundles is large, and their relationships both diffuse and complex, the following list gives a fairly clear idea of the principal relationships which have been described in *Anas, Columba, Falco, Cacatua, Pratincola, Chlorostilbon, Chrysolampis, Merula* and *Passer*. These tracts pass either to or from the tectum as total or majority projections, or, alternatively, as partial projections with other fibre bundles. A detailed and discursive review of the older data was provided by Kappers *et al.* (1936).

1. The dorsal and ventral parts of the posterior commissure;
2. The lateral part of the dorsal supra-optic decussation;

3. The ventral supra-optic decussation;
4. The dorsal part of the septo-mesencephalic tract;
5. The strio-tegmental tract;
6. The dorsal and ventral tecto-bulbar tracts;
7. The tractus tecto-thalamicus cruciatus;
8. The dorsal, ventral and ventrolateral tecto-thalamic tracts;
9. The tecto-spiriform tract;
10. The thalamo-frontal tract.

Relationships between the tectum and various structures at more anterior levels of the brain have now been elucidated by electro-physiological methods. By similar means the isthmo-optic nucleus (*q.v.*) has also been implicated in a retinal centrifugal system. These connections are discussed further below. It will suffice to say at this point that such anterior structures include the rotund nucleus in the thalamic region, the thalamo-frontal tract, the ectostriatum and certain components of the Wulst. One proposed pathway runs from the central grey stratum of the tectum to the rotund nucleus and from there to the ectostriatum via the lateral half of the thalamo-frontal tract. A polysynaptic pathway has then been suggested linking the ectostriatum with the dorsal hyperstriatum, the overlying accessory hyperstriatum and the corticoid areas of the Wulst (*q.v.*). (Karten, 1965; Karten and Revzin, 1966; Phillips, 1966; Revzinand Karten, 1967; Cohen, 1967; Zeigler, 1963). Apart from these works the important role which is played by the optic commissures in facilitating the integration of visual information from the two sides was demonstrated by Mello *et al.* (1963) using electrical stimulation of the optic nerve and surgical transection of commissural fibres.

The posterior commissures are a constant feature of vertebrate brains and situated within the rostral part of the mesencephalon. Overlying the central grey and situated below the tectal commissures they are sometimes considered to represent a single anatomical unit along with these last. Kappers *et al.* (1936) described one component as actually investing the pretectal and spiriform nuclei in some species. In mammals most of the fibres are said to originate in the large celled nucleus of Darkschewitsch, a nucleus of unknown function which in birds stretches from the internal surface of the interstitial nucleus to a more dorso-lateral position surrounded by the central grey. The tecto-bulbar system is also frequently described as a major outflow. A dorsal component leaves the stratum album centrale in association with other fibres *en route* to the oculomotor nucleus. After decussating it follows the longitudinal fasciculus. A large ventral component also passes through the tegmental region and then, after a partial decussation, contributes to the descending fibres within the ventro-lateral region of the brain-stem.

# D. Responses to Peripheral Stimulation

## (i) *Peripheral Photic Stimuli*

The neuro-anatomical descriptions would naturally lead one to expect a fairly close relationship between visual function and measurable electrical activity within the optic tectum. This was demonstrated experimentally by Peters *et al.* (1958). In the newly hatched chick direct stimulation of the eyes, using flashes of light, results in characteristic on, off and on-off responses within the optic nerve. A comparable pattern of activity follows at tectal levels. On-off responses can be detected within the contralateral lobe at the onset and cessation of sustained stimulation at 60 cyc/sec. Individual electrical responses also follow both single flashes and sequential presentations at frequencies of 5, 10, 20 and 40/sec. Such effects cannot be detected at all prior to the 17th day of incubation but are present, although less consistent, during the later phases of embryonic life. However, a generalized acceleration of embryonic development during its early stages following exposure to light, which was reported by Siegel *et al.* (1969), suggests that there may be some diffuse sensitivity at earlier phases of development.

Paulson (1965) studied the gradual maturation of specific tectal responses during the later pre-hatching period. He was able to demonstrate that the observable voltage changes alter considerably if the monitoring electrode is moved distances such as 0·2 mm. He suggested from his data, which are presented in Table 43, that the overall voltage may decline with age following maximal levels at 27 days incubation. Clearly this is not in itself conclusive.

## TABLE 43

*Average changes in the electrical activity of the optic tectum during late embryonic development.*

| Age of embryo days | Latency in msec | Duration in msec | Voltage in mV |
|---|---|---|---|
| 23 | 18·0 | 25·0 | 75 |
| 26 | 18·1 | 28·8 | 116 |
| 27 | 15·3 | 24·7 | 135 |
| 28 | 10·6 | 21·1 | 65 |
| 30 | 10·4 | 22·3 | 100 |

From Paulson (1965).

## (ii) *Direct Stimulation of the Optic Nerve*

Using the technique which they had previously described in connection with their studies on the mammalian superior colliculus (Bishop and O'leary, 1942a,b), O'leary and Bishop (1943) explored the optic tectum of both geese and ducks with a mobile electrode whilst single shocks were applied to the contralateral optic nerve. In this way they were able to record a series of post-synaptic potentials and they observed that the appropriate pre-synaptic fibres had differing thresholds and conduction rates. The plot of the amplitude of a given spike against the depth in the tectum at which it was detected showed that there was a negative maximum within the cellular layers and a positive maximum more centrally. They were not able to obtain any evidence of post-synaptic conduction but rather concluded that there was a generalized polarization of all the cellular regions. The peripheral pole of each polarized element was negative relative to the central pole during periods of activity.

An apparent correlation between the distribution of the potentials and the histological structure of the tectum suggested that the potential differences might be ascribable to the dendrites of those cells which are situated in a post-synaptic relationship to optic tract axons. The further potentials, which one would expect to find associated with the post-synaptic axons themselves, seemed to be masked by those of greater amplitude and duration within the cellular layers. They considered that the general hemispherical shape of the gross tectal components on each side served to accentuate the amplitude of positive potentials when using an indifferent electrode of fixed location. As a result the potentials of highest amplitude were recorded from the ventricular region in which they supposed there was no active tissue. However, they were able to present evidence which suggested, or at least supported the contention that those neurons which are post-synaptic to larger and more rapidly conducting optic tract fibres occupy a shallower position within the tectum than those which are functionally related to slower fibres. The earlier responses which can be obtained from such superficially situated units were isolated by employing shocks of varying strength and making simultaneous recordings from electrodes critically placed within the shallower cell layers. Neurons lying above this level, and nearer to the tectal surface, then cause a positive deflection. Those lying at deeper levels give a slightly later and negative one. The resulting diphasic form of such overlapping waves has some of the characteristics which one expects in a conducted response but is actually explicable in terms of such successive responses within differing neuronal strata. It was as a result of these studies that they suggested similar explanations for the superior colliculus, the lateral geniculate body and indeed the cerebral cortex of the cat.

Some 11 years later Cragg *et al.* (1954) found that the voltage difference

which can be measured between a micro-electrode in the tectum and an indifferent one on the skull exhibited very little spontaneous variation. The
peak amplitude was usually less than 100 $\mu$V. Following direct stimulation of
the contralateral optic nerve there was first of all a brief spike and then this
was followed by two waves of longer duration. The first spike occurred at
0·65 msec, was positive and less than 100 $\mu$V. The earliest of the two longer
lasting waves appeared a msec later and declined to zero at 2·5 msec. During
this period the surface was positive and the peak amplitude was 600 $\mu$V. The
second of the pair of long waves was of opposite polarity and began as soon
as the first disappeared. Lasting for from 10 to 30 msec it had a peak amplitude
of 400 $\mu$V.

When the voltage differences between a radial pair of micro-electrodes were
recorded at different depths the responses occurring after stimulation of the
optic nerve had a time course which resembled that of a monopolar recording.
Furthermore a maximum amplitude was detectable at a depth of 500 $\mu$ below
the pia, and the values declined above and below this level. Only very small
voltage gradients characterized the deeper tectal strata. From the thicknesses
which Cragg et al. cited for different tectal components it is apparent that the
critical level of 500 $\mu$ lies at the edge of the radial fibre layer within the
superficial plexiform layer:

   (a) Optic nerve layer, 400 $\mu$
   (b) Superficial plexiform layer, 130 $\mu$
   (c) Radial fibre layer, 600–700 $\mu$
   (d) Deep plexiform layer, 400–500 $\mu$
   (e) Central fibre layer, 250–450 $\mu$
   (f) Periventricular layer, 50–80 $\mu$

They decided that it actually coincided with the presence of many cell
bodies of the radial neurons and as a result of this concluded that it was
related, not to the axonal components, but to the dendrites and cell-bodies.

Tangential voltage gradients which contrasted with such radial ones, were
also recorded. Such gradients were maximal when aligned in the direction
along which the optic fibres run over the tectal surface, but recording from
pairs of electrodes sited at right angles to this direction gave very little change
of voltage. Placing the recording probes at varying depths revealed that there
is a reversal of the polarity of small voltage gradients between 0 and 250 $\mu$
below the surface. A comparable reversal of large voltage gradients occurred
at 1000–1250 $\mu$. Furthermore the gradients were minimal at depths of 500–
750 $\mu$ which corresponds to the level at which the radial ones were maximal.
No easy correlation was apparent between the density of the cell bodies of
tangential neurons and the recorded voltages.

Cragg et al. interpreted all these results in terms of the visual pathway and
the histological structure of the tectum. The initial brief spike which occurs

0·65 msec after stimulation of the optic nerve could be elicited for a few minutes after the death of the animal. They concluded that it was presumably indicative of the arrival of synchronized impulses at the tectum. The distance of 2·3 cm which separated the site of stimulation from the tectal monitor conforms with the available data for the speed of transmission within the optic nerve. The later responses involving large tectal gradients were attributed to the autochthonous neurons. They concluded that if the tectum contained neurons which had a 3 msec decay time for their synaptic potential, and an after-potential lasting 100 msec, then a synchronized burst of impulses in the optic nerve would evoke synaptic and after-potentials in many tectal neurons, which, when integrated together, would generate two waves of potential change. These would be similar in both their temporal and polarity characteristics to those which they recorded.

The radial voltage gradients were associated with a layer of neurons whose dendrites although extending both upwards and downwards were principally orientated in a radial direction. Theoretically the superficial processes which are situated nearest to the optic afferent fibres would be depolarized by impulses within these before those parts of the neurons which are more deeply embedded in the radial layer. This would result in a negative to positive downward voltage gradient during the first wave of responses and would correspond to that observed. The tangential gradients were positive to negative amongst the dendrites which were situated in a superficial position relative to the cell-bodies. Below this they were negative to positive among the cell-bodies and dendrites in the direction of the afferent optic fibres. As investigations on spinal motor neurons had shown that the outside becomes negative relative to distant indifferent tissue during the synaptic potential, Cragg *et al.* concluded that those dendrites receiving part of the current flowing from the cell body during its synaptic potential will become positive relative to their surroundings. Stimulated neurons would then have positive dendrites and negative cell bodies during the first response wave. Neurons which are situated closer to the chiasma may well be stimulated before more distant ones. This would give rise to superficial positive to negative tangential voltage gradients and deeper negative to positive ones. Hence Cragg and his co-workers concluded that the pattern of stimulation proceeds through the layer of tangential neurons along the line of the optic fibres, as well as downwards within the radial layer. The dendrites of stimulated neurons become positive and their cell bodies negative during the first wave of the tectal response (see Fig. 99).

Furthermore recent studies of the field potential profile resulting from stimulation of the optic nerve head were carried out by Holden (1968a). He used micro-electrodes which had been radially inserted into the tectum. The results of his work show a general similarity to those of the foregoing workers

but the specific parameters are somewhat different. The first detectable response within the superficial layers was a graded negative wave (the N-wave)
which had a latency of 2·3 msec and lasted for 4–5 msec. As in the case of the
results obtained by Cragg *et al.* there was an abrupt reversal of polarity at a
depth of 500 $\mu$. This was designated the reversal zone (R-zone) and had a
diphasic positive negative potential. Following further penetration it became
transformed into a graded positive wave which persisted throughout both the
deeper tectal laminae and also the ventricular region. This "P-wave" had a
latent period of 2·0 msec which was therefore comparable with that of the
superficial N-wave.

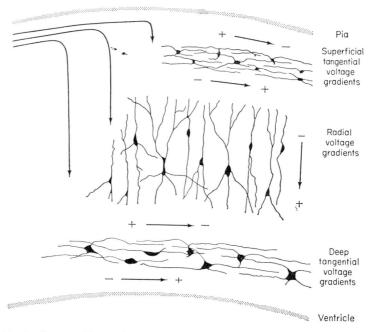

FIG. 99. A diagram illustrating the spread of impulses through the three main
layers of tectal neurones. (Cragg *et al.*, 1954).

The relationship between the stimulating current and the amplitude of the
resulting wave forms showed that there was a threshold below which no N-
wave was evoked. Above this value the amplitude grew as a sigmoid function
of the stimulatory current, although becoming independent at an upper limit
of 5 mA. The actual current range over which there was a smooth growth of
current varied from specimen to specimen and was clearly dependant upon
the distance which separated the stimulatory electrodes within the nerve head.
The growth in the observed amplitude of the N-wave was never discontinuous

and there was no evidence for quantal fluctuations at values which were close to the threshold. The amplitude profile of the P-wave was similar. This is to be expected if, as Holden suggested, it is the passive source to the N-wave sink. Some variations did, however, occur and then gave the P-wave an amplitude which was twice that of the N-wave. These could reflect the heterogeneity of the N-wave generators.

In many tracts the rise time and duration of the N-wave were independant of amplitude. The rise time of the P-wave was longer, and its total duration shorter, than in the case of the N-wave. The observation of depth profiles made it clear that there were two sinks and these were separately located within the N-, and R-zones. Lesion marking showed that the R-zone lay within the stratum griseum et fibrosum superficiale so that the spatial distribution of tissue negativity was more or less co-extensive with the anatomical distribution of optic nerve terminals.

In a paper which was published at the same time (Holden, 1968b) he gave details of the unitary responses within the lateral tectum. The spikes of the optic nerve fibres could be recognized by their conformation, fixed latency, brief recovery times and location within the superficial layers. Their action potentials were either triphasic and had a prominent second phase, or, alternatively, monophasic positive. Holden considered that the nerve contains small myelinated fibres which conduct at 5·3–8·0 m/sec. The fibre spikes were limited to the N-, and R-zones and usually preceded any tectal N-waves. Of the 156 post-synaptically firing cells which he recorded, 140 fired a single spike following each stimulus applied to the nerve head. The remaining 16 gave repetitive responses to a single stimulus.

Evidence of afferent inhibition was possibly provided by a period of background silence after a spike had been triggered by the retina. It was observed in most of the cells which exhibited background firing and he suggested that it could also be the result of post-hyperpolarization following the somatic spike. A more conclusive demonstration of afferent inhibition was obtained when increasing the amplitude of the stimulus prolonged the resulting silent period. As examples of this we may cite the fact that a 5 mA stimulus at the nerve head gave a single spike at a tectal cell and that this was followed by a 35 msec silent period. A 10 mA stimulus also gave a single spike but in this case the subsequent silence lasted 55 msec. This proved to Holden's satisfaction that a trans-synaptic inhibitory constraint was involved over and above any possible results of after-hyperpolarization.

Cells which were located within the N-zone were fired most easily during the 3 msec interval corresponding to the rising phase of the tectal wave. By comparison the cells in the P-zone seemed different. They were not fired during this interval and he concluded that discharges within this latter zone were the result of triggering by tectal inter-neurons. Complementarily some cells located in the N-zone had a recovery time of less than 5 msec. This was never the case in the P-zone. It is apparent that a very distinctive pattern of responses results from radial penetration into the tectum and that this is a direct reflection of its histological structure.

## E. Telencephalic Potentials which are Evoked by the Stimulation of the Tectum

During several experiments Revzin and Karten (1967) stimulated the central grey layer of the tectum (stratum griseum centrale) whilst they simultaneously monitored the forebrain for activation potentials. They hoped in this way to identify those telencephalic structures which are involved in forward projections from the tectum. They found that the spatial distribution of the responses was closely similar to, if not identical with, the area which is activated following stimulation of the rotund nucleus (*q.v.*). Predominantly associated with the ectostriatum they were most numerous within its posterior half, and its medial and dorsal two-thirds. The responses themselves only differed from those which can be evoked by rotundal stimulation in their lesser amplitude and longer duration. Potential changes could be detected within the ectostriatum some 2–2·5 msec later than those arising from stimulating electrodes within the rotund nucleus. Since the latency of the responses which occur within this nucleus after tectal stimulation is itself 2–2·5 msec it would seem to be highly probable that the tecto-ectostriate pathways involve a synaptic link within the diencephalon and more specifically the nucleus rotundus. This clearly corroborates the old descriptions of tectal outflow to this area. Comparable results were also obtained by Phillips (1966). He found that stimulation of the tectum was followed after 2–3 msec by the appearance of 150 $\mu$V potentials in the rotundum. The persistence of some components of such evoked potentials in the tectal-rotundal circuits, whilst others disappeared at repetition rates of 100/sec, suggested to him that both direct and polysynaptic connections exist.

## F. Partial Extirpation of the Optic Tectum

The effects which both unilateral and bilateral lesions of the tectum had upon visual discrimination were surveyed by Cohen (1967). The lesioned preparations showed no obvious sensory or motor dysfunction except that those with unilateral damage not unnaturally used the eye which was ipsilateral to it for orientation. During instrumental discrimination training such preparations took significantly longer to achieve a given level of ability by comparison with normal birds.

The lesions generally involved between 15–20% of the optic nerve fibre layer (stratum opticum) and few extended further down than the periventricular fibrous layer. Nevertheless there were a few cases in which some damage was done to the magnocellular component of the nucleus isthmi, and also to the most lateral aspect of the lateral mesencephalic reticular formation. Cohen was quite emphatic that these additional focal lesions made no obvious difference to the resultant behaviour. In the case of unilateral preparations there was a positive correlation between the actual extent of the lesion and the number of trials which were needed in order to achieve a given degree of discrimination training. There was no such correlation in bilateral preparations.

As would be expected, in view of the fact that some 80–85% of the tectum remained intact, this impairment of discrimination learning was never severe. Given the small overall size of the lesion and the absence of any gross visual disturbance, Cohen concluded that tectal involvement in the mediation of intensity discrimination must extend further than simply being the principal site at which the optic tract terminates. However, he did admit that the lesions had destroyed the areas of projection for the central part of the retina. Any such additional tectal functions in, for example, integration, would conform with the known complexity of integration and the suggestion that it occurs at a relatively central focus (Maturana and Frenk, 1963, 1965). The results do not, however, exclude the additional possibility that other visual pathways are involved in intensity discrimination. After decussating some fibres from the retina pass to other brain regions. The principal alternative routes are:

(a) to the complex of small thalamic nuclei such as the anterior lateral, lateral geniculate, external and superficial synencephalic;
(b) via the basal optic root to the ectomammillary nucleus;
(c) to the tectal grey (Cowan *et al.*, 1961).

However, Cohen followed Karten's suggestion that the principal ascending visual pathway includes the direct projection to the optic tectum, and then the long suspected route via a tecto-thalamic pathway ending at the rotund nucleus, followed by fibres passing to the ectostriate region from which projections reach the hyperstriatal regions. The studies of Hodos and Karten (*loc. cit*, Chapter 11) and Zeigler (1963) have demonstrated that lesions to both the rotund nucleus and the hyperstriatal components (*q.v.*) result in deficient discrimination. Cohen's study adds the rather basic tectal component and such a tectal–rotundal–ectostriate–hyperstriate pathway seems to be clearly implicated in visual discrimination performance. Inter-relationships of this system with others is indicated by Brown's (1965) studies on vocalization.

# IV. THE LATERAL MESENCEPHALIC NUCLEUS

## A. Neuro-anatomical Studies

Besides the optic centres, the nuclei of cranial nerves III and IV, and the mesencephalic nucleus of the trigeminal, the other well-defined midbrain structures are the nucleus isthmi, the isthmo-optic nucleus, the red nucleus, the pretectal complex at the mesencephalic-diencephalic transition and the lateral mesencephalic nucleus (Figs 94 and 95). As has already been intimated (Section I above), this last is, at least in part, the probable homologue of the torus semicircularis. Broadly speaking its component cells often resemble those of the periventricular grey. Situated near to the mid-point of the optic ventricle the nuclear cells themselves are surrounded by a sheath of more peripheral cells. Amongst its long established anatomical connections are included axons from the trigeminal sensory nucleus and, more especially, from the vestibulo-cochlear regions. Other fibres appear to relate it to the lateral lemniscus, the isthmo-mesencephalic tract and the optic tectum. Special tracts whose presence was established during the early years of the twentieth century were generally assumed (see Huber and Crosby, 1929; Kappers et al., 1936; Kappers, 1947) to permit the coordination of visual signals with ocular movements. In recent years it has, however, been clearly implicated in the ascending auditory pathways.

The nucleus is fairly clearly differentiated into two components and their relationship may be solely topographical. Certainly it is the dorsal part that is clearly related to hearing. This overlies the ventral part (Fig. 95) from which it is separated in the hindermost regions by the intercollicular nucleus. Further forward this last assumes a more definitive medial position and both ventral and dorsal components of the lateral mesencephalic nucleus are juxtaposed. The two are both underlain by the lateral mesencephalic reticular formation, whilst the dorsal surface of the dorsal part is close to the central white stratum of the tectum.

A comparison of the size of the dorsal component in 27 avian species was presented by Cobb (1964). As a measure of the relative sizes of the "torus" and the optic lobe he calculated the following simple index:

$$\text{per cent} = \frac{\text{volume of auditory torus}}{\text{volume of optic lobe}} \times 100$$

The results of such calculations are contained in Table 44.

Even a cursory inspection of the data in Table 44 will show that this ratio varies from 8·7% in *Steatornis* and 8·2% in *Otus*, to 2·0% in the loon *Gavia*. The "torus" can therefore vary from one-eleventh to one-fiftieth the size of the optic lobe. The largest single index was obtained for the left lobe of the

# TABLE 44

*A comparison of the relative sizes of the dorsal part of the lateral mesencephalic nucleus and the optic lobe in 27 avian species.*

|  | Ratio of volume of torus to optic lobe in per cent |
|---|---|
| Passeriformes | |
| *Mimus polyglottus* | 3·4 |
| *Serinus canaria* | 4·9 |
| *Cyanocitta cristata* | 5·0 |
| Piciformes | |
| *Dendrocopos pubescens* | 4·2 |
| Coraciiformes | |
| *Megaceryle alcion* | 3·3 |
| Micropodiformes | |
| *Chaetura pelagica* | 6.7 |
| Caprimulgiformes | |
| *Steatornis caripensis* | (7·5 right side) (8·7 left side) |
| *Caprimulgus vociferus* | 5·3 |
| *Phalaenoptilus nuttallii* | 4·3 |
| Strigiformes | |
| *Otus asio* | 8·2 |
| *Aegolius acadicus* | 7·4 |
| Cuculiformes | |
| *Coccyzus americanus* | 3·5 |
| Psittaciformes | |
| *Melopsittacus undulatus* | 3·0 |
| Columbiformes | |
| *Columba livia* | 3·8 |
| Lariformes | |
| *Catharacta skua maccormicki* | 3·1 |
| *Larus argentatus* | 3·5 |
| Ralliformes | |
| *Rallus limicola* | 5·1 |
| Galliformes | |
| *Meleagris gallopavo* | 2·5 |
| Falconiformes | |
| *Falco sparverius* | 2·7 |
| Anseriformes | |
| *Anas platyrhynchos* | 4·9 |
| Ardeiformes | |
| *Butorides virescens* | 4·4 |
| Pelecaniformes | |
| *Phalacrocorax pelagicus* | 4·1 |

Table 44 (*contd.*)

| | Ratio of volume of torus to optic lobe in per cent |
|---|---|
| Procellariiformes | |
| *Oceanites oceanicus* | 4·6 |
| Podicipitiformes | |
| *Podiceps auritus* | 2·85 |
| Colymbiformes | |
| *Gavia immer* | 2·0 |
| Sphenisciformes | |
| *Pygoscelis adelinae* | 2·8 |
| Casuariiformes | |
| *Dromiceius novae-hollandiae* | 5·4 |

From Cobb (1964).

oil-bird (8·7 %) although the average value for the two sides was lower (8·1 %). It is extremely interesting that there should be a bilateral disparity in this species and one is forced to speculate as to whether it may be related to the need to introduce temporal and spatial asymmetries when using echo-location (see Chapter 5, Section III and Chapter 6, Section V). A similar large value was also obtained for both the strigiform genera *Otus* and *Aegolius*, but the values for the remaining caprimulgiform genera *Caprimulgus* and *Phalaenoptilus* were lower. The average values for these two orders were therefore 7·6 % and 6·3 % respectively. As one might expect the value closest to that of the *Steatornis* was that of *Otus asio* and, with the exception of the swiftlet *Chaetura pelagica* which practices echo-location and had a value of 6·7 %, no other species approached these values.

The figures also suggest that amongst the oscine passeriforms (song-birds) there are tolerably well-developed "tori". It is not, however, immediately clear why the rails and ducks should have higher indices than is the case amongst other aquatic forms such as the Colymbiformes (Gaviiformes auctt), Podicipitiformes and Sphenisciformes. Cobb suggested that the only obvious explanation would lie in predominantly nocturnal feeding habits.

At the time that they were published these data were not easily explicable. Whilst they can easily be fitted into any scheme in which the dorsal part of the lateral mesencephalic nucleus is a station on the ascending auditory pathways, they are difficult to explain if this is not the case and Erulkar (1955) had been unable to elicit regular and significant activation potentials by the presentation of peripheral click stimuli. However, the subsequent work of Harman and Phillips (1967) and Karten (1966) has corrected this situation.

As a result one can now ascribe the relatively large size of the torus in echolocating genera to its auditory involvement. Furthermore Boord (1968) has been able to show quite conclusively that following experimentally induced lesions of the magnocellular, angular and laminar nuclei degenerated fibres from the lateral part of the magnocellular nucleus, the medial part of the angular nucleus and from the laminar nucleus, project to the homolateral superior olivary nucleus, cross the raphe in the trapezoid body, ascend in the contralateral lateral lemniscus, distribute to the ventral and latero-ventral nuclei of the lateral lemniscus, and that at least third order axons from the laminaris terminate in the dorsal part of the lateral mesencephalic nucleus. This supports the suggestion that the *torus semi-circularis* is the avian homologue of the central nucleus of the mammalian inferior colliculus and also, incidentally, that the laminar nucleus is the homologue of the medial superior olivary nucleus in mammals.

## B. Electro-physiological Investigations

The data which were obtained by electro-physiological studies of chick brain potentials showed that a number of mesencephalic sites are activated by peripheral auditory stimulation (Harman and Phillips, 1967). As noted above Erulkar (1955) had only been able to detect sporadic potentials in the lateral mesencephalic nucleus and even when they occurred these were only one-twentieth of the size of those he observed in the nucleus isthmi complex. In contrast Harman and Phillips found that the torus was usually responsive and, on all the occasions when it did give a positive result, the recording electrode was situated in the large celled dorsal part. Ipsilateral clicks evoked potentials which had amplitudes varying from 10 to 100 $\mu$V, and latencies of 4–6 msec. Subsequently a slower wave of activity, with an amplitude of between 50–250 $\mu$V, followed after 6–10 msec, and there were sporadic even slower waves with lower maximal amplitudes of 10–25 $\mu$V after intervals of 10–16 msec.

Contralateral click stimuli elicited 10–25 $\mu$V waves which peaked at 6 msec. When such stimuli were presented to both ears at once there were, as might be expected, complex compound situations. After 3 msec a 40 $\mu$V wave was observed. This was followed at 5 msec by a 10 $\mu$V wave and after 8 msec by a 90 $\mu$V one. The shorter initial latency which they observed in these cases of binaural presentation was explained by Harman and Phillips in terms of the spatial summation of crossed and uncrossed fibres. The effect of different repetitive stimulating frequencies was to give decreases in the response amplitude as the frequency of presentation was increased. The large evoked potentials were, however, still present at stimulus frequencies in excess of 100/sec. With frequencies lower than 10/sec the amplitude of the responses waxed and waned. Increased loudness gave a distinctly larger response and, if very loud stimuli were followed by soft ones, there was sometimes an additional response which was otherwise absent.

Further confirmation of these results was later obtained by Biedermann-Thorson (1968). This author showed that the neurones from which one could record a sustained response fell into three categories. These were (i) inhibitory only, (ii) excitatory in a central range of frequencies with inhibition in adjacent frequency regions and (iii) predominantly inhibitory with one or more relatively small excitatory regions. Seven of the 45 neurones which he studied only exhibited a transient response, had high thresholds or were restricted to a narrow band of frequencies. The observed responses to click presentation were variable in the case of different units and, in a single unit, they were often dependent upon the repetition rate. This may go some way towards explaining Erulkar's failure to detect any sustained and repeatable evoked responses. In their paper on the unit responses which can be detected in the medullary cochlear nuclei Stopp and Whitfield (1961) found that 20% of the neurones at that level could be inhibited but not excited by tones (see Chapter 6, Section V F). They noted that this was in fact the main difference between such neurones in the pigeon and their mammalian homologues. It appears that this may also be true of these more anterior foci, with the added qualification that even those neurones which can have an excitatory response also appear to show inhibitory responses to a greater extent than those of the cat. Very few units showed excitation without some inhibition, and in almost 50% of the sample it was the inhibitory response ranges which predominated. Twenty-five per cent of the units were actually in the totally inhibitory grouping. The characteristic frequencies varied from 400 to 2500 cyc/sec and in all cases the inhibition increased as the stimulus intensity was raised. This finally culminated in the total suppression of firing for the duration of the tone. With the exception of those whose spontaneous discharge rate was too low to be detected, the remainder of the units did respond with increased firing when presented with certain frequencies. Nevertheless they were all also inhibited by at least certain frequencies.

There is, however, a striking difference between the rates of spontaneous discharge which can be observed in the cochlear nuclei of the medulla on the one hand, and in the dorsal part of the lateral mesencephalic nucleus on the other. Rates of 100/sec were common in the former (Stopp and Whitfield, 1961) and 10/sec was considered low. In contrast almost half of the neurones in the lateral mesencephalic nucleus had discharge rates which failed to exceed 5/sec. Furthermore, although the minute by minute rate in the cochlear nuclei varied, that in the torus usually remained constant when averaged over 2–3 sec to permit some degree of compensation for the low overall values. Corroborative information for torus involvement in hearing is suggested by lesions which eliminated vocalization, and stimulations which elicited it, in the red-winged blackbird (Brown, 1965a,b, 1969).

# V. THE NUCLEUS ISTHMI COMPLEX

## A. General Considerations

The relative positions of the two principal parts of the nucleus isthmi are shown in Fig. 95. The magnocellular component is rather more extensive than the parvocellular one and underlies it for much of its length (see Karten and Hodos, 1967 for the pigeon). In the more rostral regions of their anatomical extent both are situated below the lateral mesencephalic reticular formation and dorso-lateral to the superior collicular brachium. As such both are well separated from the smaller isthmo-optic nucleus which lies above its tract and in close proximity to the lateral extremity of the trochlear decussation. Both the cytology and the fibre disposition of the isthmi complex were recently studied by Showers and Lyons (1968) using *Gallus, Numida, Meleagris, Passer, Mycteria, Cathartes, Anas* and *Phasianus*. As a result of the position which is assumed by the optic lobes relative to both diencephalic and telencephalic structures the location of the isthmi components, together with the dorsal and ventral parts of the lateral mesencephalic nucleus, varies somewhat. In the chick the optic lobes are almost directly lateral to the tegmental and diencephalic foci. In *Passer* they take up a more ventral position, whilst in the guinea fowl, *Numida*, they are more dorsal and caudal.

Prior to 1924 the nucleus isthmi complex was tacitly assumed to be the avian homologue of the mammalian medial geniculate complex (see Kappers *et al.*, 1936). However, Showers and Lyons follow Marburg (1924) in suggesting that it represents scattered reticular cells along the mammalian lateral lemniscus. In such a reticular nucleus the correlation of information derived from disparate origins within the tectum, lateral lemniscus, tegmentum, cerebellum and forebrain would be reasonable. Such correlations are almost certainly involved during the exercise of control functions like those suggested by the experimentally induced hippus of Section V B below. The entire cytological complex is then potentially related, by both its topographical position and its connections, to the acoustic and visual systems. Similar considerations also apply to the neighbouring semi-lunar nucleus which lies to the external side of the ventral part of the brachium conjunctivum. As the micro-cellular component of the isthmi complex becomes both smaller and more ventrally situated in its anterior part, the two are closely apposed.

The principal part of the nuclear complex therefore consists of two subdivisions (Craigie, 1928; Jungherr, 1945; Showers and Lyons, 1968). In the hen it appears in transverse sections as a dorsally flattened oval structure and begins at the level of the posterior commissure. In broad terms it lies in a plane which passes through the pretectal, sub-pretectal and spiriform nuclei (*q.v.*). Just behind this region it is clearly divided into its small-celled and large-celled parts. Further back the underlying magnocellular contribution remains approximately the same size and becomes rather archiform, with the concave face uppermost. In the front part of this region the small-celled partner lies within the umbo of this crescent but as its own size increases it protrudes out and takes up an elliptical shape. In their caudal regions both partners tend to rotate around one another and the parvocellular moiety decreases in size and disappears slightly in front of the magnocellular one. In this region they can extend up and come into close proximity to the isthmo-optic nucleus. In different genera they may or may not disappear at levels which pass through this latter nucleus, the larger celled unit persisting to more posterior levels in the pigeon. In *Numida*, where the optic ventricle is placed more laterally than in *Gallus* and the dorsal part of the lateral mesencephalic nucleus is medial not ventral to it, the nucleus isthmi is also involved in this shift of position. Although it maintains essentially the same relationship with the tectal hemispheres as in *Gallus* its position is rather more caudal and ventral.

The dorso-lateral rotation at the hind end also occurs in *Anas*, *Phasianus*, *Mycteria*, *Cathartes* and *Passer*. In *Anas* and *Mycteria*, however, the isthmi complex is rather behind the lateral mesencephalic nucleus. Histologically a similar picture is common to all of them. In both the hen and guinea-fowl the large-celled component comprises medium to large, deeply staining, multipolar neurones. The small cells of the other part are more closely packed together. Kappers *et al.* (1936) drew attention to the rows of cells which radiate from the concave to the convex side of the ventral unit. Showers and Lyons reported that these were very evident in both *Numida* and *Gallus*, and prominent in *Anas*, *Phasianus*, *Mycteria* and *Cathartes*. The cells of the semilunar nucleus are similar in appearance to those of the parvocellular component of the principal isthmi nucleus. There is, however, some variation in their staining characteristics. Typically the component cells are surrounded by the fibres of the lateral lemniscus and have their maximum extent within the ventro-lateral tegmentum. Extending back towards the caudal limits of the tectum and isthmo-optic nucleus the cells within these posterior regions tend to be more chromophilic when stained by classical methods. In *Mycteria* such deeper staining cells are rather smaller than the rest. In his study of the neuronal structures contributing to the ascending auditory pathway Boord (1968) found no evidence that any degenerating axons projected to either the isthmi or semilunar nuclei from the medullary auditory stations.

Following all types of lesions to the angular, magnocellular and laminar nuclei some ascending fragmented auditory axons border on the semilunar and the parvocellular part of the isthmi *en route* for the dorsal part of the lateral mesencephalic nucleus but none enter them. Preterminal degeneration was also absent within the isthmi.

## B. Lesions of the Isthmi Complex

Using stereotaxic fixation of the head Showers and Lyons (1968) made localized, direct current lesions. The lesion sites also included foci within the tectum, tegmentum and lateral mesencephalic nuclei; the current parameters involved variable voltages up to 350 V, were in the 1·0 mA range and lasted from 15 to 60 sec. After the death of the preparations they used Marchi and Nauta Gygax techniques to verify the previously described connections such as tecto-isthmo-tectal, isthmo-cerebellar, internuclear and lemniscal. They were also able to demonstrate degenerating fibres passing between the isthmi complexes of the two sides via the tegmentum and the dorsal bundles of the ventral supra-optic commissure. These had terminations at large reticular cells. Other short fibres connect the nuclei with the oculomotor nuclear-complexes and terminated in both the ipsilateral and contralateral Edinger–Westphal segments (Section II B). The bilateral connection involved passage through the ventral part of the posterior commissure.

One group of fibres entered the basal caudal ramus of the septo-mesencephalic tract and passed, via the basal frontal ramus, into the lateral prosencephalic fibre bundles. These distribute to both the paleostriatal and caudal neostriatal regions. However, Powell and Cowan (1961, see Showers and Lyons) did not observe any retrograde degeneration within the isthmi complex following total telencephalic ablation. Erulkar (1955), amongst others, had nevertheless reported marked responses in both the isthmi complex and the caudal neostriatum following the presentation of peripheral auditory stimuli. Finally a dorsal distribution of degenerating fibres ran towards the hippocampal region (q.v.).

In the living preparation the overt result of such unilateral lesions within the isthmi complex was a sustained hippus. This pupillary instability persisted for 7–30 days and was not seen after lesion of either the lateral mesencephalic nucleus or the tegmentum. Furthermore it was analogous to that obtained by direct electrical stimulation of the nucleus isthmi.

## C. Activation of the Isthmi Complex by Peripheral Stimuli

Besides monitoring the potentials which are evoked within the torus semicircularis following the presentation of click stimuli Harman and Phillips (1967) also emulated Erulkars earlier work and demonstrated that various loci within the nucleus isthmi complex are also activated. In the region of the contralateral semi-lunar nucleus the responses had a latency of 2–4 msec and an amplitude of 20–40 $\mu$V. Ipsilateral clicks evoked potentials of greater amplitude but longer latency. These appeared between 5–6 msec after click presentation and varied from 80 to 350 $\mu$V. Just above this anatomical level, and somewhat towards the mid-line, latent periods of 4–6 msec were followed by responses with an average amplitude of 160 $\mu$V.

The small celled component of the principal isthmi groupings was readily activated by both ipsilateral and contralateral clicks which conforms with what Erulkar had observed. He found that the latencies were variable, depended upon the area involved, and were generally somewhat longer than those of the medial and lateral regions of the semi-lunar nucleus—some 4–5 msec instead of 3–4. Harman and Phillips found that potentials of 25–40 $\mu$V occurred within the lateral areas after a latent period of 6–9 msec. Within the medial regions the evoked potentials were much more complicated. The most stable component was a 50–400 $\mu$V wave which occurred some 10 msec after presentation of the stimulus. When the magnocellular part responded the latencies were shorter but it by no means always did so. As with the large evoked responses within the torus those within the semi-lunar nucleus persisted even when the presentation frequency exceeded 100/sec, and with frequencies of less than 10/sec they waxed and waned.

## D. Direct Electrical Stimulation of the Isthmi Complex

Apart from their lesion studies, which were referred to in Section V B above, Showers and Lyons (1968) also undertook direct electrical stimulation of the various component structures within the isthmi complex. This showed that the consistent responses to such excitation involved the production of an unstable pupil (hippus), and that this effect is independent of the incident illumination (see Muskens, 1929, 1930). The hippus can be compared with, and contrasted to, the results of tectal stimulation. Following electrical stimulation of the lateral tectal surface there is a horizontal deviation of the eye coupled with pupillary dilation. The medial surface gives a vertical deviation accompanied by pupillary constriction. In the case of the isthmi complex, when the pupil is initially constricted then this constriction is at first sustained following stimulation. Before the end of the 5 sec stimulation period it will, however, dilate. A rapidly alternating sequence of dilation and constriction then follows during the remainder of the period of stimulation. This is hippus. Complementarily a pupil which is somewhat dilated at the onset of the current will maintain this position for a while and then also enter an analogous series of dilations and constrictions.

Increases of the voltage and amperage of the current give rise to several types of somatic movement and also lachrymation. Such movements include a horizontal convergent deviation of the eyes together with the closure of the nictitating membrane. These responses were consistently evoked by the stimulation of anterior and medial regions. Lachrymation was also a consistent concomitant of stimuli applied to posterior and medial regions. Movements of the neck and lower extremities were typically those which one associates with deglutition and walking. They occurred following high intensity stimulation of two out of the three sites which habitually produced convergent deviation of the eyes. During such responses there were no overt signs of change in the heart rate, ventilation rate or the body temperature.

The pupillary effects confirm the suggestions of Jegorow (1887). During the periods of hippus the changes which result in dilation or constriction are, like those of normal pupillary activity, complete in an average time of $\frac{1}{6}$ sec. Anatomical investigations suggest that the sphincter muscle of the iris is innervated by the oculomotor nerve, the dilator muscle by the ophthalmic division of the avian trigeminal (Zeglinski, 1885; Lillie, 1908; Carpenter, 1911; Koppanyi and Sun, 1926). Preganglionic fibres originate in the accessory oculomotor nucleus of Edinger–Westphal. These small axons pass to the ciliary ganglion along with the oculomotor nerve, and post-ganglionic fibres then traverse the short ciliary nerves to the iris sphincter (Showers and Lyons, 1968). We are still ignorant of the origin of those pre-ganglionic fibres attributed to the trigeminal nerve.

As tectal stimulation can also elicit pupillary movements and give both dilation or constriction, Showers and Lyons concluded that these effects are mediated by the numerous short tecto-isthmal fibres. Both the tecto-isthmal fibres and those originating within the Edinger–Westphal nucleus may provide the explanation for the pupillary unrest which occurs during changes in the amount of incident light. They are also almost certainly involved in the comparable changes which accompany the approach of an investigator. Furthermore the position of the isthmi complex, lying as it does in the course of the lateral lemniscus, would suggest that it is exposed to influences from other sensory modalities. In view of the observations of Erulkar (1955), together with those of Harman and Phillips (1967), it is of particular interest that pupillary unrest can follow the onset of loud noises, especially if this is sudden. This response is presumably related to the activation potentials which follow auditory stimuli.

The fact that the first result of stimulation is a brief period of pupillary stabilization which precedes the onset of hippus, led Showers and Lyons to suggest that the actual influence which is exerted by the nucleus isthmi is an intensification of the normal pupillary responses. These involve a homeostatic relationship between constriction on the one hand, and dilation on the other. The effect of isthmal lesions like those described in Section V B above is then to upset this balance which is maintained by discharges within circuits which also involve both the tectum and neostriatum (see Chapter 14, Section IX).

However, the actual integrative action can also, it seems, be far more wide-spread. In the place of such delicate muscular coordinations Phillips (1964), in his studies on "wildness", stimulated sites within the complex and observed gross activities such as crouching and running with sleeked feathers. In the nearby dorsolateral mesencephalic nucleus these responses were replaced by attack reactions directed at the preparations neighbours. It is rather odd that the results are so disparate!

# VI. THE ISTHMO-OPTIC NUCLEUS

## A. General Considerations

In both the chicken and guinea fowl the isthmo-optic nucleus appears rostrally at a plane which cuts the hind end of the oculomotor, or the front of the trochlear nucleus. It is a small, oval group of cells in the chicken, situated in the dorso-lateral tegmentum and just medial to the optic tectum. Its relationships with the posterior parts of the isthmi complex will be clear from Section V A above. Passing backwards it shifts medially and becomes associated in its ventro-lateral region with the lateral, dorsal lemniscal nucleus of Jungherr (1945). It disappears caudally at the level of the trochlear nucleus. In both *Gallus* and *Numida* the component cells are medium sized, deeply staining and multipolar. Around the periphery they form a compact ring but are more scattered at the focal centre. A similar type of organization characterizes the nucleus in *Anas, Phasianus, Cathartes* and *Passer*. However, according to Showers and Lyons (1968) in the ibis, *Mycteria americana*, the structure is absent and is replaced by a dorsal, and poorly differentiated, extension of the semi-lunar nucleus. As its identification is largely dependant upon topographical representation it is probably better to assume that the two fuse.

Showers and Lyons were able to confirm the various fibres which had been reported previously by Craigie (1928), Huber and Crosby (1929), Sanders (1929) and Cowan *et al.* (1961). Within the isthmo-optic tract there are optic tract axons, together with others which link the nucleus to the tectum, and both the oculomotor and trochlear nuclei. Cowan and Wenger (1968) have described the origin of all these fibres and also the isthmo-optic nucleus itself in the embryonic chick. The nuclear cells increase in size and a delicate fibre plexus develops during the period from the 10th to 13th day of incubation. The plexus is derived in part from outgrowths from the component cells of the nuclear mass and in part from the ingrowth of afferents from the tectum. These pass to the nucleus via the tecto-isthmal tract. Subsequently cell loss coupled with further growth of the fibres transforms the nucleus from a compact mass into a complex folded structure. Using experimentally induced anophthalmia they were also able to show that partial ablation of the developing eye does not necessarily result in the disappearance of the nucleus. However if the damage to the eye is extensive isthmo-optic hypoplasia does result.

## B. Centrifugal Fibres in the Avian Visual System

As was emphasized in Chapter 7, Section VII, one of the interesting developments in the physiology of sensory systems which took place during the late 1950s and early 1960s was the demonstration of the important part that centrifugal mechanisms play in the regulation of receptor sensitivity, and also in the modulation of the activity of main relay nuclei. Such factors have already been mentioned in relation to hearing, and a general review of the early work on all such systems was provided by Livingstone (1959). In their work on the central projections of the pigeon retina Cowan *et al.* (1961) were interested in the possibility that the isthmo-optic nucleus and its associated fibre tract formed such a regulatory system. Their attention was initially directed to this possibility by the rather unusual nature and time-course of the Nauta degeneration within the isthmo-optic tract, and also by a number of features of the cellular changes which take place in the nucleus itself after excision of the eye. These characteristics seemed to argue against the changes being trans-neuronal in origin. However, the critical evidence, which would establish the nucleus as a source of centrifugal fibres to the retina, could only be provided by showing that localized lesions within it led to the axonal degeneration of fibres that passed through the isthmo-optic tract to the retina (Cowan and Powell, 1962). Evidence of this kind was provided by their subsequent paper (Cowan and Powell, 1963) and by McGill (1964) who produced evidence for such centrifugal fibres and also showed that the fibres from the tectum to the isthmo-optic nucleus end in a discrete and localized pattern. As a result the dorsal half of the tectum is connected to the ventral half of the nucleus, and the anterior half of the tectum to the medial half of the nucleus.

Cowan and Powell produced lesions in the brain stem of five pigeons and in the optic tectum of a further three. In four out of the former group these lesions involved the isthmo-optic nucleus. Fortunately for a considerable part of its length the isthmo-optic tract remains quite distinct from the remainder of the visual pathways. Even after entering the chiasma its constituent fibres remain in a fairly compact bundle so that in Nauta Gygax preparations it is comparatively easy to follow the degenerating fibres. These enter the contralateral optic nerve and proceed to the contralateral retina. There is no evidence for a centrifugal projection to the ipsilateral eye. However, comparable degenerating fibres were also invariably found along the margins of visual relay nuclei such as the lateral geniculate and external. It was difficult to determine whether such fibres actually terminated in these nuclei, or whether they were fibres of passage and really associated with the neighbouring isthmo-optic tract. They emphasized the considerable significance from a functional point of view of the possibility, which was suggested by Wallenberg (1898), that the isthmo-optic nucleus is connected either directly, or indirectly by way of collaterals, to the primary relay nuclei. Were this the case then the centrifugal system would be capable of acting at both the periphery, and also at more central levels within the visual pathway.

Throughout most of their course the fibres are of a larger diameter than those in the optic tract, but, in front of the chiasma, their size decreases. Taking into account the rather small total number of cells in the nucleus itself, which from direct counts seemed to be rather less than 10,000, Cowan and Phillips suggested that this decrease in the axonal diameter might reflect branching. From the optic nerve head they pass into the optic nerve layer and are then distributed to all parts of the retina. It was possible that fewer degenerating fibres were present on the medial side of the eye but the actual significance of this was difficult to assess because the optic nerve layer itself is thinner on that side. On the other hand the high density of degenerating fibres which are detectable at sites close to the nerve head is due to the presence there of fibres radiating out to all retinal areas. In view of some uncertainty about the actual terminal sites they cited the results of Ramon y Cajal (1889) who had concluded that centrifugal fibres end exclusively on amacrine cells at the inner aspect of the bipolar cell layer. Such a site of termination would be of considerable importance in view of the complex relationships which exist between such cells (Dowling, 1968; Polyak, 1941; see Chapter 9, Section IV E). In the latter case it would permit interaction between the centrifugal fibres and the visual receptors by way of recurrent influence at the level of the bipolar and ganglion cells. Furthermore, in view of Kidd's (1962) electron-microscopic data on axo-axonal synapses within the inner plexiform layer, which he associated with amacrine cells, Cowan and Powell (1963) suggested that the fibre output might be responsible for a presynaptic inhibitory effect within the intra-retinal relays. Subsequently Maturana and Frenk (1965) concluded that in the pigeon retina the fibres end within the inner nuclear layer in which connections between receptors occur. Nicholls and Koelle (1968) gave a further consideration to this when they observed that in pigeon and other cone retinas there was a very ordered pattern of staining for acetylcholinesterase in the inner plexiform layer. Cells corresponding to the amacrine cells were the major source of positive activity while horizontal cells gave evidence of non specific cholinesterases. Although the isthmo-optic nucleus does stain for acetylcholinesterase the fibres in the retina which are derived from it are diffusely branched and they thought that it was unlikely these contributed significantly to their results. They noted, however, similarities to the situation in the efferent cochlear bundle and its peripheral terminations.

The results of these lesion experiments also showed that the isthmo-optic tract is an important source of tectal afferents to the isthmo-optic nucleus. It was not possible to exclude an additional origin in the lateral mesencephalic nucleus as suggested by Huber and Crosby (1929), and they were at some pains to direct attention to the possibility of yet other afferents originating from sources such as diencephalic relay nuclei on the visual pathway.

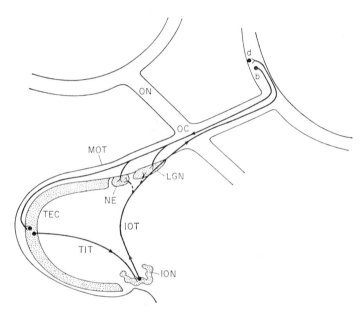

Fig. 100. A diagrammatic representation of the retinal projection upon the optic tectum and some of the diencephalic relay nuclei; the connection between the tectum and the isthmo-optic nucleus and the centrifugal pathway from this nucleus to the retina. The possible collateral projection of the centrifugal fibres to the lateral geniculate nucleus and the external nucleus is represented by broken limbs. The letters b and d refer to the ganglion and bipolar cell layers respectively (Cowan and Powell, 1963).

In conclusion Cowan and Powell speculated as to whether each part of the avian retina was reciprocally connected, via the tectum, with that particular part of the isthmo-optic nucleus from which it receives centrifugal fibres. Such a relationship would permit the fibres to serve as the limb of a reflex arc analogous to that supplying muscle spindles. Maturana and Frenk (1965), who concluded that two types of terminal are associated with the centrifugal

14

system, also suggested that they could comprise a mechanism for the localized centrifugal control of retinal function. (Fig. 100).

## C. Electrical Stimulation of the Isthmo-optic Nucleus

In his studies of the activation potentials which follow stimulation of the optic nerve head Holden (1968c) recorded from output cells within the isthmo-optic nucleus. He attempted to demonstrate three things. First of all that electrical stimulation of the optic nerve head activates the isthmo-optic nucleus antidromically and the retino-tectal system orthodromically. Secondly that the response properties of isthmo-optic events can be distinguished from those of tectal events; and thirdly that the actual output cells are within the nucleus. He was able to show that isthmo-optic penetration by the monitoring probe can be recognized by a distinctive A-wave that follows retinal stimulation, and by the isolation of antidromically activated cells. The properties of the output cells enables them to be distinguished from directly activated axonal spikes and trans-synaptically activated tectal cells.

The isthmo-optic A-wave had a prominent positive-going phase with a latency of 2–2·5 msec, a peak of 3·0 msec and a duration of the order of 3·0 msec. It was a distinctly graded response and could follow repetition frequencies up to 500/sec. Strictly localized in depth it occurred at just those positions where antidromically activated somata were encountered, and was clearly caused by such antidromic activation. It could be distinguished from the P-wave of the optic tectum, against whose medial border the isthmo-optic nucleus is apposed, by both its more spike-like conformation and its shorter duration. It was usually encountered rather abruptly after a traverse through the cerebellum and never emerged after a descent through the N-, and R-zones of the optic tectum. What is more, such a strict localization in depth is never observed in the case of the tectal P-wave. As there was good agreement between the peak of the A-wave and that of the latency histogram, Holden took this as an indication that the A-wave simply represents the field potential resulting from invasion of the somata within the nucleus. When he arranged for orthodromic discharge and simultaneous antidromic activation evidence for collisions showed that during orthodromic firing the impulse travels centrifugally towards the retina.

Holden (1968d) then investigated the synaptic actions upon the nucleus. Following the stimulation of the lateral tectum a synaptic field potential, the S-wave, can be recorded. This cannot follow repetitive stimulation in excess of 30/sec, is depressed for long periods after the A-wave, and shows frequency potentiation at 10/sec. The actual early discharge corresponds in its timing to this S-wave and presumably represents monosynaptic activation via the tecto-isthmo-optic tract. Later bursts of firing must result from more complex input routes. Surveying these results Holden concluded that 10% of the cells within the nucleus are fired synaptically from the retina. The short latency firing occurs at 5–7 msec, and is often blocked by prior antidromic invasions. The cells which discharge after a longer latent period usually do so at 80–110 msec. However, even the earliest firing can involve the serial activation of units within the retino-tectal and tecto-isthmo-optic systems. He speculated on the possibility that the initial, short latency firing represents activation of the system by a trigger stimulus, or trigger stimuli, within the visual field. The longer latency firing could then represent a command to the isthmo-optic nucleus in conjunction with the programme for motor actions. It appeared that whilst excitatory synapses were certainly activated by the isthmo-optic tract, inhibitory ones could be made to discharge from both the optic nerve head and the lateral tectum. The origins of these last discharges need not, therefore, be mutually exclusive. They could arise within the nucleus, say from inhibitory interneurones, or possibly from axon collaterals of inhibitory output cells. Alternatively they might originate from inhibitory tectal cells whose axons run within the tecto-isthmo-optic tract. A careful watch was kept for Renshaw type repetitive firing but no evidence for this type of inter-neurone was detected.

## VII. THE RED NUCLEUS

This cell grouping is the avian equivalent of the tegmental motor nucleus of other vertebrate classes, and, of course, the mammalian structure of the same name. Although rather more complex than its reptilian homologue it lacks the marked differentiation into micro-cellular and macro-cellular components that one sees in the mammalian nucleus. It is situated to the medial side of the oculomotor nuclear complex, is closely associated with components of the medial reticular formation, and is rather medio-dorsal to the quinto-frontal tract (q.v.), the ansa lenticularis and the ventral part of the deep mesencephalic nucleus (nucleus mesencephalicus profundus). For many years it has been considered to give rise to the rubro-bulbo-spinal tract which passes ventral and medial to the median longitudinal fasciculus. Amongst its other connections it numbers a high proportion of axons from the superior cerebellar peduncle or brachium conjunctivum (Kappers et al., 1936). These pass across the mid-line and then synapse around large cells in the nucleus. Indeed, although the small cells predominate in the red nucleus of man, it is the magnocellular component that is the phylogenetically older and this forms the bulk of the structure in classes other than the mammals. In view of such relationships, and its importance as an extra-pyramidal cell station in mammals, it has generally been assumed to be involved in locomotor control. Its position can be seen in the longitudinal section of Fig. 94.

## VIII. THE PRETECTAL NUCLEI

Aside from the foregoing nuclear cell masses there are, in the avian mesencephalon, interpeduncular nuclei which are associated with the habenulo-peduncular tract; two others which are homologous with those of the reptilian longitudinal fasciculus, and an ectomammillary nucleus. The interpeduncular and ectomammillary components of each side are ventrally disposed, limited below by the lower edge of the brain, and respectively situated to the medial and lateral side of the fibres of the oculomotor nerve. In the somewhat indeterminate transition region between the diencephalon and mesencephalon, at levels where the nucleus ruber is disappearing and the rotund nucleus has yet to appear, there is another group of cell masses which are generally lumped together under the collective name of pretectal nuclei. The development of both forebrain and cerebellum, the size of the eyes and, according to Kappers, the effects of neurobiotactic influences, make this region of the avian brain somewhat difficult to interpret. Portmann and Stingelin (1961), together with a number of other workers, therefore considered both mesencephalon and diencephalon together as a single entity.

The pretectal nucleus itself, along with the sub-pretectal and spiriform nuclei, are all classically ascribed to a mesencephalic origin. They have been known, at least in part, since the work of Edinger (see 1908) and Turner (1891), were all considered by a number of the older anatomists such as Edinger and Wallenberg (1899), Rendahl (1924), Huber and Crosby (1929), Papez (1929) and Craigie (1930), were reviewed in detail by Kappers *et al.* (1936) and to a lesser extent by Cordier (1954) and Kappers (1947). Details of their precise disposition in terms of stereotaxic coordinates are given by Tienhoven and Juhasz (1962) in the case of *Gallus*, and Karten and Hodos (1967) for the pigeon. The tectal grey matter of Huber and Crosby (1929) is usually associated with them. Together they comprise a relatively erect dorso-ventral block of nuclei with the sub-pretectal at the base, the mediol and lateral spiriform nuclei at more intermediate levels, and both the pretectal and medial pretectal dorsally. This is overlain medially by the area pretectalis and laterally by the isthmo-optic and septo-mesencephalic tracts.

The avian pretectal nucleus itself is a chromophilic cell grouping which, as is clear from the foregoing, lies above the spiriform nucleus and medial to the tectum. It has a relatively intimate connection with the sub-pretectal nucleus via the pretectal-sub-pretectal tract that in sections passes more or less straight down at the side of the spiriform nucleus. It was in fact the sub-pretectal that Kappers *et al.* suggested as the avian homologue of the reptilian pretectal. For many years the presence of fibres which pass to and from both the pretectal nucleus and the optic tectum, via the posterior commissure and the dorsal septo-mesencephalic tract, was taken as being indicative of a close functional relationship.

Like the pretectal nucleus the spiriform components, which are situated in the more intermediate position and also invested by fibres of the posterior commissure, are composed of chromophilic cells. These appear to have multiple connections. Both major components have been identified in a number of genera such as *Columba*, *Chrysolampis*, *Chlorostilbon* and *Gallus*, but they are possibly less distinctive in *Apteryx*. In the trochiliform genera the component cells are, however, somewhat diffuse and give the impression of a greater number of sub-units.

The final member of this pretectal group is the principal precommissural nucleus. This is generally situated in a somewhat more anterior position. Whilst its posterior region is inserted medially between the medial and lateral spiriform components, it runs forward, enlarges, and then lies as a large nuclear contribution below the diencephalic rotund nucleus. In this region it is limited on its medial side by the tecto-thalamic tract and laterally by micro-cellular parts of the mesencephalic lentiform nucleus. A well-defined cell mass it maintains axonal connections similar to those of both the spiriform and sub-pretectal nuclei that, broadly speaking, replace it behind. It also receives fibres from the ventral supra-optic decussation as this passes *en route* from the diencephalon to the optic tectum.

# IX. THE MEDIAN LONGITUDINAL FASCICULUS

The median longitudinal fasciculus has been referred to on a number of occasions in the foregoing chapters. It originates in this intermediate transitional zone separating the definitive diencephalon and mesencephalon. Passing backwards it traverses the medullary areas in which the efferent cranial nuclei are located and continues in the ventral funiculus of the spinal cord. In the pigeon it is at first situated to the side of the dorsal and ventral parts of the oculomotor nuclear complex, further back it replaces these and assumes a medial position. This is maintained throughout the medullary region and, as the nucleus raphe recedes, it extends ventrally (Figs 50, 57). Lesion experiments which were carried out during the first three decades of the century, coupled with silver tracings, suggest that anteriorly it carries the secondary vestibular fibres which pass forward to the spiriform, oculomotor and trochlear nuclei, as well as to the mesencephalic reticular formation. Contralateral descending axons pass to the definitive hypoglossal nucleus and to the ventral horn which also receives some homolateral fibres. Such descending pathways constitute the interstitial spinal and the commissuro-medullary tracts of Muskens (1929, 1930).

# X. MESENCEPHALIC CARDIOVASCULAR CONTROL

Using adult specimens of *Anas boschas* Feigl and Folkow (1963) were able to demonstrate two areas that gave striking cardiovascular effects. One of these was in the diencephalon, the other in the mesencephalon, and both were close to the mid-line. Following electrical stimulation at the mesencephalic site there was a rise in the blood pressure, a fall in heart rate, and an intense vaso-constriction within the muscle bed. As soon as the period of stimulation began the bird stopped breathing and continued to remain apneic until the stimulation ended. If this state of apnea was prevented by using artificial respiration then the cardiovascular responses changed. No bradycardia developed and the muscular vaso-constriction was either greatly reduced or, alternatively, absent altogether. However, such artificial respiration was ineffective if the gas mixture contained more than 10% carbon dioxide and the relative effects of both hypercapnia and anoxia during these studies were nearly the same as those observed during diving. In this respect it is worth noting that a state of muscular vasoconstriction is widely reported as occurring during diving in mammals (see, for example, Scholander *et al.*, 1962).

# XI. MONOAMINES IN THE MESENCEPHALON

## A. General Principles

Using a highly sensitive and specific fluorescence method Fuxe and Ljunggren (1965) investigated the monoamine distribution at a number of levels within the upper brain stem. They considered that both the green and the yellow fluorescence which they detected were quite specific and indicated the presence of a primary catecholamine and 5-hydroxytryptamine respectively. The green fluorescence was largely restricted to thin nerve fibres that had numerous, strongly fluorescent varicosities. The axonal segments which separated such varicosities could be as little as $0.2\,\mu$ in diameter and were sometimes barely visible. Such fibres were found scattered throughout the brain stem but were often aggregated within certain nuclei. In this case the non-fluorescent nuclear cells were enclosed by them. Fuxe and Ljunggren were quite clear that these represented terminals of the catecholamine type. A few yellow fluorescent fibres were also observed which, although very thin, had the same general appearance. The colour was of greater intensity following monoamine oxidase inhibition. These fibres they referred to as of the 5-hydroxytryptamine type and considered that the relatively small number which were observed was probably a reflection of their small diameter.

Groups of nerve cells, which were identified by using toluidine blue or gallocyanin-chromalum, also exhibited specific green or yellow fluorescence within their cytoplasm. These had the same general appearance as the monoamine containing nerve cells of the mammalian brain. In general two distinct categories could be distinguished *viz.*:

(a) Nerve cells that developed a weak to strong green fluorescence in both normal animals and also in preparations which had been treated with monoamine oxidase inhibitors. These they designated cells of the catecholamine type.

(b) Nerve cells that exhibited a weak to medium yellow fluorescence in normal animals, and in which the coloration was markedly increased after treatment with the inhibitors. These they designated as cells of the 5-hydroxytryptamine type.

## B. Catecholamine Cells and Fibres

A very large number of the very fine, varicose terminals were present within the superficial layers of the optic tectum. In contrast there were only a few in both the deeper layers and also the periventricular region. A large number were detected in the superficial parvocellular nucleus, the nucleus of the septo mesencephalic tract. This was also the case in lateral parts of the spiriform nucleus. However, the medial components of the spiriform nucleus, together with the external and lateral mesencephalic nuclei and the sub-pretectal nucleus contained very few. The principal part of the nucleus isthmi complex (nucleus isthmi pars principalis) had even less, whilst the pretectal nucleus itself was unlike any other area of the brain stem and was totally devoid of any fluorescent material. Whilst a similar absence characterized the outer part of the isthmo-optic nucleus a very small number of green fluorescent terminals occurred within its inner layers.

Terminals located within the tegmentum were again usually fine. Although they were particularly common in two well defined areas, others could also be seen scattered in small numbers throughout the mesencephalic reticular formation. One of the two markedly positive areas was situated approximately 2 mm to the side of the median longitudinal fasciculus at about the level of the trochlear nucleus. The other was within the ventro-lateral part of the reticular formation, and was especially clear at the level of the oculomotor nucleus. The abundant terminals in this region were often intimately associated with both cell bodies and processes of fluorescent and non-fluorescent cells. The red nucleus, and also the oculomotor, trochlear and interpeduncular nuclei all contained a few such terminals and rather more occurred within the ectomammillary nucleus. A number were present in both the region of the optic chiasma and within the oculomotor nerve just after it emerged from the brain. Fuxe and Ljunggren suggested that such fibres might innervate the vessels which supply the nerve itself but could not rule out the alternative possibility that they were definitive afferent or efferent fibres passing to or from the brain.

A small number of medium-sized cells which were situated just behind the trochlear nerve and 2 mm to the side of the median longitudinal fasciculus gave a medium to strong green fluorescence. It was clear that catecholamine containing terminals were intimately connected to them. Most lay in the ventro-lateral part of the tegmentum and the area concerned extended from the level of the trochlear nucleus, between the red and ectomammillary nuclei, to the level at which the oculomotor nerve emerges. Just in front of this aggregate, and arising from it, a large tract of green fluorescent nerve fibres ran forward within the median forebrain bundle in the lateral hypothalamic region. Some terminated at this level but the remainder followed a latero-dorsal path towards the forebrain, as was reported by Bertler et al. (1964). Fuxe and Ljunggren assumed that these had their origin in mesencephalic catecholamine cells.

## C. Cells of the 5-Hydroxytryptamine Type

In nialamide treated animals Fuxe and Ljunggren (1965) found that the yellow fluorescence, which was normally very weak, was considerably increased. It was also detectable within axons which otherwise seemed to lack it in normal animals. Some such axons, which passed forward with the green fluorescent ones within the medium forebrain bundle, arose from small to medium sized, round to oval cells. They were particularly obvious following pre-treatment with reserpine. In the lower part of the nucleus of the median longitudinal fasciculus two parallel rows of cells stood out just to the side of the mid-line. In fact almost all the cells of this 5-hydroxytryptamine type were restricted to the mesencephalon. Practically none occurred in diencephalic regions such as the thalamus and pre-optic area, although a few were present within the posterior hypothalamus. Data on cholinesterase and phosphatase are contained in Fig. 96.

## References

Azzena, G. B. and Palmieri, G. (1967). Trigeminal monosynaptic reflex arc. *Exp. Neurol.* **18**, 184–193.

Bechterev, W. (1883). Zur Physiologie des Korpergleichgewichtes. Die Funktion der zentralengrauen Substanz des dritten Hernventrikels. *Pfluger's Arch.* **31**, 479–530.

Bechterev, W. (1909). "Die Funktionen der Nervencentra." Fischer, Jena.

Bellonci, J. (1888). Ueber die centrale Endigung des Nervus opticus bei den Vertebraten. *Z. wiss. Zool.* **47**, 1–46.

Bertler, A., Falck, B., Gottfries, C. G., Ljunggren, L. and Rosengren, E. (1964). Some observations on adrenergic connections between mesencephalon and cerebral hemispheres. *Acta Pharmacol. Tox.* **21**, 283–289.

Biedermann-Thorson, M. (1968). Auditory responses of neurons in the lateral mesencephalic nucleus (inferior colliculus) of the Barbary dove. *J. Physiol.* **193**, 695–705.

Bishop, G. H. and O'leary, J. (1942a). The polarity of potentials recorded from the superior colliculus. *J. Cell. Comp. Physiol.* **19**, 289–300.

Bishop, G. H. and O'leary, J. (1942b). Factors determining the form of the potential record in the vicinity of the synapses of the dorsal nucleus of the lateral geniculate body. *J. Cell. Comp. Physiol.* **19**, 315–331.

Boord, R. L. (1968). Ascending projections of the primary cochlear nuclei and nucleus laminaris in the pigeon. *J. comp. Neurol.* **133**, 523–530.

Boyce, R. and Warrington, W. R. (1898). Observations on the anatomy, physiology and degenerations of the nervous system of the bird. *Proc. Roy. Soc. London.* **64**, 176–179.

Brandis, F. (1893). Untersuchungen uber das Gehirn der Vögel. II. *Arch. mikros. Anat.* **41**, 623–649.

Brouwer, B. (1918). Klinisch-anatomisch Untersuchungen uber den Oculomotoriuskern. *Z. neur.* **40**, 152–193.

Brown, J. L. (1965a). Vocalization evoked from the optic lobe of a song bird. *Science* **149**, 1002–1003.

Brown, J. L. (1965b). Loss of vocalization caused by lesions in the nucleus mesencephalicus lateralis of the red-winged blackbird. *Amer. Zool.* **5**, 693.

Brown, J. L. (1969). The control of avian vocalization by the central nervous system. *In* "Bird Vocalisations" (Hinde, R. A., ed.), pp. 79–96. Cambridge University Press.

Buddenbrock, W., von (1953). "Vergleichende Physiologie. II. Neurophysiologie", pp. 396. Birkhauser, Basle.

Carpenter, F. W. (1911). The ciliary ganglion of birds. *Folia neuro-biol.* **5**, 738–756.

Cate, J., Ten (1936). Physiologie des zentralnervensystem der Vögel. *Ergebn. d. Biol.* **13**, 93–173.

Cate, J., Ten (1965). The nervous system of birds. *In* "Avian Physiology" (Sturkie, P. D., ed.), pp. 697–751. Baillière, Tindall and Cassell.

Cobb, S. (1959). On the angle of the cerebral axis in the American Woodcock. *Auk,* **76**, 55–59.

Cobb, S. (1964). A comparison of the size of an auditory nucleus (n. mescencephalicus lateralis, pars dorsalis) with the size of the optic lobe in twenty species of bird. *J. comp. Neurol.* **122**, 271–279.

Cobb, S. and Edinger, T. (1962). The brain of the emu (*Dromaeus novaehollandiae,* Lath.), *Breviora* **170**, 1–18.

Cohen, D. H. (1967). Visual intensity discrimination following unilateral and bilateral tectal lesions. *J. comp. physiol. Psychol.* **63**, 172–174.

Cordier, R. (1954). Le système nerveux centrale et les nerfs cerebro-spinaux. *Traité de Zoologie.* **12**, 202–332.

Cowan, W. M., Adamson, L. and Powell, T. P. S. (1961). An experimental study of the avian visual system. *J. Anat.* **95**, 545–563.

Cowan, W. M. and Powell, T. P. S. (1962). Centrifugal fibres to the retina in the pigeon. *Nature, Lond.* **194**, 487.

Cowan, W. M. and Powell, T. P. S. (1963). Centrifugal fibres in the avian visual system. *Proc. roy. Soc. B.* **158**, 232–252.

Cowan, W. M. and Wenger, E. (1967). Cell loss in the trochlear nucleus of the chick during normal development and after radial extirpation of the optic vesicle. *J. Exp. Zool.* **164**, 267–280.

**14***

Cowan, W. M. and Wenger, E. (1968). The development of the nucleus of origin of centrifugal fibres to the retina in the chick. *J. comp. Neurol.* **133**, 207–240.

Cragg, B. G., Evans, D. H. L. and Hamlyn, L. H. (1954). The optic tectum of *Gallus domesticus*: a correlation of the electrical responses with the histological structure. *J. Anat.* **88**, 292–307.

Craigie, E. H. (1928). Observations on the brain of the humming bird (*Chrysolampis mosquitus* Linn., *Chlorostilbon caribaeus* Lawr.). *J. comp. Neurol.* **48**, 377–481.

Craigie, E. H. (1930). Studies on the brain of the Kiwi (*Apteryx australis*). *J. comp. Neurol.* **49**, 223–357.

Craigie, E. H. and Brickner, R. M. (1927). Structural parallelism in the midbrain and 'tweenbrain of teleosts and of birds. *Proc. Kon. Akad. Wet. Amsterdam* **30**, 695–704.

Dowling, J. E. (1968). Synaptic organization of the frog retina. *Proc. Roy. Soc. B.* **170**, 205–228.

Dunnebacke, T. H. (1953). The effects of the extirpation of the superior oblique muscle on the trochlear nucleus in the chick embryo. *J. comp. Neurol.* **98**, 155–177.

Edinger, L. (1908). "Vorlesungen uber den Bau der nervosen Centralorgane des Menschen und der Thiere." Vogel, Leipzig.

Edinger, L. and Wallenberg, A. (1899). Untersuchungen uber das Gehirn der Tauben. *Anat. Anz.* **15**, 245–271.

Edinger, T. (1926). The brain of *Archeopteryx*. *Ann. Mag. Nat. Hist.* **9**, 151–156.

Edinger, T. (1951). The brains of the Odontognathae. *Evolution* **5**, 6–24.

Erulkar, D. S. (1955). Tactile and auditory areas of the brain of the pigeon. An experimental study by means of evoked potentials. *J. comp. Neurol.* **103**, 421–458.

Feigl, E. and Folkow, B. (1963). Cardiovascular responses in diving and during brain stimulation in ducks. *Acta Physiol. Scand.* **57**, 99–110.

Ferrier, D. (1879). "Die Funktionen des Gehirnes." Braunschweig.

Flourens, P. (1824). "Récherches Expérimentales sur les Propriétés et les Fonctions du Système Nerveux dans les Animaux Vertébrés." Crecot, Paris.

Fujita, S. (1964). Analysis of neuron differentiation in the central nervous system by tritiated thymidine autoradiography. *J. comp. Neurol.* **122**, 311–327.

Fuxe, K. and Ljunggren, L. (1965). Cellular localization of monoamines in the upper brain stem of the pigeon. *J. comp. Neurol.* **125**, 355–382.

Gehuchten, A., van (1893). "Anatomie du système nerveux de l'homme." Leuven, van In.

Graf, W. (1956). Caliber spectra of nerve fibres in the pigeon (*Columba domestica*). *J. comp. Neurol.* **105**, 355–360.

Hamdi, F. A. and Whitteridge, D. (1954). The representation of the retina on the optic tectum of the pigeon. *J. exp. Physiol.* **39**, 111–118.

Harman, A. L. and Phillips, R. E. (1967). Responses in the avian midbrain, thalamus and forebrain evoked by click stimuli. *Exptl. Neurol.* **18**, 276–286.

Holden, A. L. (1968a). The field potential profile during activation of the avian optic tectum. *J. Physiol.* **194**, 75–90.

Holden, A. L. (1968b). Types of unitary response and correlation with the field potential profile during activation of the avian optic tectum. *J. Physiol.* **194**, 91–104.

Holden, A. L. (1968c). Antidromic stimulation of the isthmo-optic nucleus. *J. Physiol.* **197**, 183–198.

Holden, A. L. (1968d). The centrifugal system running to the pigeon retina. *J. Physiol.* **197**, 199–219.

Huber, G. C. and Crosby, E. C. (1929). The nuclei and fibre paths of the avian diencephalon, with consideration of telencephalic and certain mesencephalic connections. *J. comp. Neurol.* **48**, 1–225.

Huber, G. C. and Crosby, E. C. (1933). A phylogenetic consideration of the optic tectum. *Proc. Natl. Acad. Sci.* **19**, 15–22.

Huber, G. C. and Crosby, E. C. (1934). The influences of the afferent paths on the cytoarchitectonic structure of the sub-mammalian optic tectum. *Psychiatr. Neurol. Bl.* **3** and **4**, 459.

Ingvar, S. (1923). On thalamic evolution. *Acta Med. Scand.* **59**, 696.

Jelgersma, G. (1896). De verbindingen van de groote hersenen by de vogels met de oculomotoriuskern. *Feestb. Nederl. Vereen. Psychiat.* 241.

Jegorow, J. (1887). Ueber den Einfluss des Sympathicus auf die Vögel pupil. *Pfluger's Arch.* **41**, 326–348.

Jungherr, E. (1945). Certain nuclear groups of the avian mesencephalon. *J. comp. Neurol.* **82**, 55–175.

Jungherr, E. (1969). "The Neuro-anatomy of the Domestic Fowl", pp. 126. *Amer. Ass. Avian Pathol.*

Kappers, C. U. A. (1912). Weitere Mitteilungen uber Neurobiotaxis. VII. Die phylogenetische Entwicklung der motorischen Wurzelkerne in Oblongata und Mittelhirn. *Folia neuro-biol.* **6**, 1–42.

Kappers, C. U. A. (1947). "Anatomie Comparée du Système Nerveux", pp. 754. Masson, Paris.

Kappers, C. U. A., Huber, G. C. and Crosby, E. C. (1936). "The Comparative Anatomy of the Nervous System of Vertebrates Including Man." Macmillan (1960 reprint. 3 vols. Hafner Press).

Karten, H. J. (1963). Ascending pathways from the spinal cord in the pigeon (*Columba livia*). *Proc. Int. Congr. Zool.* **2**, 23.

Karten, H. J. (1965). Projections of the optic tectum in the pigeon (*Columba livia*). *Anat. Rec.* **151**, 369.

Karten, H. J. (1966). Efferent projections of the nucleus mesencephalicus lateralis pars dorsalis (MLD) in the pigeon (*Columba livia*). *Anat. Rec.* **154**, 365.

Karten, H. J. and Hodos, W. (1967). "A stereotaxic atlas of the brain of the pigeon (*Columba livia*)." Johns Hopkins University Press, Baltimore.

Karten, H. J. and Revzin, A. M. (1966). The afferent connections of the nucleus rotundus in the pigeon. *Brain Research* **2**, 368–377.

Kidd, M. (1962). Electron microscopy of the inner plexiform layer of the retina in the cat and the pigeon. *J. Anat.* **96**, 179–187.

Knowlton, V. Y. (1964). Abnormal differentiation of embryonic avian brain centres associated with unilateral anophthalmia. *Acta anat.* **58**, 222–251.

Kolliker, A. von (1896). "Handbuch der Gewebelehre der Menschen." Aufl. 6., Bd. 2. Engelmann, Leipzig (1889–1902).

Koppanyi, T. and Sun, K. H. (1926). Comparative studies on the pupillary reaction in tetrapods. III. The reactions of the avian iris. *Amer. J. Physiol.* **78**, 364–367.

Kschischkowski, C. (1911). Chemische Reizung des Zweihugels bei Tauben. *Zbl. Physiol.* **25**, 557–566.

Lillie, F. R. (1908). "The Development of the Chick", pp. 427. Henry Holt, New York.

Livingstone, R. B. (1959). Central control of receptors and sensory transmission systems. Chapter 31 in *Amer. Phys. Soc.* "Handbook on Neurophysiology", vol. 1. Williams and Wilkins, Baltimore.

de Long, G. R. and Coulombre, A. J. (1965). Development of the retino-tectal topographic projection in the chick embryo. *Exptl. Neurol.* **13**, 351–363.

Longet, F. A. (1842). "Anatomie et Physiologie du Système Nerveux de l'homme et des Animaux Vertébrés." Paris.

Manni, E., Azzena, G. M. and Bortolani, R. (1965). Jaw muscle proprioception and mesencephalic trigeminal cells in birds. *Exptl. Neurol.* **12**, 320–328.

Marburg, O. (1924). "Handbuch der Neurologie des Ohres", vol. 1, pp. 290. Urban and Schwarzenberg, Berlin and Vienna.

Martino, G. (1926). Contributo alla conoscenza della funzione dei lobi ottici nel columbo. *Arch. di Fisiol.* **24**, 282–292 (see also *Boll. Soc. Biol. sper.* **1**, 239–242).

Maturana, H. R. and Frenk, S. (1963). Directional movement and horizontal edge detectors in the pigeon retina. *Science* **142**, 977–979.

Maturana, H. R. and Frenk, S. (1965). Synaptic connections of the centrifugal fibres in the pigeon retina. *Science* **150**, 359–361.

McGill, J. I. (1964). Organization within the central and centrifugal pathways in the avian visual system. *Nature, Lond.* **204**, 395–396.

Mello, N. K., Ervin, F. R. and Cobb, S. (1963). Intertectal integration of visual information in the pigeon; electrophysiological and behaviour observations. *Bol. Inst. Estud. med. biol. Mexico* **21**, 519–533.

Mesdag, T. M. (1909). "Bijdrag tot de ontwikkelings-geschiedenis van de structuur der hersenen bij het kip." Dissertation, Groningen.

Munzer, E. and Wiener, H. (1898). Beitrage zur Anatomie und Physiologie des Zentralnervensystems der Taube. *Monatschr. Psychiatr.* **3**, 379–406.

Muskens, L. J. J. (1929). The tracts and centres in the pigeon dominating the associated movements of the eyes (and other moveable parts) in the sense of lateral deviation in the horizontal and of rotation in the frontal plane. *J. comp. Neurol.* **48**, 267–292.

Muskens, L. J. J. (1930). On tracts and centres involved in the upward and downward associated movements of the eyes after experiments in birds. *J. comp. Neurol.* **50**, 289–331.

Nicholls, C. W. and Koelle, G. B. (1968). Comparison of the localization of acetylcholinesterase and non-specific esterase in mammalian and avian retinas. *J. comp. Neurol.* **133**, 1–16.

Nikolaides, R. (1914). Untersuchungen uber die Regulierung der Atembewegungen der Vögel. *Arch. f. Physiol.* 553–564.

Noll, A. (1915). Uber das Sehvermogen und das Pupillenspiel grosshirnloser Tauben. *Arch. f. Physiol.* 350–372.

O'leary, J. L. and Bishop, G. H. (1943). Analysis of potential sources in the optic lobe of duck and goose. *J. cell. comp. Physiol.* **22**, 73–87.

Papez, J. W. (1929). "Comparative Neurology." Crowell and Co., New York.

Paulson, G. W. (1965). Maturation of evoked responses in the duckling. *Exptl. Neurol.* **11**, 324–333.

Perlia, D. (1898). Ueber ein neues Opticuscentrum beim Huhne. *Arch. f. Ophthalm.* **35**, 20.

Peters, J. J., Vonderahe, A. H. and Powers, T. H. (1958). Electrical studies of functional development of the eye and optic lobes in the chick. *J. Exp. Zool.* **139**, 459–468.

Phillips, R. E. (1964). "Wildness" in the mallard duck. Effects of brain lesions and stimulation on escape behaviour and reproduction. *J. comp. Neurol.* **122**, 139–155.

Phillips, R. E. (1966). Evoked potential study of the connections of the avian archistriatum and neostriatum. *J. comp. Neurol.* **127**, 89–100.

Polyak, S. (1941). "The Retina." University of Chicago Press.

Popa, G. and Popa, F. (1933). Certain functions of the midbrain in pigeons. *Proc. Roy. Soc. Lond. B.* **113**, 191–195.

Portmann, A. and Stingelin, W. (1961). The central nervous system. *In* "The Biology and Comparative Physiology of Birds" (Marshall, A. J., ed.), vol. 2, pp. 1–36. Academic Press, New York and London.

Ramon y Cajal, S. (1889). Sur la morphologie et les connexions des éléments de la retine des oiseaux. *Anat. Anz.* **4**, 111–121.

Ramon y Cajal, S. (1891). Sur la fine structure du lobe optique des oiseaux et sur l'origine réelle des nerfs optiques. *J. intern. Anat. Physiol.* **8**, 337–366.

Ramon y Cajal, S. (1909–11). "Histologie du Systeme Nerveux de l'homme et des Vertébrés." Malione, Paris.

Rendahl, H. (1924). Embryologische und morphologische Studien uber das Zwischenhirn beim Huhn. *Acta Zool.* **5**, 241–334.

Renzi, P. (1863–64). Saggio di fisiologia sperimentale sui centri nervosi della vita psic. *Schmidts Jahrb.* **123**, 151–161.

Revzin, A. M. and Karten, H. (1967). Rostral projections of the optic tectum and nucleus rotundus in pigeons. *Brain Research* **3**, 264–275.

Ris, F. (1899). Ueber den Bau des Lobus opticus der Vögel. *Arch. mikros. Anat.* **53**, 106.

Rogers, K. T. (1957). Ocular muscle proprioceptive neurons in the developing chick. *J. comp. Neurol.* **107**, 427–435.

Rogers, K. T. (1960). Studies on chick brain of biochemical differentiation related to morphological differentiation. III. Histochemical localization of alkaline phosphatase. *J. Exp. Zool.* **145**, 49–60.

Rogers, K. T., DeVries, L., Kepler, J. A., Kepler, C. R. and Speidel, E. R. (1960). Studies on chick brain of biochemical differentiation related to morphological differentiation. II. Alkaline phosphatase and cholinesterase levels and onset of function. *J. Exp. Zool.* **145**, 49–60.

Sanders, E. B. (1929). A consideration of certain bulbar, midbrain, and cerebellar centres and fibre tracts in birds. *J. comp. Neurol.* **45**, 155–221.

Scholander, P. F., Hammel, H. T., Lemessurier, H., Hemmingsen, E. and Garey, W. (1962). Circulatory adjustments in pearl divers. *J. appl. Physiol.* **17**, 184–190.

Showers, M. J. C. and Lyons, P. (1968). The avian nucleus isthmi and its relation to hippus. *J. comp. Neurol.* **132**, 589–616.

Schrader, M. E. G. (1892). Uber die Stellung des Groshirns im Reflexmechanismus des Zentralnervensystems der Wirbeltiere. *Arch. exp. Path.* **29**, 55–118.

Schupbach, P. (1905). "Beitrage zur Anatomie und Physiologie der Ganglienzellen Zentralnerven system der Taube." Dissertation. Berne.

Siegel, P. B., Isakson, S. T., Coleman, F. N. and Huffman, B. J. (1969). Photoacceleration of development in chick embryos. *J. Comp. Biochem. Physiol.* **28**, 753–758.

Starck, D. (1955). Die endokraniale Morphologie der Ratiten, besonders der Apterygidae und Dinorthidae. *Morph. Jahrb.* **96**, 14–72.

Stefani, A. (1881). Alcuni fatti sperimentali in contribuzione alle fisiologia dell' encefalo dei colombi. *Sep.-Abdr. Ferrara.*

Stopp, P. E. and Whitfield, I. C. (1961). Unit responses from the brain stem nuclei in the pigeon. *J. Physiol.* **158**, 165–177.

Tienhoven, A., van and Juhasz, L. P. (1962). Chicken telencephalon, diencephalon, and mesencephalon in stereotaxic coordinates. *J. comp. Neurol.* **118**, 185–197.

Turner, C. H. (1891). The morphology of the avian brain. *J. comp. Neurol.* **1**, 39–92.

Valkenberg, C. T. van (1911). Zur vergleichenden Anatomie des mesencephalen Trigeminusanteils. *Folia neuro-biol.* **5,** 360–418.

Vorgas-Pena, B. (1932). Contributo alla conoscenza dei rapporti tra sensibilita cutanea ed. attivita dei lobi ottici nel colombo. *Boll. Soc. ital. Biol. sper.* **7,** 762–765.

Wallenberg, A. (1898). Die secundare Acousticusbahn der Taube. *Anat. Anz.* **14,** 353–369.

Wallenberg, A. (1904). Neue Untersuchungen uber den Hirnstamm der Taube. *Anat. Anz.* **24,** 357–369.

Weinberg, E. (1928). The mesencephalic root of the fifth nerve. A comparative anatomical study. *J. comp. Neurol.* **46,** 249–405.

Zeglinski, N. (1885). Experimentelle Untersuchungen uber der Irisbewegung. *Arch. Physiol.* 1–37.

Zeigler, H. P. (1963). Effects of endbrain lesions upon visual discrimination learning in pigeons. *J. comp. Neurol.* **120,** 183–194.

# 11 The Diencephalon

## I. INTRODUCTION

The diencephalon or "between brain" corresponds to the third ventricle and its surrounding structures and is closely associated with autonomic control. During ontogeny it is derived from the hind part of the prosencephalon and in the early phases of embryonic life it lies in front of the cranial end of the notochord. It therefore probably represents an extension of the sensori-correlative centres beyond the limits of the ancestral, metamerically segmented, myotomal region. On *a priori* grounds one would not expect it, in this case, to show those imprints of both metamerism and branchiomerism which can be detected in more posterior structures. The lateral walls of the embryonic neural tube enlarge during development and give rise to the thalami which are large nuclear masses on either side. In contrast the diencephalic roof retains its epithelial character throughout much of its extent and forms the ependymal roof of the third ventricle. The sulcus limitans of His, which provides the topographical boundary to both the alar and basal plaques, curves downwards after entering the region and is widely considered to end at the pre-optic recess. This is a small evagination of the third ventricle situated in front of the optic chiasma. During ontogeny most vertebrates have another sulcus, the median, which runs from the sulcus limitans to the side of the ventricle, and separates the dorsal and posterior region from a rostro-ventral one. These comprise the so-called dorsal and ventral thalami. The former is a great diencephalic receptive area whose phylogenetic history is closely associated with that of the ascending lemniscal systems. Its relative development is usually a fair indication of that undergone by the forebrain. In contrast the ventral structures form an efferent centre.

Posterior to the diencephalic choroid plexus lies the habenular complex or epithalamus. Typically this is separated from the remaining diencephalic components by the sub-habenular sulcus. Also the *anlage* of the pineal and parapineal eyes appear on the diencephalic roof during ontogeny. The parapineal is of little significance in birds but the pineal has important control functions in relation to both hypophysial and gonadal activity. As a result of the massive topographical displacements consequent upon forebrain, tectal and cerebellar development the pineal is forced backwards. It is a relatively obscure dorsal component in many avian brains. Nevertheless it seems to share with the hypothalamic nuclei the control of automatic activity.

Benoit (1964) summarized those experiments which show that the gonadotrophic functions of the pars distalis are, in the duck, under hypothalamic
control. The long established fact that the hypothalamus controls neurohypophysial secretion, coupled with the known importance of hypothalamic
hypophysiotropic releasing factors in mammals (Guillemin, 1967), suggests
that such detailed hypothalamic control also occurs in birds.

Ten Cate (1936) reviewed the older experimental results and the effects which
diencephalic damage or ablation appeared to have on body temperature,
nystagmus and the skeletal musculature. Rogers (1918, 1919) found that
both the components of nystagmus disappeared below 25° and 30°C respectively, and that the thalamus was closely involved in the maintenance of

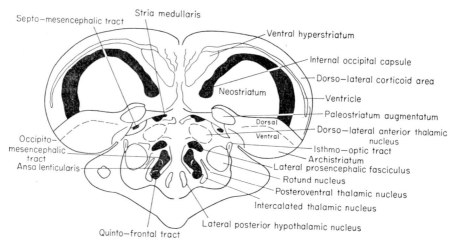

FIG. 101. A generalized section of the avian diencephalon and overlying structures.

homeothermy. Following telencephalic extirpation the body temperature
remained within the normal range. If such ablation is coupled with thalamic
cautery the body temperature falls to that of the surroundings. Rogers (1920)
also found that there was some thalamic control over what he considered were
vasomotor centres. Van Dijk (1930, 1931, 1932a, 1932b) postulated that the
thalamus was antagonistic in its effects to efferents in thoracic roots 1–5,
whilst Langworthy (1926), and also Bremer and Ley (1927), demonstrated
that it apparently had some involvement in postural control. After mesencephalic transection his preparations lay on their side with legs flexed. As was
stated in Chapter 6, Section III C Saalfeld (1936) described a specific panting
centre in the diencephalon. Transection of the region abolished panting.
Respiration rates fell from about 500/min to 60 but could regain their former
level after injections of the stimulant lobeline which was assumed to have an

effect on more posterior and probably medullary foci (see Sturkie, 1965; Richards, 1970 a,b).

## II. PREPARATIONS WITH COMPLETE TELENCEPHALIC ISOLATION

Amongst the older works only that of Thauer and Peters (1938) deals with preparations in which total forebrain extirpation left the thalamic region more or less intact. Several such preparations survived for up to 2 years and, although most of them were reported to show some lack of spontaneity, one underwent an almost complete recovery. Its spontaneous reactions were good, it could peck and eat its food, and very intensive observations were necessary to reveal that it had some optical agnosia and therefore had a slight difficulty in recognizing complicated objects. Much more recently Akermann *et al.* (1962) produced preparations with forebrain isolation by using high energy protons to achieve local irradiation lesions.

Ideal lesions which destroyed all links with the telencephalon were achieved in one group of preparations and located in the anterior ventral paleostriatum. Such lesions resulted in the destruction of all the neural tissues in the central part of the area irradiated, and in some preparations this necrotic area was surrounded by an irregular border some 0·5–1·0 mm wide in which the neurones were in varying stages of degeneration. Those cases which are of interest at this point had lost the paleostriatum, the nuclei of the ventromedial forebrain and the septal area. However there were no signs of damage to either the thalamus or hypothalamus. In these preparations both spontaneous eating and spontaneous drinking disappeared on the first post-operative day. As a result the birds lost weight. However the swallowing reflex was normal and peas placed in the bill were easily swallowed. The escape responses also disappeared at once and the preparations never flew spontaneously after 1–4 days. Nevertheless if they were thrown into the air they were able to fly normally, to avoid obstacles and also land perfectly. They were also able to avoid obstacles when they were induced to walk, but when not disturbed they sat motionless with feathers ruffled and eyes closed. None exhibited any of the characteristic behaviour patterns associated with fighting or courting. Direct stimulation of the thalamic and hypothalamic regions by way of implanted electrodes activated them. Under these circumstances they exhibited both ipsiversive and contraversive deviation of the head and trunk, coupled with raising and lowering of the head, and undertook circling movements. It was possible to elicit walking, eye closure and pupillary contraction. In some cases such stimulation resulted in the erection of the neck feathers, preening and defaecation. Although these head and locomotory movements simulated the normal patterns of attention and searching the preparations remained largely unaffected by external stimuli such as food and water. It would therefore appear that in the presence of activating stimulation the post-telencephalic structures can coordinate the visual and motor patterns involved in flight and also all the important reflexes. Tuge and Shima (1959) have shown that the paleostriatum appears to be necessary for conditioned reflex formation see (Chapter 14).

## III. THE THALAMUS

### A. General Considerations

Allowing for the effects of crowding then, together with other diencephalic structures, the avian thalamus is comparable with that of extant Diapsida. The most useful early work on neuroanatomy is that of Edinger and Wallenberg (1899), but other studies include those of Bellonci (1888), Perlia (1898), Singer and Munzer (1890), Kappers (1921), Ingvar (1923), Groebbels (1924), Rendahl (1924), Craigie (1928) and Huber and Crosby (1929). The principal genera which were studied by these workers were *Gallus*, *Columba*, *Chrysolampis*, *Chlorostilbon* and *Passer*, together with various Anseriformes and Psittaciformes. Amongst recent workers Karten (1963) investigated the wider relationships and found that in his preparations, which had undergone spinal hemisection at $C_{12}$–$C_{14}$, degenerating axons could be seen penetrating the tectal and posterior commissures and entering the thalamus. At that level they terminated diffusely in the ventral portions of the anterior and posterior dorsolateral nuclei. Occasional degenerating fibres could also be seen in the dorsomedial nucleus, but none could be clearly identified in others such as the ovoidal or rotundal. The positions of such nuclei can be seen in Figs 94 and 101.

The nuclei of the thalamic region can be very large and the rotundal is one of the most conspicuous and well-described. It shows a disproportionate development by comparison with the situation in diapsid reptiles. This was, in the past, widely ascribed to the outstanding importance of both the vestibular and cochlear regions in animals which were bipedal and also flew.

### B. The Rotund Nucleus

#### (i) *Neuro-anatomical Studies*

The rotund nucleus has been widely accepted as the avian homologue of the ventral thalamic nuclei in mammals. It was identified by Stieda (1869) and, apart from Bellonci (1888) and Turner (1891), many of the subsequent workers agreed with the implied homologies (Edinger, 1908; Kappers, 1921; Groebbels, 1924; Craigie, 1928, 1930, 1931; and Huber and Crosby, 1929).

All the conclusions of such early workers have been reviewed by Kappers *et al.* (1936), Kappers (1947), Cordier (1954) and Portmann and Stingelin (1961). In spite of the very definite shape which is implied by its name the actual appearance of the nucleus varies somewhat from genus to genus and sometimes assumes an irregular or plano-convex outline. The triangular nucleus is closely adherent to its dorsal surface for much of its extent whilst ventrally it overlies the tecto-thalamic tract. Slightly medial to this last is the postero-ventral thalamic nucleus. Throughout much of its length the lateral aspect of the rotundal is applied to the principal precommissural nucleus which was mentioned in connection with the pretectal region. From its medio-dorsal position between the medial and lateral spiriform nuclei this curves round to form a curved plate lying ventro-laterally in relation to the rotund nucleus. On the medial surface a further discrete, but often closely associated, diagonal cell mass, forms the sub-rotund nucleus. In the pigeon this lies obliquely across below the ovoidal nucleus in its hindmost regions and then assumes a more diagonal position further forward. It is therefore closely apposed to the occipito-mesencephalic tract, the ovoidal nucleus, the ansa lenticularis and, at its ventral extent, the postero-ventral thalamic nucleus. Posteriorly both the rotund and sub-rotund nuclei can be more or less confluent with this postero-ventral nucleus in, for example, Trochiliformes and Psittaciformes.

The large multipolar cells of the rotund nucleus itself are associated anatomically with a number of axonal bundles. Included amongst these are the bulbo-thalamic tract, the lateral thalamo-frontal tract, the tecto-thalamic tract and the dorsal tecto-thalamic tract. Close functional relationships between the tectum and rotund nucleus were suggested by the stimulation studies of Phillips (1966) and Karten (1965). This intimates that it is a major relay station within the central visual pathway. (See Chapter 10, Section III E.)

## (ii) *Lesions of the Rotund Nucleus*

After training 16 pigeons to peck at one of two discs on to which visual stimuli were projected, Hodos and Karten (1966) investigated the effect of bilateral electrolytic lesions within the rotund nucleus. Preparations with such lesions exhibited severe deficits in their visual discrimination performance. They were grouped into four sets depending upon the location and extent of their particular lesions. The first group of four specimens had some direct damage to the nucleus itself together with extensive direct damage to other neighbouring regions. All the preparations had undergone bilateral destruction, with or without cell loss, of at least 75% of the nucleus and as the result of marked post-operative shrinkage, more medial structures had also undergone some lateral displacement.

The effects of electrode passage through overlying structures and the fore-brain were investigated using the lesions in a second, or dorsal thalamic, group of preparations. These had no retrograde degenerations within the nucleus itself. A third set had little *direct* damage to the rotund nucleus but, as the result of the destruction of either afferent or efferent pathways, there was a considerable amount of postmortem histological evidence for cellular deterioration within it. Hodos and Karten termed these the secondary nucleus group. In two further specimens where the rotundo-ectostriate tracts

## TABLE 45

*Summary of the behaviour data obtained both before and after lesions to the nucleus rotundus etc.*

Problems: 1, brightness; 2, bars; 3, solid triangles; 4, dotted triangles.
+, reached 10% error level of performance.
−, failed to reach 10% error level in performance.

| Lesion group | | Pre-operative performance at problems 1 2 3 4 | Post-operative performance at problems 1 2 3 4 | Per cent saving | Per cent loss |
|---|---|---|---|---|---|
| Direct | 1 | + + + − | + − − − | | 100 |
| nucleus | 2 | + + + − | + − − − | | 100 |
| rotundus | 3 | + + − + | + + + − | | 17 |
| | 4 | + + + − | − − − − | | 100 |
| Secondary | 1 | + + + − | − − − − | | 100 |
| nucleus | 2 | + + + − | + − − − | | 100 |
| rotundus | 3 | + + + − | + + + − | | 100 |
| Dorsal | 1 | + + − + | + + − + | 73 | |
| thalamus | 2 | + + − + | + + + − | 69 | |
| | 3 | + + + − | + + + − | 100 | |
| | 4 | + + + − | + + − + | 73 | |
| Sham | 1 | + + + − | + + + + | 100 | |
| operates | 2 | + + + − | + + + − | 100 | |
| etc. | 3 | + − + + | − + + + | 100 | |
| | 4 | + + + − | + + + − | 100 | |
| | 5 | + + + − | + + + − | 90 | |

From Hodos and Karten (1966).

were more or less destroyed there was appropriate bilateral cell loss. In another preparation the tecto-thalamic tracts were destroyed. These had been identified as the main, and possibly the only ascending projection. It was scarcely surprising in this case that following their extirpation there was gliosis along the entire length of the tract and within the nucleus. The remaining birds formed a heterogeneous group of controls. They included sham operates and preparations with basal forebrain, septo-mesencephalic tract, septal and posterior mesencephalic ablations. The results of discrimination tests on all of these varied preparations and specimens are summarized in Table 45. A bird which required the same number of trials to reach criterion after the operation as it had done previously was scored as showing a 0% saving. If the number of post-operative trials was less than previously needed it was scored as a saving, if more, it was scored as a loss.

It is fairly clear from the data of Table 45 that bilateral lesions to both the nucleus itself, and also to its afferent or efferent tracts, result in severe deficits which involve both pattern and brightness discrimination. However a prolonged period of post-operative training gave a gradual return to pre-operative levels in most of the preparations. The observed losses could not, therefore, be attributed to overt blindness. In any case they had no difficulty in finding food and water and did not require further retraining to actually peck at the disc itself to obtain food. Furthermore the pupillary reflexes appeared to be normal (see Chapter 10, Section V). It was by no means clear whether the observed deficits should be attributed to an impaired storage of visual data, or to an altered perception of visual stimuli. In general it was the performance in brightness discrimination tests which showed the fastest post-operative improvement. Although pattern discrimination did improve, it did so at a far slower speed. There were also some indications that those problems which were most difficult to learn in the first place were the ones which showed the greatest post-operative loss. Whether the subsequent learning involved other and alternative sites within the brain, and whether the post-operative improvement was actually a function of the retraining, or just reflected the passage of time after the operation, are questions that remain unanswered.

Those preparations which had undergone lesions in regions situated above the nucleus exhibited a transient regression in the brightness discrimination. This was, however, trivial by comparison with the deficits which were seen in the *rotundus* group. They also showed a rather noticeable deterioration in their ability to carry out the more difficult pattern discriminations. Although this was neither as severe nor as prolonged as that seen in the rotundus groups it was considerably more severe than that of the controls. Nevertheless, in spite of this, their final post-operative performance still showed an average saving of 79%. It is fairly clear that such data conform with, and corroborate

Karten's (1965) suggestion that the rotund nucleus is a major relay station within the tecto-telencephalic projections of the central visual pathway. They are also themselves substantiated by the visual intensity and pattern discrimination deficits after ectostriatal lesions (Hodos and Karten, 1970; Karten and Hodos, 1970). As a result Hodos and Karten were strongly tempted to equate it with those mammalian thalamic nuclei, such as the posterio-lateral and caudal pulvinar, which are said to receive tectal projections.

## (iii) *Electro-physiological Investigations of Rotundal Relationships*

In his studies of the ascending pathways which are associated with auditory and tactile stimuli Erulkar (1955) explored a considerable amount of the rotund nucleus. In no such experiments were any electrical responses recorded from thalamic sites. Tactile stimulation over the whole body surface, and direct electrical stimulation of the sciatic nerve, also failed to evoke any responses. Although his attention was principally directed towards the rotund nucleus, his probes had to pass through the dorsal group of nuclei and all of these, including the parvocellular superficial nucleus (= the nucleus of the septo-mesencephalic tract) remained silent in the face of tactile stimulation. His exploring electrodes were actually sited at 0·5 mm intervals and the possibility remained that he had missed some very small tactile centre. However this possibility was totally excluded in the case of both the rotund nucleus and also the dorsal thalamic group. He therefore concluded that these did not comprise relay stations within the tactile system and that the rotundus is not the avian homologue of the mammalian ventral thalamic group. This is corroborated by both the lesion experiments of (ii) above, and more recent electro-physiological investigations.

Stimulation of the rotundus has demonstrated the close functional relationships which exist between it and the tectum. Potentials of 75 $\mu$V can be recorded in the tectum after a latent period of 1–3 msec. In the complementary situation stimulating the optic tectum is followed after 2–3 msec by 150 $\mu$V potentials at sites around the rotund nucleus. As the first response peak persists whilst the other disappears at repetition rates of 100 pulses/sec Phillips (1966) suggested that both direct and poly-synaptic connections exist. In fact photic stimulation regularly evokes short latency responses within the nucleus. Phillips also took the opportunity of confirming Erulkar's results and showed that no such activation follows tactile or auditory stimulation.

Turning to further ascending projections Revzin and Karten (1967) systematically explored the telencephalon for slow potentials and evoked unit discharges whilst simultaneously stimulating the rotundus. The largest evoked potentials were detected within the ectostriatum and the immediately adjacent areas of the intermediate neostriatum. Large potentials which differed slightly from these were also recorded within the hyperstriatic lamina, in that part of the ventral hyperstriatum which overlies the ectostriate components, and in the lateral regions of the paleostriatum. As far as these last are concerned it seems unlikely that they represent a major projection to the paleostriatum itself because much of the activity occurred in fibres which were traversing the part of the paleostriatum augmentatum which is both below, and closely adjacent to, the ectostriate regions. These correspond to the lateral part of the thalamo-frontal tract which is the long established rostral projection from the rotundus. Small and variable, long latency responses also occurred in the ectostriatum after stimulation of the thalamic regions above the rotundus but short latency ones were absent. If the situation was reversed and the stimulating current applied to the appropriate telencephalic regions no large responses occurred in the rotundus. Following stimulation of the ectostriatum small rotundal potentials did occur but these only comprised inconstant spike-like waves which could well represent antidromic discharges along rotundo-ectostriate fibres. Revzin and Karten therefore concluded that there was no reciprocity within the rotundo-telencephalic system of the pigeon. As a similar distribution of ectostriate potentials also followed tectal stimuli they extended this conclusion to encompass the whole tecto-rotundo-ectostriate system. Furthermore the elaborate telencephalic projections of the rotund nucleus which were described by Huber and Crosby (1929) on the basis of anatomical investigations are in no way substantiated by these electrophysiological experiments.

## C. The Ovoidal Nucleus

It will probably be clear from Section B above that the ovoidal nucleus has a fairly close relationship with the sub-rotundus in the middle to forepart of the formers extent. In this region it lies medial to the triangular nucleus, from which it is separated by the obliquely orientated thalamo-striatic tract. Further back it is situated above the occipito-mesencephalic tract and below the dorso-medial, dorso-intermedial and dorso-lateral, posterior thalamic nuclei. In this region the sub-rotund nucleus is more horizontally disposed and the fibres of the ovoidal nuclear tract run almost vertically downwards to the ansa lenticularis and quinto-frontal tract.

According to Karten (1968) Cobb has shown that there is a close relationship between the size of the ovoidal nucleus and that of the dorsal part of the lateral mesencephalic nucleus (*q.v.*) which is the probable homologue of the torus semicircularis. In Chapter 10, Section IV A it was shown that this last is of a large size in relation to the optic tectum in *Steatornis* and *Otus*, and, although relatively smaller, still large in the echo-locating swiftlets. In *Steatornis caripensis* and the owl *Cryptoglaux acadia acadia* the ovoidal nucleus is also large. Indeed in these two species its size is closely comparable with that of the rotund nucleus. The studies on the lateral mesencephalic nucleus implicated it in the auditory system. The same is true in the case of the ovoidal nucleus which remains the suggested, but unproven, homologue of the mammalian medial geniculate body.

As it had previously been suggested that efferent projections from the ovoidal nucleus passed within the tract of the same name Karten (1968) searched there for degenerating axons after nuclear extirpation. Only a very small number could be found and even these could not be traced backwards with any great degree of certainty. Although, many small, silver granules were present on both sides of the dorsal part of the lateral mesencephalic nucleus there was no more definite evidence for projections to it. Far more conclusive results were associated with rostral connections to the telencephalon.

After both complete and incomplete nuclear extirpation it became abundantly clear that the fibres passing towards the forebrain hemispheres assemble at the side of the ovoidal nucleus. They then contribute to the medial part of the lateral prosencephalic fasciculus. It is in the lateral part of this that the rotundo-ectostriate fibres of Section III B (iii) above pass forward. As seen in degeneration studies the rostrally directed axons from the ovoidal nucleus then pass dorsally, in front of the anterior commissure, and into the hind part of the hemispheres. The majority seemed to pass through the mid-part of the paleostriatum primitivum and the paleostriatum augmentatum (*q.v.*). They then crossed the dorsal medullary lamina to enter the neostriatum. Here they entered the medial part of the thalamo-frontal tract, which Karten refers to as the internal occipital capsule, and ended in great density on its intercalated cells. Although many degenerating fibres were visible in the lateral part of the caudal neostriatum only insignificant terminations occurred there. In contrast there were massive axonal and terminal degenerations within a clearly demarcated medial area of the caudal neostriatum which is both adjacent to, and continuous with, the tract.

This sharply defined area of small cells is closely coincident with field L of Rose (1914). In front it is separated from the hyperstriatic lamina by a layer of larger cells situated on its mediodorsal side. The anterio-lateral margin is apposed to the medial limits of the ectostriatum and, indeed, a few terminations were detected in that nucleus. A reciprocal study in which lesions were produced in the caudal neostriatum gave complementary data. There was extensive retrograde degeneration and gliosis in the ipsilateral ovoidal nucleus. However as the hemisphere lesions invariably included some damage to both the hyperstriatum and the paleostriatum these results can not be taken as definitive corroboration.

## D. Other Thalamic Nuclei

A number of other nuclei have been mentioned in the foregoing sections and it will be useful to briefly outline their relationships. The area which lies between the septo-mesenphalic tract and the brain mid-line includes a well-defined group of structures which are sometimes grouped together as the dorsal nucleus. More specifically they are the dorsolateral and dorsomedial thalamic nuclear groups. The lateral component abuts on the nucleus of the septo-mesencephalic tract itself, the superficial parvocellular nucleus. Together both dorsal units constitute an aggregation which is as well-recognized as the rotund nucleus (Portmann and Stingelin, 1961). Their positions can be seen by reference to Figs 101 and 103. Certain workers have indicated that the appearance of the whole complex varies somewhat as between members of the Anseriformes, Apterygiformes, Columbiformes, Galliformes, Psittaciformes, Trochiliformes and Passeriformes. For example in *Passer* the chromophilic cells which constitute the dorso-lateral unit are clearly associated into two aggregations. No such division occurs in the Kiwi, and in the humming birds which Craigie studied (*loc. cit.*) the whole complex presents a unified appearance. As in the case of the rotund nucleus (Section III B (i) above) the principal tracts which are associated with them in older anatomical works are the tectothalamic, and both lateral and medial thalamo-frontal tracts. The presence of activation potentials following the presentation of peripheral click stimuli implies that they are concerned with ascending auditory impulses but the long latencies which are involved render it unlikely that they are principal intermediate relay stations. However, once again, one can note that the existence of alternate poly-synaptic routes within the brain may well add to the precision with which sources of sonic stimuli can be located, and their content analysed (see Chapter 5, Section III). In this case it is unwise to assume that foci with less direct ascending pathways are relatively unimportant in hearing. In view of the accuracy with which birds can locate others which are singing, such alternative routes to the main auditory centres may be of critical importance.

As was implied above the most lateral part of the dorsal diencephalic margin bears, throughout its length, the superficial parvocellular nucleus which retains an intimate association with the septomesencephalic tract (Fig. 101). Anteriorly it is either round or elliptical when seen in transverse sections. This appearance merges into a triangular or laminar form further back and it has been described as fusing with another laminar cell mass, the lateral nucleus, in its more anterior region (Kappers *et al.*, 1936). It is certainly fair to say that in the region of this last-named structure both, together with the magnocellular component of the dorsolateral anterior thalamic nucleus, can be in very close proximity. These adult relationships appear to be reflected during ontogeny and in *Columba* the lateral nucleus was said to be confluent with its neighbour until the 10th day of incubation. The older authors described it in *Apteryx* and various members of the Anseriformes, Galliformes, Psittaciformes, Trochiliformes and Passeriformes. The shape seems to vary from taxon to taxon but in all cases its fibre connections have been said to indicate a functional relationship with its partner and, by means of collaterals, with the isthmo-optic tract and the lateral habenular nucleus.

The so-called geniculate region of the avian brain is controversial. One suggestion which has been widely canvassed would make it the avian homologue of the lower part of the mammalian lateral geniculate body. In mammals this structure receives a high proportion of the faster conducting fibres associated with the optic tract. In birds there is a band of markedly chromophilic cells surrounded on their ventro-lateral side by more diffusely scattered and smaller cells. Appearing at the level of the tract which is associated with the ovoidal nucleus it is, at that level, not far distant from the mid-line and close to the basal optic root. This component is frequently referred to as the ventral part of the lateral geniculate nucleus. Further forward a second component is present immediately to the side of the first. It underlies both small-celled and large-celled parts of the mesencephalic lentiform nucleus, and is referred to as the principal part of the lateral geniculate. Its longitudinal extent is limited and it is the ventral part which runs forward to end rostrally just in front of the magnocellular part of the dorsolateral anterior thalamic nucleus. As it passes forward it moves to an oblique lateral position under the ventro-lateral thalamic nucleus. Karten (1963) found that it received some ascending fibres from the spinal cord. Due to the compression of the avian mesencephalon and diencephalon all these thalamic cell masses are closely juxtaposed. Furthermore an additional series of relatively scattered cell groups is distributed along the lateral forebrain bundle. These are subject to considerable topographical variation. Sometimes they are associated into discrete endo-peduncular nuclei; at other times they are more diffuse. When they are associated together the overall histological appearance of the resulting structure is very similar to that of the neighbouring paleostriatum primitivum.

## E. Auditory Stimulation of the Thalamic Region

Influenced by the anatomical suggestions of Papez (1929) Erulkar extended his investigation of the central effects of click stimuli to include the ovoidal nucleus. As his attempts to elicit activation potentials at this focus were uniformly unsuccessful, he was forced to conclude that neither it, nor other thalamic nuclei, were involved in the ascending auditory pathways. It is such results that are reflected in Schwartzkopf's (1968) ambivalent position about supra-medullary auditory relay stations. However the more recent electro-physiological studies carried out by Harman and Phillips (1967) are quite conclusive and corroborate the lesion experiments which were outlined above. They found that following the presentation of peripheral click stimuli both ipsilateral and contralateral thalamic potentials could be detected. These were usually simple monophasic effects and, if other components were observed, they were inconsistent, faded rapidly following repetitive stimulation, and had a markedly lower amplitude than the principal response.

In general the highest amplitudes occurred within the dorsolateral and dorsomedial thalamic nuclei but these had longer latencies than those in the ovoidal nucleus. Following the presentation of click stimuli to the contralateral ear there was a 50–120 $\mu$V wave in the lateral part of the area and this had a latency of 10–16 msec. Ipsilateral stimuli gave slightly different results with 20–170 $\mu$V amplitudes after latent periods of 12–20 msec. Responses from medially situated areas were usually more complex than those which were obtained in the lateral area. The initial responses varied from 50 to 100 $\mu$V in amplitude and had latencies of 14–16 msec. Later components of such potentials had smaller amplitudes varying between 10 and 60 $\mu$V, and followed at 20–30 msec. Those which resulted from ipsilateral clicks tended to have amplitudes in the lower parts of the range and to follow after longer latent periods than those evoked by contralateral stimuli.

In contrast to the foregoing the evoked potentials which could be detected within the ovoidal nucleus were less complex and had shorter latent periods. The amplitudes generally ranged from 50 to 80 $\mu$V and they followed some 10–12 msec after presentation of the stimulus to the ear. At its lateral edge ipsilateral clicks resulted in waves of 10–100 $\mu$V after latencies of 9–12 msec. In the region lying directly below the ovoidal nucleus, and either within or in close proximity to its tract, peripheral stimuli presented to either ear were followed after 6–11 msec by waves with an amplitude of 30–150 $\mu$V. Even further down, and where the mesencephalon meets the supra-optic commissure, yet shorter latencies were recorded. In this region, which Harman and Phillips assumed from lesion experiments to be afferent to the ovoidal nucleus, the potentials had amplitudes of 65–175 $\mu$V, and latencies of 4·5–7 msec. Such responses faded rapidly with stimulation frequencies of 10–20/sec and none could be detected at frequencies in excess of 20/sec (see also Biedermann-Thorson, 1970a,b).

Although the rotund nucleus itself exhibited no activity, responses which were broadly similar to those of the ovoidal nucleus could be observed in regions which were either medial to it, or below it. These short latency responses never had amplitudes in excess of 40–50 $\mu$V. Other even smaller responses were also recorded sometimes. In these cases the amplitudes fell between 10 and 30 $\mu$V and the latencies were comparable with some of those in the dorsomedial and dorsolateral nuclei. They generally involved periods of 14–17 msec.

Within the diencephalic components of the lateral forebrain bundle responses to contralateral stimuli had latencies of 8–12 msec and amplitudes of 30–100 $\mu$V. On some occasions analogous potentials also occurred after ipsilateral presentation. Repetitive clicks up to a limit of 8/sec all elicited such responses but as the frequency rose above this level they disappeared. Indeed the second of a pair of sequentially presented clicks elicited no response if the interval separating it from the other was less than 10 msec. All these results are relevant to Chapter 14, in connection with those ascending auditory pathways which evoke responses in the caudal neostriatum.

## IV. THE EPITHALAMUS

Above all the foregoing structures, and dorso-medial to the septo-mesencephalic tract (Fig. 101), there are two relatively small habenular nuclei and an underlying sub-habenular nucleus. During early embryonic life the more lateral of the two is a collection of small cell-bodies which is traversed by axons and situated amongst the component fibres of the medullary stria. Although it is initially quite distinct from the medial unit some of the latters cells can subsequently overlie it so that the medial nucleus can be divided into dorsal and ventral components. Both structures appear posteriorly close to the level at which the posterior commissure disappears and at which the pretectal-subpretectal tract occurs (q.v.). They then run forward. The lateral does not persist for any great distance in front of the parvocellular part of the mesencephalic lentiform nucleus but the medial continues in front of the anterior end of the posterio-lateral hypothalamic nucleus.

The compound medullary stria was considered by the older anatomists to provide the majority of the axons which reach the habenular complex. Within it fibres which originate in the septal regions and elsewhere form a more or less well-defined septo-habenular tract. A reported connection with the taenial nucleus constitutes the taenio-habenular tract, and there are also fibres linking the epithalamic region with both the pre-optic and archistriatal regions. Other axons pass between the habenular nuclei, the tectum and the interpeduncular nucleus.

## V. THE HYPOTHALAMIC REGION

The overall appearance and disposition of the avian hypothalamic nuclei is not unduly dissimilar from those of reptiles. The pre-optic nucleus has long been known to have at least two components and to extend up to the level of the endopeduncular nuclei. The paraventricular nucleus was described as confluent with it in the ratite birds, but separate in the carinates. Kappers (1947) summarized the diverse earlier descriptions. Afferent projections to the large-celled paraventricular component were considered to include septo-paraventricular fibres as well as others from the anterior dorso-medial nucleus of Huber and Crosby (1929) and Craigie (*loc. cit.*). The efferent connections, hardly elucidated at all prior to the publication of Wingstrand (1951), were thought to resemble those of reptiles. A number of other cell-groupings were also thought to have an autonomic involvement, to be related to spinal autonomic relays, and to be concerned with peripheral homoiothermal control.

Posteriorly, and situated in a position close to the midline, the lateral and medial posterior hypothalamic nuclei appear along with the lateral hypothalamic nucleus in transverse sections. All three are situated medio-ventrally in relation to the quinto-frontal tract, the ansa lenticularis and the occipito-mesencephalic tract. Directly above the posterio-medial component lies the magnocellular periventricular nucleus. Passing forward the posterio-lateral disappears and the lateral plus the posterio-medial are underlain by the dorsal and ventral supra-optic decussations. In the pigeon the lateral nucleus (*sensu stricto*) persists into more anterior regions and overlies the optic chiasma. As the supra-optic decussation is reduced the others are replaced by the anterio-medial hypothalamic nucleus and the large-celled periventricular nucleus moves ventrally. Even nearer to the rostral pole the chiasma is overlain by the supra optic nucleus which has first of all an oblique dorso-medial to latero-ventral position, and then, in the forward part of its extent, lies horizontally. Its dorsal tip is closely apposed to the large celled paraventricular nucleus to the side of which is the medial pre-optic nucleus. When the supra-optic nucleus has assumed a horizontal orientation in transverse sections the anterior pre-optic nucleus forms a diffuse mass lying above it and ventro-medial to the septo-mesencephalic tract. The relationships of the supra-optic and related cell masses are considered in greater detail below (see Chapter 12, Section III). Accounts of the inter-relationships of a variety of non-nervous factors are provided by Sturkie's review of avian physiology (1965) and the wealth of articles on endocrine integration.

# VI. GENERALIZED CONTROL FUNCTIONS OF THE DIENCEPHALON

## A. Respiratory Control

Following the older authors which were cited in Section I above, a number of workers besides Akerman *et al.* (1962) have investigated diencephalic functions using the standard electro-physiological techniques. An overall picture which is comparable with that suggested by Akerman *et al.* was provided by Vedyaev (1963, 1964). Using 55 stimulatory foci within the thalamic and telencephalic areas he found that the responses could be broadly assigned to three distinct categories. First of all there were some, primarily motor, mixed effects. He thought that these were often of the "through path" type and resulted in reactions which did not involve a change in the ventilation rate even when the stimulation level was gradually increased. In all, these made up 28% of the total results and were particularly frequent when the anterior and ventral thalamic regions were stimulated. Secondly an intensification of respiratory movements occurred in 58% of the experiments. This was noticed in relation to various thalamic nuclei and the responses were of two principal types. In simple cases the movements were accelerated during the actual period of stimulation but returned to the normal rate when it ceased. In contrast markedly diphasic responses also occurred on occasions. In these the period of acceleration which accompanied stimulation was followed by a subsequent period of quiescence during which ventilation seemed to be inhibited. It is clearly difficult to interpret this solely in terms of diencephalic activity in view of the apnea which can follow hyperventilation. However, more definitive inhibitory effects appeared to accompany stimulation of the posterior thalamic regions and also both the paleostriatum augmentatum (= mesostriatum) and the strio-cerebellar tract. Such inhibition was observed in 14% of the total cases and resulted in an arrest of respiration for 5–7 sec. However, in view of the conclusions which have, in recent years, superseded the earlier conceptions of discrete control foci within the brain, Vedyaev was loathe to assume that any strictly localized reflex areas existed. Instead he suggested that the units which are involved in central respiratory control are diffuse. Furthermore a comparison with conditions in lagomorphs suggested that inhibitory mechanisms are less well represented in the avian brain, at least within the areas which he studied.

## B. Thermo-regulation

During the nestling period the young of altricial birds undergo a striking change from their initial poikilothermy to the adult homoiothermy. This has, of course, been of paramount importance in the evolution of parental behaviour. According to Dawson and Evans (1957) the establishment of functional nervous control over the processes of heat production is complete by the 5th and 6th days after hatching in *Spizella pusilla pusilla* and *S. passerina passerina*. However, the actual speed with which such effective temperature control develops when the young are experiencing moderate environmental temperatures varies from species to species. In some it actually precedes, at least in part, the growth of feathers. In the house wren it occurs at 9 days; in the red-backed shrike, *Lanius collurio*, at 10 days; a similar period was cited for *Melopsittacus undulatus*, whilst 11 days was suggested for the wryneck *Jynx torquilla* (Baldwin and Kendeigh, 1932; Kendeigh, 1939; Boni, 1942; and Dawson and Evans, 1957).

There is some evidence that the direct cooling of the avian brain, or the stimulation of cutaneous cold receptors, gives an increased level of heat production. By means of ablation studies Rogers and Lackey (1923) demonstrated such effects to their own satisfaction and found that they were absent after thalamic extirpation. Following his own long series of experiments Rogers (1928) also reported that both shivering and an increase in body temperature followed if the intact thalamus was cooled. Peripheral studies in which Randall (1943) inserted a cooling tube into the cloaca of 7-day-old chicks have also been used as evidence for central temperature control since, although no cutaneous temperature change occurred, shivering was elicited. Clearly, as Whittow (1965) pointed out, such results are equivocal. The actual deep temperature at which such shivering appears is variable. If the specimens are first of all exposed to experimentally induced hyperthermia the subsequent deep body temperature which results in shivering can be lower than normal. Howell and Bartholomew (1962) also demonstrated similar responses in nestlings of the red-tailed tropic bird and the red-footed booby. Exposure to a cold environment quite clearly produces shivering and this before there is any detectable drop in the deep body temperature. In consequence they thought that cutaneous cold receptors were predominantly involved in the mediation of the resulting reflexes. Such suggestions are not in themselves particularly conclusive. It would be surprising if local receptors were not the first to be involved following exposure to low temperatures whether at the skin surface or in the cloaca. Nevertheless bilateral vagotomy seems to prevent such increases in heat production in the pigeon (King and Farner, 1961) and thereby implicates the vagus as an integral pathway mediating the "reflexes".

The converse situation has also been investigated. Thalamic extirpation did not produce thermal polynea in the pigeon according to Rogers and Lackey (1923), and Sinha (1959) located an area in the mesencephalon that abolished any existing panting. Bilateral lesions within the hypothalamic region were also found to have a comparable result by Feldman *et al.* (1957). Behavioural responses to heat disappeared in preparations with such lesions. Randall (1943) asserted that in the absence of an increase in temperature at cephalic thermoregulatory centres birds, unlike mammals, cannot be induced to pant by elevating the cutaneous temperature. He based this on experimental situations in which a circulation of cold water around the neck region of a 7-day-old chick was accompanied by a simultaneous increase in the peripheral temperature. Yeates *et al.* (1941) always observed an increase in cloacal temperature prior to the onset of panting although El Hadi (1969) has described certain exceptions to this. Richards (1970a, b) reported that at hermal delay of about 10 min occurs between the brain and colon in fowls subjected to rapid changes of ambient temperature and that panting never precedes the rise of hypothalamic temperature. He therefore viewed with suspicion the reported peripheral initiation of gular flutter in tropic birds and boobies. The various effects of bilateral vagotomy upon such activities were discussed in relation to the nervous control of respiration (see Chapter 6, Section III (C)). The significance of species differences and also the specific roles of vagal pathways which are involved in thermal polynea remain to be elucidated. Further examples of hyperthermia after diencephalic, and in this case hypothalamic, lesions are cited in association with experimental aphagia below and many natural responses are cited in Sturkie (1965).

In recent years Richards (1970b) found that, whilst panting certainly resulted from square wave stimulation of the fowl mesencephalon somewhat dorsal to the red nucleus (see also Akermann et al., 1960), it was not possible to elicit it in this way from the anterior hypothalamus in areas where such stimulation gives positive results in mammals. This in spite of the fact that lesion studies suggest that such an area exists (Feldman et al., 1957; Kanematsu et al., 1967; Lepkovsky et al., 1968). Additional confirmatory evidence which contradicts the electro-physiological studies was also provided by McPhail (1969). He found that in the ring dove panting sometimes ensued during cholinergic stimulation of the hypothalamus. The involvement of the hypothalamus, and more particularly the anterior region, in the organization of panting would then seem to occur in birds as in mammals. In some species polynea is apparently dependant upon afferent impulses transmitted to the respiratory centres by the vagus nerves, in others vagotomy has little or no effect and the central mechanisms alone seem capable of maintaining the rhythm. Forebrain hemisphere involvement in respiratory activity is also suggested by the data of Chapter 14, Section VIII (C).

## C. Cardio-vascular Responses

In their study of the cardio-vascular responses which accompany, and occur during, diving, Feigl and Folkow (1963) found that stimulation of an area lying 1–2 mm to the side of the mid-sagittal diencephalic plane gave a marked increase in the blood-flow to muscles. Characteristically there was a slight tachycardia and such increases in heart rate were not unnaturally associated with rises in blood pressure. Contemporaneous respiratory changes varied. Atropine blocked the vasodilatory effects but left both the blood pressure and tachycardia unaffected. This would conform with the involvement of a cholinergic dilator mechanism. Together with the sympathetic excitation observed in the other components it compares fairly closely with the defence reaction of, say, a cat. In this context it is worth remembering that under certain conditions a diving bird will dive when alarmed and in these circumstances a diving reflex would be appropriate as a defence reaction. On other occasions overt flight results in which case the cardio-vascular pattern which was observed by Feigl and Folkow would possibly be more relevant.

## D. The Intake of Food and Water

Preparations which have been exposed to a number of operative manipulations subsequently fail to eat and drink normally. A number of such conditions have already been mentioned, and others will be dealt with in association with the forebrain hemispheres. More specifically Feldman *et al.* (1957) found that bilateral lesions within the medial and lateral hypothalamic regions caused aphagia. This was accompanied by disturbances of the general level of activity, water metabolism, heat regulation, reproduction and the homeostatic balance of body composition. Some of these aphagic preparations subsequently resumed feeding spontaneously but 12 failed to do so. In another group of preparations tube-feeding was followed by weight increases and gastro-intestinal function appeared to be normal. However, the importance of diencephalic structures in autonomic regulation is shown by the fact that two of them died during a hot-spell in which their body temperature reached 43·5°C. One of the remaining four also failed to regulate its temperature, did not pant, spread its wings or ruffle its feathers. When tube-feeding was discontinued such preparations did not resume pecking to any great extent although some did make incipient pecks at both food and other objects, and two drank water. Cases of experimental hypophagia are mentioned in Section VI E below.

During periods of exposure to a hot environment heat loss is almost entirely a product of moisture evaporation. This, together with the hyperthermal effects just noted, would lead one to expect on, *a priori* grounds, that the areas controlling such responses might be functionally or topographically related. When studying the central control of body temperature, feeding and water intake in pigeons Akerman *et al.* (1960) evoked polydipsia by electrical stimulation of the hypothalamus. Stimulation of the pre-optic area gave consistent results and complements the adipsia which Feldman *et al.* observed in their lesioned preparations. Actually the literature contains a number of accounts of supposedly low water intake in birds under normal conditions. Wolf (1958) asserted that quails and parrots scarcely ever need water, a suggestion which runs quite counter to my experience. He also cited a case of a hen which lived for 6 weeks with neither food nor water (!) and pointed out that poultry were the only components of Hédin's caravan through the Taklamakan desert which seemed to remain undisturbed. He also states that

15

members of the Falconiformes, as opposed to granivores, never drink and concluded that many desert birds such as thrushes, desert larks and ostriches are independent of a supply of drinking water. Such factors are discussed in Sturkie (1965).

Inter-relationships between hunger and thirst were considered in terms of the central areas by McFarland (1965a,b, 1967) and McFarland and L'Angellier (1966). Adopting a black box approach he suggested that the reciprocal effect of the feeding and drinking systems is not symmetrical. Thirst decreases during 2 days of food deprivation and increases rapidly to supra-normal levels during the subsequent recovery period.

## E. Control of Ovulation and Lay

A number of studies have been carried out on the control of ovulation and lay. Ralph and Fraps (1960) investigated the effect of injecting progesterone into the brain of hens and hypothalamic involvement in the control of ovi-position was suggested by the effects of lesions which were produced in fowls by Yasuda (1957) and Ralph (1959), and by the effects of certain barbiturates (Fraps and Case, 1953). Focal ablation of regions in the neighbourhood of the pre-optic area, and also in this area itself, caused extensive interference with the laying of those eggs that were already present in the oviduct when the operations were carried out. Although they could be, and were eventually, layed, they were usually retained in the oviducts for several days. Small electrolytic lesions in the median part of the diencephalon of actively ovulating hens prevented ovulation itself regardless of the actual site which was de-stroyed (Ralph and Fraps, 1959a, b). However, the duration of this anovulatory period varied and could be controlled by altering the location of the lesion. When this was in the pre-optic hypothalamus, the supra optico-hypophysial tract or the dorso-caudal thalamus, the mean length of the periods was about twice as long as after lesions to central diencephalic areas.

Preparations which had undergone such lesion operations were frequently hypophagic, an observation which conforms with the experimental aphagia and adipsia mentioned in Section VI D above. This condition lasted for a few days and Ralph and Fraps thought that it was significant that the larger follicles become atretic by the second day of starvation in normal hens. Earlier results had suggested that lesions to areas which are not particularly concerned with ovulation can cause an anovulatory phase lasting up to 15 days. Such nutritional factors would account for the lack of ovulation in preparations with lesions outside the ventro-median area and no irreparable damage to either this area or the neural structures which are specifically con-cerned with gonadotrophic hormones. They would not be expected to have any effect on ovulation after the initial phase of hypophagia. More lengthy interruptions were more likely caused by damage to the specific neural structures concerned with gonadotrophic activity. They concluded that the extent of the period of anovulation probably reflects both the scope and significance of such damage.

As this and other work had implicated the ventro-median region of the pre-optic hypothalamus in the control of normal ovulation Ralph and Fraps (1959a) utilized electrolytic lesions to determine whether or not the hypothalamus was concerned in progesterone induced ovulation. Only damage to the paraventricular nucleus, the ventro-median region of the pre-optic area, or to fibre tracts passing back from it, prevented the ovulation that would normally follow progesterone administration. Furthermore if they were to exert such an overt inhibitory effect they had to be produced within 2 hours of administering the steroid. If they were produced after a longer interval they had no effect on such induced ovulation although they either abolished or delayed subsequent natural ovulations. The ventro-median area could therefore be either the actual site of progesterone excitation or, alternatively, merely a link in the neural complex which mediates its effects. Ralph and Fraps suggested that the destruction of the site would preclude the release of gonadotrophic hormone which normally follows progesterone administration. The actual timing relationships agree with the results of Rothchild and Fraps (1949). These had suggested that excision of the pituitary within 2 hours of injecting progesterone prevented ovulation. Tienhoven (1955, 1961) had also shown that a cholinergic component of progesterone-induced ovulation is only operative for up to 2 hours after the administration of the steroid.

More recently Opel (1963, 1967) carried out experiments which suggested that premature oviposition of the terminal egg of a series $(C_t)$ follows stimulation of the pre-optic areas. This appears to conflict with the results of Juhasz and Tienhoven (1964) in which electrical stimulation of the telencephalon caused a *delay* in the oviposition of the first egg of a sequence $(C_1)$. More precise investigations showed Opel that premature oviposition not only followed electrical stimulation but could be elicited by simply inserting the electrodes into any of several sites. As far as the discrepancies with Juhasz and Tienhoven's results are concerned he felt that major differences in both the technique employed and the brain areas which were examined, together with known differences in the hormonal control of the $C_1$ and $C_t$ lay, precluded any definitive comparisons.

He found that piqure, the mere insertion of the electrodes, and full electrical stimulation of the pre-optic brain both produced similar effects on the $C_1$ lay, the associated $C_t$ ovulation and its lay. He therefore pooled the data and they are presented in Table 46 as effects of piqure. During any interval of time such piqure advances the $C_1$ lay but the greatest incidence (56%) of premature ovipositions followed treatment between 9 p.m. and midnight. The frequency then fell off with the advancing time of operation. Such results contrast with the earlier report because in that case the observed frequencies of premature lay *increased* with the advancing time of treatment between 9 p.m. and 9 a.m. and 100% results occurred between 6 a.m. and 9 a.m. In addition to this piqure operations which were performed prior to 3 a.m. also induced a significant number of delayed $C_t$ ovulations. The period of the delay varied from 3 to 6 hours in 6 hens, and 24 to 48 hours in 15 others. Opel considered that the delays reflected interference with the release of luteinizing hormone. Nine birds which were treated prior to 3 a.m. never ovulated the $C_t$ egg. At autopsy they exhibited atresia of the large ovarian follicles and such atresia also occurred in 16 preparations which had, nevertheless, undergone either normal or delayed ovulation.

When $C_t$ lay was not affected by the experimental treatment it either occurred at the expected time (31 birds), or was delayed by 1–13 hours (7 birds). No statistically significant correlations could be detected between $C_1$ lay, ovulation and $C_t$ lay. Piqure of either the thalamus or the hemispheres had no disturbing effect on $C_1$ lay.

## TABLE 46

*The effects of piqure of the preoptic brain regions on $C_1$ oviposition, the associated $C_t$ ovulation and also $C_t$ lay. Those figures in the columns which have different superscripts are significantly different.*

| Time of piqure | Hours before period of $C_1$ lay | Number of hens | No. of $C_1$ ovipositions premature | No. of $C_t$ ovulations delayed | No. of $C_t$ ovipositions premature |
|---|---|---|---|---|---|
| 6–9 p.m. | 12–15 | 16 | 5[b] | 6[a] | 2[a] |
| 9 p.m.–12 midnight | 9–12 | 16 | 9[a] | 8[a] | 1[b] |
| 12–3 a.m. | 6–9 | 16 | 6[b] | 7[a] | 0[c] |
| 3–6 a.m. | 3–6 | 16 | 5[b] | 3[b] | 2[a] |
| 6–9 a.m. | 0–3 | 16 | 2[c] | 2[b] | 2[a] |
| Unoperated controls | — | 100 | 2[d] | 11[b] | 2[c] |

From Opel (1967).

The effects on $C_t$ lay were time dependent (Table 47). No displacement was detectable following piqure operations which were undertaken between 9 and 11 p.m., but $C_t$ lay was advanced in a significant number of hens when such stimulation was administered between 6 and 8 a.m.

The actual method by which brain piqure, involving simple mechanical injury, can produce these responses remains obscure. Clearly a number of hypotheses can, however, be put forward in terms of injury effects. Opel himself concluded that there were major differences in the effects on $C_1$ lay on the one hand, and on $C_t$ lay on the other. Besides the differing sequence of events after operations at different times, $C_t$ lay was advanced as the result of piqure at all the brain sites which he tested. In contrast $C_1$ lay was only advanced when the area involved was in the pre-optic region (Table 47). He suggested that one possible explanation for these differences could be a variation in the brain sensitivity to piqure. This could in itself be dependant upon the levels

## TABLE 47

*The effects of the site and time of brain piqure on $C_1$ and $C_t$ lay.*

| Time of piqure | Brain site | Incidence of premature lay | |
| | | $C_1$ | $C_i$ |
| --- | --- | --- | --- |
| 9–11 p.m. | Preoptic region | 5/10 | 3/10 |
| | Thalamus | 0/10 | 0/10 |
| | Telencephalon | 0/10 | 0/10 |
| 6–8 a.m. | Preoptic region | 3/10 | 10/10 |
| | Thalamus | 0/10 | 6/10 |
| | Telencephalon | 0/10 | 2/10 |
| Unoperated controls | | 2/100 | 2/100 |

From Opel (1967).

of circulating ovarian steroids. The negative feedback relationship of steroids to hypothalamic and hypophysial centres has been widely implicated in the control of cyclical ovulation and oviposition. The experiments which were carried out by Rothchild and Fraps (1949) suggested that $C_1$ lay is influenced to a very marked extent by secretions from both the last ruptured, and also the currently ovulating ovarian follicles. $C_t$ lay, which occurs on a day when there is no ovulation, is primarily timed by the secretions of the ruptured follicle. If this is the case then higher blood steroid levels, and a possibly consequential lowered threshold of brain sensitivity, may account for the differences which piqure has on $C_1$ lay and $C_t$ lay. The piqure evoked oviposition should therefore be readily blocked by appropriately timed steroid injections. Finally Opel suggested that the zone of the pre-optic brain which had previously been implicated in ovulation, may also be a primary centre in the systems controlling oviposition. However, the mechanism of the two processes may not be the same. Neurohypophyseal involvement is discussed in Chapter 12, Section VI B.

## VII. THE EFFECTS OF TESTOSTERONE PROPIONATE ON COCKS WHICH HAVE UNDERGONE HYPOTHALAMIC LESIONS ETC.

### A. General Considerations

Following lesions of the ventro-medial hypothalamic region, especially in the neighbourhood of the median eminence (*q.v.*), cockerels had impairments of their gonadotrophin production and/or release. Snapir *et al.* (1969) observed that the combs of such preparations atrophied, and their external appearance was similar to that of capons (*sensu stricto*). Injections of physiological saline had no subsequent effects on this regression of the secondary sexual characters. However, following the administration of testosterone propionate the atrophied combs achieved a size which compared favourably with those of normal birds although some differences persisted and such combs appeared bright red and glistened rather more than those of unoperated birds. Alongside this renewed comb growth the androgen administration also restored the libido of the preparations but no seminal fluid could be elicited. Kordon and Gogan (1964) had previously reported complementary experiments. They had implanted glass canulae containing testosterone crystals at various loci. After 14 days in continuous light all the birds then exhibited gonadal stimulation, except those whose stereotaxic implants were situated in close proximity to the ventro-median complex. These showed testicular regression. As a result they concluded that the ventro-median nuclear components were sensitive to testosterone levels and that one, at least, was involved in the negative feed-back system which is responsible for cutting off gonadotrophin production. More detailed studies on such androgen implantations are given in Section B(ii) below.

Aberrations of the secondary sexual characters are not the only overt results of functional castration by ventro-medial hypothalamic lesions. The other parameters were summarized by Snapir *et al.* Normally the male bird stores very little fat. Following direct hypophysectomy Nalbandov and Card (1942) found that cockerels became obese. Chernick (1960) had formerly suggested that the fat content of avian adipose tissues is involved in a homeostatic relationship that reflects changes from lipolytic to lipogenic activity in conjunction with the proximity of feeding time, the body's needs and the amount of activity. Allied concepts had been widely accepted by pathologists for some years earlier. In association with this it was logical to assume that some mechanism prevents the accumulation of excess, or large quantities of fat in the adipose tissues of normal cockerels. Using *in vitro* experiments Gibson and Nalbandov (1966) obtained data which led them to suggest that the obesity of hypophysectomized preparations was caused by an impaired mobilization of fat from the body's depot deposits. As the synthesis of lipids from glucose within the adipose tissue was greatly diminished by hypophysectomy they ruled out a direct increase in the total fatty acids. *In vivo* work supported this hypothesis as injections of pituitary extracts raised the plasma concentration of free fatty acids in such hypophysectomized preparations. This was a rapid response which argued for a simple mobilization rather than increased synthesis. Furthermore they thought that the effect was direct and not mediated via target organs because the thyroid, adrenal cortex and testes etc., are atrophied in such preparations and their involvement in the rising plasma concentrations would preclude such a rapid response.

At about the same time Lepkovsky and Yasuda (1966) demonstrated that in cocks ventro-medial hypothalamic lesions involving sites close to the median eminence produced both functional castration and its attendant obesity. There were, however, some marked differences in the observed symptoms in hypophysectomized and lesioned preparations. Whilst hypophysectomized specimens decreased their food intake the hypothalamic preparations did not seem to. Indeed, in view of Section VI D it is interesting that many became hyperphagic. Subsequently Lepkovsky *et al.* (1967) showed that injections of testosterone propionate are followed by an increase in the plasma levels of free fatty acids in cockerels but not in hens. This they interpreted as a fat mobilization analogous to that postulated by Gibson and Nalbandov, however, it clearly implicates the target organs in such responses.

Returning to the work of Snapir *et al.* four out of the six preparations which had had ventro-medial hypothalamic lesions followed by injections of physiological saline subsequently became hyperphagic. This was also true of two preparations out of the six which had received testosterone propionate. Such specimens were in differing periods of either a "dynamic" or "static" phase of hyperphagia. They then deposited abnormally large amounts of fat in their carcasses even if their overall dietary budget was controlled so that it was the exact equivalent of that in control birds. However, Snapir *et al.* reported that all the lesioned preparations, whether in a hyperphagic or normophagic state, decreased their food intake during the actual treatment period. As this did not happen in the controls their food intake during this time exceeded that of the experimental animals. Nevertheless no clear results, which would enable one to give a synoptic account of the effect of exogenous androgen on feeding behaviour and its hypothalamic control, emerged because of the vast fluctuations of food intake in the experimental group. One can merely state that adiposity, although also subject to considerable variations, was greater in the saline treated group than in the steroid treated one. Unfortunately as the food intake of the latter was lower this is again a confused picture. From the data Snapir *et al.* cautiously suggested that hypothalamic lesions may activate at least two mechanisms which lead to changes in the degree of adiposity. There is first of all an impaired gonadotrophin release. This leads to an increased fat content in the adipose tissues comparable with that following treatment of normal birds with oestrogens (Lorenz, 1954; Nalbandov, 1953; Sturkie, 1958), and also that of hypophysectomized preparations. There is also an actual rise in food intake. Together these two effects, functional castration and increased food intake, determine the actual degree of adiposity.

## B. Hormonal Effects on Mating Behaviour

### (i) *Introduction*

The behaviour of castrates and the changes which are consequent upon hormone therapy are well-known and have long been part of our lay experience. Erpino (1969) gives a detailed account of the situation in the pigeon and full details of work on other species is summarized in the relevant chapters of Sturkie (1965). More precise investigations of the possible focal effects of such hormones within the brain are scarce. The results which Barfield obtained are therefore of paramount importance and discussed below. It is, however, worth recalling some less specific contributions and noting that activation of such systems by peripheral sensory information is dealt with elsewhere. (See Chapter 12, Section VI B.)

Using the canary Warren and Hinde (1959, 1961) found that large doses of oestrogen stimulate nest-building in the female. Similar observations were made on the ring-dove by Lehrman and Brody (1961) and Lehrman (1961, 1963). On the other hand Noble and Wurm (1940) were unable to detect any such response when studying the black-crowned night heron, *Nycticorax nycticorax*. Orcutt (1967) found that the administration of diethyl stilboestrol to female *Agapornis roseicollis* would also stimulate the preparation of nest material but only after the female had reached a certain minimum age. Using exogenous oestrogen it was then possible to initiate such behaviour at least 2 weeks before it would normally occur.

As the result of Lehrman's work on the induction of parental behaviour by progesterone there followed an extensive reconsideration of the function of prolactin (Riddle, 1963). As the discoverer of this substance some decades earlier Riddle was still inclined to ascribe to it a primary position in the hierarchical organization of factors which cause parental behaviour. Lehrman (1963) acknowledged the importance of Riddle's work. He felt that the criticisms which the latter levelled arose from the impression that the work of Lehrman (1958) devalued the role of prolactin. Whilst he was at pains to emphasize that he doubted whether prolactin would work in the absence of progesterone as a mediator, he was quite clear that prolactin was the primary substance in nature. The difficulties seemed to disappear as a result but more recently Murton *et al.* (1968) have carried out experiments that give some indication of the expected complexity of the hormone involvement in breeding behaviour. They considered that in, for example, the male pigeon, there is a cycle which is successively dependent on a changing hormone balance. The initial phases of courtship where aggressive components are much in evidence depends upon high follicular stimulating hormone/androgen titres. This phase is followed by one in which oestrogen becomes dominant and leads to nest demonstration. Oestrogen released from the testis tubules may be facilitated by androgen. At the end of this oestrogen phase of behaviour follicular stimulating hormone becomes more involved and leads to nest building. This sequence is then followed by progesterone secretion and the onset of incubation.

The release of prolactin itself has been shown to be under hypothalamic control in birds by Kragt and Meites (see Meites and Nicoll, 1966). In an extensive study these last two authors showed that an extract equivalent to one parent hypothalamus, and prepared on the day that the young hatched, produced a two-to four-fold increase in the release of prolactin from the pituitary of young birds. Telencephalic extracts had no such effect, even when obtained from the same source birds. It is interesting that attempts to use rat and pigeon extracts on the young of the other species were uniformly unsuccessful. Meites and Nicoll (1966) understandably concluded from this that the factors which lead to prolactin release are different in these two species. Especially since hypothalamic extracts from other avian sources such as *Gallus* do stimulate prolactin release in the pigeon. Furthermore Nicoll had reported that the addition of a neutralized acid extract of homologous hypothalamic tissue to cultures of the pars distalis of the tri-coloured blackbird, *Agelaius tricolor*, maintained prolactin secretion at a high level. Telencephalic extracts again gave negative results. One can conclude from this that adenohypophysial secretion is under hypothalamic control and recall the hypothalamic hypophysiotropic substances of mammals (Guillemin, 1967).

## (ii) *Pre-optic Androgen Implantations*

Barfield (1969) investigated the effects of androgen implantations in the fowl and other studies have been reported on the male ring-dove, *Streptopelia rissoria* (Barfield, 1967; Hutchinson, 1967, 1970a,b). Complementarily Komisaruk (1967) described the suppression of male courtship by progesterone implantations which activated incubation behaviour instead. In Barfield's (1969) experiments 59 capons received a stereotaxic implant of testosterone propionate in the forebrain regions. Copulatory activity was then shown by 14 of 19 birds which had pre-optic implants although very little courtship or

15*

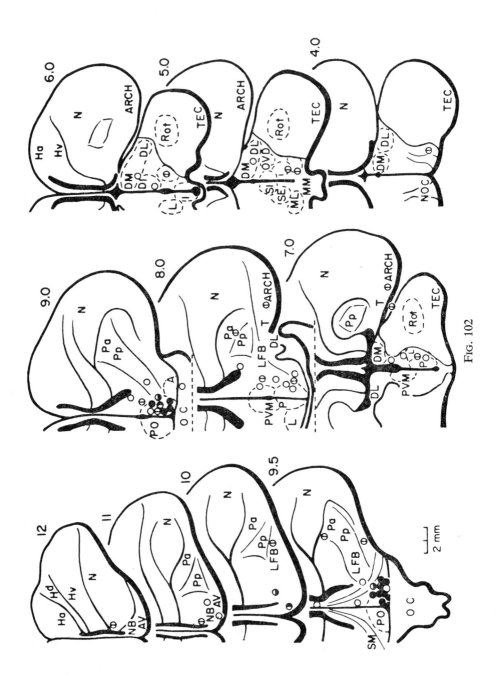

Fig. 102

aggressive behaviour occurred. None of the 40 birds which received implants elsewhere in the brain showed any such copulatory activity. Pre-optic implantation of cholesterol had no such results either and Barfield concluded that circulating autochthonous androgen normally acts on the pre-optic region alone to activate copulation but not courtship or aggression.

High androgen titres had previously been shown to be associated with all these behaviour patterns by Davis and Domm (1942). Contrariwise Guhl (1950) showed that oestrogens activate copulatory activity but suppress both courtship and aggression. Their data had therefore implied a marked distinction between the mechanisms which control these usually coordinated components of the copulatory, courtship and aggression complex a fact that agrees with the cycle of Murton *et al.* The actual behavioural patterns which are involved were outlined by Davis and Domm (1942), Wood-Gush (1954, 1955), Fisher and Hale (1957), Guhl (1962) and Kruijt (1964). Copulation follows a sequence of actions. These include courtship display, an approach, mounting and treading. During courtship the male performs the *waltz* as he side-steps around the female. The outer wing is lowered, the primaries scratch the ground and are struck by the feet as these move past them. *Tid-bitting* is a solicitation display which resembles a call to food. The male pecks at the ground, often scratches with his feet and emits a repetitive vocalization. In the *approach* he grabs the hen's comb, or neck feathers, with his beak and starts to place one foot on her back. A *mount* involves standing on either her back or out-stretched humeri, and *treading* refers to a rhythmical up-and-down

---

FIG. 102. Survey of transverse sections through the forebrain and midbrain showing placements of implants. The numbers next to each section refer to the anterioposterior stereotaxic coordinates in millimeters: the higher numbers are anterior. Areas enclosed by broken lines show sites of major nuclei and fibre tracts are shown by unbroken lines. Symbols: single implants–●, copulate: ◖, mount only; ○, no sex behaviour; bilateral implants◐, copulate; ◌, no sex behaviour. (Barfield, 1969).

Abbreviations for Figs 102–104: A, see HAM; AC, anterior commissure; Arch, archistriatum; AV, area ventralis; Cer. cerebellum; DL, M, n. dorsolateralis, medialis; DSO, supraoptic decussation; Ha, d, v, hyperstriatum accessorium, dorsale, ventrale; HAM, n. hypothalamicus anterior medialis; HPM, n. hypothalamicus posterior medialis; I, n. hypothalmic inferior; L: LHT, n. hypothalamicus lateralis; LFB, lateral forebrain bundle; M. ML. M, n. mammilaris lateralis, medialis; N, neostriatum; NB, n. diagonal band of Broca; NOC, oculomotor nerve; OC, optic chiasma; OVD. n.ovoidalis; P. see HPM; Pa. p., paleostriatum augmentatum, primitivum; PC, posterior commissure; PI, n. paramedianus intermedius; PO; POM-PPM, n. praeopticus medialis-paraventricularis magnocellularis (preoptic nuclear complex). PS. Position of pituitary stalk. PVM. n. paraventricularis magno-cellularis; ROT, N. rotundus: SE. i, stratum cellularis externum, internus; SM, tractus septomesencephalicus; SO, n. supraopticus; TEC, optic tectum.

movement of the feet which occurs while mounted and culminates in the depression of the tail bringing both male and female cloacas into contact to facilitate sperm transport.

The distribution of Barfield's implants is shown in Fig. 102. Those close to the midline are represented in Fig. 103 and those in the pre-optic region in Fig. 104. Their associated behaviour patterns are indicated by the symbols. Generally speaking the bilateral implantations represented the crudest localizations. The two associated with copulation were situated at opposite ends of the area which included all other placements. One bilateral focus which did

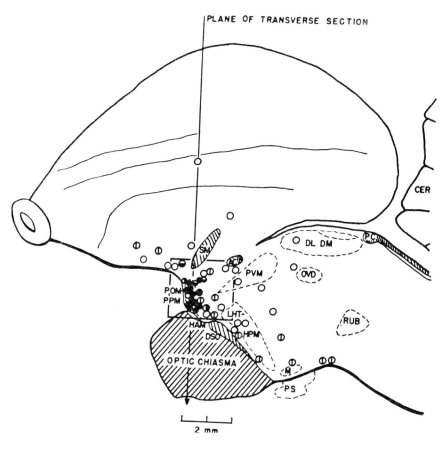

FIG. 103. Para-sagittal reconstruction demonstrating implant positions 0·5–1·5 mm from the midline. The plane of the transverse sections of Fig. 102 is drawn through the pre-optic area. The lines drawn around the hypothalamic-preoptic region indicate the area represented in Fig. 104. Cross-hatched areas indicate fibre tracts. Other symbols and abbreviations as in Fig. 102. (Barfield, 1969).

not result in copulatory behaviour lies within the pre-optic complex. Three
others in surrounding regions were also negative. However, most of the sites
which are figured were revealed by unilateral implantation and the majority
of the implants situated in the pre-optic or adjacent areas induced copulatory
activity. The two principal exceptions lie at its basal and postero-dorsal
margins. Of the 14 birds which showed complete copulatory activity 12 had

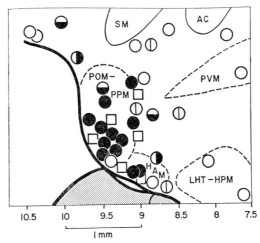

FIG. 104. The anterior hypothalamo-pre-optic region of the fowl brain in para-
sagittal section. The lower scale gives anterio-posterior coordinates in millimeters.
Symbols show where tips of implants lay. Open square, cholesterol implants.
Other symbols as for Fig. 102. (Barfield, 1969).

single implants. *Waltzing* was only seen in two of these 14 and in one of these
it occurred 10 days after overt copulatory activity had ceased. Three sites at
the periphery were associated with mounting alone and it is interesting that
these three animals also *waltzed*.

The latent period which elapsed between the time of implantation and the
appearance of the behavioural components varied from 4 to 17 days. The
longest periods occurred in those capons which had received rather peripheral
implantations. Using as controls capons which had received testosterone propionate systemically Barfield demonstrated that the frequency of copulation
was very similar in the two groups. The results are contained in Table 48. A
marked contrast was seen, however, in the incidence of *waltzing*. Eight out of
the 11 controls exhibited it against only 2 out of the 14 implanted birds.

The sexual behaviour of those capons which had received injected testosterone was in fact similar to that of normal cockerels. Both groups approached
females with an intenseness of movement, and they often emitted a growling
vocalization. They *circled, waltzed* and *tid-bitted* both before and after copulation. The implanted capons merely stood near to the female and often mounted

with absolutely no prelude at all. Although they would, on occasions, circle the hen, these movements were of low intensity and any concomitant vocalizations were rare. None showed obvious signs of *tid-bitting*. Often the female would act aggressively and as a result implanted capons only succeeded in mounting when the hens were receptive. In contrast the injected controls were never driven off by such pugnacious behaviour.

A measure of the actual amount of implanted androgen which entered the systemic circulation was provided by the degree of comb growth. This enabled

## TABLE 48

*A comparison of copulatory and courtship behaviour in implanted and injected capons as shown by the data from the four tests representing maximum reproductive behaviour for each animal.*

| Measure | Group | | P |
|---|---|---|---|
| | Implanted (N = 14) | Injected (N = 11) | |
| Frequency of copulation | 2·3 ± 0·4 | 2·6 ± (0·4) | > 0·20 *t* test |
| Occurrence of waltzing | 2 | 8 | < 0·005 chi-squared |
| Frequency of waltzing | 0·25 (N = 2) | 2·3 (N = 8) | |

From Barfield (1969).

## TABLE 49

*The comb growth of implanted capons three weeks after implantation.*

| Group | N | Mean increase ± SEM (in cm²) | P |
|---|---|---|---|
| A. Copulating; single implant | 12 | 7·9 ± 1·3 | |
| B. Not copulating; single implant | 25 | 7·8 ± 0·9 | |
| C. Not copulating; Bilateral implant | 18 | 20·3 + 1·7 | P < 0·01 |

From Barfield (1969).

Barfield to obtain a measure of the systemic hormonal effects in both copulating and non-copulating birds. The results are shown in Table 49. The mean increase in comb size amongst copulating birds with a single implant was practically identical to that in their non-copulating analogues. The lack of any apparent correlation between the systemic activity associated with implants and the induction of copulatory behaviour was further emphasized by comparing the bilaterally implanted non-copulating birds with the singly implanted copulators (Table 49). The increase in comb size was significantly greater in the doubly implanted non-copulators. $(P<0.01.)$

The average duration of these behavioural effects following testosterone propionate implantations was of the order of a month. Peak comb size occurred at 30 days and there was subsequently a regression. It would therefore appear that the cessation of copulatory activity followed closely upon the decline in systemic effects. This fits the general pattern in both bird and mammal castrates. Young (1961) summed this up and concluded that copulatory behaviour diminishes or ceases a short time after either castration or withdrawal of the supporting hormone therapy. It is exceptionally interesting that in his comparable work on the ring dove Barfield (1967) did not find the differentiation of aggressive and sexual systems which would compare with these observations on capons. He suggested (1969) that this is because the dove is not a sexually dimorphic species, and, that it may also reflect the differences in social structure which exist between the two species.

## VIII. ELECTRICAL STIMULATION AND SOCIAL BEHAVIOUR

### A. Electrical Stimulation and Courtship

Direct electrical stimulation of various diencephalic sites also results in courtship behaviour (Akerman, 1965a). Electrodes situated in the region of the pre-optic area and in neighbouring sites such as the septum, anterior hypothalamic and anterior paraventricular regions all evoked various components. Fifty pigeons gave elements of the bowing action and of these only two were females. A rather large number responded primarily with a complete and intense action. The head and body were erected at the onset of the stimulus and crop movements were initiated. The bird then looked around in a searching manner and began to walk in circles with ruffled crown feathers and uttering the cooing call. In some specimens the tail feathers were spread out. This behaviour was performed in periods separated by short pauses during which they stood still and looked around attentively. A mixed behaviour pattern was also evoked from the lateral border of the preoptic periventricular nucleus masses. Here the bowing response was succeeded by pecking. It is of interest that a less intense bowing action was elicited by stimulation of hemisphere foci. This emphasizes the close functional integration of such regions with the preoptic area.

Nest demonstration and displacement preening are also closely associated with courtship. The nest demonstration is manifested by small nodding movements and a special vocalization, the nest call. At high activation these components are performed in a lying position and the tip of the wing closest to the mate is vibrated. Intermediate stages between ordinary preening and rapid movements of the bill towards the scapulae—displacement preening—are frequent in courtship. The nest demonstration was induced in 14 birds, 8 of which were males. Displacement preening occurred in four males and three females. In the case of the females the vocalization was softer and more rattling. Such effects were elicited with greatest intensity from the anterior paraventricular masses. The most complete response followed stimulation of the medial pre-optic area at a threshold of 0·02 mA. Once again the fact that certain lateral neostriatal sites gave similar results suggests a close functional inter-relationship.

## B. Diencephalic Stimulation and Defensive Behaviour

Besides his observations on bowing, nest demonstration and displacement preening Akerman (1965b) evoked certain defensive actions by diencephalic stimulation. These were very similar to both the normal defence and escape reactions. One group of birds reacted by adopting a motionless erect posture. All movements ceased at the onset of stimulation even if they had been alert previously. With continuing stimulation they crouched a little and looked forward. The tail was lowered, feathers were spread out, and the feathers of both neck and body were ruffled. Looking to one side the preparations emitted the growling call and raised the opposite wing. Powerful blows, first of all with the non-erected wing and then with both, were delivered into empty space. These were accompanied by the short "wao" call and small jumps. There were no tendencies to escape in such cases. Post-mortem examination revealed that the electrode tips were situated medially in the magnocellular portion of the paraventricular nuclei and the anterior hypothalamus. A few loci which elicited similar responses also occurred in the posterior part of the paleostriatum near to the archistriatum, and within the archistriatum. Preparations which exhibited less complete threat reactions had electrode sites at the border between the magnocellular paraventricular nucleus and the external cellular stratum or in the posterior paleostriatum. These stood still or walked about slowly with an erect wing and emitting the growling call but did not deliver any blows with the wings. A similar pattern of behaviour was also shown by two specimens which were stimulated in the anterior ventral area. Wing erection was lacking from the responses of some specimens which had electrodes in the medial part of the paleostriatum. In these latter cases the vocalizations were not of the characteristic growling type but rather a hybrid noise including components of this combined with elements of the cooing call that normally accompanies the bowing action of courtship.

Stimulation in another group at first released restlessness instead of "freezing". They ran about stopping now and then, and made rapid turning movements of the head. With higher current strengths the running gave way to walking with erect wing and emitting the growling call. Blows with the wings were also observed sometimes. The responsible sites were in the lateral regions of the preoptic area and in the external cellular stratum of the para-ventricular grey. Finally regions in the posterior part of the pre-optic area, the external cellular stratum and the hypothalamus could elicit plain escape reactions. When stimulated the birds tried to run or fly away swiftly almost at once, even at a low current strength. Rapid glances to both sides, crouching or cringing could intervene in the brief intervening period. Similar escape responses were also elicited at hemisphere levels in areas of both paleostriatum and archistriatum which lay more laterally than those producing defence actions. The observed thresholds for such avoidance reactions were lower than those of the defensive ones and the highest (0·3 mA) was obtained in the archistriatum.

These two different patterns of reaction therefore seem to be controlled by a closely interacting system in the forebrain. The field involved would appear to extend from the hypothalamus, the diencephalic paraventricular grey and the preoptic area via the ventral area, paleostriatum and the border area between the septal nuclei, up into the archi-, neo- and hyperstriatum. A zonation in this pattern of distribution is also likely. The defence and escape reactions seem to be closely linked with the hypothalamus and the adjacent grey with a medio-lateral gradient of defence–escape. The primarily more aggressive behaviour patterns such as bowing and fighting are represented in front of, and medial to, the defence and escape reactions. As a result there seem to be both medio-lateral and rostro-caudal gradients of representation.

## IX. MONOAMINES IN THE DIENCEPHALON

Together with their work on mesencephalic monoamines (see Chapter 10, Section XI) Fuxe and Ljunggren (1965) also obtained data relating to the differential distribution of such chemicals in various diencephalic foci. Within the epithalamic region the medial habenular nucleus appears as a generally dark area when viewed through the fluorescence microscope. However, a particularly concentrated fluorescence was detectable in the zone of contact between the anterior tela choroidea and the lateral habenular nucleus. This last contained fine, green fluorescent, varicose fibres. A number of comparable terminals had previously been seen in the sub-habenular nucleus by Kuhlenbeck (1937). The overlying epiphyseal components had a marked yellow fluorescence in their parenchymal cells suggesting the presence of 5-hydroxytryptamine, but no sign of any fluorescence of either type could be detected in the choroid areas.

Within the thalamic nuclear masses only scattered varicose fibres occurred in both the anterior and posterior dorso-medial nuclei, the dorso-lateral and the lateral. On the other hand both the ovoidal nucleus and the lateral geniculate complex had very abundant fibres which exhibited a green fluorescence, and this was also true of some areas which lay above and to the side of the rotund nucleus. A similarly large quantity was detected in the periventricular zone, in the dorsal thalamic regions, in the anterio-ventral area, and in the nucleus of the dorsal supra-optic decussation. No cells with a yellow fluorescence occurred in any of these regions or elsewhere in the thalamus.

The septal area had a large number of green fluorescing fibres reminiscent of the situation which was described by Bertler *et al.* (1964) for "striatal" components. Within the pre-optic region the principal concentrations were within the periventricular and anterior pre-optic nuclei. The bed nuclei of both the anterior and pallial commissures received a large number of such fibres although only a few actual cell bodies were seen to fluoresce, and these only after pre-treatment with reserpine.

Both the paraventricular and posterio-medial hypothalamic nuclei had numerous terminals. These seemed to make contact with many of the included cells. In particular this resulted in a dense fibrillar network within the posterio-medial nucleus. Elsewhere some local variations gave various areas with less intense results and in the lateral part of the posterior hypothalamic nucleus the number of catecholamine terminals was far less than that in the medial part. The periventricular areas outside the pre-optic region contained fewer terminals than in that region itself. Approximately comparable conditions occurred in the inferior hypothalamic and supra-optic nuclei. In contrast the zone which is nearest to the ependyma along with both the anterio-lateral and the anterio-medial hypothalamic nuclei did not fluoresce at all.

A detailed comparison of adrenalin and noradrenalin levels in the hypothalamus and other brain regions was provided by Juorio and Vogt (1967). Their data are contained in Table 50. These results, which refer to the fowl, give adrenalin concentrations in excess of those obtained for the pigeon although the levels of nor-adrenalin were similar in the two species, and they found that the hypothalamic dopamine concentration amounted to about 10% of the nor-adrenalin levels. The hypothalamus had higher nor-adrenalin concentrations that anywhere else in the brain and it is interesting that the figures of Table 50 show a high proportion of adrenalin in both this, and other nor-adrenalin rich areas. The percentage methylation which was detectable was also, at 40, the highest value obtained in birds or mammals, and this agrees with the data reported previously by Callingham and Cass (1965). The existence of nor-adrenergic fibres in these regions which are functionally associated with both the hypothalamo-hypophyseal and the extra-hypophyseal neuro-secretory systems resembles the situation in mammals.

## TABLE 50

*The values obtained for adrenalin and nor-adrenalin concentrations in the hypothalamus and other brain regions of the fowl, expressed as μgrams per gram of tissue.*

| Brain region | Nor-adrenalin | Adrenalin |
|---|---|---|
| Anterior part of basal nucleus | 0·37 | — |
| Posterior part of basal nucleus | 0·39 | — |
| Remainder of basal complex | 0·23 | — |
| Thalamus | 0·88 | 0·64 |
| Hypothalamus | 1·42 | 1·01 |
| Optic lobes | 0·45 | 0·35 |
| Cerebellum | 0·23 | 0·07 |
| Medulla | 0·21 | 0·02 |

From Juorio and Vogt (1967).

## References

Akerman, B. (1965a). Behavioural effects of electrical stimulation in the forebrain of the pigeon. (i) Reproductive behaviour. *Behaviour* **26**, 328–338.

Akerman, B. (1965b). Behavioural effects of electrical stimulation in the forebrain of the pigeon. (ii) Protective behaviour. *Behaviour* **26**, 339–350.

Akerman, B., Andersson, B., Fabricius, E. and Svensson, L. (1960). Observations on central regulation of body temperature, and food and water intake in the pigeon (*Columba livia*). *Acta Physiol. Scand.* **50**, 328–336.

Akerman, B., Fabricius, E., Larsson, B. and Steen, L. (1962). Observations on pigeons with prethalamic radiolesions in nervous pathways from the telencephalon. *Acta Physiol. Scand.* **56**, 286–298.

Baldwin, S. P. and Kendeigh, S. C. (1932). Physiology of the temperature of birds. *Scient. Publ. Cleveland Mus. Nat. Hist.* **3**, 1–196.

Barfield, R. J. (1967). Activation of sexual and aggressive behaviour by androgen implants in the brain of the male ring dove. *Amer. Zool.* **7**, 800 (abst.).

Barfield, R. J. (1969). Activation of copulatory behaviour by androgen implanted into the preoptic area of the male fowl. *Hormones and Behaviour* **1**, 37–52.

Bellonci, J. (1888). Ueber die centrale Endigung des Nervus opticus bei den Vertebraten. *Z. wiss. Zool.* **47**, 1–46.

Benoit, J. (1964). The structural components of the hypothalamo-hypophyseal pathway in birds, with particular reference to photostimulation of the gonads. *Ann. N.Y. Acad. Sci.* **117**, 23–34.

Bertler, A., Falck, B., Gottfries, C. G., Ljunggren, L. and Rosengren, E. (1964). Some observations on adrenergic connections between mesencephalon and cerebral hemispheres. *Acta Pharmacol. Toxic.* **21**, 283–289.

Biedermann-Thorson, M. (1970a). Auditory evoked responses in the cerebrum (field L) and ovoid nucleus of the ring dove. *Brain Res.* **24**, 235–246.

Biedermann-Thorson, M. (1970b). Auditory responses of units in the ovoid nucleus and cerebrum (field L) of the ring dove. *Brain Res.* **24**, 247–256.

Boni, A. (1942). Uber die Entwicklung der Temperatur regulation bei verschiedenen Nesthockern. *Schweiz. Arch. Ornithol.* **2**, 1–56.

Bremer, F. and Ley, R. (1927). Récherches sur la physiologie du cervelet chez la pigeon. *Arch. internat. Physiol.* **28**, 58–95.

Callingham, B. A. and Cass, R. (1965). Catecholamine levels in the chick. *J. Physiol.* **176**, 32–33.

Cate, J., Ten (1936). Physiologie des zentralnervensystem der Vögel. *Ergebn. Biol.* **13**, 93–173. (See also 1965, below.)

Cate, J., Ten (1965). The nervous system in "Avian Physiology" (Sturkie, P. D., ed.), pp. 697–751. Baillière, Tindall and Cassell.

Chernick, S. S. (1960). "Symposium of the Czechoslovakian Academy of Science", pp. 165–178.

Cordier, R. (1954). Le système nerveux-centrale et les nerfs cerebro-spinaux. *Traité de Zoologie* **12**, 202–332.

Craigie, E. H. (1928). Observations on the brain of the humming bird (*Chrysolampis mosquitus* Linn., *Chlorostilbon caribaeus* Lawr.). *J. comp. Neurol.* **48**, 377–481.

Craigie, E. H. (1930). Studies on the brain of the Kiwi (*Apteryx australis*). *J. comp. Neurol.* **49**, 223–357.

Craigie, E. H. (1931). The cell masses in the diencephalon of the humming bird. *Proc. Kon. Akad. Wet. Amsterdam* **34**, 1038–1050.

Davis, D. E. and Domm, L. V. (1942). The influence of hormones on the sexual behaviour of the fowl. *In* "Assays in Biology", pp. 171–181. University of California Press.

Dawson, W. R. and Evans, F. C. (1957). Relation of growth and development to temperature regulation in nestling field and chipping sparrows. *Physiol. Zool.* **30**, 315–327.

Dijk, J. A. van (1930). The effect of stimulation of the cervical sympathetic cord upon the function of cross-striated muscle in the pigeon. *Arch. Néerl. Physiol.* **15**, 126–137.

Dijk, J. A. van (1931). The influence of stimulating the diencephalon upon the contractions of striped muscle in the pigeon. *Arch. Néerl. Physiol.* **16**, 567–573.

Dijk, J. A. van (1932a). On the nature of the diencephalic stimuli affecting rhythmical muscle contractions in the bird. *Arch. Néerl. Physiol.* **17**, 495–503.

Dijk, J. A. van (1932b). On the conduction of impulses set up in a birds 'tween brain causing changes in its muscle activity. *Arch. Néerl. Physiol.* **17**, 504–524.

Edinger, L. (1899). Untersuchungen uber das Gehirn der Tauben. *Anat. Anz.* **15**, 245–271.

Edinger, L. (1908). "Vorlesungen uber den Bau der nervosen Centralorgane des Menschen und der Thiere." Vogel, Leipzig.

Edinger, L. and Wallenberg, A. (1899). Untersuchungen uber die vergleichende. Anatomie des Gehirns. 5. Das Vorderhirn der Vögel. *Abhandl. Senckenb. nat. Gesellsch. Frankfurt am Main* **20**(4), 343.

El Hadi, H. (1969). "The acid base status of the fowl during thermal polynea." Dissertation, London, quoted by Richards (1970).

Erpino, M. J. (1969). Hormonal control of courtship in the pigeon (*Columba livia*). *Anim. Behaviour.* **17**, 401–405.

Erulkar, D. S. (1955). Tactile and auditory areas of the brain of the pigeon. An experimental study by means of evoked potentials. *J. comp. Neurol.* **103**, 421–458.

Feigl, E. and Folkow, B. (1963). Cardiovascular responses in diving and during brain stimulation in ducks. *Acta Physiol. Scand.* **57**, 99–110.

Feldman, S. E., Larsson, S., Dimick, M. K. and Lepkovsky, S. (1957). Aphagia in chickens. *Amer. J. Physiol.* **191**, 259.

Fisher, A. E. and Hale, E. B. (1957). Stimulus determinants of sexual and aggressive behaviour in male domestic fowl. *Behaviour* **10**, 309–323.

Fraps, R. M. and Case, J. F. (1953). Premature ovulation in the domestic fowl following administration of certain barbiturates. *Proc. Soc. Exp. Biol. Med.* **82**, 167–171.

Fuxe, K. and Ljunggren, L. (1965). Cellular localization of monoamines in the upper brain stem of the pigeon. *J. comp. Neurol.* **125**, 355–382.

Gibson, W. R. and Nalbandov, A. V. (1966). Lipid mobilization in obese hypophysectomized cockerels. *Amer. J. Physiol.* **211**, 1345–1351.

Groebbels, F. (1924). Untersuchungen uber den Thalamus und das Mittelhirn der Vögel. *Anat. Anz.* **57**, 385–415.

Guhl, A. M. (1950). Heterosexual dominance and mating behaviour in chicks. *Behaviour* **2**, 106–120.

Guhl, A. M. (1962). The behaviour of chicks. *In* "The Behaviour of Domestic Animals" (Hafez, E. S. E., ed.), pp. 491–530. Baillière, Tindall and Cox., London.

Guillemin, R. (1967). The adenohypophysis and its hypothalamic control. *Ann. Rev. Physiol.* **29**, 313–348.

Harman, A. L. and Phillips, R. E. (1967). Responses in the avian midbrain, thalamus and forebrain evoked by click stimuli. *Exptl. Neurol.* **18**, 276–286.

Hodos, W. and Karten, H. J. (1966). Brightness and pattern discrimination deficits in the pigeon after lesions of the nucleus rotundus. *Exp. Brain Res.* **2**, 151–167.

Hodos, W. and Karten, H. J. (1970). Visual and pattern discrimination deficits after lesions of the ectostriatum in pigeons. *J. comp. Neurol.* **140**, 53–68.

Howell, T. R. and Bartholomew, G. A. (1962). Temperature regulation in the red-tailed tropic bird and the red-footed booby. *Condor.* **64**, 6–17.

Huber, G. C. and Crosby, E. C. (1929). The nuclei and fibre paths of the avian diencephalon, with a consideration of telencephalic and mesencephalic connections. *J. comp. Neurol.* **48**, 1–225.

Hutchinson, J. B. (1967). Initiation of courtship by hypothalamic implants of testosterone propionate in castrated doves (*Streptopelia risoria*) *Nature, Lond.* **216**, 591–592.

Hutchinson, J. B. (1970a). Influence of gonadal hormones on the hypothalamic integration of courtship behaviour in the Barbary dove. *J. Reprod. Fert. Suppl.* **11**, 15–41.

Hutchinson, J. B. (1970b). Differential effects of testosterone and oestradiol on male courtship in Barbary doves (*Streptopelia rissoria*). *Anim. Behaviour* **18**, 41–51.

Ingvar, S. (1923). On thalamic evolution. *Acta med. Scand.* **59**, 696.

Jegelsmar, G. (1896). De verbindingen van de groote hersenen by de vogels met de oculomotoriuskern. *Feestb. Nederl. Vereen. Psychiatrie* p. 241.

Juhasz, L. P. and Tienhoven, A., van (1964). Effect of electrical stimulation of the telencephalon on ovulation and oviposition in the hen. *Amer. J. Physiol.* **207**, 286–290.

Juorio, A. V. and Vogt, A. (1967). Monoamines and their metabolites in avian brain. *J. Physiol.* **189**, 489–518.

Kanematsu, S., Kii, M., Sonada, T. and Kato, Y. (1967). Effects of hypothalamic lesions on body temperature in the chicken. *Jap. J. vet. Sci.* **29**, 95–104.

Kappers, C. U. A. (1920–21). "Vergleichende Anatomie des Nervensystems." Bohn, Haarlem.

Kappers, C. U. A. (1947). "Anatomie Comparée du Système Nerveux", pp. 754. Masson, Paris.

Kappers, C. U. A., Huber, G. C. and Crosby, E. C. (1936). "The Comparative Anatomy of the Nervous System of Vertebrates Including Man." Macmillan (1960 reprint. 3 vols. Hafner and Co., New York).

Karten, H. J. (1963). Ascending pathways from the spinal cord in the pigeon. (*Columba livia*). *Proc. Int. Congr. Zool.* **2**, 23.

Karten, H. J. (1965). Projections of the optic tectum in the pigeon (*Columba livia*). *Anat. Rec.* **151**, 369.

Karten, H. J. (1968). The ascending auditory pathway in the pigeon (*Columba livia*). II. Telencephalic projections of the nucleus ovoidalis thalami. *Brain Res.* **11**, 134–153.

Karten, H. J. and Hodos, W. (1970). Telencephalic projections of the nucleus rotundus in the pigeon (*Columba livia*). *J. comp. Neurol.* **140**, 35–52.

Kendeigh, S. C. (1939). The relation of metabolism to the development of temperature regulation in birds. *J. Exp. Zool.* **82**, 419–438.

King, J. R. and Farner, D. S. (1961). Energy metabolism, thermoregulation and body temperature. *In* "The Biology and Comparative Physiology of Birds" (Marshall, A. J., ed.), vol. 2, pp. 215–288. Academic Press, London and New York.

Komisaruk, B. R. (1967). Effects of local brain implants of progesterone on reproductive behaviour in ring doves. *J. comp. Physiol. Psychol.* **64**, 219–224.

Kordon, C. and Gogan, F. (1964). Localization by microelectrode implantation of hypothalamic structures responsible for testosterone feedback in the duck. *C.R. Soc. Biol. Paris* **158**, 1795–1798.

Kruijt, J. P. (1964). Ontogeny of social behaviour in the Burmese Red Junglefowl (*Gallus gallus spadiceus*). *Behaviour* Suppl. **12**, 1–201.

Kuhlenbeck, H. (1937). The ontogenetic development of the diencephalic centres in a bird's brain (chicken) and comparison with the reptilian and mammalian diencephalon. *J. comp. Neurol.* **66**, 23–27.

Langworthy, O. R. (1926). Abnormalities of posture and progression in the pigeon following experimental lesions of the brain. *Amer. J. Physiol.* **78**, 34–46.

Lehrman, D. A. (1958). Effect of female sex hormones on incubation behaviour in the ring dove (*Streptopelia risscria*). *J. comp. Physiol. Psychol.* **51**, 142–145.

Lehrman, D. A. (1961). Hormonal regulation of parental behaviour in birds and infra-human mammals. Chapter 21. *In* "Sex and Internal Secretions" (Young, W. C., ed.), 3rd ed. Williams and Wilkins, Baltimore.

Lehrman, D. A. (1963). On the initiation of incubation behaviour in doves. *Anim. Behaviour.* **11**, 433–438.

Lehrman, D. S. and Brody, P. (1961). Does prolactin induce incubation behaviour in the ring dove? *J. Endocrinology* **22**, 269.

Lepkovsky, S., Dimick, M. K., Furuta, F., Snapir, N., Park, R., Narita, N. and Komatsu, K. (1967). Response of blood glucose and plasma free fatty acids to fasting and to injection of insulin and testosterone in chickens. *Endocrinology* **81**, 1001–1006.

Lepkovsky, S., Snapir, N. and Furuta, F. (1968). Temperature regulation and appetitive behaviour in chickens with hypothalamic lesions. *Physiol. Behav.* **3**, 911–915.

Lepkovsky, S. and Yasuda, M. (1966). Hypothalamic lesions, growth and body composition of male chickens. *Poultry Sci.* **54**, 582–594.

Lorenz, F. W. (1954). Effects of oestrogens on domestic fowl and applications in the poultry industry. *In* "Vitamins and Hormones" (Harris, R. S., and Thimann, K. V., eds), pp. 235–275. Academic Press, London and New York.

McFarland, D. J. (1965a). The effect of hunger on thirst motivated behaviour in the Barbary dove. *Anim. Behaviour* **13**, 286–292.

McFarland, D. J. (1965b). Control theory applied to the control of drinking in the Barbary dove. *Anim. Behaviour* **13**, 478–492.

McFarland, D. J. (1967). Phase relationships between feeding and drinking in the Barbary dove. *J. comp. Physiol. Psychol.* **63**, 208–213.

McFarland, D. J. and L'Angellier, A. B. (1966). Disinhibition of drinking during satiation of feeding behaviour in the Barbary dove. *Anim. Behaviour.* **14**, 463–467.

McPhail, E. M. (1969). Cholinergic stimulation of dove diencephalon; a comparative study. *Physiol. Behav.* **4**, 655–657.

Meites, J. and Nicoll, C. S. (1966). Adenohypophysis; prolactin. *Ann Rev. Physiol.* **28**, 57–88.

Murton, R. K., Thearle, R. J. P. and Lofts, B. (1968). The endocrine basis of breeding behaviour in the feral pigeon (*Columba livia*). (i) Effects of exogenous hormones on the pre-incubation behaviour of intact males. *Animal Behaviour* **17**, 286–306.

Nalbandov, A. V. (1953). Endocrine control of physiological function. *Poultry Sci.* **32**, 88–103.

Nalbandov, A. and Card, L. E. (1942). Effect of hypophysectomy of growing chicks upon their basal metabolism. *Proc. Soc. Exp. Biol. Med.* **51**, 294–296.

Nalbandov, A. and Card, L. E. (1943). Effect of hypophysectomy of growing chicks *J. Exp. Zool.* **94**, 387–413.

Noble, G. K. and Wurm, M. (1940). The effect of testosterone propionate on the black crowned night heron. *Endocrinology* **26**, 837–850.

Opel, H. (1963). Delay in ovulation in the hen following stimulation of the preoptic brain. *Proc. Soc. Exp. Biol. Med.* **113**, 488–492.

Opel, H. (1967). Effects of brain piqure on the first and second ovipositions of the hens 2 egg sequence. *Proc. Soc. Exp. Biol. Med.* **125**, 627–630.

Orcutt, F. S. Jnr. (1967). Oestrogen stimulation of nest material preparation in the peach faced love-bird (*Agapornis roseicollis*). *Anim. Behaviour* **15**, 471–478.

Papez, J. W. (1929). "Comparative Neurology." Crowell and Co., New York.

Perlia, D. (1898). Ueber ein neuers Opticuscentrum beim Huhne. *Arch. Ophthalmol.* **35**, 20.

Phillips, R. E. (1964). "Wildness" in the mallard duck. Effects of brain lesions and stimulation on escape behaviour and reproduction. *J. comp. Neurol.* **122**, 139–155.

Phillips, R. E. (1966). Evoked potential study of the connections of the avian archistriatum and neostriatum. *J. comp. Neurol.* **127**, 89–100.

Portmann, A. and Stingelin, W. (1961). The central nervous system. *In* "The Biology and Comparative Physiology of Birds" (Marshall, A. J. ed.), vol. 2. pp. 1–36. Academic Press, London and New York.

Ralph, C. H. (1959). Some effects of hypothalamic lesions on gonadotrophin release in the hen. *Anat. Rec.* **134**, 411–427.

Ralph, C. H. and Fraps, R. M. (1959a). Long terms effect of diencephalic lesions on the ovary of the hen. *Amer. J. Physiol.* **197**, 1279–1283.

Ralph, C. H. and Fraps, R. M. (1959b). Effect of hypothalamic lesions on progesterone-induced ovulation in the hen. *Endocrinology* **65**, 819–824.

Ralph, C. H. and Fraps, R. M. (1960). Induction of ovulation in the hen by injection of progesterone in the brain. *Endocrinology* **66**, 269–272.

Randall, W. C. (1943). Factors influencing the temperature regulation of birds. *Amer. J. Physiol.* **139**, 56–63.

Rendahl, H. (1924). Embryologische und morphologische Studien uber das Zwischenhirn beim Huhn. *Acta Zool.* **5**, 241–334.

Revzin, A. M. and Karten, H. (1967). Rostral projections of the optic tectum and nucleus rotundus in pigeons. *Brain Res.* **3**, 264–275.

Richards, S. A. (1970a). The biology and comparative physiology of thermal panting. *Biol. Rev.* **45**, 223–264.

Richards, S. A. (1970b). A pneumotaxic centre in avian brain. *J. Physiol. Lond.* **207**, 57–59P.

Riddle, O. (1963). Prolactin or progesterone as key to parental behaviour, A review. *Anim. Behaviour* **11**, 419–432.

Rogers, F. T. (1918). The relation of lesions of the optic thalamus of the pigeon to body temperature, nystagmus and spinal reflexes. *Amer. J. Physiol.* **45**, 553.

Rogers, F. T. (1919). Studies on the brain stem. I. Regulation of body temperature in the pigeon and its relation to certain cerebral lesions. *Amer. J. Physiol.* **49**, 271–283.

Rogers, F. T. (1920). Studies on the brain stem. IV. On the relation of the cerebral hemispheres and thalamus to arterial blood pressure. *Amer. J. Physiol.* **54**, 355–374.

Rogers, F. T. (1928). Studies on the brain stem. XI. The effects of artificial stimulation and of traumatism of the avian thalamus. *Amer. J. Physiol.* **86**, 639–650.

Rogers, F. T. and Lackey, R. W. (1923). Studies on the brain stem. VII. The respiratory exchange and heat production after destruction of the body temperature regulating centres of the thalamus. *Amer. J. Physiol.* **66**, 453–460.

Rose, M. (1914). Uber die cytoarchitektonische Gliederung des Vorderhirns der Vögel. *J. Psychol. Neurol. Leipzig.* **21**, 278–352.

Rothschild, I. and Fraps, R. M. (1949). The induction of ovulating hormone release from the pituitary of the domestic fowl by means of progesterone. *Endocrinology* **44**, 141–149.

Saalfeld, F. E. von (1936). Untersuchungen über das Hacheln bei Tauben. *Z. vergleich. Physiol.* **23**, 727.

Schwartzkopf, J. (1968). Hearing in birds. *In* CIBA symposium, "Hearing in Vertebrates", pp. 41–59.

Singer, J. and Munzer, E. (1890). Beitrage zur Anatomie des Zentralnervensystems, insbesondere des Ruckenmarkes. *Denkschr. d. k. Akad. Wiss. Wien. Math-Nat.* **57**, 569.

Sinha, M. P. (1959). Observations on the organization of the panting centre in the avian brain. *21st Int. Congr. Physiol. Sci.* Buenos Aires.

Snapir, N., Nir, I., Furuta, F. and Lepkovsky, S. (1969). Effect of administered testosterone propionate on cocks functionally castrated by hypothalamic lesions. *Endocrinology* **84**, 611–619.

Stieda, L. (1869). Studien über das zentrale Nervensystem der Vögel und Säugetiere. *Ztschr. wiss. Zool.* **19**, 1.

Sturkie, P. D. (1958). A survey of recent advances in poultry physiology. *Poultry Sci.* **37**, 495–509.

Sturkie, P. D. (1965). "Avian Physiology", pp. 766. Baillière, Tindall and Cassell, London.

Thauer, R. and Peters, G. (1938). Sensibilität und Motorik bei lange uberlebenden Zwischen Mittelhirntauben. *Pfluger's Arch.* **240**, 503–526.

Tienhoven, A., van (1955). The duration of stimulation of the fowls anterior pituitary for progesterone induced LH release. *Endocrinol.* **56**, 667–674.

Tienhoven, A., van (1961). Endocrinology of reproduction in birds. *In* "Sex and Internal Secretions" (Young, W. C. ed.). Williams and Wilkins, Baltimore.

Tuge, H. and Shima, I. (1959). Defensive conditioned reflex after destruction of the forebrain in pigeons. *J. comp. Neurol.* **111**, 427–446.

Turner, C. H. (1891). The morphology of the avian brain. *J. comp. Neurol.* **1**, 39–92.

Vedyaev, F. P. (1963). Role of striatal and thalamic structures in the central nervous system of birds in control of respiration and functional characteristics. *Fiziol. Zh. USSR.* **49**, 666.

Vedyaev, F. P. (1964). Role of striatal and thalamic structures in the central nervous system of birds in control of respiration. *Fed. Proc.* **23**(5), part II.

Warren, R. P. and Hinde, R. A. (1959). The effects of oestrogen and progesterone on the nest-building of domesticated canaries. *Anim. Behaviour* **7**, 3–4.

Warren, R. P. and Hinde, R. A. (1961). Does the male stimulate oestrogen secretion in the female canary? *Science* **133**, 1354–1355.

Whittow, G. C. (1965). Regulation of body temperature. *In* "Avian Physiology" (Sturkie, P. D., ed.), pp. 766. Baillière, Tindall and Cassell.

Wingstrand, K. G. (1951). "The Structure and Development of the Avian Pituitary." Gleerup, Lund.

Wolf, A. V. (1958). "Thirst". Thomas, Springfield, Illinois.

Wood-Gush, D. G. M. (1954). The courtship of the brown leghorn cock. *British J. Anim. Behaviour* **2**, 95–102.

Wood-Gush, D. G. M. (1955). The behaviour of the domestic chicken. A review of the literature. *Anim. Behaviour* **3**, 81–110.

Yasuda, M. (1957). The functional relationship between the diencephalon and the reproductive organs in the fowl. *Jap. J. zootech. Sci.* **28**, 69–75. (In Japanese.)

Yeates, N. T. M., Lee, D. H. K. and Hines, H. J. (1941). Reactions of domestic fowls to hot atmospheres. *Proc. R. Soc. Qd.* **53**, 105–128.

Young, W. C. (1961). The hormones and mating behaviour. *In* "Sex and Internal Secretions" (Young, W. C., ed.), 3rd ed., pp. 1173–1239. Williams and Wilkins, Baltimore.

# 12 Hypothalamo-hypophyseal, Neuro-hypophyseal and Pineal Systems

## I. INTRODUCTION TO HYPOPHYSEAL STUDIES

The outstanding role which is played by the hypothalamo-hypophyseal systems in vertebrate endocrine control is well-known. Similarly the extensive literature, including many review articles, on the adenohypophysis means there is little need to provide a detailed account of functional studies on this structure which is, in any case, not of neural origin. Many references to such studies are contained in papers cited in the following pages. Without any doubt at all the outstanding account, which laid the foundations for all subsequent anatomical and physiological work, is that of Wingstrand (1951). More recently Tixier *et al.* (1962) and Tixier and Assenmacher (1962) have demonstrated that all the typical secretory cell types occur in the avian pars distalis and there is also in addition a cell grouping which is peculiar to birds.

The hypothalamus and neurohypophysis have now been directly implicated in mammalian anti-diuretic and oxytocic functions for a long period. Their involvement in a wide variety of other responses, which vary from thyroid control to alarm reactions, is both more recent and more controversial. In contrast to this situation the actual neurosecretory pathways in birds were not described until a later date than those of other vertebrates. In 1945 Scharrer and Scharrer directed attention to the miserable lack of any histological evidence for such neurosecretion within the avian hypothalamo-neurohypophyseal system. However, with the development of the Gomori chrome-haematoxylin and paraldehyde-fuchsin staining techniques all the components which are common to other vertebrates were subsequently observed in a wide variety of birds.

Wingstrand (1951, 1953, 1966) and Sturkie (1965) have given detailed descriptions of the overall relationships and histogenesis in a number of genera. Other works which include studies on the histology, histogenesis and fine structure are those of Assenmacher (1957, 1958), Bargmann and Jacob (1952), Benoit (1964a,b), Benoit and Assenmacher (1955, 1966), Duncan (1956), Farner *et al.* (1962), Fujita (1955a,b, 1957), Kobayashi and Farner (1960), Kobayashi *et al.* (1961), Legait (1955, 1957, 1959), Legait and Legait (1958), Oksche (1962) and Oksche *et al.* (1959). The results of many of these workers were reviewed by Oksche (1962), Gabe (1966) and various contributors to Harris and Donovan (1966). One factor which led to the late appearance of these basic descriptions was almost certainly the marked cyclical changes that take place in the characteristic morphological and histo-chemical units. The perikaryons of the hypothalamo-neurohypophyseal system undergo an annual cycle during which the neuro-secretory material available for histochemical purposes undergoes a decrease and then a sub-sequent increase in concentration. The presence of hypothalamic hypophysio-trophic substances in mammals, coupled with the results of injecting hypo-thalamic extracts into young birds, suggest that much hypothalamic control over the adenohypophysis of birds remains to be elucidated (see Guillemin, 1967 and Dodd *et al.*, 1971).

## II. THE GENERAL STRUCTURE OF THE ADENOHYPOPHYSIS

Although we have explicitly excluded a detailed account of the adenohypo-physeal structures from the scope of this book (see above), a brief description of its basic appearance helps to elucidate the disposition of the neurohypo-physis (see Fig. 105). We may therefore summarize the overall arrangement succinctly before proceeding with the consideration of the pars nervosa and allied systems. Although there are some variations in the morphology Wing-strand (1951, 1966) concluded that these are of little importance in the case of the pars distalis. Amongst the Galliformes and Anseriformes it is flattened and somewhat elongated. More variation is seen amongst passeriform genera and the arrangement which characterizes *Xolmis*, *Elaenia* and *Leptasthenura* is almost confined to that order. In these genera the pars distalis is compressed rostro-caudally. A similar compressed appearance was observed in another tyrannid, *Sayornis phoebe*, by Painter and Rahn (1940). Wingstrand suggested that it may be characteristic of primitive passerines and one is reminded of the accompanying aberrations of the inter-carotid anastomosis (Chapter 2, Section II). The only birds other than members of the sub-order Mesomyodes which have this type of organization are swifts and, to a less extent, the hum-ming bird *Sephanoides*. All birds, other than domestic varieties of ducks and hens, appear to have relatively constant species-specific adenohypophyseal organization and Wingstrand found the compressed form in all his specimens of *Apus*, *Sephanoides*, etc.

The vestigial remains of the connection between the pars distalis and its initial site of origin in the oral epithelium is frequently represented by a thread-like projection, the epithelial stalk. A continuous string of cells linking these two areas was only observed in adult *Micropus*, but fairly extensive structures were also present in some specimens of *Sephanoides*, *Spheniscus magellanicus*, *Diomedea epomophora*, *Anser indicus*, *Melanitta nigra*, *Enicognathus leptor-hynchus*, *Xolmis pyrope* and *Emberiza citrinella*. The cells are, in such cases, usually attached to the ventral or caudo-ventral surface of the pars distalis, but amongst anseriform genera were found predominantly at the posterior end of the gland.

The avian pars tuberalis includes three principal units. The pars tuberalis (*sensu stricto*) is a thin layer of cells lying in the pia mater and associated with a portal zone. This consists of strings of epithelial cells which connect it to the pars distalis and is continuous with the pars tuberalis interna. This is a paired structure and is intimately fused to the pars distalis. Histologically it appears as a transitional region which is intermediate in appearance between the two. Wingstrand noted that it was particularly well developed in *Larus, Lyryrus, Cygnus, Melanitta, Coragyps, Phalacrocorax, Procellaria* and *Spheniscus.* The pars tuberalis itself, although universally present, is of a very variable form. In the Procellariiformes it comprises several cell layers and the portal zone is a large and compact glandular body. In contrast the trochiliform genus *Sephanoides* has a portal system which only consists of a few cells on each side. Indeed the pars tuberalis is not easily distinguished in this order. Most other birds fall between these two extremes. Although the portal zone of *Phalacrocorax* and *Coragyps* resembles that of the Procellariiformes, that of *Columba* and many anseriform species is relatively poorly developed. There are, however, considerable variations between different specimens of a single species.

The widespread origin of secretagogic stimuli influencing these centres was summarized by Benoit (1964). For example, in the duck gonadotrophic activity is under the control of hypothalamic centres regulated by nervous stimulation of ocular origin. Retinal autonomic cells of Becher, tuberal nuclei, neurosecretory fibres which end in the median eminence, portal vessels and intra-hypophyseal secondary capillary networks were all involved in his opto-hypothalamo-hypophyseal pathway which mediates his opto-sexual reflex. Work on the pineal would extend such control to that structure as well. However, it seems unlikely that the magnocellular nucleus, which Benoit also cited, is related to any retinal pathway as it is in the ascending auditory pathway but the known stimulatory function of the song of a potential mate (Section VI B) would agree with its inclusion in such afferent pathways to the hypophysis.

## III. HYPOTHALAMO–NEUROHYPOPHYSEAL RELATIONS

### A. The Paraventricular and Supra-optic Nuclei

The avian paraventricular nucleus comprises two cell groups. The more median of these is situated close to the ventricle and stretches from the level of the hind part of the supra-optic nucleus to that of the pallial commissure. The lateral group overlies the supra-optic decussation in front and then runs back to a level which is just in front of the pallial commissure. Both have mutual inter-connections by relatively isolated neuro-secretory perikaryons. The unity of the whole complex has been emphasized by a number of authors. There are also other banks of cells connecting the paraventricular masses to the supra-optic nucleus. This occupies a transverse position at the level of the optic chiasma and to the side of the preoptic recess of the third ventricle. Laterally some of its components can reach the brain surface. Incidentally it is worth noting that anatomically the functionally hypothalamic region extends further than is suggested by the usage of neurologists and physiologists so that the relationships of the structures are sometimes somewhat confused. Wingstrand (1951) reviewed the nomenclature of the region and further accounts are contained in Assenmacher (1958), Legait (1959) and Kobayashi *et al.* (1961).

## B. The Median Eminence

The median eminence is that part of the tuber cinereum which is related to the pars tuberalis of the hypophysis. The tuberal region is much more elaborate in fishes than in other classes. This has been ascribed to the close topographical relationships which it has with the buccal region and its consequent involvement in gustatory activity. With the greater prominence of olfaction in air-breathing vertebrates it has undergone regression. The eminence is the ventral or rostro-ventral region of the floor of the third ventricle and lies between the optic chiasma and the infundibular stalk, of which it is a continuation. The ventral surface is smooth in certain avian genera but can be traversed by furrows in others. These mark the presence of vessels and in general it maintains a very intimate relationship with the capillaries of the hypophyseal portal system. The vascular relationships of the whole region are depicted in Fig. 106. The eminence is rather thin in genera such as *Coragyps* and *Columba*, only contains a few internal glial cells and the capillaries run on its smooth surface. In *Anser*, however, it is thick, surrounded by a deep tubero-infundibular sulcus, and contains many free glial cells or pituicytes. In this case the capillaries are buried within deep furrows.

In transverse sections of the avian median eminence three component layers can be detected. After Wingstrand these components are referred to as the glandular, fibre and ependymal layers. More recently Kobayashi *et al.* (1961) referred to the glandular component as the palisade layer, and Oksche (1962) distinguished between two principal zones. These are an external, including reticular and palisade layers, and an internal zone comprising ependymal and fibre layers. The ependymal cells are occasionally ciliated and constitute a simple epithelial lining to the ventricle. Processes which are directed towards the external surface traverse the fibrous layer, ramify in the palisade layer, and disappear in a dense network of fibres from which there originate conical vascular podia which are supported by the thin reticular membrane separating the nervous tissue from underlying capillaries. Large glial cells which also occur in the ependymal layer have processes that again follow this path towards the capillaries. Within the fibre layer the fibres of the hypophyseal tract are interspersed with glial cells. These are smaller than those of the ependymal layer and their processes pass towards both the blood vessels and the ventral surface.

The palisade or glandular layer contains component fibres of the tubero-hypophyseal tract, neuro-secretory axons, processes from the ependymal and neuroglial cells of the fibre and ependymal layers, and also some morphologically aberrant neuroglial cells (Stutinsky, 1957). The neuro-secretory axons arise from perikaryons which also give rise to the hypothalamo-neurohypophyseal tract although their exact relationships with both the supra-optic and paraventricular tracts is controversial. Their contents differ in their overt staining characteristics. In the anterior regions of the median eminence there is a large quantity of material which stains with chrome-haematoxylin and paraldehyde-fuchsin but this is more or less absent posteriorly. The transition between these two zones varies in its relative abruptness from species to species. Fibres which contain the product have been traced through a characteristic and looped path in both *Anas boschas* and hens. This looping, which occurs in the region of the hypophyseal portal system, has not been seen elsewhere in birds (Kobayashi *et al.*, 1961; Oksche, 1962), but neuro-secretory material is always detectable, and often in large quantities, near to the vessels where some axons terminate.

Kobayashi *et al.* reported that electron-microscopic studies revealed the ependymal and neuroglial cell processes, together with the terminal components of axons. These contained four types of vesicles and granules:

1. Synaptic vesicles with an average diameter of 390 Å;
2. Ovoid vesicles with a diameter of about 490 Å;
3. Electron dense neuro-secretory granules with diameters that ranged from 600 to 1000 Å;
4. Other larger vesicles with diameters comparable with those of the granules.

A careful examination of electron-micrographs showed that the overall pattern of distribution varies and that there are other vesicles which are intermediate in appearance between (1) and (2). One wonders whether those of (4) are developmental precursors of the more electron dense structures in (3). Some terminals, which are probably not neuro-secretory in function, only contained synaptic vesicles. Others contained a mixture of electron opaque granules and electron transparent vesicles.

Within the hypophyseal portal plexus the primary capillaries have a thick basement membrane. This projects into the interior of the median eminence as a series of folds. Some of the axons which contain synaptic, and also larger, vesicles may end in direct contact with this basement membrane but they are rarely seen to actually penetrate it. Generally, however, both ependymal and neuroglial cell processes are interposed between the membrane and the terminal branches of nerve cells. Within the fibre layer the supra-optico-hypophyseal tract fibres also contain elementary granules analogous to those of (3) above. Vesicular structures are, however, found less frequently than in the glandular layer. Electron-micrographs reveal numerous ependymal cell processes which have been cut transversely or obliquely, and also a large number of axonal sections.

The infundibular stalk, which is a ventral continuation of the substance of the median eminence, has a similar composition. In this region the external capillary network is less dense and the relative development of the glandular and fibre layers differs. As the axons are concentrated into a relatively small space the glandular layer is relatively reduced and the fibre layer highly developed. In the vicinity of the pars nervosa the stalk hollows out to form a tube whose dorsal and ventral walls have a similar structure. Although the glandular layer shows little evidence of neurosecretory material such material, together with moderately large Herring bodies, is present throughout the entire extent of the supraoptico-hypophyseal tract. A comparison with the gonad control functions of other regions is provided by Gogan *et al.* (1963).

## C. The Hypothalamo-hypophyseal Tract

The majority of the processes which originate in the neurosecretory perikaryons appear to contain neurosecretory products. Many workers have nevertheless commented on the difficulties which they experience in tracing the axon paths on the basis of the presence of this product (see, for example, Wingstrand, 1951; Arizona and Okomoto, 1957; Legait, 1959; Kobayashi *et al.*, 1961). This is particularly true when studying the proximal components of the supraoptico-hypophyseal tract. In spite of this difficulty it has now been established that both supraoptico-, and paraventriculo-hypophyseal tracts exist, pass round the supra-optic decussation, and converge towards the hypophysis. The fibres may or may not constitute a compact fasciculus which lies in a plane behind the optic chiasma.

The original detailed description of the innervation of the neurohypophysis was provided by Wingstrand (1951). Using the pigeon he showed that coarse fibres in the fibre layer of the median eminence have a predominantly longitudinal orientation. They can be followed into the neural lobe, where they form a thick layer near to the lumen, and then spread diffusely into more peripheral regions. In closer proximity to their sites of origin in the hypothalamus the pattern is more complicated. At the points of junction between the pars nervosa and the stem they are concentrated into the small tract which was alluded to above, and this tube shaped tract surrounds the stem lumen. Anteriorly, at the junction between the stem and the median eminence, the tract opens and continues in the fibre layer of the eminence itself. The whole grouping of hypophyseal tract fibres includes a number of components with differing initial paths. (See Fig. 107.)

At the base of the stem a number of fibres turn dorsally. They run along the hind surface of the hypothalamus and make up the posterior hypophyseal tract. A diffuse plexus of thin fibres lying outside the fibre layer of the eminence, and in the proximal parts of the stem, forms the superficial plexus. This contains a number of transverse fibres but is largely formed by fibres from the tuber nuclei and comprises the tubero-hypophyseal tract. The front part of the eminence is characterized by the decussation of many fibres which takes place just behind the chiasma. Not all fibres do decussate, however, and uncrossed fibres, with some crossed ones, form the supraoptico-hypophyseal tract. This originates, as its name implies, in the supra-optic nucleus, and passes down along the post-optic slope. The remainder of the decussating fibres pass to the pre-optic area and comprise the anterior hypophyseal tract.

## IV. THE NEUROHYPOPHYSIS

The neuro-secretory material which is elaborated in the supra-optic and paraventricular nuclei within the hypothalamus accumulates in the pars nervosa of the hypophysis and also in the glandular zone of the median eminence. The relationships of the pars nervosa are shown in Fig. 105.

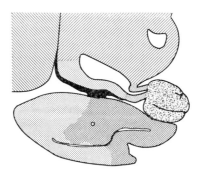

FIG. 105. The pituitary in *Perdix perdix*. The different parts of the pars distalis and pars nervosa are represented by the different size of the dots and the different density of dots respectively. (After Wingstrand, 1966).

Experimental transection of the infundibular stalk in *Anas* is followed by an extensive and almost complete regression of the neurohypophysis (Assenmacher, 1958). The voluminous neuroma which subsequently develops at the site of transection acquires a very rich vascular supply and can be shown to contain a large quantity of neurosecretory material. Evidence from elsewhere suggests that this product, which is made visible by using chrome-haematoxylin and paraldehyde-fuchsin techniques, is in all probability a carrier complex in which one or both of the principal neurohypophyseal hormones are bound to a protein substrate. The hormones are peptides of low molecular weight which readily combine with the large molecule of the protein neurophysin. They are discussed further in Section V, below. Either the vasopressor or the oxytocic principle can be isolated from hypophyseal extracts by varying the pH at which the complex is precipitated. In birds Scharrer's concept is clearly appropriate and the neurosecretory cells form a connecting link between the nervous system and the endocrine glands. It is worth remembering in this context that Willmer (1960) suggested that these may represent nephro-coelomoduct homologues which, having lost the direct excretory and reproductive roles that they played in ancestral forms, have still retained control over both ionic and reproductive activities.

The micro-anatomy of the avian pars nervosa is subject to some variation. Wingstrand (1951) distinguished four anatomical categories:

1. In Procellariiformes, many Galliformes and also Strigiformes, the posterior lobe comprises a series of hollow buds. These have relatively large lumina which maintain a direct communication with the infundibular recess. They are considered to be primary structures and secondary tissue of infundibular stalk origin is relatively scarce. When seen in transverse section it is clear that the original bilateral symmetry has been lost during development and few pituicytes are visible.

2. The second type occurs in a wide variety of birds. As such it characterizes the Struthioniformes, Ardeiformes, Columbiformes, Falconiformes, Psittaciformes, Piciformes, Micropodiformes and Passeriformes. At the base of the lobe there is a central lumen which is extended laterally as two diverticula. These can undergo a further sub-division. The walls of the tubes are thick and the bulk of the tissue is of secondary origin.

3. The third category has an even greater quantity of secondary tissue. As a direct result of this the pars nervosa of *Perdix*, *Larus* and the Anseriformes has a compressed appearance. The lumen of the infundibular recess is limited to the base of the neurohypophysis, and there are only two lateral diverticula on the rostro-lateral surfaces.

4. The fourth type of organization occurs in the Sphenisciformes. An even greater reduction in the lumen of the infundibular recess reduces it to a narrow slit which is bordered by ependymal cells. The buds which were derived from

16

the primitive sac have fused and the superficial limiting membranes have vanished. Only the connective tissue septa remain evident.

In general the ependymal cells of the pars nervosa are more compact than those of the median eminence. Their expansions rest upon the superficial membranes of the primitive walls, or, alternatively, on the basement membranes of the blood vessels. There is a considerable amount of variation in both the size and numbers of the pituicytes. They are most numerous in the

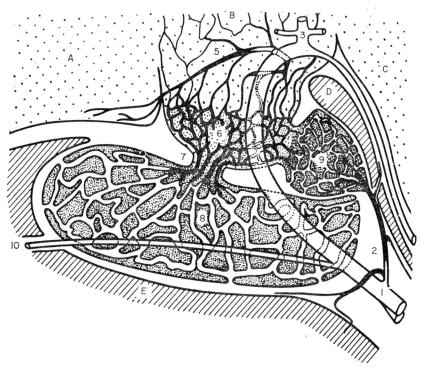

Fig. 106. The blood supply to the avian hypophysis.
A. Optic chiasma; B, Diencephalon; C, Medulla oblongata; D, Dorsum sellae; E, Base of skull; 1, Internal carotid artery; 2, Inferior hypophyseal artery; 3, and 4. Rami of internal carotid; 5, Infundibular artery; 6, Primary capillaries; 7, Portal vessels; 8, Secondary capillaries; 9, Capillary network; 10, Ophthalmic artery. (Wingstrand, 1951).

representatives of categories 3 and 4 above. Wingstrand pointed out that the zones which are in contact with the intima of the pia of the primitive walls, and also the peri-vascular regions, are organized in a manner which is reminiscent of the glandular zone within the median eminence.

The ultimate terminations of the neuro-secretory fibres are close to the basal membranes of the capillaries where they are interspersed with the processes of neuroglial cells. In all cases that part of the fibres which is contained within the pars nervosa is rich in neuro-secretory material. This seems to accumulate in the vicinity of neuroglial components. Duncan (1956), Kobayashi *et al.* (1961) and the Legaits (1958) demonstrated by means of ultra-structural studies that the neuro-secretory axons are surrounded at their ends by the cytoplasmic processes of pituicytes. A close inspection of photographs showed, in optimally situated sections, that the fine axonal processes are applied to neuroglial cells rather than to the basement membranes of blood vessels. At these points their ultra-structure compares with that of the median eminence although Kobayashi and his co-workers emphasized significant

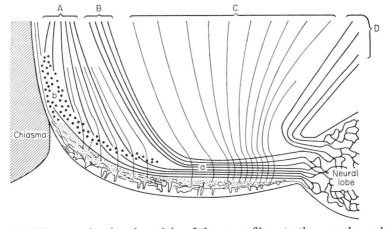

FIG. 107. Diagram showing the origin of the nerve fibres to the neurohypophysis of the pigeon. (*Columba livia*). (Wingstrand, 1951).

A, tractus hypophyseus anterior; B, tr. supraoptico-hypophyseus; C, tubero-hypophyseus; D, tr. hypophyseus posterior; a, fibre layer of the eminentia (tr. hypophyseus); b, decussation of the tr. hypophyseus anterior; c, decussation of the superficial eminentia plexus, which is indicated with delicate lines.

differences in the dimensions of the numerous neuro-secretory granules. Those of the median eminence ranged in diameter from 600 to 1000 Å. In the pars nervosa larger specimens extended the range to 1750 A. Various workers (e.g. Oksche *et al.*, 1959; Kobayashi and Farner, 1960; Kobayashi *et al.*, 1961) have suggested that certain paraldehyde-fuchsin positive inclusions which exist in avian pituicytes are not related to neuro-secretory functions.

## A. Histogenesis of the Neuro-secretory Pathway

The development of the hypothalamo-neurohypophysial pathways was described by Wingstrand in his original papers (1951, 1953) and by Grignon (1956). *In Gallus* the characteristic organization of the median eminence is apparent by the 7th day of incubation. At the 9th day the infundibular stalk is invaded by fibres of the supraoptico-hypophyseal tract. Using the chrome haematoxylin staining technique some definitive neurosecretory material becomes detectable at the 13th or 14th days. At this time it is present in both the neurohypophysis and the perikaryons of the hypothalamic supra-optic nucleus. Once it has appeared in these foci it increases rapidly in concentration but is still not detectable in the paraventricular nucleus until the 16th day. Alongside the hypertrophy of these supraoptic and neurohypophyseal components there is a concomitant hypertrophy of the cells in the glandular layer of the anterior part of the median eminence.

Wingstrand (1953) assayed the anti-diuretic properties of hypothalamo-hypophyseal extracts made from chick embryos at this stage of development. From the 10th day of incubation onwards such extracts have marked anti-diuretic effects when tested by the Brunn method. Some anti-diuretic activity is already present on the 9th day but none can be detected in younger embryos. This discrepancy in which chemical extracts produce effects from the 10th day but no histochemical tests are positive prior to the 13th day was thought by Wingstrand (1953) to result from one of two alternative reasons. Either the amount present before the 13th day is too small to be detected by light microscopy or, the hormones are present in the absence of their chrome-haematoxylin positive carrier.

A similar pattern of development has also been described in *Anas*. Here the neurosecretory material is detectable in both the supraoptic nucleus and the hypophysis at the 14th day of incubation but no positive staining can be detected within the cells of the paraventricular nucleus for a further 24–48 hours. The anterior region of the median eminence is fully developed histologically by the 26th day, 2 days prior to hatching, and gives positive results for chrome haematoxylin staining at the same time as the supra-optic nucleus (Assenmacher, 1957, 1958).

## V. AVIAN NEUROHYPOPHYSEAL HORMONES

As oxytocin has a vaso-depressor affect in fowls this has been the basis of a long-established method of bio-assay (see Dodd *et al.*, 1966). That chicken hypophysis contains pressor, milk-ejection, avian depressor, oxytocic and anti-diuretic principles, or at least properties, has also been known for a long time (Herring, 1913; Hogben and de Beer, 1925; Lawder *et al.*, 1934). Heller (1945) demonstrated the anti-diuretic activity of the pituitaries in both *Anas* and *Columba*, although this was less than in mammals. In the pigeon the ratio of frog water balance activity to anti-diuretic activity was 48:1 which conforms with the pharmacological and chromatographic data of Munsick *et al.* (1960). Ishii *et al.* (1962a,b) also demonstrated frog bladder activity in the eminence of *Zosterops palpebrosa japonica*. An analysis of the oxytocic and frog bladder activity of *Columba* led them to make the suggestion that whilst only arginine vasotocin was present in the eminence it occurred alongside oxytocin in the pars nervosa. Arginine vasotocin was tentatively identified in both pigeons and white-crowned sparrows by Heller and Pickering (1961) and Sawyer (1966).

CyS–Tyr–*Ileu*–Glu(NH$_2$)–Asp(NH$_2$)–CyS–Pro–*Arg*–Gly(NH$_2$)

Munsick *et al.* (1960) investigated the pharmacological properties of the neurohypophysis of *Gallus*. The extracts which they obtained differed from those of mammalian origin. Their effect on isolated rat uterus was noticeably greater in the presence of magnesium, and had a greater milk ejection activity in the rabbit. They concluded that oxytocin was absent. Furthermore the fowl extracts had more marked effects on fowl oviduct than mammalian extracts did. Careful comparisons showed that such oviducts were more responsive to arginine vasopressin than to the lysine analogue, that oxytocin exhibited moderate activity and that 3-phenylalanine oxytocin was virtually inactive. They concluded that both an 8-arginine ring and an oxytocic ring favoured fowl oviduct activity. The use of 8-arginine oxytocin, an analogue which was synthesized by Katsoyannis and du Vigneaud (1958), showed it to be extremely active and it had the same ratio of fowl oviduct to rat vasopressor activity as extracts of fowl neurohypophysis. Chromatographic analysis of the natural fowl extracts gave two active fractions. One of these had a migration pattern which was comparable with that of oxytocin, the other with synthetic arginine vasotocin. This last fraction also had similar biological properties to arginine vasotocin, was destroyed by trypsin, and probably contained either arginine or lysine. Further tests showed that both the fowl extract and arginine vaso-tocin had identical properties when assayed on frog water balance, isolated frog bladder and fowl anti-diuretic activity. Chauvet *et al.* (1960) finally isolated arginine vasotocin from avian neurohypophyses and Ishii *et al.* (1962a, b) subsequently showed that it was present on its own in the eminence and with oxytocin in the pars nervosa.

Arginine vasopressin was also isolated by Chauvet *et al.* (1960) but Sawyer (1966) was reluctant to accept this as evidence for its widespread natural occurrence, especially since the detailed chromatographic studies of Heller and Pickering (1961) and Munsick *et al.* (1960) appeared to rule out the presence of large quantities. Clearly this would not completely preclude its presence were it not for the fact that Chauvet and his co-workers added equine neurophysine to their preparations. In the absence of further evidence once must therefore consider arginine vasopressin a mammalian chemical. Alongside these results oxytocic properties of neurohypophyseal extracts from chick, pigeon and white-crowned sparrow were also observed by a number of these authors and Boissonas *et al.* (1961) showed that shortening the peptide chain reduced avian depressor activity. Indeed certain synthetic analogues have a greater depressor response than the naturally occurring oxytocin does.

# VI. NEUROHYPOPHYSEAL CONTROL FUNCTIONS

## A. Osmoregulation

The osmoregulatory functions of the mammalian pars nervosa, coupled with the significance of osmoreceptor cells in the supra-optic nucleus, have now been known for a long time. It was therefore intellectually satisfying when Wingstrand (1951, 1953) and Legait (1955, 1959) confirmed analogous functions for the avian hypothalamus and neurohypophysis. As was stated in Section IV, Wingstrand also showed that the time of appearance of anti-diuretic activity during embryonic life is, with some slight qualifications, co-incident with the appearance of chrome-haematoxylin positivity in the neuro-hypophysis, median eminence and supra-optic nuclei.

Legait demonstrated that osmoregulatory activity resulted in changes within the hypothalamo-hypophyseal system of Rhode Island hens. These were comparable with the changes reported from other vertebrates. A water free dietary regime gives a marked diminution in the neurosecretory material which can be detected within the perikaryons of the supra-optico-hypophyseal tract and in the pars nervosa. Furthermore it is no longer possible to trace the extra-hypophyseal pathways which lead to the ventricle and to the habenular regions, and the capillaries within the hypothalamic region appear congested. It may be recalled that some controversy exists about the water requirements of birds and Wolf (1958) cites observations which date from Aristotle, to Tancredi in the seventeenth century, Callenfels in the nineteenth, and Krogh in the early decades of the twentieth. However, experimental investigations of the Japanese quail, *Coturnix coturnix*, together with *Taeniopygia castanotis* and *Zonotrichia leucophrys* (Oksche *et al.*, 1959; Farner *et al.*, 1962) showed that the administration of hypertonic salt solutions in place of water resulted in all the symptoms of hypothalamic hyper-activity. The concentration of the neuro-secretory product in the infundibular process decreases and that in the median eminence increases. (See also Table 51.)

## B. The Female Genital Tract

The actual inter-relationships which exist between the avian neurohypophysis and the female genital tract are not clearly understood. Fitzpatrick (1966a) summarized much of the literature. Injections of extracts of the posterior pituitary result in the premature expulsion of eggs in both galliform and columbiform genera (Burrows and Byerly, 1942; Burrows and Fraps, 1942 and the classic work of Riddle, 1921). Vasopressin is more effective than oxytocin but the natural avian 8-arginine vasotocin is more effective than either. Stalk section and hypophysectomy, which gave polydipsia and polyurea, did not result in any appreciable variation in either oviposition time or egg formation according to Shirley and Nalbandov (1956). They cautiously concluded that this result argued against neurohypophyseal involvement in the control of oviposition but that the release of an oxytocic principle from the hypothalamus could not be totally excluded (see Chapter 11, Section VI). It should also be remembered that the neuromata which form proximal to the point of transection can assume the secretory functions.

In more recent years the neural lobe has been implicated in the control of egg-laying by the work of Tanaka and Nakajo (1960, 1962). They demonstrated that the oxytocic potency of neurohypophyseal extracts made from birds which were killed within 5 min of egg-laying was significantly less than that derived from birds sacrificed an hour prior to the expected time of oviposition, or, indeed, at other points in the ovarian cycle. In view of the chemical characteristics of avian hormones which were outlined above, this has been attributed to low arginine vasotocin levels. The results are of particular importance because they were far and away the clearest demonstration of a decrease in oxytocic-like activity under these circumstances, and contrasted very markedly with the less obvious changes which accompany parturition in mammals. It therefore seems probable that the results of preoptic stimulation, which altered the timing of the $C_1$ and $C_t$ egg of a sequence, may involve the release of neurohypophyseal oxytocic factors although the precise causal relationships have yet to be elucidated. Fitzpatrick (1966b) also went on to suggest that the depletion of aldehyde-fuchsin positive material which accompanies the testicular growth spurt in birds (Farner et al., 1962) may implicate neurohypophyseal factors in the release of gonadotrophin from the pars distalis.

The peripheral stimuli which lead to natural ovulation come from widely varying sources. One may conclude that such influences converge upon the hypothalamus, adenohypophysis and neurohypophysis. As an example one can cite the effect of auditory stimuli in ring doves and budgerigars. Female ring doves normally lay eggs after spending several days in association with a male (Harper, 1904; Lehrman et al., 1961). Erickson and Lehrman (1964), together with Matthews (1939), showed that such stimulation to ovulate is not solely a product of direct contact as it can follow separation by a glass screen. Lehrman and Friedman (1969) also demonstrated that placing female birds in sound isolation chambers resulted in ovarian follicular regression. If they were then allowed to hear the sounds picked by a microphone placed in a breeding room the size of the ovarian follicles increased by a factor of two. Using tape-recordings Brockway (1965, 1969) allowed female budgerigars which were kept in total darkness to hear male calls and obtained comparable results. *Soft warble*, a vocalization which is closely associated with male pre-copulatory activity, was found to have a profound stimulatory effect on the ovarian follicles. Other components of the male song had no such effect. Furthermore analogous results were obtained with male birds (Brockway, 1964) which underwent gross testicular development if they could either see or hear breeding pairs.

## VII. HYPOTHALAMO-HYPOPHYSEAL CHANGES DURING THE REPRODUCTIVE CYCLE

An extensive literature exists which describes the control of gonadal cycles in birds (see, Gabe, 1966; Wolfson, 1966; Lofts and Murton, 1968; and Lehrman, 1965). Following the annual breeding season many species enter a period of photo-refractoriness during which there is gonadal regression and the light conditions which otherwise stimulate the gonads to develop are ineffective. Subsequently the photo-period which is associated with the breeding season then induces gonadal activity. In view of the intimate relationships between the hypothalamus and hypophysis, together with the control of gonadal development by release of gonadotrophins from the adenohypophysis, it is not surprising that the annual cycle is accompanied by variations in hypothalamic activity.

In his contributions Legait (1955, 1959) reported that an increase in neuro-secretory activity followed the administration of oestrogens. Injections of hexoestrol appeared to produce nuclear hypertrophy in both the supra-optic and paraventricular nuclei, together with an increase in the amount of chrome-haematoxylin positive material throughout the neuro-secretory pathways. Progesterone injections produced less striking results. Although they appeared to provoke some activity in the perikaryons there was no subsequent accumulation of neuro-secretory material at any focus within the hypothalamo-hypophyseal systems. The effects of oestrogen at this level are probably direct. Some idea of the mechanism of such differential activity is given by the results of Lovelock *et al.* (1963). They found that a marked affinity for thermal electrons characterizes those steroids with the side-chain

$$CO \cdot CH:CR_1R_2.$$

Those steroid hormones with the highest absorption coefficients were members of the adrenocortical complex. The androgens and progestogens absorb electrons to a moderate extent, on the other hand oestrogens have no particular affinity for available thermal electrons. They postulated that, as the cytochrome sequence and other intermediates of oxidative metabolism function as an ordered series of electron traps, the thyroid and steroid hormones may exert their influence by their own capacity to act as traps. The fact that oestrogens do not appear to absorb electrons led to the further suggestion that they act as antagonists to other electron trapping steroids by competitive displacement from the active sites of the cell.

In view of the reaction of the supra-optic and paraventricular foci to injected steroids it is not very surprising that analogous events accompany the annual and other reproductive cycles. Throughout the period of sexual inactivity in the hen the perikaryons of origin are moderately active and there is a high concentration of chrome-haematoxylin positive material in the hypophyseal tracts. *Rosary figures* and Herring bodies are plentiful in both the infundibular stalk and the supra-optico-hypophyseal system. At the egg-laying phase there is some slight degree of nuclear hypertrophy in both supra-optic and paraventricular regions. This is accompanied by a decrease in the hormonal material which is accessible to staining. With the onset of brooding massive nuclear hypertrophy occurs. This is seen first of all in the para-ventricular components and a few days later in the supra-optic nucleus. The hypertrophy is associated with a marked drop in the concentration of the neuro-secretory material in the supra-optico-hypophyseal tract, the median eminence and the infundibular process. Subsequently a further rest period lasts from the 7th until the 10th day. On the next day nuclear hypertrophy takes place within the perikaryons of the hypothalamo-hypophyseal pathways and there is yet another drop in the chrome-haematoxylin positive material in all regions apart from the glandular layer of the median eminence. This second phase of nuclear hypertrophy and its attendant phenomena ends at the 20th day. The cycle is then complete and the appearance of all the components returns to that characteristic of the beginning of the cycle. During moulting the histological characters are comparable with those of the sexual rest period.

The reproductive cycle of most birds is of course based on yearly phases. The seasonal pattern of activity in the hypothalamo-hypophyseal system which accompanies the annual cycle has been intensively investigated in *Zonotrichia leucophrys gambelii*. As in other genera the testicular cycle involves weight changes and these range from 2 mg in winter to more than 600 mg at the end of the spring. The increase in day length controls the spring moult, increased feeding, deposition of migratory fat reserves and also the migration itself. Farner *et al.* (1960), Kobayashi and Farner (1960), Laws (1961) and Oksche *et al.* (1959) established that simultaneous changes occur in both the hypothalamic and hypophyseal regions. Autopsies which were carried out in June showed that neither the supra-optic nucleus nor the median eminence contained very much neuro-secretory material. During August and

16*

September there followed a gradual increase in paraldehyde-fuchsin positive material throughout the system. The change was less obvious within the infundibular stalk as this contained fairly high concentrations even in June. From the middle of November until the spring, during the so-called refractory period, the concentration of paraldehyde-fuchsin positive material underwent a gradual reduction, although a slight interruption in this pattern occurred during January. However, experimentally induced day length changes show that during the refractory period photoperiods of from 8 to 20 hours produce no overt variations in the concentration of neuro-secretory material. Outside this phase of zero responsiveness an increase in the day length causes a drop in the concentration of neuro-secretory products in the median eminence together with concomitant testicular growth. When the testicular development is complete the concentration in the glandular layer of the eminence begins to increase again. If castration is performed at differing points in this cycle the results vary. Autumn castrations have no effect on the concentrations within the eminence. However, by comparison with normal birds, January castrates have higher concentrations in both the supra-optic nucleus and the median eminence.

Farner *et al.* (1962) postulated that the release of gonadotropin from the pars distalis of the hypophysis was under the control of a factor which is itself produced within the perikaryons of the supra-optic nucleus, and possibly also the paraventricular nucleus although this is poorly developed in *Zonotrichia*. This chemical has little or no affinity for paraldehyde-fuchsin and can only be detected as a result of its agglomeration into a particulate form, of polymerization, or fixation on to a paraldehyde-fuchsin positive carrier. Clearly the situation is, however, highly complex and we are far from establishing the definitive inter-relationships within the hypothalamic, pars nervosa and pars distalis systems. A number of additional points of interest are also available.

During the period that birds, kept on an 8-hour photoperiod during the winter, are exposed to 20-hour photoperiods in the summer, there is a significant increase in the acid phosphatase activity in the supra-optic nuclei and in the median eminence (Kobayashi and Farner, 1960). A statistically significant increase in such activity has also been found in the anterior lobe in *Zonotrichia albicollis* (Wolfson and Kobayashi, 1962; Kobayashi *et al.* 1961). Although the precise function of such phosphatase activity was not known Farner *et al.* (1964) assumed that it was implicated in the use of high energy phosphate bonds during the transport and secretory phases. There is also a contemporaneous increase in catheptic-proteinase activity (Table 51). They assumed that this was correlated with the breakdown of neuro-secretory material, a suggestion that fitted the data on this material. Although attempts at demonstrating comparable changes in the neuro-hypophysis were less successful in *Z. leucophrys* they were clear in *Z. albicollis* (Wolfson and

Kobayashi, 1962). Winget *et al.* (1967) reported other changes. They studied the effect of light on the activity of acid and alkaline phosphatase and also cholinesterase activity within the diencephalic, hypophyseal and blood plasma derivatives. In general light increased such activity except in the case of hypothalamic alkaline phosphatase and pituitary cholinesterase. As a result of this they inferred that visible light is a *Zeitgeber* for the enzyme systems concerned.

### TABLE 51

*A summary of the effects of photo-stimulation and dehydration, or osmotic stress, on the hypothalamo-hypophyseal system of the white crowned sparrow.* (*Zonotrichia leucophrys gambelii*).

|  | Supra-optic nucleus | Median eminence | Neural lobe |
| --- | --- | --- | --- |
| Photo-stimulation | Neuro-secretory cells more active. Acid phosphatase activity increased. Catheptic protein-ase activity unchanged | Decrease in alde-hyde fuchsin positive material in the zona externa. Acid phosphatase and catheptic pro-teinase activity increased | No definitely detectable changes |
| Dehydration or osmotic stress | Neurosecretory cells more active. Acid phosphatase activity increased | No definitely detectable changes | Depletion of alde-hyde fuchsin posi-tive material. Increase in acid phosphatase activity |

From Farner *et al.* (1964).

## VIII. THE PINEAL ORGAN

### A. Introduction

The gross anatomy, fine structure and physiological functions of the pineal derivatives in all vertebrate classes have recently been the subject of an extensive and well-informed review by Wurtman *et al.* (1968). This has been supplemented by Ralph (1970) who has indicated the tenuous nature of many of the statements relating to avian pineals. In general terms the organ system undergoes the well-known and gradual transformation from a functional third eye to a sacciform structure which was for very many years thought to serve no function. In both birds and mammals it has now been shown to exert an influence on ovarian and oviducal development, on testicular development, and on both thymus and adrenocorticoid activity. Recent works which are concerned with the avian organ are those of Homma *et al.* (1967), Singh and Turner (1967) and Saylor and Wolfson (1967).

In his semi-popular review of vertebrate pineals Kelly (1962) concluded that birds generally have a single structure although some evidence of para-pineal components is detectable in species such as the swan. Quay (1965), when reviewing both avian and mammalian pineal organization, emphasized that during the foregoing 20 years only about five articles had been published which considered the situation in birds. Previous to that period the available papers are still scarce and largely relate to embryological studies. He suggested that, in broad terms, our knowledge had scarcely advanced beyond the level which was current at the time of Studnička's review (1905).

Cobb and Edinger (1962) have also discussed the absence of a parietal sense organ in birds. Wetzig (1961) studying 10 embryonic and 3 post-embryonic development stages in *Larus* found that any hypothesis of mechanical sup-pression of the stalked organ was untenable. However, the club shape with the largest circumference distal suggests that the actual form is governed by the available space. Cobb and Edinger drew attention to the absence of parietal eyes from many Archosauria and justifiably concluded that its loss occurred during pre-avian phylogeny at a period prior to the appearance of the Jurassic avian genera.

## B. Developmental Studies

The gross development of the avian epiphysis was described by Krabbe (1955) who studied representatives of the Sphenisciformes, Galliformes, Anseriformes, Charadriiformes, Lariformes, Ardeiformes, Psittaciformes, Trochiliformes, Strigiformes and Passeriformes. A small prominence on the diencephalic roof is visible at an early stage and lies between the posterior and habenular commissures. It soon becomes a little hollow sac, and later rather club-shaped. A narrow proximal stalk joins the dilated distal region to the brain and the internal cavity is continuous with the third ventricle. Sub-sequently a large number of hollow "buttons" develop from the outer wall so that in a number of species the organ assumes a "coralline" appearance. Such excrescences tend to be globular in the more proximal regions and pro-gressively more columnar as one passes distally. Krabbe was quite clear that their hollow interiors were not derived from the main central canal during ontogeny but were secondary cavities which arose within the buttons them-selves. This type of development occurred in *Spheniscus*, *Ardea*, *Larus*, *Charadrius*, *Phasianus*, *Pica*, *Parus* and *Hirundo*.

In other genera various differences occurred. Amongst the Strigiformes the entire epiphysis remains rudimentary throughout embryonic life. In a 6·5 mm embryo of *Strix aluco* it was only a small prominence. However, it later assumed the appearance of a small stick and traces of a lumen could be detected in the proximal regions. Amongst the Trochiliformes embryos of both *Chlorostilbon* and *Eupetoma* had a pear-shaped sac, within which lay a free round body, and the entire epiphysis lacked any connection with the roof of the diencephalon. In *Cygnus* the sacciform epiphysis of 10 mm embryos was narrow proximally and its hollow interior remained in communication with the third ventricle. However, a 30 mm embryo had what Krabbe described as an *extraordinary* pineal. Although the main part resembled that of other species it bore two slender rod-shaped prolongations at the distal end. These were themselves bifurcated and terminated in two large buttons. Later these two distinct rods appeared to fuse and in a 44 mm embryo there was a single rod bearing a distal swelling. In embryos which measured 170 mm this parietal corpuscle was detached from the main part of the pineal and lying as a free body within the connective tissue of the dura and the as yet unossified skull. Subsequent work failed to confirm these studies.

A somewhat less aberrant structure occurred in the budgerigar. Here the epiphysis consisted of a long hollow tube which lacked both ramifications and buttons. At its distal extremity a small furrow divided it into two lateral prominences and inside it small keel-shaped eminences were visible on each side. In a 23·5 mm embryo the entire organ was traversed by a central canal whilst the distal end curved backwards and formed a primitive eye just below the epidermis. Within the enlargement forming this eye the internal canal was both dilated and eccentric with a small anterior lenticular part lying in front of a more voluminous region. At later stages of development this epiphyseal organization disappears. The whole *anlage* is short and conical, has a slightly spherical surface and a stalk of reduced length. At such late stages the epiphysis of *Melopsittacus* is massive and its outer surface is closely apposed to the overlying dural membrane.

## C. Adult Anatomy and Histology

To supplement the data which then occurred in the literature Quay (1965) gave descriptions of the adult pineal anatomy in a number of species which he had himself studied. These included material from both captive and wild sparrows, from chickens, and from a series of marine and coastal species predominant amongst which were *Phalacrocorax pelagicus*, *P. penicillata*, *Ereuntes mauri*, *Cataptrophorus semipalmatus* and *Larus delawarensis*. The relatively diverse appearance of the pineals in such varied genera was not easily correlated with the known physiological or adaptive characteristics. Furthermore Ralph and Lane (1969) were unable to find any correlation between changes in the appearance of the pineal of male and female sparrows and either the time of year or gonad size.

The largest known avian pineals are those of the cursorial ratite genera. In *Dromiceius novaehollandiae*, the emu, it is 10 mm long and weighs 100 mg after formalin fixation. Although such a large size immediately suggests that its relative development is related to the overall body weight this relationship is not completely linear if studied in all birds. Quay contrasted the situation in his material of *Phalacrocorax pelagicus* and *Cataptrophorus semipalmatus* which had similar brain and body sizes.

In general the epiphyses of birds are consolidated structures and composed of epithelial parenchyma which is encapsulated by connective tissue. This last is continuous with that part of the meningeal region lying between the forebrain and cerebellum. During the early post-hatching period Ralph and Lane (1969) found that that of the sparrow had an initial, loose, and highly folliculated structure. Throughout the first 90 days of life this was gradually transformed into a compact and more solid organization and area measurements indicated that an overall size decrease took place between 30 and 180 days. Such striking changes led them to conclude that the function during early post-embryonic life may differ from that in the adult. This would, in fact, agree with the results of Saylor and Wolfson (1967, 1968) who observed a transitory, post-natal, progonadotropic phase in young, female, Japanese quail.

The epithelial parenchyma seen under the light microscope is a derivative of the diencephalic neuro-ectoderm and, following the very early suggestions of Studnička (1905), is generally divided into three types. There is first of all a saccular configuration which is principally described in species with small pineals. Secondly there can be a structure consisting of open tubules and follicles, or, thirdly, a solid lobular parenchyma with either closed follicles or clusters of cells. This type has been found predominantly in birds possessing large epiphyses. However, in view of their results on young sparrows Ralph and Lane thought that the ascription of bird pineals to particular categories in cases where the age was unknown was unwise. Furthermore, as might be expected from these results, considerable variations exist and it is very often not particularly easy to decide which of the three categories is appropriate for a particular species or specimen. Even in pineals with a largely solid and lobular structure some patent tubular canals of modified ependymal cells may penetrate through it and remain in continuity with the pineal recess. For example in the pelagic cormorant the distal part consists of a relatively solid parenchyma but tubules and follicular structures predominate in the basal region. The formation of the various closed vesicles or follicles during ontogeny was discussed by Jullien (1942a,b), Romieu and Jullien (1942a,b) and Spiroff (1958).

Nerve cells were considered to be absent from the avian pineal by Stammer

(1961) but, in view of the diversity of the structure in different species, Quay (1965) preferred to leave the question open. In particular he cited some rather nerve-like cells which he observed in both the willet and the western sandpiper. Whatever conclusion one draws from such results undoubted fine nerve fibres and terminals are abundant in the peri-follicular and perivascular areas (Quay, 1965; Stammer, 1961) and a structure described as resembling a chemoreceptor nerve ending occurred within the epiphyseal stalk region of Stammer's ducks. In fact Quay and Renzoni (1963) in an earlier publication had described large neurons with fibre connections from smaller ones in members of the Passeriformes. The axons of the larger cells were then said to make up the pineal tract which could be traced down the epiphyseal stalk as thick bundles. In some species these could even be followed to the habenular commissure and in *Regulus calendula* they disappeared at that level.

Indeed the work of Quay and Renzoni (1963) seemed to confirm the suggestions of van der Kamer (1949) who suggested that some avian pineal ependymal cells had characteristics which are usually associated with sensory cells. They found evidence of both secretory and definitive sensory cells. However Oksche and Harnack (1965a) using electron-microscopy concluded that no such sensory cells existed. Using pigeons, chicken and sparrow material they studied the fine structure and observed many granules and vesicles which had diameters ranging from 0·1 to 0·25 $\mu$ in diameter, and also club-like perivascular terminals with granular inclusions and filaments. In his comparable study of the magpie Collin (1966a,b) drew attention to two contrasting components. One of these comprised ependymal cells, the other rudimentary sensory cells. He considered that the presence of both internal and external segments in these units justified a comparison with the constituents of the parietal organs of lower vertebrates and also the retina *sensu stricto*. However, the reduced size of the external segment, which is present at hatching and persists until at least the 20th day of life, warranted the use of the term "rudimentary". During embryonic development important transformations occur at the level of these units. Normally they have an electron-transparent appearance which is not dissimilar to that of ependymal cells. However, in a given epiphysis a certain percentage undergo a change and become more electron-dense. This transformation is associated with changes in the sub-cellular organelles.

Throughout the electron transparent phase the polymorphic mitochondria are limited by the traditional double membranes, the internal component of which gives rise to the cristae. In electron micrographs the intervening material is at first relatively opaque but subsequently becomes transparent and also enlarges. At the same time the cristae assume a peripheral position and Collin said that 2, 3 or 4 mitochondria become attached to each other by the external lamella of their double membrane. A number of vesicles are also present at the level of the sensory cell ellipsoid. The gross electron dense appearance is partially a reflection of the numerous and widespread ribosomes, whilst the Golgi apparatus is suggestive of intense secretory activity. Other organelles include lysosomal-like structures which contain a fine granular material. These were abundant in both the internal segment and ependymal cells, and also present within the free cells of the lumen. Their overall appearance and larger size distinguished them from the other electron dense granules which were associated with the Golgi system. Collin concluded that these changes, which result in a transformation from a clear to a dense cytoplasm, are an onto-genetic recapitulation of the phylogenetic trend from a sensory to a glandular structure. They were not synchronous in all cells of a given specimen and it is of particular interest that they parallel very closely the changes which follow the administration of gonadotrophins.

Both light and electron-microscopic investigations show that there are numerous capillaries, fibroblasts, histiocytes and mast cells within the stroma. According to Quay (*loc. cit.*) the mast cells vary a lot even within a single species and are particularly numerous in the stalk of the sparrow. In addition to such components elderly specimens of *Ardea cinerea* and *Anser albifrons* have epiphyseal concretions (Stammer, 1961) which seem to have a rather more irregular appearance than the acervuli of ungulate and primate pineals. Lymphoid tissue is also a more conspicuous pineal component in birds than it is in mammals. That of the chicken was described by Romieu and Jullien (1942a) and also Spiroff (1958). The latter author found lymphoid patches in close proximity to the walls of superficial blood vessels as early as 4 days after hatching. They then increased in number to achieve a maximal level at 2–3 months and then retrogression occurred from the 4th month onwards, although some such lymphoid cells persist into adult life. Quay (1965) also reported the almost invariable presence of a nodule within either the capsule or the stroma at 12 months.

The parenchymal cells themselves, which have been classified as neuroglial, resemble ependymal cells in both their appearance and position. An electron microscope study was provided by Oksche and Harnack (1965b). Over and above the pineal ependymocytes *sensu stricto* two other distinct groupings exist (Romieu and Jullien, 1942a; Spiroff, 1958). These have been termed hypendymocytes and pineocytes. Cells comparable with these were classified by Quay into three categories:

1. Cells with small, round, acidophilic nuclei;
2. Cells with their elongated nuclei disposed radially in relation to the lumen and with small nucleoli.
3. Cells with larger nuclei and a large nucleolus.

Outside and also between the follicular or tubule walls there are scattered cells with a large round nucleolus in a very large nucleus. As they have certain similarities to nerve cells, and their processes stain with protargol, Quay (1965) regarded them as potential nerve cells, a less definitive position than he had taken earlier. Some cells lining the tubules and follicles in species such as the cormorant have numerous granular and globular inclusions. Those with large globules appear to be related to certain of the cells lining the pineal recess. Other, possibly secretory cells, are sometimes in close proximity to either the lumina or capillaries. The staining characteristics of these pineal components were studied by Quay (1965) and also Beattie and Glenny. In *Gallus* the nuclei of the parenchymal cells which make up the majority of the gland are stained very strongly by Feulgen. The most positive results were obtained in areas within the periphery of the *cor*. Some cells which line the tubules and follicles stain with the periodic acid Schiff technique and are resistant to amylase digestion. Alcian blue gives distinctly positive results, particularly in connective tissue cells and less obviously around the sheaths of the vascular rete. Both neutral and acid mucopolysaccharides are therefore present at sites throughout the organ. Both the ninhydrin and diazotization reactions also give sufficient positive results to indicate the presence of biologically active amines, and melatonin is indicated within the areas near to the sinuses.

The total vascular supply of the avian pineal was described by Beattie and Glenny (1966). The major source of arterial blood is the posterior meningeal artery which usually arises from the cranial branch of the left internal carotid or the posterior cerebral artery (see page 24). Within large spaces in the neighbourhood of the epiphysis this meningeal artery has a complex anastomosis and thereby forms the circus vasculosus which bathes some lobules of the pineal stalk in blood. The pineal rete *sensu stricto* comprises a further network of both vascular cavities and fine vessels which divide the gland into lobes of varying size. Very narrow vessels lie in close association with the *cor secretorium*. The venous out-flow appears to emerge from the anterio-dorsal region as the left and right epiphyseo-meningeal veins. These curve laterally and downwards along the posterior and medial surfaces of the telencephalon to a point near the base of the posterior cerebral artery.

## D. Melatonin and Pineal Function

Claims that changes occur in the appearance and disposition of pineal cells in association with differing levels of endocrine activity have been put forward at a number of times over many years. Such suggestions remained controversial, and, indeed, largely ignored, until relatively recently (see, for example, Desogus, 1924, 1926; Spiroff, 1958). Wetzig (1961) suggested that the contents at the luminal ends in many of the cells may reflect apocrine secretory activity. However, one of the basic discoveries in recent decades was the identification of the characteristic pineal indole, melatonin. This is a pale yellow, crystalline material most of which can only be found within the pineal gland itself. It is more abundant in birds than mammals (Axelrod *et al.*, 1964) and the concentration within hen pineal appeared to be some 10 times greater than that in any mammal then examined. This conforms with equally high concentrations of the enzyme which is responsible for its synthesis.

Structurally the melatonins are 5-methoxy N-acetyltryptamine and the synthesizing enzyme is hydroxyindole-O-methyl transferase (HIOMT). Using the pineal glands of white Leghorn hens Axelrod *et al.* produced homogenates in cold water and then incubated them with S-adenosylmethionine-methyl-$^{14}$C, and N-acetylserotonin. The radioactive product had an identical $R_f$ value to that of melatonin and the enzymatic activity per milligram of pineal tissue was twice as high as the highest values which had previously been established for mammals—those of Primates.

Subsequently groups of hens were kept in differing light régimes. These varied from continuous darkness to continuous light. Assays for HIOMT which were carried out at the end of this period are represented in Table 52. There was a highly significant decrease in the activity of the melatonin forming enzyme when the birds were kept in darkness for 5 days and, complementarily, a significant increase occurred in those which were maintained in continuous light. No comparable changes took place in the levels of monoamine oxidase activity. Another interesting result which emerged was that the overall weight of the pineals was greater in the specimens which had been kept in a light régime than in those kept in continuous darkness. This provides a marked contrast to the situation in rats.

## TABLE 52

*The effects of light on the melatonin synthesizing enzyme in the hen pineal gland.*

| Lighting conditions | Pineal weight (in mg) | HIOMT (per pineal) | HIOMT (per mg) | MAO (per pineal) | (MAO per mg) |
|---|---|---|---|---|---|
| Diurnal | 5·0±0·57 | 44·4±4·0 | 8·9±0·8 | 9·0±0·8 | 1·8±0·17 |
| Constant light | 4·9±0·62 | 56·9±4·2 | 11·4±1·8 | 9·6±1·5 | 2·0±0·45 |
| Constant darkness | 3·0±0·34 | 17·4±3·4 | 5·8±0·7 | 10·1±3·2 | 3·3±0·77 |

From Axelrod, Wurtman and Winget (1964).

Diurnal cycles in the amount of presumptive melatonin within the pineals of *Coturnix coturnix* and *Steganura paradisea* are shown in Figs 108–110. Ralph *et al.* (1967) found that when subjected to differing periods of darkness the melatonin content of the pineal varies during the dark phase and the light phase. Quay (1966) also found 24-hour rhythms of both 5-hydroxy-tryptamine and 5-hydroxyindole acetic acid concentrations in the pineal. However Lynch and Ralph (1970) found that the pineal melatonin rhythm persisted when chickens were kept under a constant light régime.

It has already been emphasized that Ralph and Lane (1969) were unable to find any correlation between the pineal appearance of either sex and either the time of year or gonad size. However, following the implication of the pineal gland as a mediator in the gonadal responses of rats and hamsters a number of workers have demonstrated similar results in birds. Singh and Turner (1967) found that birds which had received injections of 50 or 100 μg of melatonin at 7 weeks of age subsequently had reduced testis, ovarian, adrenal and thymus weights. There seemed to be no corresponding change in the thyroid secretion rate. Saylor and Wolfson (1967) investigated the effect

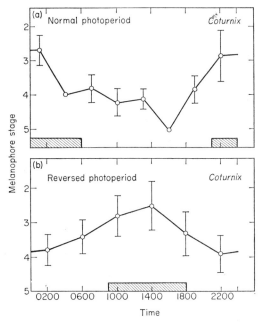

Fig. 108. The concentration of melatonin in *Coturnix*. (Ralph *et al.*, 1967) for comparison with Figs 109–110.

which pineal extirpation exerted on the achievement of sexual maturity in *Coturnix*. As a result they concluded that the pineal normally contributes to the sudden and rapid ovarian development which precedes laying. In the absence of a pineal both this maturation and the onset of lay are delayed by a period of several days. As a simultaneous lowering of ovarian, oviducal and

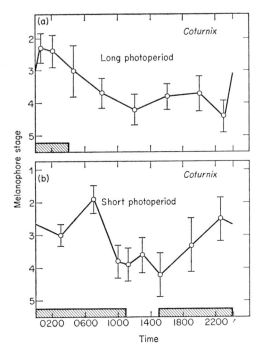

FIG. 109. The concentration of melatonin in *Coturnix*. (Ralph *et al.*, 1967).

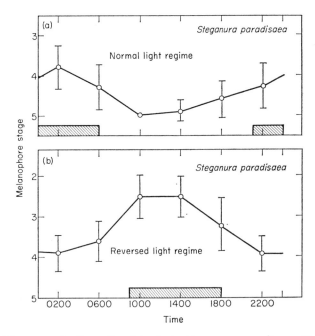

FIG. 110. The concentration of melatonin in *Steganura paradisaea*. (Ralph *et al.*, 1967).

adenohypophyseal weights occurred they suggested that the observed differences may reflect an alteration in the levels of gonadotrophin secretion. However, whatever the exact relationships between pineal and pars distalis function may be, the effects of pinealectomy are only transitory. Indeed they seem to be largely restricted to a brief period prior to the onset of sexual maturity during which the pineal appears to have a pro-gonadotropic influence.

Other workers have also noted a similar age-dependant effect. Shellabarger (1952) found that the pineal in male White Leghorns is pro-gonadotropic during the first 20 days of life. It then seemed to enter an anti-gonadotropic phase between the 40th and 60th days. During both these periods bovine pineal extracts reversed the results of pinealectomy. The overt results under experimental conditions were a decrease in testis weight of pinealectomized cockerels at the 19th day, an increase at 40 days, and no effect at 90 days. As Saylor and Wolfson (1967) found that removal of the superior cervical ganglia in mature female *Coturnix* caused a cessation of egg production for up to 13·2 days they suggested that this too may be involved with the pineal but that, unlike the situation in mammals, it exerted a gonad stimulatory not a gonad inhibitory influence. A more diffuse pineal control of the bodies activities was suggested by the observations of Gaston and Menaker (1968). They showed that under constant environmental conditions it contributes to the maintenance of locomotor activity. It was not, however, necessary for entrainment to light cycles. They tentatively concluded that the pineal does not merely have a direct control over the circadian cycles of activity, because if this was the case then pinealectomy should render the preparations either continuously active or inactive. In fact such preparations were rhythmically active in a light-dark régime, with inactivity in the dark phases, and continuously active when kept in permanent darkness. Although they intimated that the pineal might be a self-sustaining oscillator which normally drives a damped oscillator that in turn controls locomotor activity, they concluded that this was not in itself the entire control mechanism. The observed entrainment to light cycles could therefore result from a direct photic response involving the hypothetical damped oscillator. However, the pineal could also be a coupling device within a multi-oscillator clock system. Two or more of the oscillating components of the overall timing machinery might then be coupled and a fixed phase relationship maintained between them by the intact pineal. In view of the numerous sensory clues which are available as potential indicators of diurnal environmental changes it would be surprising if one, two or even three primary control systems were in themselves of totally overriding importance. Auditory, thermal and visual receptors are all potential gauges of such environmental changes and their central pathways potential lines of transmission. In view of the overt experimental evidence for their effects on endocrine control all their primary and secondary centres offer themselves as alternatives to pineal integration.

## References

Arizona, H. and Okomoto, S. (1957). Comparative neurologic study on the hypo-thalamo-hypophysial neurosecretory system. *Med. J. Osaka Univ.* **8**, 195–228.

Assenmacher, J. (1957). Évolution de la neurosécretion hypothalamo-neuro-hypophysaire au cours de l'organogenèse du Canard domestique. *C.R. Soc. Biol., Paris* **151**, 1301–1304.

Assenmacher, J. (1958). Récherches sur la controle hypothalamique de la fonction gonadotrope prehypophysaire chez le Canard. *Arch. Anat. micro.* **47**, 447–572.

Axelrod, J., Wurtmann, R. J. and Winget, C. M. (1964). Melatonin synthesis in the hen pineal gland and its control by light. *Nature, Lond.* **201**, 1134.

Bargmann, W. and Jacob, K. (1952). Uber Neurosekretion im Zwischenhirn von Vögeln. *Z. Zellforsch.* **36**, 556–562.

Beattie, C. W. and Glenny, F. H. (1966). Some aspects of the vascularization and chemical histology of the pineal gland in *Gallus. Anat. Anz.* **118**, 396–404.

Benoit, J. (1964a). The structural components of the hypothalamo-hypophyseal pathway in birds with particular reference to photostimulation of the gonads. *Ann. N.Y. Acad. Sci.* **117**, 23–34.

Benoit, J. (1964b). The role of the eye and the hypothalamus in the photostimulation of gonads in the duck. *Ann. N.Y. Acad. Sci.* pp. 204–216.

Benoit, J. and Assenmacher, I. (1955). Le controle hypothalamique de l'activité prehypophysaire gonadotrope. *J. Physiol. Paris* **47**, 427–567.

Benoit, J. and Assenmacher, I. (1966). Photosensitivity of the superficial and deep nerve receptors during gonado-stimulation by visible radiation in the prepubertial Pekin duck. *C.R. Acad. Sci., Paris* **262D**, 2750–2752.

Boissonas, R. A., Guttman, S., Berde, B. and Konzett H. (1961). Relationships between chemical structures and the biological properties of the posterior pitui-tary hormones and their synthetic analogues. *Experientia* **17**, 377–391.

Brockway, B. F. (1964). Social influences on reproductive physiology and ethology of budgerigars (*Melopsittacus undulatus*). *Anim. Behaviour.* **12**, 493–501.

Brockway, B. F. (1965). Stimulation of ovarian development and egg-laying by male courtship vocalizations in budgerigars. *Anim. Behaviour* **13**, 575–578.

Brockway, B. F. (1969). Roles of budgerigar vocalization in the integration of breed-ing behaviour. *In* "Bird Vocalisations" (Hinde, R. A., ed.), pp. 131–158. Cam-bridge University Press.

Burrows, W. H. and Byerly, T. C. (1942). Premature expulsion of eggs by hens following injection of whole posterior pituitary preparations. *Poultry Sci.* **21**, 416–421.

Burrows, W. H. and Fraps, R. M. (1942). Action of vasopressin and oxytocin in causing premature oviposition in domestic fowl. *Endocrinology* **30**, 702–705.

Chauvet, J., Lenci, M-T. and Acher, R. (1960). Presence de deux vasopressines dans la neurohypophyse du poulet. *Biochim. Biophys. Acta.* **38**, 571–573.

Cobb, S. and Edinger, T. (1962). The brain of the emu (*Dromaeus novaehollandiae*, Lath.). *Breviora* **170**, 1–18.

Collin, J. P. (1966a). Sur l'évolution des photorecepteurs rudimentaire epiphysaires chez Pie (*Pica pica* L.). *C.R. Soc. Biol., Paris* **160**, 1876–1880.

Collin, J. P. (1966b). Étude préliminaire des photorecepteurs rudimentaires de l'épiphyse de *Pica pica* pendant la vie embryonnaire et post-embryonnaire. *C.R. Acad. Sci., Paris*, **263D**, 660.

Desogus, V. (1924). La pineale negli ucelli normale e cerebrolesionati. *Riv. Biol.* **6**, 495–504.

Desogus, V. (1926). Contributo allo studio della pineale e dell'ipofisi degli ucelli in stato maternita. *Monit. Zool. ital.* **37**, 273–282.

Dodd, J. M., Follett, B. K. and Sharp, P. J. (1971). Hypothalamic control of pituitary function in submammalian vertebrates. *In* "Advances in Comparative Physiology and Biochemistry". **4.** Academic Press London and New York, pp. 114–224.

Dodd, J. M., Perks, A. M. and Dodd, M. H. (1966). Physiological functions of the neurohypophysial hormones in submammalian vertebrates. *In* "The Pituitary Gland" (Harris, G. W. and Donovan, B. T., eds), vol. 3, pp. 578–623. Butterworths, London.

Duncan, P. (1956). An electron microscope study of the neurohypophysis of a bird: *Gallus domesticus. Anat. Rec.* **125**, 457–463.

Erickson, C. J. and Lehrman, D. S. (1964). Effect of castration of male ring doves upon ovarian activity of females. *J. comp. physiol. Psychol.* **58**, 164–166.

Farner, D. S. (1961). Comparative physiology: photo-periodicity. *Ann. Rev. Physiol.* **23**, 71–96.

Farner, D. S., Kobayashi, H., Oksche, A. and Kawashima, S. (1964). Proteinase and acid phosphatase activities in relation to the function of the hypothalamo-hypophyseal neurosecretory systems of photostimulated and of dehydrated white-crowned sparrows. *Progr. Brain Res.* **5**, 147–155.

Farner, D. S. and Oksche, A. (1962). Neurosecretion in birds. *Gen. comp. Endocrinol.* **2**, 113–147.

Farner, D. S., Oksche, A., Kobayashi, H. and Laws, D. F. (1960). Hypothalamic neurosecretion in the photoperiodic testicular response of birds. *Anat. Rec.* **137**, 354.

Farner, D. S., Oksche, A. and Lorenzen, L. (1962). Hypothalamic neurosecretion and the photo-periodic testicular response in the white-crowned sparrow, Zonotrichia Leucophyrs Gambelii. *In* "Neurosecretion" (Heller, H. and Clark, R. B., eds), pp. 187–197. Academic Press, London and New York.

Fitzpatrick, R. J. (1966a). The posterior pituitary gland and the female reproductive tract. *In* "The Pituitary Gland" (Harris, G. W. and Donovan, B. T., eds), vol. 2, pp. 453–504. Butterworths, London.

Fitzpatrick, R. J. (1966b). The neurohypophysis and the male reproductive tract. *In* "The Pituitary Gland", vol. 2, pp. 505–516. Butterworths, London.

Fujita, H. (1955a). Die histologische Untersuchung des Hypothalamus-hypophysen-system der Vogel. *Arch. histol. Jap.* **9**, 109–114.

Fujita, H. (1955b). Ontogenetische Studien uber den Bau und das Neurosekretions-und Nisslbild des hypothalamisch-hypophysaren system des Haushuhnes. *Arch. histol. Jap.* **9**, 213–224.

Fujita, H. (1957). Electron microscopic observations on the neurosecretory granules in the pituitary posterior lobe of the dog. *Arch. histol. Jap.* **12**, 165–172.

Gabe, M. (1966). "Neurosecretion", pp. 872. Pergamon Press.

Gaston, S. and Menaker, M. (1968). Pineal function; the biological clock in the sparrow. *Science* **160**, 1125–1127.

Gogan, F., Kordon, C. and Benoit, J. (1963). Effects of lesions in the median eminence on gonad stimulation in the duck. *C.R. Soc. Biol. Paris* **157**, 2133–2136.

Grignon, G. (1956). "Développement du complexe hypothalamohypophysaire chez l'embryon de poulet." Soc. Impress, Typograph, Nancy.

Guillemin, R. (1967). The adenohypophysis and its hypothalamic control. *Ann. Rev. Physiol.* **29**, 313–348.

Harris, G. W. and Donovan, B. T. (eds) (1966). "The Pituitary Gland", 3 vols. Butterworths, London.

Harper, E. H. (1904). The fertilization and early development of the pigeon's egg. *Amer. J. Anat.* **3**, 349–386.

Heller, E. J. (1945). The effect of neurohypophyseal extracts on the water balance of lower vertebrates. *Biol. Rev.* **20**, 147–158.

Heller, H. and Pickering, B. T. (1961). Neurohypophyseal hormones of non-mammalian vertebrates. *J. Physiol.* **155**, 98–114.

Herring, B. T. (1913). Further observations upon the comparative anatomy and physiology of the pituitary body. *Quart. J. exp. Physiol.* **6**, 73–108.

Hogben, L. T. and de Beer, G. R. (1925). Studies on the pituitary. VI. Localization and phyletic distribution of active materials. *Quart. J. exp. Physiol.* **15**, 163–176.

Homma, K., McFarland, L. Z. and Wilson, W. O. (1967). Response of the reproductive organs of the Japanese quail to pinealectomy and melatonin injections. *Poultry Sci.* **46**, 314–319.

Ishii, S., Hirano, T. and Kobayashi, H. (1962a). Preliminary report on the neuro-hypophysial hormone activity in the avian median eminence. *Zool. Mag. Tokyo* **71**, 206–211.

Ishii, S., Hirano, T. and Kobayashi, H. (1962b). Neurohypophysial hormones in the avian median eminence. *Gen. comp. Endocrin.* **2**, 433–440.

Jullien, G. (1942a). Sur l'origin et la formation des vesicules closes dans l'épiphyse des Gallinacé. *C.R. Soc. Biol., Paris* **136**, 243–244.

Jullien, G. (1942b). Sur les formations vesiculaires de la gland pineale des oiseaux. *Bull. Mus. Hist. Nat. (Marseilles)*, **2**, 163–170.

Kamer, J. C. van der (1949). Over de ontwikkeling, de determinatie, en de betekenis van de epiphyse en de paraphyse van de amphibia. Thesis. Arnhem.

Katsoyannis, P. G. and du Vigneau, V. (1958). Arginine vasotocin, a synthetic analogue of the posterior pituitary hormones containing the ring of oxytocin and the side chain of vasopressin. *J. biol. Chem.* **233**, 1352–1354.

Kelly, D. E. (1962). Pineal organs: photoreception, secretion and development. *Amer. Sci.* **50**, 597–625.

Kobayashi, H., Bern, H. A., Nishioka, R. S. and Hyado, Y. (1961). The hypothalamo-hypophyseal neurosecretory system of the parakeet *Melopsittacus undulatus. Gen. comp. Endocrinol.* **1**, 545–564.

Kobayashi, H. and Farner, D. S. (1960). The effect of photoperiodic stimulation on phosphatase activity in the hypothalamo-hypophyseal system of the white-crowned sparrow *Zonotrichia leucophrys gambelli. Z. zellforsch.* **54**, 275–306.

Krabbe, K. H. (1955). Development of the pineal organ and a rudimentary parietal eye in some birds. *J. comp. Neurol.* **103**, 139–150.

Lawder, A. M. de, Tarr, L. and Gelling, E. M. K. (1934). The distribution in the chicken's hypophysis of so-called posterior lobe principles. *J. Pharmacol.* **51**, 142–143.

Laws, D. F. (1961). Hypothalamic neurosecretion in the refractory and post-refractory periods, and its relationship to the rate of photoperiodically induced testicular growth in *Zonotrichia Leucophrys Gambelii. Zeit Zellforsch.* **54**, 275–306.

Legait, H. (1955). Etude histophysiologique et experimentale du système hypothalamo-neurohypophysaire de la poule Rhode Island. *Arch. micr. Anat.* **44**, 323–343.

Legait, H. (1957). Les voies extra-hypothalamo-neurohypophysaire de la neuro-sécretion diencephalique dans le serie des vertébrés. "2nd Intern. Symp. Neurosecretion" (Bargmann, W., Hanstrom, E. and Scharrer, B., eds), pp. 42–51. Springer, Berlin.

Legait, H. (1959). "Contribution a l'étude morphologique et expérimentale du système hypothalamo-hypophysaire de la poule Rhode Island." Soc. Impress, Typograph, Nancy.

Legait, H. and Legait, E. (1958). Rélations entre les noyaux hypothalamiques neurosécretoires et les régions septale et habenulaire chez quelques oiseaux. In "Pathophysiologia Diencephalica" (Curry, S. B. and Martini, L. eds), pp. 143–147. Springer, Berlin.

Lehrman, D. S. (1965). Interaction between internal and external environments in the regulation of the reproductive cycle of the ring dove. In "Sex and Behaviour" (Beach, F. A., ed.), Wiley, New York.

Lehrman, D. S., Brody, P. N. and Wortis, R. P. (1961). The presence of the mate and of nesting material as stimuli for the development of incubation behaviour and for gonadotrophin secretion in the ring dove. Endocrinology 68, 507–516.

Lehrman, D. S. and Friedman, M. (1969). Auditory stimulation of ovarian activity in the ring dove. Anim. Behaviour 17, 494–497.

Lofts, B. and Murton, R. K. (1968). Photoperiodic and physiological adaptations regulating avian breeding cycles, and their ecological significance. J. Zool. 155, 327–394.

Lovelock, J. E., Simmonds, P. G. and Vandenhauvel, W. J. A. (1963). Affinity of steroids for electrons with thermal energies. Nature. Lond. 197, 249–251.

Lynch, H. J. and Ralph, C. L. (1970). Persistent rhythm of pineal melatonin in chicks under constant conditions. Amer. Zool. 10, 491–492.

Matthews, L. H. (1939). Visual stimulation and ovulation in pigeons. Proc. Roy. Soc. B. 126, 557–560.

Munsick, R. A., Sawyer, W. H. and Dyke, H. B. von (1960). Avian neurohypophysial hormones, pharmacological properties and tentative identification. Endocrinology 66, 860–871.

Oksche, A. (1962). The fine nervous, neurosecretory and glial structure of the median eminence in the white-crowned sparrow. In "Neurosecretion" (Heller, H. and Clark, R. B. eds), pp. 199–208. London and New York. Academic Press.

Oksche, A. and Harnack, A. von (1965a). Vergleichende Electronmikroskopische Studien am Pinealorgan. Progr. Brain Res. 10, 237–258.

Oksche, A. and Harnack, A. von (1965b). Rudimentary sense cell structures in the pineal organ of the chicken. Naturwissenschaften 52, 662.

Oksche, A., Laws, D. F., Kamemoto, F. S. and Farner, D. S. (1959). The hypothalamo-hypophyseal neurosecretory system of the white-crowned sparrow, Zonotrichia leucophrys gambelii. Z. Zellforsch. 51, 1–42.

Painter, B. T. and Rahn, H. (1940). A comparative histology of the pituitary. Anat. Rec. 78, abstract 112.

Quay, W. B. (1965). Histological structure and cytology of the pineal organ in birds and mammals. Progress. Brain. Res. 10, 49–84.

Quay, W. B. (1966). Rhythmic and light induced changes in levels of pineal 5-hydroxy indoles in the pigeon. Gen. comp. Endocrinol. 6, 371–377.

Quay, W. B. and Renzoni, A. (1963). Studio comparativo e sperimentale sull struttura e citologia della epifisi nei Passeriformes. Riv. Biol. 56, 363–407.

Ralph, C. L. (1970). The structure and alleged function of avian pineals. Am. Zoologist. 10, 217–235.

Ralph, C. L., Hedlund, L. and Murphy, W. A. (1967). Diurnal cycles of melatonin synthesis in bird pineal bodies. *Comp. Biochem. Physiol.* **22**, 591–599.

Ralph, C. L. and Lane, K. B. (1969). Morphology of the pineal body of wild house sparrows in relation to reproduction and age. *Canad. J. Zool.* **41**, 1205–1208.

Riddle, O. (1921). A simple method of obtaining premature eggs from birds. *Science* **54**, 664–666.

Romieu, M. and Jullien, G. (1942a). Caracteres histologiques et histophysiologiques des vesicules epiphysaire des gallinaces. *C.R. Soc. Biol., Paris* **136**, 628–630.

Romieu, M. and Jullien, G. (1942b). Evolution et valeur morphologique des vesicules closes de la glande pineale des oiseaux. *C.R. Soc. Biol., Paris* **136**, 630–632.

Sawyer, W. H. (1966). Neurohypophysial principles of vertebrates. *In* "The Pituitary Gland" (Harris, G. W. and Donovan, B. T., eds), vol. 2, pp. 307–329. Butterworths, London.

Saylor, A. and Wolfson, A. (1967). The avian pineal gland. *Science* **158**, 1478–1479.

Saylor, A. and Wolfson, A. (1968). Influence of the pineal gland on gonadal maturation in the Japanese quail. *Endocrinology* **51**, 152–154.

Seharrer, E., and Sharrer, B. (1945). Neurosecretion. *Physiol. Rev.* **25**, 171–181.

Shellabarger, C. J. (1952). Pinealectomy vs. pineal injection in the young cockerel. *Endocrinology* **51**, 152–154.

Shirley, H. V. and Nalbandov, A. V. (1956). Effects of neurohypophysectomy in domestic chickens. *Endocrinology* **58**, 477–483.

Singh, D. V. and Turner, C. W. (1967). Effect of melatonin upon thyroid hormone secretion rate and endocrine glands of chicks. *Proc. Soc. Exp. Biol. Med.* **125**, 407–411.

Spiroff, B. E. N. (1958). Embryonic and post-hatching development of the pineal body of the domestic hen. *Amer. J. Anat.* **103**, 375–401.

Stammer, A. (1961). Untersuchungen uber die Struktur und die Innervation der Epiphyse bei Vögeln. *Acta. Biol. N.S. Acta. Univ. Szeged.* **7**, 65–75.

Studnička, F. K. (1905). Die Parietalorgane. *In* "Lehrbuch der vergleichenden mikroskopischen Anatomie" (Oppel, A., ed.), vol. 5, pp. 1–254. Fischer, Jena.

Sturkie, P. D. (1965). "Avian Physiology", pp. 776. Baillière, Tindall and Cassell, London.

Stutinsky, F. (1957). Récherches morphologiques sur le complexe hypothalamo-neurohypophysaire. *Bull. Micro. appl. mem. hors. series.* **2**, 1–90.

Tanaka, K. and Nakajo, S. (1960). Oxytocin in the neurohypophysis of the laying hen. *Nature, Lond.* **187**, 245.

Tanaka, K. and Nakajo, S. (1962). Participation of neurohypophysial hormone in oviposition in the hen. *Endocrinology* **70**, 453–458.

Tixier, V. A., Herlant, M. and Benoit, J. (1962). The prehypophysis of the pekin male duck through an annual cycle. *Arch. Biol. Liege* **73**, 317–368.

Tixier, V. A. and Assenmacher, I. (1962). Effect of reserpine and permanent light on prehypophysis of male duck. *C.R. Soc. Biol., Paris* **156**, 37–43.

Wetzig, H. (1961). Die Entwicklung der Organe des Zwischenhirndaches (Epiphyse und Plexus choroideus anterior), bei der Sturmmove, *Larus canus*, L. *Morph. Jahrb.* **101**, 406–431.

Willmer, E. N. (1960). "Cytology and Evolution" (2nd ed., 1970). Academic Press, London and New York.

Winget, C. M., Wilson, W. O. and McFarland, L. Z. (1967). Response of certain diencephalic, pituitary and plasma enzymes to light in *Gallus domesticus*. *Comp. Biochem. Physiol.* **22**, 141–147.

Wingstrand, K. G. (1951). "The Structure and Development of the Avian Pituitary." Gleerup, Lund.

Wingstrand, K. G. (1953). Neurosecretion and anti-diuretic activity in chick embryos with remarks on the sub-commissural organ. *Ark. Zool.* **6**, 41–67.

Wingstrand, K. G. (1966). Comparative anatomy and evolution of the hypophysis. *In* "The Pituitary Gland" (Harris, G. W. and Donovan, B. T., eds), vol. 1, pp. 58–126. Butterworths, London.

Wolf, A. V. (1958). "Thirst", pp. 536. Thomas, Springfield, Illinois.

Wolfson, A. (1966). Environmental and neuroendocrine regulation of annual gonadal cycles and migratory behaviour in birds. *Recent. Progr. Hormone Res.* **22**, 177–244.

Wolfson, A. and Kobayashi, H. (1962). Phosphatase activity and neurosecretion in the hypothalamo-hypophysial system in relation to the photoperiodic gonadal response in Zonotrichia *albicollis. Gen. comp. Endocrinol. Supplement*, **1**, 168–179.

Wurtman, R. J., Axelrod, J. and Kelly, D. E. (1968). "The Pineal", pp. 199. Academic Press, London and New York.

# 13 Anatomy and Histology of the Forebrain Hemispheres and Related Systems

## I. INTRODUCTION

The telencephalic hemispheres are, together with the laterally displaced optic tectum, the overt characteristic features of the avian central nervous system. The hemispheres themselves correspond to the dorsal sensory alar plate during development and are separated from the diencephalon by a transverse velum. Kappers (1922) described four longitudinal regions in the chick embryo consisting at an early stage of a ventro-medial basal eminence, ventral basal ganglia, the latero-dorsal paleopallium and the medio-dorsal archipallium. He concluded that by the 5th day of incubation the early vestiges of the striatum were present and that in front of the foramen of Munro traces of both neostriatum and hyperstriatum occurred on the lateral walls. They comprised a juxta-ventricular matrix with a less dense cellular region. He ascribed a larger ventral protrusion, which was characterized by three cell layers, to the paleostriatum and noted that it overlay the ventral thalamic eminence at the anterior pole of the diencephalon.

Since the time of this original work by Kappers further detailed studies have been made by Kuhlenbeck (1938), Kallén (1953, 1955) and Haefelfinger (1958). The generalized sequence of events is described in Chapter 4, Section VIII and the embryonic regions together with their adult derivatives are summarized in Table 20. Although there is a marked difference in the temporal relationships of myelination between nidicolous and nidifugous genera (see Chapter 16) the general sequence of events is common to all birds and the typical avian characteristics tend to develop gradually between the 7th and 12th day. Prior to this time the structure is comparable with that of other vertebrates. The growth of the ventricular wall reduces the ventricle itself to a medial and caudo-lateral slit whilst the paleostriatal, neostriatal and hyperstriatal complexes are progressively elaborated. By the 15th day both ordinal and familial variations are apparent. Amongst the nidifuges the myelin sheaths are then visible but in nidicoles their development is long-delayed and takes place during the post-hatching period.

As in all amniotes the neopallium develops in association with the four primary regions and extends laterally and posteriorly to contribute a significant fraction of the hemispheres. However, although this is a "surface" structure in mammals it has a massive organization in birds and an intimate relationship with the "striatal" regions. This marked difference between the mammalian and avian structures is strongly emphasized in many modern mammals, and particularly the Primates, by the increase in area which gives rise to the gyri and sulci. Similar developments also occur in less representative genera such as *Tachyglossus* and *Zaglossus*. In contrast to this in the avian brain it undergoes a thickening in the peri-ventricular grey. For a long time it has been assumed, without much justification, that this difference is a result of the precocious and extensive concrescence of the avian cranial sutures and the relatively limited intra-cranial space which results. On this reasoning the lateral pallial region has therefore only limited scope for development and gives rise to massive hyperstriatal components. This argument was also put forward by Kappers (1947) to explain the comparatively high density of the component cells.

It is worth emphasizing at this point that the nomenclature which is associated with the avian telencephalon is unfortunate. For historical reasons the components are referred to as paleostriatum, archistriatum, ectostriatum, neostriatum, ventral, dorsal and accessory hyperstriatum, together with cortical regions such as the hippocampus, parahippocampus and peri-amygdaloid regions. This obscures the actual origins and suggests homologies which are rather misleading. From Table 20 it can be seen that there are the following principal homologies with the mammalian brain and that the dorsal and accessory hyperstriatum, together with the intervening intercalated nucleus when this is present, are ontogenetically related to each other and not to the ventral hyperstriatum which is the embryological congener of the neostriatum.

| | |
|---|---|
| Paleostriatum primitivum | ? Globus pallidus |
| Paleostriattum augmentatum | ? Putamen-caudate |
| Archistriatum | |
| Ectostriatum | |
| Neostriatum | |
| Ventral hyperstriatum | |
| Dorsal hyperstriatum | ⎫ |
| Accessory hyperstriatum | ⎬ Neopallium |
| Parahippocampus | ⎭ |
| Hippocampus | Archipallium |

The avian hemispheres are therefore laminated structures. The upper three layers of dorsal hyperstriatum, intercalated nucleus and accessory hyperstriatum are overlain to a varying degree by superficial corticoid structures and together comprise the "Wulst". Directly below this the ventral hyperstriatum, neostriatum and paleostriatum are arranged in a more or less orderly fashion whose variations will be considered below. A brief synonymy of the variety of terms which exist in the literature is contained in Table 53.

Apart from the work of Kuenzi (1918) there are relatively few comparative works on the overall appearance of the avian hemispheres. The subject was touched on by Bumm (1883), Turner (1891), Edinger (1908) and Dennler (1921). Additional information is also provided by the works of Craigie, which are referred to in subsequent sections, and by Schaposchnikow (1953) but the outstanding account is that of Stingelin (1958). The latter author was at some pains to integrate the inter-relationships between overall structure and internal nuclear organization. Kuenzi (1918) drew attention to the importance of telencephalic development in determining the form of the entire brain. This varies from the broad condition of, say, *Pernis*, *Micropus* and *Corvus*, to the relatively narrow brain of *Phoenicopterus*. There has been a fair amount of discussion about the influence which the eyes have upon this

## TABLE 53

*A brief outline of the principal synonymies of the telencephalic nomenclature in various works. Note particularly the unfortunate extent of the term hyperstriatum in the older work.*

| Current terminology | Kappers *et al.* (1936) Kappers (1947) | Kuhlenbeck (1938) | Rose (1914) | Edinger *et al.* (1903) |
|---|---|---|---|---|
| Accessory hyperstriatum | Hyperstriatum accessorium | Nucleus diffusus dorsalis | B | Cortex frontalis |
| Intercalated nucleus | Nucleus intercalatus hyperstriati suprema | Nucleus diffusus dorsolateralis | A | Frontal mark |
| Dorsal hyperstriatum | Hyperstriatum dorsale | Nucleus epibasalis dorsalis, pars superior | C | |
| Superior frontal lamina (LFS) | Nucleus intercalatus hyperstriati superioris | | | |
| Ventral hyperstriatum (dorso-ventral) (ventro-ventral | Hyperstriatum ventrale (dorso-ventrale) (ventro-ventrale) | Nucleus epibasalis dorsalis pars inferior | D. $D_1$ | Hyperstriatum. |

Table 53 (*contd.*)

| | | | | |
|---|---|---|---|---|
| Frontal neostriatum | Neostriatum frontale | Nucleus epibasalis centralis, pars medialis | $G_1$ | |
| Intermediate neostriatum | Neostriatum intermediale | Nucleus epibasalis centralis, pars posterior | $G, G_2$ | |
| Caudal neostriatum | Neostriatum caudale | | $L, G_3$ | |
| Parolfactory lobe | | | | Parolfactory lobe |
| Basal nucleus | Nucleus basalis | Nucleus epibasalis ventrolateralis | R | Mesostriatum laterale |
| Ectostriatum | Ectostriatum | Nucleus epibasalis centralis accessorium | S | Ectostriatum |
| Archistriatum | Archistriatum | Nucleus epibasalis caudalis | K | Epistriatum |
| Paleostriatum augmentatum | Paleostriatum augmentatum | Nucleus basalis | H | Mesostriatum |
| Paleostriatum primitivum | Paleostriatum primitivum | Nucleus entopeduncularis | J | Nucleus entopeduncularis |

variation. Kuenzi concluded that although some broad brains, such as those of *Pernis* and *Micropus*, are associated with large eyes, and the narrower brains of *Phoenicopterus* and *Lorius* with relatively small eyes, no general and consistent pattern emerges. *Alca* and *Charadrius* with relatively large eyes do not have particularly broad brains whilst complementarily *Sula*, *Pelecanus*, *Turdus* and *Corvus*, with what Kuenzi considered had rather low eye indices, do. It would certainly be somewhat unexpected if any such generalization was of very wide application but Kuenzi drew attention to the fact that in a given order the appearance is often fairly uniform. A further consideration of this will emerge from the following sections but one can note that this statement is a fair reflection of the conditions in, say, the Falconiformes, Strigiformes and Columbiformes. Within the Psittaciformes Kuenzi thought that it was less true and contrasted the Loriidae and Psittacidae with the Cacatuidae. Such differences which are epitomized by the conditions in a variety of genera can arise from a number of causes.

The dorsal furrows which traverse the hemisphere in various directions were also described at an early stage although their full significance only emerged with the work of Haefelfinger (1958) and Stingelin (1958). Kuenzi differentiated three types. He thought that none were present in the Casuariiformes, Struthioniformes, Upupidae, Charadriidae, Tyrannidae, Laniidae, Hirundinidae, Motacillidae, Cinclidae and Sturnidae. His first type originated in the hind or middle third of the hemispheres and described a gentle arc to terminate close to the olfactory bulb. It was most clearly differentiated amongst members of the Tinamidae, Odontophoridae, Rallidae, Ciconiidae, Columbinae, Treroninae and Laridae. With slight modifications it also occurred in the Alcidae, Ardeidae, some Cuculidae, the Rhamphastidae, Alcedinidae, Cypselidae, Turdidae, Tanagridae, Fringillidae and Ploceidae. The second type possessed a rather similar disposition but originated nearer to the hind end of the hemispheres and did not extend as far forward as the olfactory bulbs. Instead it bent towards the mid-line some distance behind the anterior region. This typified the Gruidae, Pelecanidae, Ibididae, Anatidae, Phoenicopteridae, Cacatuidae, Loriidae, Psittacidae and Picidae. The third type of furrow differed from both the foregoing as it was directed obliquely or transversely and not aligned along the longitudinal axis of the brain. Originating close to the mid-line and within the middle third of the hemispheres it described a wide arc to the lateral borders, down and round to the olfactory bulbs. As such it was present in the Sulidae, Bucerotidae, Bubonidae and Strigidae. Kuenzi considered that the vallecula, demarcating the limits of neopallial and striatal structures, was particularly well represented amongst the Gruidae, Pelecanidae, Ardeidae, Ciconiidae, Ibididae, Alcedinidae, Bucerotidae, Strigidae, Cacatuidae, Loriidae, Psittacidae and Picidae. He seemed unaware of the presence of a definitive and well marked Wulst in the cursorial ratites. The detailed analysis of the internal nuclear components enables one to appreciate the significance of such observations.

## II. THE OLFACTORY BULBS AND TRACTS

### A. Peripheral Structures

The degree of development undergone by the olfactory bulbs varies considerably from order to order, and to a less extent from family to family and genus to genus. This reflects their respective positions within a macrosmatic-microsmatic series. The anatomical evidence for avian olfactory function was summarized by Bang (1960). On the basis of comparative anatomy one can produce good morphological evidence that inherent and learned responses to chemical stimuli are of vital importance to a number of species. Bang observed that it was rather unfortunate, in view of the wide disparity in the size and degree of development of the olfactory organs, that experimental investigations usually involved the study of species such as the pigeon which are relatively poorly equipped. This is certainly the basis for the frequently reiterated statement that the avian chemical sense is either minimal or lacking. However, even in pigeons Michaelsen (1959) and Calvin (1960) have demonstrated odour discrimination.

As far as the peripheral olfactory epithelium is concerned Wesołowski (1967) concluded that in geese and turkeys there is no essential difference from that of other vertebrates. Following the terminology of Andres he demonstrated that blastemic cells can be distinguished from sensory cells. The olfactory cells themselves contain an enhanced number of small granular inclusions and rod-shaped mitochondria by comparison with the supporting cells. Large and strongly innervated organs occur in the turkey vulture, *Cathartes aura*, the Trinidad oil-bird *Steatornis caripensis*, and both the Laysan and black-footed albatrosses, *Diomedea immutabilis* and *D. nigripes*. These four species clearly represent both very divergent feeding habits and ordinal positions. The vulture is a diurnal carrion eater which constructs solitary nests in relatively sequestered hiding places. The oil-bird feeds at night on fruits and is a colonial nester in the darkness of caves. In contrast to both of these the two *Diomedea* species feed largely on crustaceans and nest in hollow scrapes on oceanic islands. From anatomical studies Bang concluded that there could be no question of degenerate or indifferent olfactory function in these species. Although the olfactory components of other species varied from modest to considerable those of these birds were outstanding. *Cathartes* may detect food odours which are rising on the thermal updraughts upon which they glide. They have been observed to overshoot carrion which one would expect them to detect visually and to then turn back in a manner which could be interpreted as following a scent beam down at an angle to the carrion. Other field observers have reported that a flock of 40–50 regularly came in on the first morning thermal and fed off offal which had actually been exposed since dawn. Yet other field observations are said to show that albatrosses were attracted, within an hour, and from distances of up to 20 miles, by a scum of bacon fat. Similarly, marked individuals completely ignored paint washings. In view of the rather different physical appearance it would be unfair to draw very definitive conclusions from such results.

However, leaving this aside, Bang considered that the nasal fossae of such

17

species had a structure which was precisely that which would be needed to filter, warm, moisten and sample the inspired air. The general arrangement provides an olfactory chamber that is enclosed on three sides and although the actual size of the conchae varies, their appearance is similar. The anterior conchae, peculiar to birds, serve principally as a baffle and thermo-regulator. The middle, or respiratory conchae are covered with mucous producing and ciliated cells, whilst the posterior or olfactory concha is different again and supports the free receptor fibrils of the olfactory nerve together with a quantity of serous glands which are peculiar to the region. In the fresh state this last is a conspicuous yellow colour and in both albatrosses and oil-birds it is hollow with its concavity in communication with the cephalic air sac system. In the vulture it is scrolled $2\frac{1}{2}$ times and has therefore got a greater overall surface area than elsewhere.

Scraping the concha revealed the underlying fibres of the olfactory nerve. These form prominent branches which are distributed over the entire convexity of the concha in *Diomedea*. However, the corresponding network of terminal branches in *Cathartes* is so thick that it is difficult to make a tidy preparation. In *Steatornis* the condition is roughly intermediate between these two. The fibrils are closer together than those of *Diomedea* but less dense than those of *Cathartes*. In each species other well-defined branches of the olfactory nerve also lie along the roof, the posterior and ventro-lateral walls, and the upper part of the nasal septum within the olfactory chamber. Cobb and Edinger (1962) reported that the concha of the emu is not scrolled thereby emphasizing the variations that occur.

Portmann (1961) recorded that the anterior concha is particularly large in Galliformes and Falconiformes, small in Podicipidiformes, Sphenisciformes, Procellariiformes and Ciconiae. It is absent from *Casuarius*, the Anhimae and *Opisthocomus*. Furthermore in *Apteryx* the posterior concha is exceptionally well developed, presents about five transverse foldings, and is large relative to the respiratory parts. In the Procellariiformes the middle part of the nasal cavity on each side of the septum bears a valve structure, the triangular valve, which opens into a tubular entrance (Technau, 1936). Mangold (1946) showed that air currents fill this structure and he suggested that it may act as a pressure detector during the sailing flight of petrels.

In the absence of any inequivocal experimental data Bang suggested that birds such as the vultures and albatrosses almost certainly respond to air-borne chemical stimuli either alone, or in combination with information provided by the auditory and visual modalities. It seemed clear that the kiwi had a well-developed sense of smell and there were even then somewhat equivocal experimental results for ducks and many song birds (Bajandurow and Larin, 1925; Zahn, 1933; Wagner, 1939). It is only fair to add that the work of Soudek and Walters which was cited by Portmann, conflicts with this and that both authors consider that birds are anosmic. More recently Tucker (1965) demonstrated that electrical activity can be recorded in the olfactory nerves of 14 avian species including the pigeon, quail and turkey vulture. In the following year Henton, Smith and Tucker (1966) showed that behaviour data relate quite closely to Tucker's electrophysiological threshold measurements. Other behaviour studies include those of Calvin (1960) and Michaelsen (1959). In one experiment two pigeons learned to discriminate, in a key-pecking situation, between no odour and those of sec-butyl acetate or iso-octane. The responses which they observed were shown to be dependent upon the olfactory pathways. In another type of investigation Neuhaus (1963) observed that significant changes in anserine respiratory patterns occur in the presence of scatol but not distilled water. He went on to suggest that the principal central connections of the avian olfactory pathways are with visceral rather than associative centres. This, he thought, would explain the apparently greater effects on "vegetative" functions rather than on the responses of skeletal muscles. Clearly in our present state of knowledge this rather obscurely worded suggestion may be a gross over-simplification. Other studies, which are similar to the field observations cited by Bang, were carried out by Stager (1964). If fumes of ethyl mercaptan were released into the air from an invisible source, *Cathartes* and *Sarcorhamphus* were clearly seen to alter their flight patterns and move in the direction of increasing odour concentration. Bang (1960) also suggested that the sense of smell may be related to breeding as well as feeding. It is certainly interesting that Wenzel and Sieck (1966) have observed ovarian and uterine hypoplasia following olfactory bulb ablation in mammals and Donovan (1966) cites a number of other papers leading to similar conclusions involving olfactory influences on ovulation. The oil ejected through the nostrils of the Tubinares seems indeed to be used in sexual rituals and may be a unique stimulus for recognizing mate and young (Cobb, 1960a, b).

## B. Central Structures

The degree of variation in olfactory bulb organization is described in a number of papers. Strong (1911) noted that the bulbs of *Dromiceius* are constricted at their bases and hence borne on short stalks. This was not the case in either *Rhea* or *Struthio*. In these two genera the organization resembled that of the majority of Carinates. Craigie (1939) also drew attention to the differential development in these forms. He agreed that they were largest in the emu and progressively smaller in rheas and ostriches. Rose (1914) observed that those of *Apteryx* were particularly large and that in *Anas* and various Lariformes, Charadriiformes and Alciformes they exceeded the size of those in other neognaths. In *Ardea* the actual size could equal that in *Anas*, and in addition they were clearly hollow. In *Gavia* the bases form a cap on the ventral and medial surfaces of the rostral pole of the hemisphere. Craigie emphasized that in both the Trochiliformes and Micropodiformes they are very reduced. Here a cap-like structure is reminiscent of the situation in some Passeriformes where the two bulbs can be either partially fused or form a single unified structure. The distribution relative to the various hemisphere components in a number of genera is shown in Figs 119–123.

A relatively recent detailed discussion of the subject is that of Cobb (1960a, b). Table 54 gives the size of both the olfactory bulb and the hemisphere in 36 species which are broadly arranged according to the increasing values of the ratio of bulb size to hemisphere size. Species which possess relatively small bulbs are listed first and those with relatively larger bulbs subsequently. Some of the species at the head of the list, such as the house sparrow, fox sparrow and canary have a single bulb. They are placed at that point because they only have small quantities of olfactory tissue although the single-fused structure may actually exceed the size of one of the paired bulbs of other species. From the list it is clear that the ratio of bulb to hemisphere diameter

## TABLE 54

*Various species of bird arranged according to the ratio of the size of the olfactory bulb to that of the hemisphere.*

| Species | Diameter of olfactory bulb (mm) | Diameter of hemisphere (mm) | Ratio of olfactory bulb to hemisphere (%) |
|---|---|---|---|
| Passeriformes | | | |
| *Parus atricapillus* | 0·8 | 12·5 | 6 |
| *Carpodacus purpureus* | 1·1 | 13·0 | 8 |
| *Passer domesticus* | 1·0 | 13·0 | 8 |
| *Hesperiphona vespertina* | 1·2 | 15·0 | 8 |
| *Passerella iliaca* | 1·5 | 13·0 | 11 |
| *Serinus canarias* | 1·5 | 12·5 | 12 |
| *Corvus brachyrhynchos* | 1·3 | 26·0 | 5 |
| *Cyanocitta cristata* | 1·0 | 16·0 | 6 |
| *Sturnus vulgaris* | 1·3 | 14·0 | 9 |
| Psittaciformes | | | |
| *Melopsittacus undulatus* | 0·8 | 13·0 | 6 |
| Piciformes | | | |
| *Colaptes auratus* | 1·5 | 18·0 | 8 |
| *Dendrocopos pubescens* | 1·5 | 15·0 | 10 |
| Pelecaniformes | | | |
| *Phalacrocorax penicillatus* | 3·0 | 29·0 | 10 |

Table 54 (*contd.*)

| | | | |
|---|---|---|---|
| Galliformes | | | |
| *Meleagris domesticus* | 2·5 | 19·0 | 13 |
| *Bonasus umbellus* | 2·0 | 14·0 | 14 |
| Falconiformes | | | |
| *Pandion haliaetus* | 3·0 | 21·0 | 14 |
| *Coragyps atratus* | 4·0 | 24·0 | 17 |
| Charadriiformes | | | |
| *Capella gallinago* | 2·0 | 14·0 | 14 |
| *Limnodromus griseus* | 2·0 | 13·0 | 15 |
| *Charadrius semipalmatus* | 1·5 | 10·0 | 15 |
| *Philohela minor* | 2·5 | 15·0 | 17 |
| Lariformes | | | |
| *Larus argentatus* | 3·0 | 19·0 | 16 |
| Columbiformes | | | |
| *Columba livia* | 2·0 | 12·0 | 17 |
| Sphenisciformes | | | |
| *Pygoscelis adelinae* | 5·0 | 30·0 | 17 |
| Ardeiformes | | | |
| *Nycticorax nycticorax* | 4·0 | 22·0 | 18 |
| Strigiformes | | | |
| *Bubo virginianus* | 4·5 | 25·0 | 18 |
| *Asio flammeus* | 3·5 | 18·0 | 19 |
| Coraciiformes | | | |
| *Megaceryle alcion* | 2·5 | 13·0 | 19 |
| Micropodiformes | | | |
| *Chaetura pelagica* | 1·5 | 8·0 | 19 |
| Anseriformes | | | |
| *Anas platyrhynchos* | 4·0 | 21·0 | 19 |
| *Cygnus olor* | 6.0 | 28·0 | 21 |
| Gaviiformes | | | |
| *Gavia immer* | 5·0 | 25·0 | 20 |
| Cuculiformes | | | |
| *Coccyzus americanus* | 1·9 | 9·0 | 21 |
| Gruiformes | | | |
| *Fulica americana* | 4·0 | 17·0 | 24 |
| Caprimulgiformes | | | |
| *Caprimulgus vociferus* | 2·5 | 10·0 | 25 |
| Podicipitiformes | | | |
| *Podiceps auritus* | 4·0 | 15·0 | 27 |
| Procellariiformes | | | |
| *Oceanites oceanicus* | 3·5 | 12·0 | 29 |
| *Diomedea nigripes* | 3·5 | 12·0 | 29 |
| *Puffinus opisthomelas* | 5·0 | 17·0 | 29 |
| *Puffinus gravis* | 6·0 | 20·0 | 30 |
| Apterygiformes | | | |
| *Apteryx australis* | — | — | 33 |

From Cobb (1960a,b).

varies between 5 and 30%. Passerines, parrots and wood-peckers head the list although *Dendrocopos* with a value of 10 is similar to the group of columbiform, galliform, falconiform, charadriiform and lariform species with intermediate ratios of 10–17. Finally there are 16 of these carinate species with large olfactory bulbs and ratios ranging from 18 to 30. This group includes members of the Ardeiformes, Coraciiformes, Cuculiformes, Micropodiformes, Caprimulgiformes, Anseriformes, Podicipitiformes and Procellariiformes. Measurements taken from Craigie (1935b) and corroborated by Cobb indicate that the kiwi has certainly got larger olfactory bulbs relative to hemisphere size than any of these other birds.

## TABLE 55

*Various orders of birds arranged according to the ratio of olfactory bulb size to that of the hemisphere.*

| Avian order | Habitat | Ratio of olfactory bulb to hemisphere (%) |
|---|---|---|
| Passeriformes | Trees and shrubs | 5–9 |
| Psittaciformes | Trees | 6 |
| Piciformes | Trees | 8–10 |
| Pelecaniformes | Water | 10 |
| Galliformes | Ground | 13–14 |
| Falconiformes | Trees | 14–17 |
| Charadriiformes | Beach, marsh and water | 14–17 |
| Columbiformes | Trees and ground | 17 |
| Ardeiformes | Marsh | 18 |
| Strigiformes | Trees | 18–19 |
| Micropodiformes | Air | 19 |
| Coraciiformes | Water | 19 |
| Anseriformes | Marsh and water | 19–21 |
| Gaviiformes | Water | 20 |
| Cuculiformes | Trees | 21 |
| Gruiformes | Plain and marsh | 24 |
| Caprimulgiformes | Air | 25 |
| Podicipitiformes | Water | 27 |
| Procellariiformes | Ocean | 29–30 |
| Apterygiformes | Ground | 33 |

From Cobb (1960a,b).

The first and last species on the list are conspicuously different. Whilst the conditions in *Apteryx* may conceivably reflect the ancestral condition it is much more likely that they are a specialization in association with its macrosmatic habits. The tree-living passerines are predominantly seed eating and Cobb suggested that olfaction is of little importance to them. Table 55 was constructed to reflect such overall habits of the orders and it is clear that the first three, being tree-living, would have to have odours carried to them over fair distances. After the intermediate group the orders which have values in excess of 19 include a marked preponderance of water birds. Leaving aside the kiwi, the genera with the largest central olfactory apparatus are all Procellariiformes living a largely oceanic existence. The two exceptions to this, Cobb's wet rule, are the Cuculiformes which live in trees, and the Caprimulgiformes, such as the whip-poor-wills, which catch insects on the wing.

Besides the positive relationship of the size of bulb to that of the hemisphere microscopic surveys suggest that a large hippocampus (*q.v.*) is present in species with large olfactory bulbs and vice versa. It also appeared that other corticoid layers are more extensive in brains with large bulbs and hippocampal regions. Thus the prepyriform area (*q.v.*), which is situated at the rostral end of the telencephalon and in close proximity to the olfactory bulb, is obviously larger in the grebe than in crows or canaries. Furthermore the parahippocampal components in the occipito-parietal regions are more extensive in gulls, grebes and coot than they are in the canary, crow, woodpecker or parakeet. Such suggestions indicate that a correlation exists between the size of the olfactory bulb and the extent of the corticoid components.

The zones which occur elsewhere in the vertebrates can be distinguished with varying degrees of ease in avian bulbs. Generally there is a wide glomerular layer, a fibrillar zone, and a layer of mitral cells which is demarcated from the succeeding granular layer by olfactory tract fibres (see Fig. 111). The unmyelinated olfactory fibres synapse after entry with the terminal dendrites of the mitral cells giving rise to the so-called glomeruli. The mitral cells form a ring within the resulting plexiform layer and their axons pass to the olfactory tract. The cells of the granular layer invest the olfactory ventricles. In many Psittaciformes and Passeriformes all these layers are reduced so that isolated mitral cells are interspersed along the outer border of the granular cells.

In recent years the situation has been investigated at the ultra-structural level by Andres (1970) using *Larus argentatus* and *Philomachus pugnax*. He concluded that the situation was directly analogous with that in turtles which have fewer myelinated axons in the plexiform layer than *Varanus niloticus*.

Peripherally the olfactory fibres are often few in number and vary in relation to the degree of macrosmatic development. Both Kappers (1920) and Kappers *et al.* (1936) concluded that there is an even greater reduction of the central pathways than of the peripheral ones. As shown in Fig. 111 for every 10–20 fila olfactoria there is only a single mitral cell. Besides this the secondary olfactory tracts are also of a relatively insignificant extent and tend to synapse in close proximity to their point of entry to the hemispheres.

In relatively macrosmatic genera the axons of the mitral cells, and also possibly the granule cells, pass as medial and lateral olfactory tracts. The older authors suggested that even in *Passer* the tiny lateral tract still runs along the ventro-lateral surface of the hemispheres to the prepyriform area (*q.v.*) and the ventral part of the basal nucleus. Other fibres enter the archistriate and neostriate tract. The definitive medial tract runs to the anterior olfactory nucleus and thereby parallels the condition in diapsid reptiles such as *Alligator*.

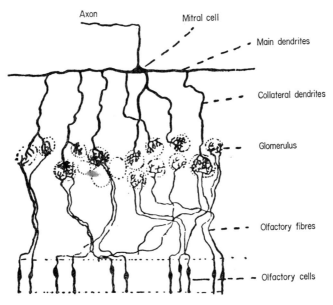

FIG. 111. Diagram illustrating the relationships of the olfactory neurons and mitral cells in birds. (From Kappers *et al*, 1936. After van Gehuchten).

The tertiary olfactory connections within the wall of the hemispheres appear to involve the large fronto-archistriato and neostriatal tract (Huber and Crosby, 1929; Kappers *et al.*, 1936; Kappers, 1947). Histologically this can be traced to a variety of origins. The resulting compound tract then penetrates the archistriatum and neostriatum and a relatively small group of axons also passes to the occipito-mesencephalic component of the anterior commissure. Connections between the archistriatum and overlying structures' have been assumed to take place via the dorsal archistriate tract and various axons which penetrate neostriatal areas. In *Apteryx* Craigie (1930) was unable to obtain any evidence for a definitive fronto-archistriatal and neostriatal tract. There was, however, an additional unit, the so-called mesostriatic tract, which joins the archistriatum and the paleostriatum augmentatum.

The older workers (see, for example, Kappers *et al.*, 1936) tended to emphasize that the relatively poor development of the olfactory bulb was accompanied by rather under-developed olfactory derivatives elsewhere. They were, however, unaware of the existence of definitive cortical areas and, as was just mentioned above, it is possible that the actual extent of these is a reflection of olfactory function. On the other hand the two sets of structures may be separate results of a common cause and it is certainly interesting that the olfactory modality appears to have an activating effect on the central control systems of other sensory modalities (see Chapter 14, Section II). As far as other olfactory derivatives are concerned they emphasized that many genera have a poorly defined nuclear mass ensheathed in olfactory tract fibres at the level of the olfactory bulb. In the sparrow the anterior olfactory nucleus, to which the medial olfactory tract passes, is a small, laminar group of cells (see Section XII C). It was originally identified by analogy with the comparable but more prominent structure in reptiles, and is more distinct in *Anas* and *Columba*. Furthermore the nucleus accumbens also figured in these discussions. It is a continuation of the paleostriatal material. Lying medial to

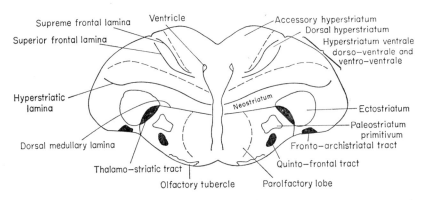

FIG. 112. A generalized transverse section of the hemispheres in birds. See also Fig. 101.

the parolfactory lobe (Fig. 112) it is below and around the lower ventricle component and ventro-lateral to the lateral septal nucleus. In more anterior regions, in which only the dorsal ventricular cavity is prominent in the pigeon, it assumes a ventral and medial position and is closely apposed to the olfactory tubercle. All such older studies were put in perspective by Craigie's data on cortical structures. These are discussed in detail below.

## III. THE LATERAL VENTRICLE

The massive nuclear development within the ventricular walls, which gives rise to the characteristic components of the avian telencephalon, more or less occludes the telencephalic cavities. The lateral ventricles are best represented in *Apteryx* and probably undergo their greatest reduction amongst various Psittaciformes. In these birds little remains apart from small spaces on the median side of the principal cell masses. Other orders exhibit stages which are intermediate between these two extremes and amongst the Trochiliformes, for example, the ventricles can be seen in transverse sections as tolerably extensive spaces. Indeed in association with the rather thick medial wall the outline of such sections is reminiscent of many reptilian conditions (Craigie, 1928).

Posteriorly there can be equally extensive spaces and in birds generally the ventricles project upwards as a dorso-lateral slit. At the occipital extremity this narrows fairly sharply and ends as a narrow cleft. This is disposed vertically and over much of its extent is only separated from the medial surface by a rather thin septum. Some idea of the distribution can be obtained from Figs 115 to 118. In contrast the ventricle of diapsid reptiles is disposed laterally with the septum and pallium forming fairly thick limiting partitions. Although gross lateral extensions can be present in birds, e.g. *Lophortyx*, it is only in the trochiliform genera such as *Chlorostilbon* and *Chrysometrius* that the overt development approaches that of the Lacertidae. A recent detailed study using Tensol impregnation has been reported by Böhme (1969). Although this particular report relates to the conditions in the hen we may look forward to future data that will no doubt clarify our knowledge of other genera.

## IV. THE HISTOLOGY OF THE HEMISPHERES

A number of workers have described the histological structure of the hemispheres. In view of the widespread interest in mammalian cerebral cytoarchitectonics during the early years of the present century it is not surprising that anatomists and histologists expected to find comparable and definitive differences between the various avian forebrain components. Today the emphasis placed on cortical cell differences in mammals is rather less and interest is directed towards other more electro-physiological lines of approach. However, there are just as clear variations in both cell size and appearance within the various nuclei which make up the birds hemispheres as there are in the mammalian cortex. The work of Stingelin (1958) amplified and corroborated the earlier studies of such workers as Edinger *et al.* (1903), Rose (1914), Craigie (1928, etc.), Huber and Crosby (1929), Kappers *et al.* (1936),

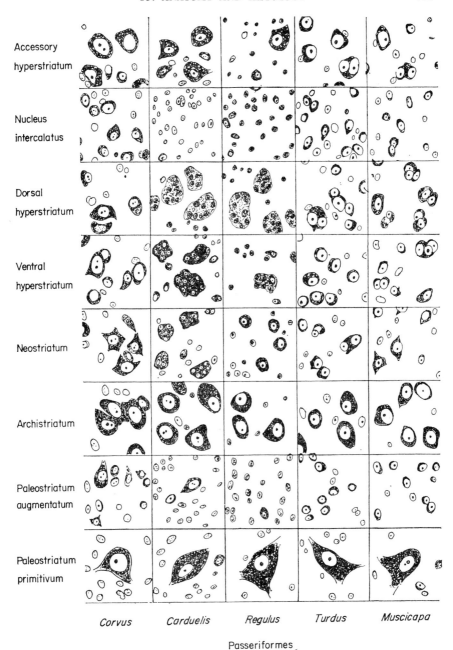

FIG. 113. Diagrammatic representation of the histological structure of the hemispheres in *Corvus*, *Carduelis*, *Regulus*, *Turdus* and *Muscicapa*. (Stingelin, 1958).

Kappers (1947) and Kuhlenbeck (1938). Stingelin concluded from his extensive comparative analysis that there are nine principal types of cell. These may be listed seriatim and are illustrated in Fig. 113.

*Type (a).* Large granular cells with a diameter of 25–50 $\mu$ and a somewhat pyramidal appearance that are restricted to the paleostriatum primitivum.

*Type (b).* Somewhat smaller and compressed but also very granular cells varying in size from 18 to 25 $\mu$ and with large clear nuclei. These occur in some cases in the magnocellular part of the archistriatum, in the accessory and ventral hyperstriatum and in the neostriatum.

*Type (c).* More polygonal cells with an overall diameter of 15–20 $\mu$ which occur in the same localities as type (b).

*Type (d).* Large cells measuring some 15–18 $\mu$ in diameter which again occur alongside type (b) but differ from both the foregoing types in their less dense granulation.

*Type (e).* Medium-sized cells with a diameter of some 10–15 $\mu$ that are very granular and generally have an oval nucleus. Their overall appearance is variable and they can be spindle-shaped, round or angular. They occur in all regions other than the magnocellular part of the archistriatum.

*Type (f).* Small strongly granular cells with a round or oval nucleus that are widespread in the telencephalon. They are somewhat smaller than those of type (e) and vary from 8 to 10 $\mu$ in diameter.

*Type (g).* Small cells which are comparable in size to the last named but have much less granulation.

*Type (h).* Minute cells varying from 4 to 8 $\mu$ in diameter with round or oval nuclei and no granulation. These are ubiquitous.

*Type (i).* This is the smallest cell type observed in the forebrain and measures from 3 to 6 $\mu$. They lack granulation but have very large amounts of chromatin.

Stingelin also drew attention to the four types of organization which such cells can exhibit. Broadly one can summarize these as involving a rather diffuse array with an overall longitudinal orientation; a marked linear arrangement; small cell clumps; and, finally, dense plasmodial clusters. According to Balcells (in Stingelin) these clusters are formed by neurones becoming enswathed within oligodendroglial derivatives. The actual appearance of a given telencephalic component is a function of the cell types which occur within it and the actual cell density. A series of measurements on comparable and uniform fields in 15 genera gave the results which are contained in Table 56. From these data it is clear that the density varies considerably in all regions. The highest values per unit area occur within the intercalated hyperstriate nucleus of *Regulus*, the lowest in the neostriatal and accessory hyperstriatal regions of *Columba*. Average values for all five components listed exceeded

80 in the Psittaciformes and Passeriformes. Comparable high values are also found in members of the Strigiformes. *Columba* has an average value of 40, whilst that of the cuckoo was also relatively low at 55. It is interesting that Stingelin did not find any overall correlation between brain size and cell density such as might be expected if the suggestions of Kappers (1947) were correct. As mentioned in Section I above the latter author considered that the

## TABLE 56

*The cell density within the neostriatum, ventral, dorsal and accessory hyper-striatum in fifteen genera.*

| Genus | Neo-striatum | Ventral hyper-striatum | Dorsal hyper-striatum | Inter-calated nucleus | Acces-sory hyper-striatum | Average |
|-------|------|------|------|------|------|------|
| *Columba* | 27 | 51 | 46 | — | 34 | 40 |
| *Anas* | 58 | 74 | 44 | 105 | 74 | 71 |
| *Vanellus* | 51 | 60 | 83 | 100 | 60 | 68 |
| *Ardea* | 69 | 58 | 56 | 68 | 54 | 61 |
| *Cuculus* | 45 | 70 | 62 | — | 45 | 55 |
| *Athene* | 82 | 69 | 61 | 160 | 62 | 87 |
| *Amazona* | 78 | 95 | 67 | 103 | 65 | 82 |
| *Melopsittacus* | 86 | 90 | 82 | 134 | 103 | 99 |
| *Apus* | 64 | 96 | 64 | — | 63 | 70 |
| *Merops* | 62 | 83 | 82 | — | 73 | 75 |
| *Muscicapa* | 50 | 61 | 53 | 80 | 66 | 62 |
| *Turdus* | 50 | 69 | 56 | 101 | 64 | 68 |
| *Corvus* | 67 | 82 | 91 | 94 | 64 | 80 |
| *Carduelis* | 153 | 169 | 140 | 169 | 125 | 151 |
| *Regulus* | 100 | 131 | 147 | 192 | 120 | 138 |

From Stingelin (1958).

cell density of the avian hemispheres was a direct consequence of the very limited volume of the cranial cavity. If one takes the two pairs of genera *Melopsittacus/Amazona* and *Carduelis/Muscicapa* one finds that in the first pair the size of the hemispheres in terms of their weight differs in the ratio 1:10. However, the cell density values are 99:82. Complementarily in the second pair the hemispheres are of more or less comparable size but the density values are 151:62. Nevertheless in spite of this certain general trends can be detected.

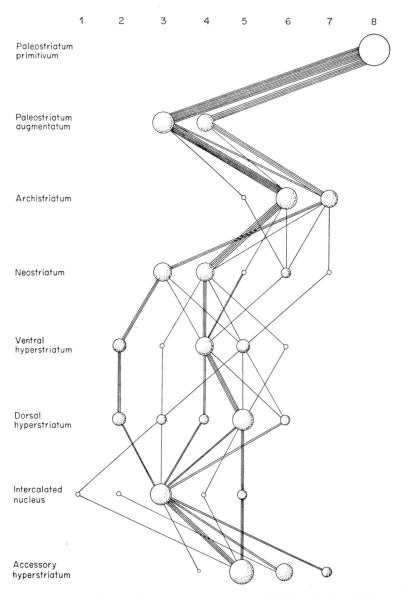

FIG. 114. A nexus diagram illustrating the structure of the hemispheres in terms of cell size and form. (After Stingelin, 1958). For further details see text.

In all the species in which a distinct intercalated nucleus intervenes between the dorsal and accessory hyperstriatum the transition from the dorsal hyperstriatum to the nucleus involves an increase in density. On the contrary passing from the nucleus to the accessory hyperstriatum one observes a decrease in the density. However, the very considerable variations of cell form and cell distribution which occur within the various components makes any wide-ranging description very difficult, if not impossible. Some idea can be obtained by allocating numerical values to the nine cell groupings (a), (b), (c), (d), (e), (f), (g), (h) and (i), so that these numbers give a measure of the particular cell size. Stingelin gave type (a) the value 8 and the subsequent types decreasing values from 7 to 1. A transition from a region of large cells such as type (a), to one with type (d) then gave a differentiation coefficient of 3. He then ascribed comparable numerical values to differences of cell density. For every 30 units difference between the regions of Table 56 he allocated a value of 1. For example in the budgerigar the values for the intercalated nucleus (134) and accessory hyperstriatum (103) will give this unit value. Thirdly he assessed the degree of cell association that exists in the various laminae. Any transition from areas with no such grouping to others with loose grouping, or to close oligodendroglial packing gives values of $\frac{1}{2}$ and 1 respectively. Summing all these results he expressed them in a nexus diagram. A simplified version of this is shown in Fig. 114. It is quite clear from this figure that the relative homogeneity of the different components varies. Whilst the paleostriatum primitivum, the paleostriatum augmentatum and the archistriatum are relatively homogeneous this is not the case for the other components. This implies that whilst the cytoarchitectonic characteristics of the paleostriate and archistriate regions remain fairly constant, those of the accessory hyperstriatum vary somewhat and there is much greater variation in both the dorsal hyperstriatum and neostriatum.

# V. THE PALEOSTRIATUM PRIMITIVUM AND AUGMENTATUM

The paleostriate regions consist of two quite distinct nuclear masses whose histology differs. The *augmentatum* surrounds the *primitivum* peripherally and is separated from the overlying structures such as the neostriatum by the dorsal medullary lamina. Works which are of particular significance in any discussion of these and subsequent cell masses are Edinger *et al.* (1903), Rose (1914), Kappers (1922, 1928, 1929, 1947), Huber and Crosby (1929), Craigie (1930, 1932), Durward (1932), Stingelin (1958), Portmann and Stingelin (1961), and the experimental investigations which are cited in Chapter 14. Clear ontogenetic information was provided by Kuhlenbeck (1938) and Haefelfinger (1958), and the stereotaxic atlases of Karten and Hodos (1967) and Tienhoven and Juhasz (1962) give accurate maps of the relative distribution in the hemispheres of the pigeon and chick. Clearly the outstanding contributions are those of Stingelin.

If one compares the situation in *Corvus* and *Amazona* one finds that both the paleostriatum primitivum and the paleostriatum augmentatum are not dissimilar to their supposed form in the hypothetical ancestral condition. This makes a very marked contrast to the changes which other telencephalic components have undergone in terms of this supposed ancestry. Nevertheless these latter changes necessarily involve rearrangements of the paleostriate regions and these will be discussed in Section XIII. Amongst the Charadriiformes the anatomical disposition of the paleostriatal areas is certainly very similar to that figured as a probable ancestral condition by Stingelin. However, as such suggestions about the original form have been made following a consideration of the conditions in modern genera this is possibly not so very

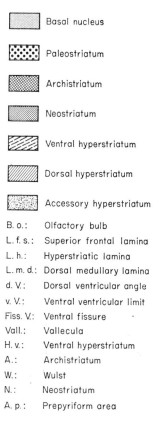

|  | Basal nucleus |
|  | Paleostriatum |
|  | Archistriatum |
|  | Neostriatum |
|  | Ventral hyperstriatum |
|  | Dorsal hyperstriatum |
|  | Accessory hyperstriatum |

B. o.:    Olfactory bulb
L. f. s.:    Superior frontal lamina
L. h.:    Hyperstriatic lamina
L. m. d.:    Dorsal medullary lamina
d. V.:    Dorsal ventricular angle
v. V.:    Ventral ventricular limit
Fiss. V.:    Ventral fissure
Vall.:    Vallecula
H. v.:    Ventral hyperstriatum
A.:    Archistriatum
W.:    Wulst
N.:    Neostriatum
A. p.:    Prepyriform area

Key to Figs. 115–124.

surprising! Anteriorly the paleostriate region ends at the ventral fissure and immediately behind the olfactory bulb (Fig. 115). The dorsal limits which are embodied by the dorsal medullary lamina are markedly convex. In *Vanellus vanellus* it does not extend so far forward as in *Charadrius*. In *Capella gallinago*, as indeed is the case in *Cuculus*, *Otus*, and *Athene* (Figs 116 and 117) the anterior and ventral boundary is directed strongly backwards.

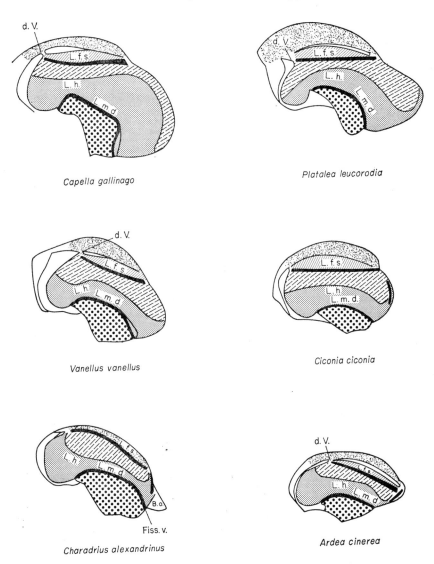

FIG. 115. The structure of the forebrain hemispheres in *Capella*, *Vanellus*, *Charadrius*, *Ardea*, *Ciconia* and *Platalea*. (After Stingelin, 1958).

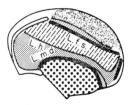

*Columba livia*

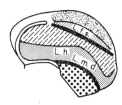

*Merops apiaster*

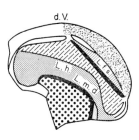

*Cuculus canorus*

Fɪɢ. 116. The structure of the forebrain hemispheres in *Columba, Merops* and *Cuculus*. (After Stingelin, 1958).

Amongst the Ardeiformes the situation varies. In *Ardea cinerea* the medullary lamina is not very strongly arched and, in sagittal sections, lies close to the horizontal plane. This is also true of a line joining the ventral and neopaleostriatic fissures which represent its anterior and posterior ends. In *Ciconia ciconia* and *Platalea leucorodia* Stingelin drew attention to a progressive increase in the degree of arching. This can approach the convex form of *Charadrius* and consequently restricts the ventral anterior-posterior extent of the paleostriate region. An analogous condition occurs in the owls *Otus*

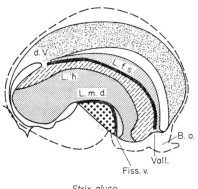

*Strix aluco*

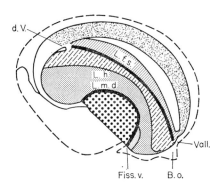

*Athene noctua*

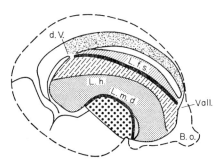

*Otus scops*

FIG. 117. The forebrain hemispheres in three species of Strigiformes. (After Stinge-lin, 1958).

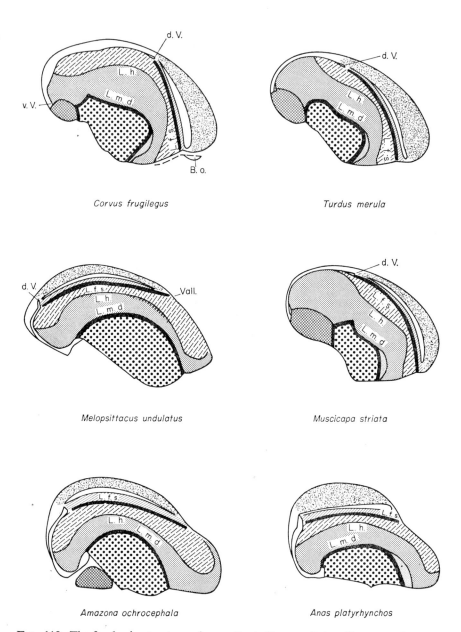

FIG. 118. The forebrain structure of some Passeriformes, Psittaciformes and *Anas*. (After Stingelin, 1958).

*scops, Athene noctua* and *Strix aluco.* On the other hand in the duck *Anas platyrhynchos* there is an opposite development. In the mid-line of the brain the long paleostriatal area forms a major component of the hemisphere floor. Such a massive form also characterizes the budgerigar and *Amazona.* Elsewhere the region can be small and orientated obliquely upwards as in the flycatcher *Muscicapa striata.*

What has been said in the foregoing paragraphs will clearly suggest that the actual appearance of the paleostriatum primitivum and the paleostriatum augmentatum, when seen in transverse sections, will vary considerably depending upon the species concerned. In some cases the primitivum appears anteriorly at the level of the olfactory bulb. Where the bulbs are situated further forward this is not the case. Broadly speaking it appears first as a relatively small contribution, situated laterally and over the quinto-frontal tract, in the pigeon. In sections from more posterior regions its size increases and it assumes a position more closely associated with the ventral surface. Its outline is variable and may be round, rhomboidal or triangular. Posteriorly it is surrounded by fibres of the lateral forebrain bundle and then tapers dorsally behind. At such levels it is frequently confluent with the endopeduncular nuclei which lie both alongside and within the bundle as it passes from the diencephalon to the telencephalon. These nuclei have an appearance which is histologically similar to that of the primitivum. As was indicated in Section IV this comprises large cells of Stingelins type (a), together with micro-cellular components (h) and (i). There are only very few examples of cell types (f) and (g).

The augmentatum surrounds the primitivum throughout much of its dorsal and lateral extent. Separated from the neostriatum, archistriatum and ectostriatum by the dorsal medullary lamina, the augmentatum very often appears C shaped as a result with the open part of the C directed towards the midline. It continues to surround the primitivum until the level of the inter-ventricular foramen is reached and then decreases in size posteriorly. Behind the forebrain bundle it has the appearance of an irregularly rounded nuclear mass lying near to the ventro-medial surface of the hemisphere. As such it can be traced back to the level of the posterior commissure in some species but disappears at the mid-point of the rotundal nucleus in sections of pigeon brain (see Karten and Hodos, 1967, for example). Rich in the micro-cellular components of Stingelin's cytological classification it is particularly well provided with cell types (h) and (i), although there are scattered examples of types (e) and (f). The identification of the augmentatum in various older works is sometimes hindered by the use of alternative names. Besides those identified in Table 53 it is worth noting that following the suggestion of Johnston (1923) who homologized it with the caudate and putamen of mammals, and that of Kappers (1947), who suggested a homology with the globus pallidus, these terms are also sometimes associated with it. Furthermore considerable ambiguity can arise from Kuhlenbeck's (1938) use of the term basal nucleus. The structure referred to by that name by other workers being his ventro-lateral epibasal nucleus. In recent years electron microscopic investigations have actually supported the usage of Kappers and suggest that the primitivum is homologous with the globus pallidus, and the augmentatum with the putamen-caudate complex (Fox *et al.*, 1966).

# VI. THE ARCHISTRIATUM

The archistriatal complex is situated within the posterior third of the hemi-spheres and at the lateral wall so that in the pigeon it comprises the tissue of the ventro-lateral angle. In longitudinal sections it can appear more or less surrounded by the neostriatum dorsally, and is applied to the hind or postero-dorsal areas of the paleostriate complex. In the sparrow and pigeon it appears in transverse sections at the level of the habenular nuclei and the archistriatal tract lies more or less vertical above it. At such levels it can be dissected by fibre bundles from the occipito-mesencephalic tract and then takes the form of bands of cells. The taenial nucleus lies medial to it and below the paleo-striatum augmentatum. Craigie (1930, 1932) included this structure as part of the archistriatal complex, a position which was foreshadowed by Huber and Crosby (1929).

The archistriatum *sensu stricto* can be clearly divided into ventral and dorsal parts. The former consists of chromophilic cells often intercalated amongst the fibres of the occipito-mesencephalic tract. The dorsal part is often a crescentic shaped mass of larger cells, is not dissimilar to the ventral part, but is separated from it by a laminated micro-cellular component. Posteriorly the entire structure becomes more compact and the percentage representation of the micro-cellular components increases. The dorsal lobe widens out, loses its crescentic appearance, assumes a progressively greater share of the whole structure and finally contributes the entire hind limits. The macro-cellular units include examples of Stingelin's cell types (b) and (c) whilst the smaller cells are those of categories (h) and (i).

The actual degree of dissection of the two parts varies from order to order but both were identified at an early stage in *Passer*, *Apteryx* and various Trochiliformes. Reference to Figs 115–118 and also Figs 123 and 124 will show that the overall topographical limits also vary considerably. In *Chara-drius* the entire grouping is situated to the side. This is a rather antero-ventral position by comparison with that in the hypothetical ancestral form and is also the case in *Vanellus vanellus*, *Capella gallinago*, *Ardea cinerea*, *Ciconia ciconia*, *Anas platyrhynchos*, *Otus scops*, *Athene noctua*, *Strix aluco*, *Melopsit-tacus undulatus* and *Amazona ochrocephala*. However, Stingelin (1958) drew attention to the rather different position in Passeriformes such as *Muscicapa striata*, *Turdus merula* and *Corvus frugilegus*. In these species it lies nearer to the mid-line and in a rather postero-dorsal position between the ventral angle of the ventricle and the paleostriatal regions. In contrast to the relatively stable distribution of this latter region in longitudinal sections of avian hemispheres the archistriatum has, for example, a rostro-ventral position in *Amazona* and a caudo-dorsal one in *Corvus*.

## VII. THE ECTOSTRIATUM

The ectostriatum has been widely recognized in avian brains since the end of the last century. It lies above the dorsal medullary lamina, which forms its basal boundary, and projects into the neostriatum. As a result it divides the neostriatal region very clearly into the medio-dorsal intermediate neostriatum and more extensive lateral components. If one looks at transverse sections of pigeon brains the ectostriatum gradually moves ventro-laterally and actually makes a small contribution to the hemisphere wall at the level where the paleostriatum primitivum is disappearing. The fronto-archistriatal tract lies to the ventral side of it at this level but further forward once more occludes it from the wall. Slightly posterior to this region the fibres of the thalamo-striatic tract intervene between the ectostriatum and augmentatum. Fibres of the thalamo-frontal tract traverse the primitivum and augmentatum to reach it and represent important forward projections from the rotundal nucleus. Some idea of the relationships can be obtained from Fig. 112. The close topographical relations between the ectostriate and neostriate regions are emphasized by Kuhlenbeck who attributed both to his central epibasal nuclear complex (see Table 53).

## VIII. THE BASAL NUCLEUS

In the pigeon the basal nucleus lies above the fronto-archistriatal tract in front of the level at which the ectostriatum disappears and the final anterior prolongation of the augmentatum is being occluded between the parolfactory lobe and the lateral hemisphere wall. Further forward it makes a significant contribution to the lateral wall of the hemisphere and in this area is rectangular in cross section. Finally, as the neostriatum above it is gradually occluded and converted to a horizontal strip by the predominant ventro-ventral ventral hyperstriatum, it gradually peters out. As this distribution is largely well out to the side of the hemisphere it is not seen in median longitudinal sections taken close to the mid-line. A somewhat more extensive disposition, together with slight topographical variations, does, however, render it visible in such sections in some genera. One may cite *Vanellus*,

*Capella, Cuculus, Otus, Athene, Strix, Muscicapa, Turdus* and *Corvus* which are represented in Figs 115–118.

## IX. THE NEOSTRIATUM

The neostriatum is one of the very extensive telencephalic components although like the others it is subject to considerable variations. Furthermore, as with the others, it has been given a number of different names in past decades and has sometimes been homologized with the putamen and caudate. Capping the paleostriatum it can extend along the entire length of the hemisphere from the anterior to the posterior pole. Its form reflects with considerable accuracy the variations of hemisphere structure and some idea of this diversity can be obtained from Figs 115 to 118; 121 to 124. A number of authors have suggested its division into three discrete regions the frontal, intermediate and caudal neostriatum (Huber and Crosby, 1929) and these terms are usually used today. The areas concerned are intimated anatomically by the passage through it of various axonal systems and also by histological differences. The experimental studies which are considered in Chapter 14 also suggest that these divisions are of considerable functional importance.

In generalized terms the frontal neostriatum comprises that area which is associated with the basal nucleus; the caudal neostriatum lies above and behind the paleostriatum; and the intermediate neostriatum separates the two. However, the frontal component also swings out sideways and occupies the ventro-lateral area of the hemispheres. It is clearly demarcated from the paleostriatum by the dorsal medullary lamina and the neo-paleostriatic fissure, and from the ventral hyperstriatum by the hyperstriatic lamina. As the intermediate neostriatum, which is equivalent to Rose's field G, $G_2$, increases in size it takes over the composition of much of the nucleus and is then replaced posteriorly by the caudal component.

Histologically the frontal neostriatum is composed of both medium-sized and large cells with a few smaller ones. They are generally larger than those of the intermediate part and are more diffusely arranged. By comparison the medium sized cells of the intermediate part can be more closely packed and in this respect it contrasts with both the frontal and caudal regions. However, as was emphasized in Section IV above, considerable intergeneric variation occurs, and in the occipital region there are a number of larger cells which are both similar to, and partially continuous with, the components of the periventricular band. In terms of Stingelin's classification the neostriatal cells fall into classes (b)–(i) and there is a tendency for them to be aggregated together. This can either result in clumps of associated cells or definite fused units in which the cells are held together by the oligodendroglial contributions. These are particularly prevalent in *Amazona, Melopsittacus* and *Carduelis.*

striatum, are less frequent if they are present at all. It can also differ from the two overlying Wulst structures in its cell density as is shown in Fig. 114. In *Ardea, Vanellus, Muscicapa, Turdus, Carduelis* and *Regulus*, for example, the intercalated nucleus has more cells per unit volume. Complementarily in *Columba, Ardea, Athene, Amazona, Carduelis, Merops, Corvus* and *Regulus* the accessory hyperstriatum has less. Further examples can be obtained by referring to Table 56.

The degree of overt differentiation between the dorsal and accessory hyperstriatum can also vary. In *Charadrius alexandrinus* the two are very closely associated and the intercalated nucleus is also rather poorly differentiated in *Columba, Cuculus, Micropus* and *Merops*. Nevertheless in most genera and species the two are clearly distinguishable and a well defined intercalated nucleus is equally distinct from both. In longitudinal sections the dorsal hyperstriatum more or less equals the size of the accessory hyperstriatum in the hemispheres of *Capella gallinago, Ardea cinerea, Ciconia ciconia, Merops apiaster* and *Otus scops*. It is, however, smaller in *Platalea leucorodia, Cuculus canorus, Anas platyrhynchos, Athene noctua, Strix aluco, Muscicapa striata* and *Turdus merula*. Furthermore as in the case of the other Wulst structures its position can vary from an area near to the occipital pole in *Anas*, to the frontal pole in the Strigiformes and Passeriformes.

## B. The Intercalated Hyperstriate Nucleus

This structure, whose characteristics have already been outlined in the previous section and in Section IV above, is situated between the dorsal and accessory hyperstriatum. As such it is associated with the supreme frontal lamina and was described in association with that structure by Kappers *et al.* (1936). In those genera or species in which it exists as a discrete entity it usually contains rather densely packed micro-cellular components. These are referable to Stingelin's cell types (f)–(i). The cells of category (i) are particularly prevalent in the case of the little owl and tawny owl, *Athene noctua* and *Strix aluco*. Its position necessarily parallels that of the other Wulst structures. Lying in an oblique dorso-ventral plane in the frontal region of *Muscicapa, Turdus, Corvus* and many other forms, it has a more horizontal disposition in *Ardea* and *Platalea*. In *Anas* the situation contrasts even more with the first-named genera and it assumes an oblique postero-ventral to anterio-dorsal orientation at the occipital pole.

## C. The Accessory Hyperstriatum

Together with the intercalated nucleus the accessory hyperstriatum was ascribed by Kuhlenbeck to his diffuse nucleus. The two were also linked together by Kallén (1953, 1955) who designated them as the principal components of his units $d^I$ and $d^{II}$. Both suggestions emphasize the lack of any intervening fibrous structure separating the two and emphasize the supreme frontal lamina which separates the intercalated nucleus from the dorsal hyperstriatum. In carinates the accessory hyperstriatum is largely composed of Stingelin's cell types (b), (c), (d), (h) and (i). Types (e), (f) and (g) are encountered less often. It differs from the intercalated nucleus, where this is present, by its more diffuse cellular arrangement (Table 56), and from the dorsal and ventral hyperstriatum together with the neostriatum by the absence of marked plasmodial aggregations. Very large representatives of category (b) cells occur in *Corvus* and *Ardea*, medium sized (e) cells in *Vanellus*. In general, however, the large and relatively dense layers of micro-cellular elements give it a characteristic appearance.

In the Casuariiformes it is a very large mass which occupies the anterior and anterio-dorsal region of the hemispheres (Dennler, 1932). By a comparison of the gross anatomy it would appear that although extensive *Struthio* it does not extend quite so far backwards as in *Casuarius* and *Dromiceius*. Craigie (1936a) concluded that it was nevertheless very similar in all three genera. Dennler noted that the vallecula was not visible from above in *Struthio* and that it also failed to extend on to the ventral surface as in the Strigiformes, being largely visible from the side. Prior to the more recent studies Craigie suggested that this was due to the relative depression of strigiform brains which gives a lateral extension so that the accessory hyperstriatum projects sideways over the more ventral regions.

Craigie (1936a) described three longitudinal divisions internally in the ostrich which he homologized with those described by Durward (1932) in *Apteryx*. The most lateral of these he designated zone 1. It was approximately triangular in cross section and its three angles extended over the dorsal and ventral surfaces of his zone 2, and over the dorso-lateral part of the neostriatum. It was the least extensive of his three sub-divisions and characterized by the presence of numerous rather large cells with a few smaller ones interspersed between them. Anteriorly it narrowed into a thin layer that curved over the cephalic pole of the hemisphere. Posteriorly it became indistinguishable from the other two zones.

Zone 2 was crescent-shaped in cross section and comprised small closely packed cells. The packing was so dense that after cresyl violet staining the whole zone could be seen with the naked eye. The appearance was further enhanced by the relatively chromophilic nature of the intercellular substance. The dorsal arm of the crescent extended over the outer surface of zone 3 to reach the dorso-medial angle of the hemisphere. The ventral component stretched across the medial region of the inner surface. Like zone 1 it also formed a layer over the cephalic extremity of the next but lost its identity further back. This situation compared fairly closely with that in the Casuarii-formes, less so with the condition in *Apteryx* (Craigie, 1935c). Zone 3 was dense, less chromophilic, and had relatively large component cells which were interspersed between smaller ones reminiscent of those in zone 2. The most extensive of the three zones it merged medially with the parahippocampal area.

The aberrant condition in *Apteryx* is discussed in association with the corticoid structures in Section XII B below. As was emphasized both in association with the dorsal hyperstriatum and the intercalated nucleus the topographical position of the accessory hyperstriatum in various carinate orders is different. Dennler (1932) considered that *Phalacrocorax* fitted his bubonid type in which, although not thick, it is broad and forms a cap over the entire rostral pole. He pointed out that amongst the Strigiformes them-selves this results in the accessory hyperstriatum being the only structure which is seen in transverse sections of the anterior end. In *Mergus* it just failed to reach the anterior end and extended well back. In *Anas* he found that it reached further back than the more ventral Wulst components. Both these conclusions conform with the data of more recent authors.

Together with the rest of the Wulst the accessory hyperstriatum also has a relatively posterior position in *Capella gallinago* and such occipital sites reflect the very considerable development of both the neostriatum and the ventral hyperstriatum at the frontal pole. Stingelin (1958) emphasized that the con-dition in *Ardea cinerea* resembles that of his hypothetical ancestor. In a series arranged as *Ardea*, *Ciconia ciconia* and *Platalea leucorodia* there is a pro-gressive relegation of all the Wulst structures to the occipital region (Fig. 115). Complementarily the ventral position of the vallecula amongst the Strigi-formes, which was observed by Dennler, is of course a reflection of the massive development of all the Wulst structures, but particularly the accessory hyper-striatum in the frontal region of the owl brain. The situation is particularly marked in *Athene noctua* and *Strix aluco*, less so in *Otus scops* (Fig. 117) and is true in the owls which I have investigated. A relatively dorso-frontal position also characterizes the Wulst in *Cuculus*, *Columba* and *Merops* but a ventrally located vallecula is most clear in the Passerines. Rostro-ventral in the Muscicapidae it is tucked well under on the ventral surface in *Turdus* although the actual position in longitudinal sections varies with the plane of the section. Such forms comprise an interesting comparison with the Psittaci-formes. With the Wulst in an occipital position the whole appearance of the hemispheres is quite different.

The Trochiliformes are of historical interest because they were the source of a vigorous and rather unjust *riposte* by Huber and Crosby in Kappers *et al.* (1936) against a barely ambiguous statement by Craigie (1932) which will be referred to below. He seems (1940b) to have taken it in good part and Kappers seems to have at least partially disassociated himself from it. Broadly it related to the traces of 3 cell laminae which Craigie had described in the region between the ventricle and dorso-lateral wall in *Lampornis*, *Cyanolaemus* and the swift genus *Chaetura*. Less evident in *Cyanolaemus* than *Lampornis* they were restricted to a relatively short zone and the accessory hyperstriatum soon thinned. It was these cell laminae which subsequently resulted in Craigie's description of the superficial cortical areas of birds.

# XII. AVIAN CORTICAL STRUCTURES

## A. Introduction

In retrospect it is unfortunate that many of Craigie's early studies on avian neuro-anatomy were focused on *Apteryx* (1929a,b, 1930). At that time, and throughout the thirties, there was some considerable controversy about the possible occurrence of cortical structures in birds. The fact that both his results and those of Durward (1932) referred to the Kiwi, which in any case has an aberrant brain, simply served to add to the confusion. As was mentioned above Huber and Crosby (Kappers *et al.*, 1936) were totally unconvinced about many of the so-called corticoid structures. Up until that time suggestions about the extent of pallial development in general had varied considerably and it was Kuhlenbeck's (1938) embryological observations which formed the basis for subsequent definitive differentiation between the Wulst structures and those underlying them. In their 1903 paper Edinger *et al.* concluded that neither the ventral quarter nor the medial region of the hemispheres had any cortex. However, they ascribed the covering of the dorsal three-quarters to "true cortex". In the dorso-medial region Rose (1914) figured what he called the entorhinal area. This corresponds to what Craigie in his later papers designated zone 3 of the hippocampus. Although Huber and Crosby (1929) retained Rose's terminology, their guarded scepticism about such cortical layers was shown by their emphasis on the fact that they were simply using the term topographically. It did not, as far as they were concerned, carry any connotations implying a homology with similarly named areas in mammals. In Kappers *et al.* (1936) they agreed that it was the homologue of the dorsal part of the hippocampus in diapsid reptiles such as *Alligator* but remained fairly dubious about the possible existence of laminated areas in the avian telencephalon. Nevertheless, even at that time, it was generally agreed that at least the components of the accessory hyperstriatum were the functional analogues of the cortex of some mammalian genera (see, for example, Kappers, 1922, 1928; Huber and Crosby, 1929; Craigie, 1930; Durward, 1932). Despite the works of Kuhlenbeck (1924, 1929) it was only after the publication of Craigie's later works (1934, 1935a,b,c, 1936a,b, 1939, 1940a,b) that the confused and highly controversial situation was elucidated. Even then, in spite of the supporting embryological work of Kuhlenbeck (1938), Crosby remained unconvinced (Crosby and Humphrey, 1939). The condition in the chick is shown in Fig. 39.

## B. Cortical Structures in Ratite Birds

Following Craigie (1940b) one can say that the avian cortex basically comprises median, dorsal and lateral areas which have close phylogenetic and topographic relationships with structures in the brain of reptiles. Such topographical differences as one observes, and these vary, result from the transformation of the neopallial primordium into a large non-cortical mass, the differing position of the Wulst and other structures, and finally the relative development of the rhinencephalon. The clearest representation and most extensive distribution of cortical areas occurs in the orders Struthioniformes, Casuariiformes and Rheiformes. They are particularly well represented in *Dromiceius*, the emu. The cortex of all other birds can be derived from this condition but involves the greater or less reduction of various components. Although it is just as well-developed in certain neognaths as it is in some paleognaths, no neognathine genus has the degree of development and differentiation which occurs in *Dromiceius*. Conversely no paleognathine genus has the cortex reduced to the extent of some neognath orders although the tinamiform genus *Rhynchotus* occupies a rather independent position. Only the small ventro-lateral portion of the parahippocampal formation of the Kiwi comprises cortex which is more highly differentiated than any area in either *Dromiceius* or other avian genera.

The pattern in *Dromiceius* involves a hippocampal area, a parahippocampal area, an anterio-lateral prepyriform area and a postero-lateral peri-amygdalar area. The last two represent the lateral cortex of reptiles. (Dart, 1934; Craigie, 1940b). The hippocampal component extends along the dorsal part of the median wall of each hemisphere, curves downwards in the mid-part of the caudal wall, and then projects forward for a short distance ventrally. In this ventral position it forms a longitudinal thickening of the wall. It can be subdivided further into four longitudinal zones (Craigie, 1935b) which, from below upwards, have been designated $H_1$–$H_4$. $H_1$ consists of a narrow plate of small dark cells which are crowded together. They are somewhat flattened in a plane parallel to the surface and therefore look as if they have been squeezed between the fibres lying both below and above them. $H_2$ is a small area, triangular in cross-section, with a single layer of larger and less chromophilic cells and a broad zonal layer. Together these two components contribute to the hippocampus of Huber and Crosby (1929).

$H_3$ is the entorhinal area of Rose (1914). From the dorsal limits of $H_2$ it expands further dorsally to a very considerable thickness and shows three distinct cell layers coupled with a superficial zonal layer. Together with $H_4$ it is, broadly speaking, the dorsal part of the hippocampus of Huber and Crosby. In $H_4$ the two deeper cell layers which are present in $H_3$ continue unchanged but the outer cell layer is replaced by a more or less clearly differentiated mass of smaller cells. These are usually both crowded together and chromophilic. The intercellular substance of this region, when viewed after the classical procedures, is also more deeply staining than that elsewhere and the degree of vascularization is greater. The whole layer has considerably more capillaries than neighbouring areas. The zonal layer gradually thins out over it and is finally obliterated by fibres of the septo-mesencephalic tract. The actual position of the cell column is usually made clear by the arrangement of the cells internal to it which are distributed in sweeping curves.

The parahippocampal area occupies the dorsal part of the median wall between the hippocampal formation and the neighbouring accessory hyperstriatum. Ventrally its cell layers are continuous with the deeper layers of the hippocampus. Dorsally they are continuous with virtually the entire thickness of the accessory hyperstriatum and there is no clear line of demarcation between the two. When traced backwards it runs up the medial part of the dorsal wall and then spreads across the whole unattached dorsal wall behind the accessory hyperstriatum. Like the hippocampal components it is further divisible. That region lying next to the hippocampal region itself is characterized by the presence of a very broad zonal layer, and rather diffuse, elongated pyramidal cells which are either arranged in oblique lines or lie in sweeping curves. This region is not very extensive and forms a transitional zone between the hippocampus and the remainder of the parahippocampus. Craigie designated it zone (a). Dorsally it is followed by two narrow subdivisions. These have a broad zonal layer and a single layer of underlying cells. The majority of these are pyramidal and, whilst they are both crowded and chromophilic in zone (b) they are less so in zone (c). It is above this last-named component that the most extensive of the para-hippocampal subdivisions occurs. Designated zone (d) by Craigie, the single cell-layer of zone (a) is replaced within it by several distinct laminae. These are at least three in number and there are sometimes four. In contrast to the increased thickness of the cellular components its zonal lamina is thinner than those of zones (a), (b) and (c). The outermost cell layer was thought by Craigie to continue over the accessory hyperstriatum as the superficial corticoid layer of Huber and Crosby. The deeper layers gradually merge with the accessory hyperstriatum itself. The distinctness of these laminae varies somewhat between the Casuariiformes, Rheiformes and Struthioniformes. In *Rhea*, although generally well marked, they are less clear in that region adjoining the accessory hyperstriatum. In *Struthio* they are not recognizable at this focus at all, and even in the more ventro-medial part of zone (d) the layers are not very clear, so that in the ostrich the area was described as comprising both somewhat laminated (d') and laminated (d'') components. Zone (c) is also less clearly differentiated in these two genera than in *Dromiceius*.

The periamygdalar and prepyriform (= anterior olfactory nucleus) areas also vary somewhat in their extent amongst the Casuariiformes, Rheiformes and Struthioniformes. The periamygdalar area is well-developed in all four genera although least so in *Rhea*. The prepyriform cortex is a distinct ventro-lateral plate at the anterior end of the hemisphere. It is both larger and more distinct in the emu than in either *Rhea* or *Struthio*. At its ventro-medial edge it adjoins the lateral margin of the olfactory tubercle which lies behind the olfactory bulb.

By comparison with all the foregoing the situation in the Kiwi is aberrant! In spite of the fact that all the areas which were described for the emu are present, and exhibit a similar cytological appearance, they are displaced. This change in their topographical relationships is the result of a marked disloca-tion of most parts of the telencephalon (Craigie, 1930, 1935b; Durward, 1932). This affects more particularly the accessory hyperstriatum which has undergone a backward migration leading to a great reduction of the more anterior and dorsal parts of the parahippocampal area. As a result it is only in the ventro-lateral region of the occipital pole of the hemisphere that this cortex is well-developed. At that point zone (d) is very thick and shows the highly developed lamination which was referred to above.

The hippocampus is less reduced dorsally than the parahippocampus but it too has a reduced anterior region. As if to compensate for this the hind end curves down over the caudal pole and then runs forward as a massive longitudinal thickening. This thickening, together with the lamination of parahippocampal zone (d), is more marked than in any other avian genus. The prepyriform area is, however, roughly comparable with that in *Struthio*. The periamygdalar area is more reduced and, unlike the condition in any other avian or indeed reptilian genera, it is isolated from the parahippocampus.

## C. Cortical Structures in Other Birds

### (i) *The Hippocampus*

The cortical development in 6 genera of the Tinamiformes was investigated by Craigie (1940a). It is of considerable interest that both the structure and the topographical relationships resembled those of the emu. With the excep-tion of the genus *Rhynchotus* there was a well-developed hippocampus which was, however, less extensive than in the foregoing birds. Zones $H_1$ and $H_2$ were both small and not distinct the one from the other. In *Rhynchotus* the overall shape of the hemispheres is different and in some respects recalls the apterygiform condition. This is associated with very little differentiation of the sub-divisions of the hippocampus.

In the genera of other orders the hippocampus presents a more uniform appearance than the parahippocampus. Sub-divisions $H_1$ and $H_2$ are indeed frequently inseparable but, this aside, all the components are invariably present. In certain genera whose brains are more removed from the supposed ancestral condition the combined unit of $H_1 + H_2$ is frequently confined to the

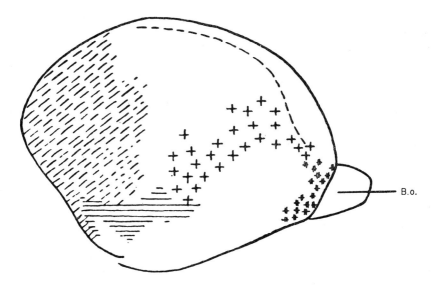

FIG. 119. The cortical structure of the hemisphere of a loon viewed from the side. The widely spaced crosses indicate the more or less vestigial sheet connecting pre-pyriform and periamygdalar areas. Vallecula is shown by the broken line; parahip-pocampal field *d* obliquely hatched; periamygdalar area is horizontally hatched; pre-pyriform area is shown by crosses with double uprights; vestigial sheet by widely spaced crosses. (From Portmann and Stingelin, 1961; after Craigie, 1940).

more caudal region where the formation swings down into the caudo-ventral wall. On the other hand there are a number of genera in which this hook is lacking. Its presence is dependent upon the posterior extent of the accessory hyperstriatum. The general conditions which were observed by Craigie are summarized in Table 57. Additional points of particular interest occur in the Strigiformes. Here the hippocampus occupies the entire dorsal region behind the level at which the accessory hyperstriatum disappears. Although $H_{1+2}$ is small anteriorly, $H_2$ becomes more distinct caudally and both $H_3$ and $H_4$ are very thick. $H_3$ also develops a very distinct inner layer of many small cells, a condition which also occurs in *Cariama*.

## TABLE 57

*A summary of the hippocampal development in some neognathine genera.*

| Species | Ventral longitudinal thickening | $H_1$ | $H_2$ | $H_3$ | $H_4$ |
|---|---|---|---|---|---|
| Gaviiformes | | | | | |
| *Gavia immer* | present | + | + | + | + |
| Colymbiformes | | | | | |
| *Colymbus nigricollis* | | | | | |
| *californicus* | slight | + | + | + | + |
| Pelecaniformes | | | | | |
| *Phalacrocorax auritus* | | | | | |
| *auritus* | none | + | + | + | + |
| Gruiformes | | | | | |
| *Cariama cristata* | none | fused | | + | + |
| Galliformes | | | | | |
| *Bonasus umbellus* | small | fused poor | | + | + |
| Anseriformes | | | | | |
| *Mergus americanus* | thick | fused and small | | + | + |
| *Anas domesticus* | thick | ? | ? | + | + behind |
| *Somateria mollissima* | | | | | |
| *borealis* | thick | + | + | + | + |
| Charadriiformes | | | | | |
| *Capella delicata* | none | + | indefinite | + | + |
| *Philohela minor* | slight | + | | + | + |
| Alciformes | | | | | |
| *Fratercula artica artica* | small | + | + | + | + |
| *Uria troile* | small | + | + | + | + |
| Falconiformes | | | | | |
| *Accipiter nisus* | ? | + | + | + | + |
| *Accipiter velox* | slender | + | + | + | + |
| Psittaciformes | | | | | |
| *Melopsittacus undulatus* | small | fused | | posteriorly | posterior |
| Strigiformes | | | | | |
| *Otus asio* | none | fused small | | Huge | Huge |
| Caprimulgiformes | | | | | |
| *Chordeiles virginianus* | ? | fused All present posteriorly | | + | + |
| Coraciiformes | | | | | |
| *Ceryle alcion* | | + distinct dorsal groove | + | + | + |

Table 57 (*contd.*)

| Species | Ventral longitudinal thickening | $H_1$ | $H_2$ | $H_3$ | $H_4$ |
|---|---|---|---|---|---|
| Micropodiformes | | | | | |
| *Chaetura pelagica* | none | fused and caudal | | $+$ | $+$ |
| Trochiliformes | | | | | |
| *Lampornis mango* | none | fused and caudal | | $+$ | $+$ |
| *Cyanolaemus elemanciae* | none | similar to, but less distinct than, *Lampornis* | | | |
| Passeriformes | | | | | |
| *Planesticus migratorius* | ? | $+$ | $+$ | ? | ? |

After Craigie (1940b).

## (ii) *The Parahippocampus*

In the parahippocampus of the Tinamiformes there is never a clear distinction between sub-divisions (b) and (c). These are represented by a narrow zone intermediate between a more or less well-developed field (a) and the multi-laminar cortex of (d). The dorso-lateral edge tends to be separated from the deeper part of the accessory hyperstriatum by a cellular zone and therefore only appears to be continuous with its thin outer layer. Craigie (1940a,b) considered that this was a secondary condition. The free caudo-lateral wall is reduced to a varying extent, but is typically much thinner than in the emu, rhea and ostrich. Under such conditions it loses its cortical structure. In *Rhynchotus* the parahippocampus parallels the condition of the hippocampus and shows very little differentiation. The entire occipital pole in this genus is, in fact, covered by a thin membranous wall and the parahippocampus does not extend on to the lateral surface.

The lateral and posterior parts of the parahippocampal region in neognathine birds are reduced and there is little evidence of a differentiated cortex. Such weakness or absence of cortical differentiation can be associated with both a reduction, or, alternatively, an increase in the thickness of the wall in the region concerned. The laminar organization is usually best developed when the wall is not too thick. It disappears when it is membranous and acellular, and also when it is massive. The medial and anterior parts remain more massive and also more frequently retain overt evidence of structural differentiation. Traces of lamination are particularly prone to occur at a region close to the hind end of the accessory hyperstriatum. Such traces are probably as weak as anywhere in the guillemot, *Uria troile*, the American downy woodpecker, *Dryobates pubescens medianus*, and in the fowl. The degree of reduction does not, therefore, follow a gradation of form in the sequence which is usually accepted as representing the phylogenetic status of the orders.

## TABLE 58

*A summary of the parahippocampal development in some neognathine genera.*

| Species | Zones | | | |
|---|---|---|---|---|
| | a | b | c | d |
| *Gavia immer* (auctt) | + | fused | | + |
| *Colymbus nigricollis californicus* (auctt) | + | fused | | + |
| *Phalacrocorax auritus auritus* | + | fused and poor | | + |
| *Cariama cristata* | + | fused | | + |
| *Bonasus umbellus* | + | — | — | + |
| *Ardea herodias herodias* | + | fused | | + |
| *Mergus americanus* | + | generally indistinct | | |
| *Anas domesticus* | + | extensive but not well differentiated | | ? |
| *Somateria mollissima* | + | fused | | + |
| *Capella delicata* | + | fused | | multi-laminate |
| *Philohela minor* | + | fused | | multi-laminate |
| *Fratercula arctica arctica* | + | + | + | + |
| *Uria troile* | + | + possibly fused | | ? |
| *Accipiter nisus* | ? | ? | ? | + |
| *Accipiter velox* | + | fused small | | + |
| *Melopsittacus undulatus* | + | — | — | + |
| *Otus asio* | + | fused and indistinct | | + |
| *Ceryle alcion* | small | small | | inseparable |
| *Planesticus migratorius* | + | fused | | + |

After Craigie (1940b).

In certain genera rather special conditions occur. One can note the similarity which exists between both the hippocampus and parahippocampus of the Micropodiformes and Trochiliformes. Although thinner in the former than in the latter, that part of the parahippocampus lying in the dorso-medial part of the hemisphere appears as somewhat compressed in both orders. Two outer cell laminae form an area which is triangular in cross-section. A deeper macrocellular layer tapers off medially, but laterally it is broadly continuous with the tissue of the thin accessory hyperstriatal components of that region. The parahippocampus of *Phalacrocorax* extends laterally as the accessory hyperstriatum gradually disappears. It then loses its lamination and becomes reduced to, or rather is replaced by, a thickish membrane caudo-laterally. In *Accipiter velox* the parahippocampal area again fades into a membrane posteriorly and this covers the entire occipital pole of the hemisphere. Comparable membranous regions characterize the hind end of the hemisphere in *Cariama*, *Chaetura* and *Cypselus*. A synopsis of the organization in 19 species is contained in Table 58.

## (iii) *The Periamygdalar and Prepyriform Areas*

In the reptilian hemisphere a lateral cortex connects the periamygdalar field to the prepyriform area (= anterior olfactory nucleus). This zone is generally absent in birds but is present in the young of cormorants and grebes, in the newly hatched emu, and, as a narrow cortical structure, in *Gavia immer* (Fig. 119). In the ratites the prepyriform area is entirely lateral in position but in the Tinamiformes it traverses across the cephalic end of the hemisphere just above the olfactory bulb. As a result it is also represented on the medial surface. In the adults of the Colymbiformes (Gaviiformes) it also extends across but does not reach the medial surface, although it does so in the Podicipidiformes. In *Phalacrocorax*, *Cariama*, *Ardea* and *Otus* it is entirely lateral and, in contrast to the condition in the Colymbiformes, there is only a very small lateral component in the Anseriformes. In these birds the "centre of gravity" is differently distributed so that the area covers the entire medial, and even medio-dorsal, surface at the cephalic pole of the hemisphere. A variety of conditions link these two extremes and again have no immediate correlation with the accepted phylogenetic sequence. Craigie (1940b) concluded that in some ancestral condition the prepyriform cortex was distributed on the lateral aspect only. The cephalic and medial positions are in this case secondary conditions and reach their apogee in the Anseriformes. The fact that it is continuous, via a lateral field, with the periamygdalar area in the young birds mentioned above would support this hypothesis.

The periamygdalar area is far more variable in its occurrence amongst adult forms than is the prepyriform area. Amongst the Tinamiformes it is poorly developed in *Calopezus* and better developed in other genera. The condition in *Nothura* compares very closely with that in *Struthio*. In some neognathine genera, for example *Accipiter* amongst the Falconiformes, its very presence is questionable. However, it is well-represented in both *Gavia* and *Phalacrocorax* and quite extensive in *Ardea, Anas* and *Columba.* In *Cariama* on the other hand, which may not be representative of the condition in Ralliformes, it is only distinct and well differentiated in the occipital region. Perhaps the most prominent development is in the coraciiform genus *Ceryle*, whilst it is a small, thin and very weak component of the cortical structures in *Otus*, and only represented by scattered cells in the budgerigar. A comparable condensation of scattered cells in the ventral edge of the free lateral wall is the only representation which is detectable in charadriiform genera such as *Capella* and *Philohela.* As a result of this it does not have the same relationships to other structures in these two genera as it does elsewhere.

From this rather indecisive and heterogeneous distribution of cortical components it is perhaps difficult to extract a unifying principle. It does, however, seem that well-developed cortical structures occur in genera in which the olfactory structures are relatively complex. In the case of the prepyriform and periamygdalar areas the genera in which they are most extensive are those with tolerably high values for the ratio of olfactory bulb size to hemisphere size (Table 54). This would imply that one can extend Cobb's "wet rule" to the cortical areas of neognathine genera.

# XIII. VARIATIONS IN THE STRUCTURE OF THE HEMISPHERES

## A. General Considerations

Following Portmann and Stingelin (1961) and Stingelin (1958) one can describe the basic type of carinate organization from which other telencephalic form is derived as follows:

(1) The hemispheres are of relatively small size.
(2) There is a relatively large olfactory bulb.
(3) There are relatively large archistriatal and paleostriatal regions.
(4) The anterior extremity of the lateral ventricle reaches in front of the so-called base line.
(5) There is a small and medially situated Wulst with no clear differentiation of the accessory hyperstriatum, nucleus intercalatus and dorsal hyperstriatum.
(6) There is no well defined ventral fissure.
(7) There is an extensive prepyriform area.

In contrast, in this scheme the evolutionary trends leading from this primary condition appear to have resulted in:

(1) An overall increase in the size of the hemisphere.
(2) A relative and overall decrease in the size of the olfactory bulbs.
(3) A similar diminution in the relative size of the archistriatal and paleostriatal regions.

(4) A posterior displacement of the dorsal limits of the lateral ventricle.

(5) A well developed ventral fissure.

(6) A curvature of the hyperstriatic and dorsal medullary laminae.

(7) A reduction of the prepyriform area.

However, it is essential to emphasize that in the emu's brain the Wulst is conspicuous. Its anterior end is almost in contact with the olfactory bulb and posteriorly it reaches back to within 4 mm of the occipital pole of the hemisphere. Its position is somewhat reminiscent of that in the pigeon but in relation to the remainder of the hemisphere it is both longer and broader. Cobb and Edinger (1962) discussed the possibility that this reflected a primitive condition but emphasized that the large cursorial ratites are anything but primitive. In fact they are highly specialized to their cursorial way of life.

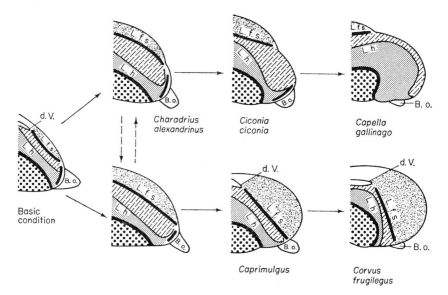

Fig. 120. Summary of the sort of evolutionary changes that neognathine forebrains may have undergone. No derivation of one species from another is intended. (After Stingelin, 1958).

Indeed it is worth noting that Cobb was consistently one of the most broadminded contributors to avian neurology, persistently refused to draw trite analogies himself, and urged others not to do so.

The result of such evolutionary trends as those suggested by Stingelin is to produce a greater concentration of phylogenetically new material in the front part of the hemispheres. However, as will have become apparent from the foregoing sections, this can take two principal and distinct forms and these are summarized in Fig. 120. The frontal development may involve the Wulst.

In this case the dorsal and accessory hyperstriatum reach down ventrally and are in a close relationship with the olfactory bulbs. Alternatively the frontal development may affect the intermediate nuclear masses. In this case the Wulst is separated from the olfactory bulbs by these structures and it is the neostriatum and ventral hyperstriatum which form the frontal pole of the hemispheres. The variations in dorsal fissure formation and the differences of surface topography which were both of such interest to Kuenzi (1918), are a direct result of these two tendencies. The archistriatum and both components of the paleostriatum are not directly involved in these developments and they retain their relatively consolidated appearance. Nevertheless it is obviously unlikely that they would be completely unaffected and they do undergo topographical changes which Portmann and Stingelin ascribe to compression. The contribution of the dorsal and accessory hyperstriatum to the frontal pole is marked in members of the Columbiformes, Coraciiformes and Caprimulgiformes. It reaches its zenith in the Strigiformes and Passeriformes. Complementarily the Wulst has a more dorsal and posterior position in many Charadriiformes, Ardeiformes and Anseriformes, and the formation of the frontal pole by neostriatal and ventral hyperstriatal tissue reaches its apogee in the Psittaciformes. In a number of orders certain genera and species do, however, have a less extreme condition and the heron *Ardea cinerea*, for example, has a degree of nuclear differentiation which is not dissimilar from that of the supposed ancestral condition.

## B. The Dorsal Aspect of the Hemispheres

Stingelin (1958) emphasized the variations which occur in both the dorsal and ventral aspects of the brain in association with the tendencies outlined above and I have been able to corroborate these in a variety of species. In the brain of some species there is a rather generalized condition which, following Stingelin, has been widely assumed to reflect the conditions in the hypothetical ancestral forms. The Wulst structures are relatively undifferentiated and occupy a medio-frontal position. Their lateral limits run from the medio-frontal to a caudo-lateral position and are defined by the vallecula. To one side, and situated between the Wulst and olfactory bulb, there is a prepyriform area. Behind this the lateral edge of the supposed ancestral brain included a small ventral hyperstriatal and a large neostriatal contribution. Posteriorly the archistriatal tissue was in a close association with the occipital pole. The actual conditions amongst neognathous birds enables them to be attributed to various positions along the suggested evolutionary series depicted in Figs 121 and 122. It is, of course, essential to emphasize that in these diagrams it is not intended to suggest that the particular species either evolved from, or gave rise to, the other species represented. They merely reflect possible changes in organization, and neuro-anatomical phases through

18*

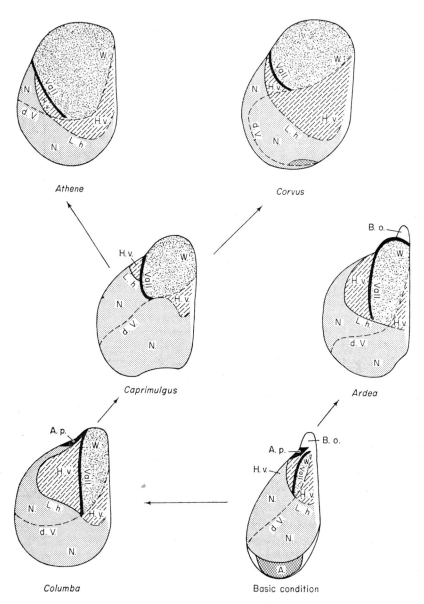

FIG. 121. A summary of similar data to Figure 120 when seen from above. (After Stingelin, 1958). The so-called developmental type A.

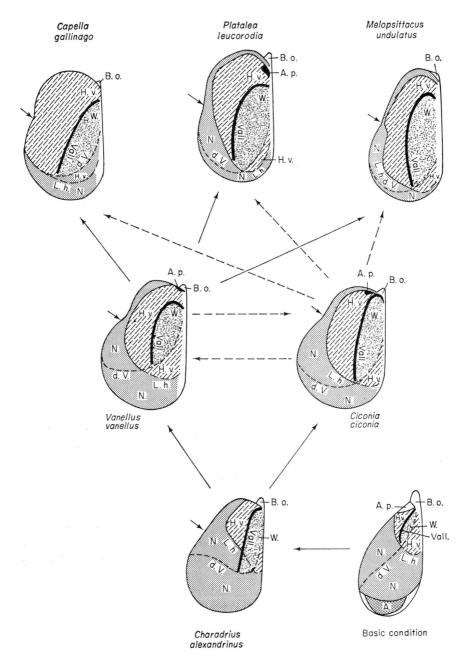

FIG. 122. The possible trends of hemisphere development as represented by some charadriiform, ciconiiform and psittaciform brains viewed in the dorsal aspect. (After Stingelin, 1958). The so-called developmental type B.

which the respective brain may have passed during phylogeny. In both types of developmental trend there are a number of obvious changes. These include a backward movement of the olfactory bulb, an increase in both the relative and absolute size of the Wulst structures, and both a lateral and caudal extension of the ventral hyperstriatum. Stingelin emphasized the progressive contribution which the dorsal and accessory hyperstriatum make to the frontal region in forms such as *Columba, Ardea, Caprimulgus, Corvus* and *Athene* (Fig. 121). In contrast the Wulst is both smaller and further back in *Charadrius, Vanellus, Ciconia, Capella, Platalea* and *Melopsittacus* (Fig. 122).

In the former series the conditions in *Columba* are envisaged as the more primitive extant situation. The relatively small Wulst has a rather elongated overall appearance reminiscent of the condition in the emu and reflecting both its length and its relative narrowness. There is some lateral expansion of the ventral hyperstriatum and the dorsal limits of the ventricle are moderately far back. In *Ardea* the situation is somewhat different from the other genera figured. Although the olfactory bulb is situated well forward, giving an appearance which is reminiscent of the ancestral form, both the greater lateral extent of the Wulst and its overall position give an impression of greater evolutionary elaboration. It will be remembered that from Cobb's wet rule the development and prominence of the olfactory bulb conform with the conditions in other aquatic, littoral or oceanic birds. At the posterior limits of the ventral hyperstriatum the hyperstriatic lamina runs directly to the mid-line and the dorsal limits of the ventricle follow a path which is broadly comparable with the situation in *Columba* although curving back in the postero-lateral region.

In contrast to this slightly disparate condition in the heron the hemispheres of the nightjar are more clearly intermediate between both the hypothetical primitive condition and *Columba* on the one hand, and that of the corvids and Strigiformes on the other. The dorsal and accessory hyperstriatum are short, somewhat rounded, and in a frontal position. The vallecula only extends a short way on to the dorsal surface and the dorsal angle of the ventricle lies markedly further forward than do those of either *Columba* or *Ardea*. These characters are emphatically accentuated in strigiform and passeriform genera such as *Athene* and *Corvus*. As has been emphasized in a number of earlier sections the Wulst of these genera is very broad and in the Strigiformes it extends further back so that, when viewed from above, the vallecula is more distinctive. In *Corvus* the line of the dorsal corner of the ventricle follows a path which is broadly comparable with that in the basic type, but in *Athene* it curves strongly forward in the lateral region.

The alternative putative developments are represented in Fig. 122. These involve a number of conditions which resemble those of *Ardea* and *Columba*. This is particularly true in the case of *Charadrius*. In this genus it will be remembered that the components of the Wulst are not always clearly differentiated. Furthermore although the ventral hyperstriatum is relatively smaller than in *Columba* its limits, as outlined by the hyperstriatic lamina, have a similar disposition when viewed from above. In *Vanellus* and *Ciconia* both the neostriatum and ventral hyperstriatum extend well back but the dorsal and accessory hyperstriatum are still rather anterior and occupy a medial position. Such organization affords some idea of the possible neuro-anatomy of forms ancestral to *Platalea, Capella* and *Melopsittacus*. In these the Wulst has a medial to posterior position, is relatively circumscribed laterally and has an elongated appearance. The ventral hyperstriatum and the neostriatum extend forward to the frontal pole and the dorsal angle of the ventricle is displaced into the occipital region.

## C. The Basal Aspect of the Hemispheres

The differences which are reflected in these variations of dorsal topography also result in modifications of deeper and more ventral regions. This has already been stated but in any case will be an obvious consequence of the differential ventral penetration of the dorsal and accessory hyperstriatum. The organization at these ventral levels is depicted in Figs 123 and 124. Broadly speaking one can envisage the supposed primary condition as one in which the paleostriate components are closely apposed to the median line and extend in an anterio-posterior direction with their front end abutting on to the hindpart of the olfactory bulb. The ventral fissure runs parallel to the lateral curvature of the hemispheres which is largely composed of neostriatum with a small archistriatal component approaching the surface posteriorly. Neither the Wulst structures nor the ventral hyperstriatum reach to the ventral surface. The subsequent evolutionary changes result in a marked anterior foreshorten-ing of the basal paleostriatum and a lateral displacement of the archistriatum. In the caprimulgiform-corvid-strigiform series of Fig. 123 the Wulst also makes a progressive contribution to these deeper levels. As would be expected from all the earlier considerations this does not happen in the *Ciconia-Melopsittacus* type of organization shown in Fig. 124.

If the pigeon is again taken as reflecting a relatively *primitive* type, that is to say one in which no great accentuation of any characteristic can be detected, a number of points emerge. Although the olfactory bulb is situated behind the level at which it is conceived as occurring in the putative basal form, the paleostriate limits retain a position close to its hind end. The sub-triangular outline of the paleostriatum is, however, rather similar to that in *Ardea* and quite unlike that in the basal form. Correlated with this the relatively large archistriatum is displaced from the mid-line and as a result occupies a postero-lateral position comparable with that of the nightjar. At the anterior extremity the Wulst just reaches the ventral level and assumes the position in

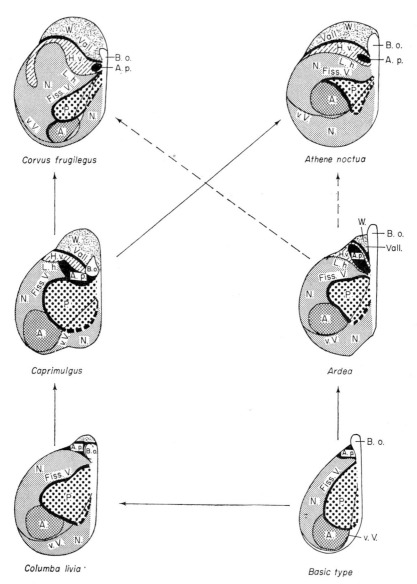

FIG. 123. Basal aspect of developmental type A (see Fig. 121). Solid lines indicate the suggestions of intermediate stages made by Stingelin (1958).

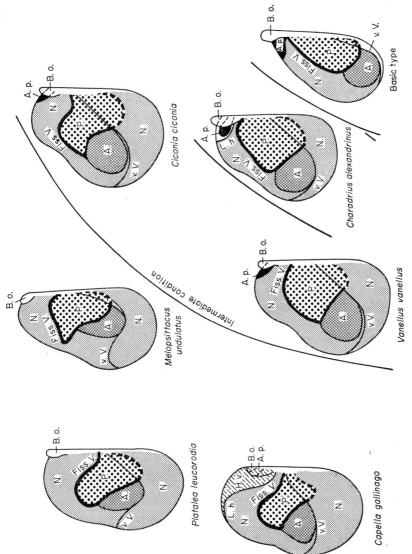

FIG. 124. Basal aspect of the hemispheres in series B of Stingelin (see Fig. 122). (After Stingelin, 1958).

front of the prepyriform area which might otherwise be occupied by the olfactory bulb. In *Ardea* the olfactory bulb is of course further forward and more prominent, but a small Wulst component occurs laterally. The presence of a small strip of ventral hyperstriatum is suggestive of the large crescentic contribution in *Caprimulgus*, *Athene* and *Corvus*. Very extensive in the latter two genera it has a backwardly projecting finger-like protrusion in *Corvus*. As would be expected from Tables 54 and 55 there is a large prepyriform area in *Ardea* and *Caprimulgus* but the portion reaching the ventral plane in *Corvus* and *Athene* is a considerably less conspicuous component. Although the outstanding characteristic of both these brains is the large size of the Wulst component a very marked difference exists in the form and position of their paleostriate regions. Retaining a pear-shaped appearance in horizontal section in *Corvus*, that of *Athene* is compressed and somewhat triangular in outline.

The contrasting conditions are represented in Fig. 124. In *Charadrius*, *Vanellus*, *Ciconia*, *Capella*, *Platalea* and *Melopsittacus* neither the dorsal nor the accessory hyperstriatum can possibly reach the ventral regions. Nevertheless the development of the neostriatum again results in a distortion of the paleostriatum whose limits are somewhat rectangular in appearance in *Capella*, *Vanellus* and *Platalea*. In view of the limited differentiation which occurs in the Wulst of *Charadrius* one must bear in mind that a massive rectangular paleostriatum may well be a closer approximation to the ancestral condition than the pear-shaped form which is generally attributed to it. Represented in the lowest region of the brain of *Charadrius*, *Ciconia* and *Vanellus*, the prepyriform area is more dorsally located in *Capella*, *Platalea* and *Melopsittacus*.

# XIV. TELENCEPHALIC FIBRE SYSTEMS

## A. Older Results

The various laminae which demarcate the regions of the hemispheres have been referred to previously. They were outlined at an early stage by Bumm (1883), Edinger *et al.* (1903) and Kappers *et al.* (1936). The dorsal medullary lamina separates the paleostriatal complex or globus pallidus, caudate and putamen from the neostriatum, ectostriatum and archistriatum. The ventral medullary lamina separates the primitivum from the augmentatum. Together they include fields of axons which contribute in part to the lateral forebrain bundles. The hyperstriatic lamina separates the neostriatum from the overlying ventral hyperstriatum and its principal component has been assumed to be the fronto-parieto-occipital tract. The two frontal laminae separate the so-called hyperstriate regions from each other. The superior separates the ventral hyperstriatum from the Wulst, and the supreme together with intercalated nucleus divides the dorsal from the accessory hyperstriatum. Short axonal projections pass to both these regions. The avian lateral forebrain bundle is very prominent. It was surveyed by Craigie (1930), Huber and Crosby (1929), Papez (1929), Kappers (1947), Kappers *et al.* (1936) and Port-

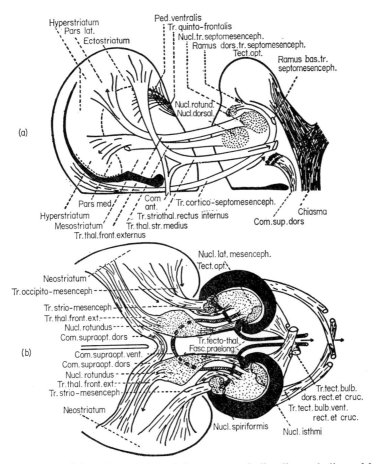

FIG. 125. Some of the relationships of the mesencephalic, diencephalic and hemi-sphere regions as indicated by early anatomical studies. (Kappers *et al.*, 1936; see also Fig. 140).

mann and Stingelin (1961). At the transition from the diencephalon to the telencephalon. There are five major axonal groupings which form the lateral to the medial position are:

1. The lateral thalamo-frontal tract.
2. The intermediate thalamo-frontal tract.
3. The strio-tegmental tract.

4. The medial thalamo-frontal tract.

5. The medial strio-hypothalamic tract.

Three further systems which figured prominently from an early stage are the quinto-frontal tract, which was referred to in association with the trigeminal nucleus, the dorsal supra-optic decussation, and the septo-mesencephalic tract or median forebrain bundle. These structures are represented in Figs. 125 and 140.

## B. Recent Observations

Many of the specimens in which the foregoing tracts were described were normal stained brain sections. Others were the results of lesion experiments. However, these all pre-date the introduction of the Nauta-Gygax staining method (Nauta and Gygax, 1951, 1954). As a result Adamo (1967) undertook a re-examination of the fibre systems in *Gallus*, *Corvus corax* and the African Lovebird, *Agapornis roseicollis*. Lesions were produced by suction and, in one case, thermo-coagulation. These indicated that in all three species the primary route for projections to the accessory and dorsal hyperstriatum, together with the intervening intercalated hyperstriatic nucleus, is the septo-mesencephalic tract. In *Corvus* the lateral forebrain bundle is also implicated. Efferents from the Wulst also seemed to pass in this latter bundle in *Gallus* and *Agapornis* although no direct information was obtained, possibly because the appropriate Wulst regions were not destroyed. Such relationships are indicated in Fig. 126.

In all three species projections both to and from all the telencephalic components were common. There were also projections via the medial forebrain bundle which linked them with the superficial parvocellular nucleus, the lateral anterior nucleus and the optic tectum. In the raven fibres passing through the lateral forebrain bundle end in the lateral geniculate nucleus, the intercalated nucleus and the tectum. In some of the specimens efferents reach the pretectal nuclei via the septo-mesencephalic tract. In the chicken and lovebird these originate in the caudal half of the Wulst. This was not the case in the raven where two specimens with *anterior* lesions had projections to this nucleus. Two chickens had degenerating fibres passing to the dorso-lateral anterior nucleus. In the raven the septo-mesencephalic tract did not project to the intercalated nucleus although the lateral forebrain bundles did. One raven had fibres projecting to the nucleus of the cruciate tecto-thalamic tract via the lateral forebrain bundle. It was, however, only in *Agapornis* that axons in the septo-mesencephalic tract reached the posterio-lateral hypothalamic nucleus. Such results broadly corroborate those of the older neuro-anatomists.

In one raven and two of the three lovebirds it was possible to distinguish fine degeneration tracks reaching to the medial and lateral pontine nuclei via

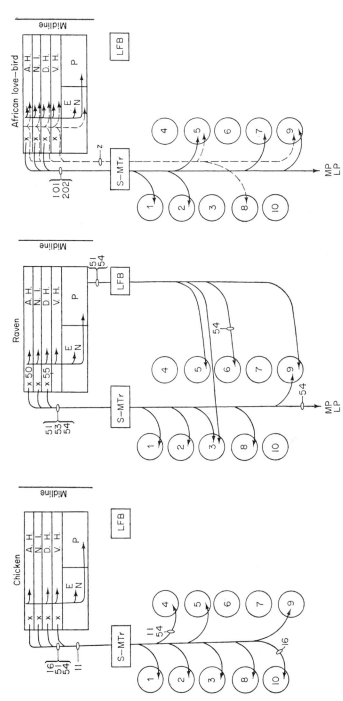

Fig. 126. Fibre relationships of the accessory hyperstriatum (A.H.), intercalated nucleus (N.I.), dorsal hyperstriatum (D.H.), ventral hyperstriatum (V.H.) and paleostriatum (P), in the chicken, raven and African love-bird. (Adamo, 1967). E, Ectostriatum; LFB, Lateral forebrain bundle; N, Neostriatum; S-MTr, Septomesencephalic tract; 1, Parvocellular superficial nucleus; 2, Anterior lateral nucleus; 3, Lateral geniculate nucleus; 4, Dorso-lateral anterior nucleus; 5, Intercalated nucleus; 6, Nucleus of cruciate tecto-thalamic tract; 7, Postero-lateral hypothalamic nucleus; 8, Pretectal nucleus; 9, Tectum; 10, The region of the posterior commissure.

Numbers in excess of 10 refer to lesions in specific preparations.

the basal caudal branch of the septo-mesencephalic tract. These presumably comprise a link within a hyperstriato-ponto-cerebellar system which is possibly comparable with the cortico-ponto-cerebellar servo systems of mammals. Should this be the case then, by analogy with mammals, it may well be an important monitor for the control and modulation of motor activity. Huber and Crosby (1929) reported projections to the diencephalon in the pigeon, and others running further back in the sparrow. The terminal sites were not elucidated. Boyce and Warrington (1898) observed connections with the mesencephalon, and Kalischer (1905) suspected that component fibres of the septo-mesencephalic tract might reach the spinal cord. The degenerated fibres that pass posteriorly and behind the tectum led Adamo to suggest that the terminal sites for fibres in the septo-mesencephalic tract may indeed be different in the species of the orders Galliformes, Psittaciformes and Passeriformes. Whether this actually reflects basic differences in the fibre organization between these orders remains conjectural. The significance of these tracts has been enhanced in recent years by studies which have shown that birds appear to "like" having these fibres stimulated (see Chapter 14, Section IV), and that they therefore comprise examples of so-called pleasure centres.

## XV. HISTOCHEMICAL INVESTIGATIONS

A number of data relating to the chemistry of the hemispheres are included in Chapter 3. We can cite here a few further histochemical results. Periodic acid Schiff staining of the neuropile was studied by Singh (1967). Earlier work by Hess and Ozello had been inconclusive but he found that in parrots the peri-neuronal cavities of the hemispheres gave strongly positive results. In his Gomori investigations Rogers (1960) found that areas indicative of alkaline phosphatase activity only occurred deep within the telencephalon where the fibre tracts collect into the peduncles. Such activity appeared on the 16th day of incubation and was concentrated in the region of the thalamo-frontal, occipito-mesencephalic and strio-mesencephalic tracts. At the 18th day the intensity of staining increased and was then particularly clear in the septo-mesencephalic tract. In the young adult these regions are less strongly positive and in all cases the cell bodies of the large neurones stand out because they show no activity, and hence lack the sulphide deposits resulting from the Gomori method. Because of the extensive areas with no activity it seemed that the hemispheres lacked the high levels seen in the medulla.

As in the case of the mid-brain and pons-medulla the acetylcholine concentration in the hemispheres is higher than that of 5-hydroxytryptamine, dopamine and nor-adrenalin. Expressed as a percentage of the acetylcholine concentration these were 38, 42 and 20% respectively (Aprison and Takahashi,

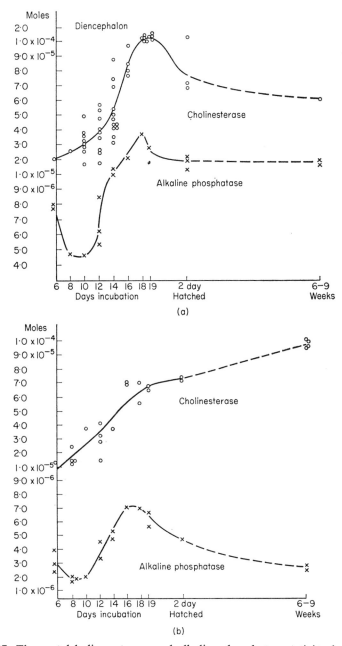

FIG. 127. The acetylcholine esterase and alkaline phosphatase activity during the development of (a) the diencephalon and (b) the forebrain hemispheres. Data expressed as moles of substrate split per hour for each milligram of tissue nitrogen. (Rogers *et al.*, 1960).

1965). The dopamine level was, however, far in excess of that in other brain regions. The high values for acetylcholine conform with Whittaker's suggestion that pigeon brain has an excess of cholinergic neurones. Eiduson (1966) found that the 5-hydroxytryptamine concentration rose from 0·05 $\mu$g/g wet weight in a 13-day embryo to 0·7 $\mu$g in the 3-day-old hatchling, and that there was normally a sharp increase which contributed to this difference some 1–3 days prior to hatching. The comparable rise in cholinesterase activity is shown in Fig. 127. Rogers et al. (1960) found that the level had risen by the 15th day and that this rise is closely comparable with the temporal characteristics of the increasing electrical activity.

Using pigeons, ducks, chickens and finches Juorio and Vogt (1967), Bertler et al. (1964) and Spooner and Winters (1966) demonstrated that a high concentration of dopamine exists within the sharply contoured region of the "basal nucleus". As they thought that this might contain some endopeduncular nuclear material they would appear to mean components of the paleostriatum (see Table 53). They concluded that this dopamine rich zone was the only one which is comparable biochemically with the corpus striatum in mammals and hence concur with Fox et al. (1966). The concentration was approximately 3 $\mu$g/g of tissue in the pigeon, duck and chicken but 7·5 $\mu$g/g in the spicefinch Lonchura punctulata. In the pigeon 0·6 $\mu$g/g of 5-hydroxyindolylacetic acid were also present which represents a higher concentration than in any other brain areas.

The use of fluorescence microscopy showed that a large anterior area had green–yellow fluorescence which recalled that of the striatum in rats. In addition there were some fluorescent fibres and varicosities. The more diffuse fluorescence disappeared after pre-treatment with reserpine and probably represented the presence of dopamine. Indeed, as in mammals, reserpine, tetrabenazine and prenylamine all decreased the concentration of brain monoamines. The influence of various such amines is considered further in Chapter 15, Section IV. In birds there was no increase in the concentration of homovanillic acid as there is in mammals, although there was a rise in 5-hydroxyindole acetic acid. Furthermore $\beta$-tetrahydronaphthylamine administration resulted in a drop in the level of all brain monoamines and their metabolites. This is an action which is quite unlike that in mammals. Chlorpromazine only affected dopamine metabolism. Noting that the histochemistry of pigeon paleostriatum was comparable with that in the caudate and putamen of rabbits Bertler et al. (1964) concluded that the dopamine must originate in the ventral mesencephalon. They therefore suggested that the adrenergic connections between these two regions are, at least in part, dopaminergic. There also remained the possibility that nor-adrenergic neurones exist in the system as nor-adrenalin is present in both the paleostriatum and neostriatum, and occurs in large quantities in the ventral mesencephalon. Nor-adrenalin positive results might, however, reflect contamination from adjacent areas.

In his comparative studies of the histochemical characteristics of various vertebrate brains Baker-Cohen (1968) observed that the paleostriatum primitivum of *Melopsittacus* is strongly positive for both succinic dehydrogenase and nicotinamide-adenine dinucleotide tetrazolium reductase. As in both lizards and alligators there was no strongly positive reaction for acid phosphatase. The neuropile of the paleostriatum augmentatum was intensely positive for succinic dehydrogenase activity, especially at its rostral end where fewer fibre bundles traverse it. The mass of neuropile was in fact the most heavily stained area of the telencephalon and only the accessory hyperstriatal and ectostriatal regions approached it in intensity. Acid phosphatase was present in a large number of weakly stained neurons in the *augmentatum* although there were also scattered *deeply* stained cells. It is these latter that are the more abundant units in turtle, alligator and mouse.

The neostriatum together with the bulk of the hyperstriatum had weak to moderate succinic dehydrogenase activity. However, a more strongly stained area of neuropile occurred in the medial neostriatum in close proximity to the fibres of the medullary lamina. The lamina hyperstriatica also showed an enhanced level of activity and both laminae exhibited browning in the acid phosphatase incubations. In neither the neostriatum nor the hyperstriatal components did any neurones show a degree of staining which was as prominent as that seen in the caiman. As in the case of the paleostriatum augmentatum some scattered neurons had a high acid phosphatase titre but most neostriatal and hyperstriatal neurons were not very active for this enzyme. This contrasts with the positive reactions of ectostriate neurons which were also strongly positive for succinic dehydrogenase. Light NAD diaphorase staining in the neuropil throughout the forebrain also highlighted the prevalence of markedly positive neurons in the ectostriatum, by contrast to their scarcity or absence in other striatal areas apart from the paleostriatum primitivum and localized areas of the archistriatum.

## References

Adamo, N. J. (1967). Connections of efferent fibres from hyperstriatal areas in chicken, raven and African love-bird. *J. comp. Neurol.* **131**, 337–356.

Andres, K. H. (1970). Anatomy and ultrastructure of the olfactory bulb in fish, Amphibia, reptiles, birds and mammals. In CIBA symposium "Taste and Smell in Vertebrates" (Wolstenholme, G. E. W. and Knight, J., eds), pp. 177–194. J. and A. Churchill, London.

Aprison, M. H. and Takahashi, R. (1965). 5-hydroxytryptamine, acetylcholine, 3,4-dihydroxy phenylethylamine and norepinephrine in several discrete areas of the pigeon brain. *J. Neurochem.* **12**, 221–230.

Bajandurow, B. J. and Larin, E. F. (1925). Contributions to the physiology of the olfactory analysor in birds. *Trudi Med. Inst. Tomsk.* **2**, 10.

Baker-Cohen, K. F. (1968). Comparative enzyme histochemical observations on submammalian brains. I. Striatal structures in reptiles and birds. *Ergebn. Anat. Entwicklungsges.* **40**, (b), 1–41.

Bang, B. G. (1960). Anatomical evidence for olfactory function in some species of bird. *Nature, Lond.* **188**, 547–549.

Bang, B. G. and Bang, F. B. (1959). A comparative study of the vertebrate nasal chamber in relation to upper respiratory inspections. *Bull. Johns Hopkins Hospital* **104**, 107.

Bertler, A., Flack, B., Gottfries, C. G., Ljunggren, L. and Rosengren, E. (1964). Some observations on adrenergic connections between mesencephalon and cerebral hemispheres. *Acta Pharmacol. Toxicol.* **21**, 283–289.

Böhme, G. (1969). Vergleichende Untersuchungen am Gehirnventrikelsystem. Das Ventrikelsystem des Huhnes. *Acta anat.* **73**, 116–126.

Boyce, R. and Warrington, W. B. (1898). Observations on the anatomy, physiology and degenerations of the nervous system of the bird. *Proc. Roy. Soc. B.* **64**, 176–179. (See also 1899, *Phil. Trans. Roy. Soc. B.* **191**, 293.)

Bumm, A. (1883). Das Grosshirn der Vögel. *Z. wiss. Zool.* **38**, 430.

Calvin, A. (1960). Olfactory discrimination. *Science* **131**, 1265–1267.

Cobb, S. (1960a). A note on the size of the avian olfactory bulb. *Epilepsia* **1**, 394–402.

Cobb, S. (1960b). Observations on the comparative anatomy of the avian brain. *Perspectives Biol. Med.* **3**, 383–408.

Cobb, S. and Edinger, T. (1962). The brain of the emu (*Dromaeus novaehollandiae*, Lath.). *Breviora* **170**, 1–18.

Craigie, E. H. (1928). Observations on the brain of the humming bird (*Chrysalompis mosquitus*, Linn., *Chlorostilbon caribaeus* Lawr.). *J. comp. Neurol.* **48**, 377–481.

Craigie, E. H. (1929a). The cerebral cortex of *Apteryx*. *Anat. Anz.* **68**, 97–105.

Craigie, E. H. (1929b). The vascularity of the cerebral cortex in a specimen of *Apteryx*. *Anat. Rec.* **43**, 209–213.

Craigie, E. H. (1930). Studies on the brain of the kiwi (*Apteryx australis*) *J. comp. Neurol.* **49**, 223–357.

Craigie, E. H. (1932) The cell structure of the cerebral hemisphere in the humming bird. *J. comp. Neurol.* **56**, 135–168.

Craigie, E. H. (1934). Multilaminar cortex in the dorsal pallium of the emu (*Dromiceius novaehollandiae*). *Psych. Neur. Blad. Amsterdam. Feestbdl. Kappers.* pp. 702–711.

Craigie, E. H. (1935a). The hippocampal and parahippocampal cortex of the emu (*Dromiceius*). *J. comp. Neurol.* **61**, 563–591.

Craigie, E. H. (1935b). The cerebral hemispheres of the Kiwi and the emu (*Apteryx and Dromiceius*). *J. Anat.* **69**, 380–393.

Craigie, E. H. (1935c). Some features of the pallium of the cassowary (*Casuarius uniappendiculatus*). *Anat. Anz.* **81**, 16–28.

Craigie, E. H. (1936a). The cerebral cortex of the ostrich (*Struthio*). *J. comp. Neurol.* **64**, 389–415.

Craigie, E. H. (1936b). Notes on the cytoarchitectural features of the lateral cortex and related parts of the cerebral hemisphere in a series of birds and reptiles. *Trans. Roy. Soc. Canada* Series 3, sec. V, **30**, 87–113.

Craigie, E. H. (1939). The cerebral cortex of *Rhea americana*. *J. comp. Neurol.* **70**, 331–353.

Craigie, E. H. (1940a). The cerebral cortex in some Tinamidae. *J. comp. Neurol.* **72,** 299–328.

Craigie, E. H. (1940b). The cerebral cortex in paleognathine and neognathine birds. *J. comp. Neurol.* **73,** 179–234.

Crosby, E. C. and Humphrey, T. (1939). Studies on the vertebrate telencephalon. *J. comp. Neurol.* **71,** 121–213.

Dart, R. A. (1934). The dual structure of the neopallium, its history and significance. *J. Anat.* **69,** 3–19.

Dennler, G. (1932). Zur morphologie des Vorderhirns der Vögel. Der Sagittalwulst. *Folia neuro-biol.* **12,** 343–362.

Donovan, B. T. (1966). The regulation of the secretion of follicle stimulating hormone, *In* "The Pituitary Gland" (Harris, G. W., and Donovan, B. T., eds), vol. 2, pp. 49–98.

Durward, A. (1932). Observations on the cell masses in the cerebral hemisphere of the New Zealand kiwi (*Apteryx australis*). *J. Anat.* **66,** 437–477.

Edinger, L. (1908). "Vorlesungen uber den Bau der Nervosen Centralorgane des Menschen und der Thiere." Vogel, Leipzig.

Edinger, L., Wallenberg, A. and Holmes, G. (1903). Untersuchungen uber die vergleichenden Anatomie des Gehirns. 5. Das Vorderhirn der Vögel. *Abh. Senckenb. nat. Ges.* **20,** 343–426.

Eiduson, E. (1966). 5-hydroxytryptamine in the developing chick brain. *J. Neurochem.* **13,** 923–932.

Fox, C. A., Hillman, D. E., Siegesmund, K. A. and Sether, L. (1966). The primate globus pallidus and its feline and avian homologues; a Golgi and electron microscopic study. *In* "Evolution of the Forebrain" (Hassler, R. and Stephen, H., eds), pp. 237–248. Thieme, Stuttgart.

Haefelfinger, H. R. (1958). "Beitrage zur vergleichenden Ontogenese des Vorderhirns bei Vögeln", pp. 99. Helbing and Lichtenhahn, Basle.

Henton, W. W., Smith, J. C. and Tucker, D. (1966). Odour discrimination in pigeons. *Science* **153,** 1138–1139.

Huber, G. C. and Crosby, E. C. (1929). The nuclei and fibre paths of the avian diencephalon, with consideration of telencephalic and certain mesencephalic structures. *J. comp. Neurol.* **48,** 1–225.

Johnston, J. B. (1923). Further contributions to the study of the evolution of the forebrain. *J. comp. Neurol.* **35,** 337–481.

Juorio, A. V. and Vogt, A. (1967). Monoamines and their metabolites in avian brain. *J. Physiol.* **189,** 489–518.

Kallén, B. (1953). On the nuclear differentiation during ontogenesis in the avian forebrain. *Acta Anat. (Basle)* **17,** 72–84.

Kallén, B. (1955). Notes on the mode of formation of brain nuclei during ontogenesis. *C.R. Ass. Anat. Reun* **42,** 747–756.

Kalischer, O. (1905). Das Grosshirn der Papageien in anatomischer und physiologischer Beziehung. *Abhandl. preuss. Akad. Wiss. Abh.* **IV,** 1–105.

Kappers, C. U. A. (1920–21). "Vergleichende Anatomie des Nervensystems." Bohn, Haarlem.

Kappers, C. U. A. (1922). The ontogenetic development of the corpus striatum in birds and a comparison with mammals and man. *Proc. Kon. Akad. Wet. Amsterdam* **26,** 135–158.

Kappers, C. U. A. (1928). "The corpus striatum, its phylogenetic and ontogenetic development and function." Levin and Munksgaard, Copenhagen.

Kappers, C. U. A. (1929). "The Evolution of the Nervous System." Bohn, Haarlem.

Kappers, C. U. A. (1947). "Anatomie Comparée du Système Nerveux", pp. 754. Masson, Paris.

Kappers, C. U. A., Huber, G. C. and Crosby, E. C. (1936). "The Comparative Anatomy of the Nervous System of Vertebrates, Including Man." (1960 reprint. Hafner, New York, 3 vols.)

Karten, H. J. and Hodos, W. (1967). "A Stereotaxic Atlas of the Brain of the Pigeon (*Columba livia*)", pp. 193. Johns Hopkins University Press, Baltimore.

Kuhlenbeck, H. (1924). Ueber die Homologien der Zellmassen im Hemispharenhirn der Wirbeltiere. *Folia anat. Jap.* **2**, 235.

Kuhlenbeck, H. (1929a). Die Grundbestandteile des Endhirns im lichte der Bauplanlehre. *Anat. Anz.* **67**, 1–51.

Kuhlenbeck, H. (1929b). Erwiderung auf die rorstehende Entgegung von M. Rose. *Anat. Anz.* **67**, 323–329.

Kuhlenbeck, H. (1938). The ontogenetic development and phylogenetic significance of the cortex telencephali in the chick. *J. comp. Neurol.* **69**, 273–301.

Kuenzi, W. (1918). Versuch einer systematischen Morphologie des Gehirn der Vögel. *Rev. Suisse. Zool.* **26**, 17–112.

Mangold, O. (1946). Die Nase der segelnden Vögel, ein organ des Stromungsinnes? *Naturwissenschaften* **33**, 19–23.

Michaelsen, W. J. (1959). Procedure for studying olfactory discrimination in pigeons. *Science* **130**, 630–631.

Nauta, W. J. H. and Gygax, P. A. (1951). Silver impregnation of degenerating axon terminals in the central nervous system. 1. Technique. 2. Chemical notes. *Stain Technol.* **26**, 5–11.

Nauta, W. J. H. and Gygax, P. A. (1954). Silver impregnation of degenerating axons; a modified technique. *Stain. Technol.* **29**, 91–93.

Neuhaus, W. (1963). On the olfactory sense of birds *In* "Olfaction and Taste" (Zotterman, Y. ed.), pp. 396. Pergamon Press, London.

Portmann, A. (1961). Sensory organs; skin, taste and olfaction. *In* "The Biology and Comparative Physiology of Birds" (Marshall, A. J., ed.), pp. 37–48. Academic Press, New York and London.

Portmann, A. and Stingelin, W. (1961). The central nervous system. *In* "The Biology and Comparative Physiology of Birds" (Marshall, A. J., ed.), pp. 1–36. Academic Press, New York and London.

Rogers, K. T. (1960). Studies on the chick brain of biochemical differentiation related to morphological differentiation. III. Histochemical localization of alkaline phosphatase. *J. Exp. Zool.* **145**, 49–60.

Rogers, K. T., DeVries, L., Kepler, J. A., Kepler, C. R. and Speidel, E. R. (1960). Studies on the chick brain of biochemical differentiation related to morphological differentiation. II. Alkaline phosphatase and cholinesterase levels. *J. Exp. Zool.* **145**, 49–60.

Rose, M. (1914). Uber die cytoarchitektonische Gliederung des Vorderhirns der Vögel. *J. Psychol. Neurol., Leipzig* **21**, 278–352.

Schaposchnikow, L. K. (1953). Der Bau des Kopfhirns der Vögel im Zusammenhange mit Besonderheiten der Funktion der Nahrungssuche. *Nachr. Akad. Nauk. U.S.S.R.* **91**, 679–682.

Singh, R. (1967). Some observations on the histochemistry of the neuropile tissue of the brain of the parrot, *Psittacula Krameri*. *Acta Anat.* **68**, 567–576.

Spooner, C. E. and Winters, W. D. (1966). Distribution of monoamines and regional uptake of D. L. norepinephrin 7-$H^3$ and dopamine 1-$H^3$ in the avian brain. *Pharmacologist* **8**, 189.

Stager, K. E. (1964). The role of olfaction in food location by the turkey vulture *Cathartes aura*. *Los Angeles County Museum. Contrib. Sci.* **81,** 1–63.

Stingelin, W. (1958). "Vergleichend morphologische Untersuchungen am Vorderhirn der Vögel auf cytologischer und cytoarchitektonischer Grundlage", pp. 123. Helbing and Lichtenhahn, Basle.

Strong, R. M. (1911). On the olfactory organs and smell in birds. *J. Morph.* **22,** 619–660.

Technau, G. (1936). Die Nasendruse der Vögel. *J. ornithol., Leipzig* **84,** 511–617.

Tienhoven, A., van and Juhasz, L. P. (1962). The chicken telencephalon, diencephalon and mesencephalon in stereotaxic coordinates. *J. comp. Neurol.* **118,** 185–197.

Tucker, D. (1965). Electro-physiological evidence for olfactory function in birds. *Nature, Lond.* **207,** 34–36.

Turner, C. H. (1891). The morphology of the avian brain. I. Taxonomic value of the avian brain and the histology of the cerebrum. *J. comp. Neurol.* **1,** 39–92.

Wagner, H. O. (1939). Untersuchungen uber Geruchsreaktionen bei Vögeln. *J. Ornithol.* **87,** 1–9.

Wenzel, B. M. and Sieck, M. H. (1966). Olfaction. *Ann. Rev. Physiol.* **28,** 381–434.

Wesołowski, H. (1967). The behaviour of mitochondria and the secretion of the olfactory epithelial cells and the olfactory glands in domestic birds. *Folia biologica* **15,** 303–324.

Zahn, W. (1933). Ueber den Geruchssinn einiger Vögel. *Ztschr. vergleich. Physiol.* **19,** 785–796.

# 14 Experimental Studies on the Hemispheres and Related Structures

## I. INTRODUCTION

It will be clear from the information presented in Chapter 13 that the hemispheres exhibit considerable variations upon a fundamentally similar anatomical plan and that besides the important relationships which result from the conspicuous horizontal laminae between the principal component cell masses, a dense network of fibres is developed in the whole hemisphere. Most of the studies which followed those of Rolando, and also attempted to elucidate the function of different areas, involved the production of either localized or widespread lesions. These were intended to demonstrate any specific focal representation of the different sensory modalities and also any localized motor control areas. The extirpation experiments which were cited in Chapter 11 showed that birds can continue their sensory and motor activities after telencephalic extirpation and particularly striking reports were also presented by Schrader (1889) and by Visser and Rademaker (1934). According to the latter authors light and hunger are the predominant factors which lead to activity in forebrain-less preparations. At most times these are characterized by quiescence. Rogers (1922) considered that the extirpation of neo-, and hyperstriatic components deprived the resulting preparations of any ability to learn, pair, or care for offspring. Kalischer (1905) suggested that the archistriatum was the principal centre involved in sensory integration at a telencephalic level. Both authors concluded that merely extirpating the foremost frontal regions of the hemispheres had no notable consequences on the overt behaviour, an operation clearly inspired by the developments of frontal lobotomy in Primates.

A very large number of other investigations which date from the nineteenth and the earliest part of the twentieth century exist in the literature. With Ten Cate (1936) one can cite the following, most of which are now largely the province of historians of science. More or less extensive extirpations were carried out by Couty (1882), Ewald (1895), Fasula (1889), Jastrowicz (1876), Jolyet (1902), Kalischer (1900), Levinsohn (1904), Lussana and Lemoigne (1871), Martin and Rich (1918), McKendrick (1873), Moeli (1879), Munk (1883), Munzer and Wiener (1898), Musehold (1878), Onimus (1871), Rosenthal (1868), Schrader (1889), Shaklee (1921, 1928a,b), Stefani (1881),

Trendelenburg (1906), Voit (1868) and Vulpian (1866). In addition Gallerani and Lussana (1891) used topical application of chemicals whilst Bickel (1898), Ferrier (1879) and Steiner (1891) used direct electrical stimulation.

In recent decades a fairly large number of rather more definitive experiments have been undertaken. These vary in the specificity of their results. Some give direct focal information, others more diffuse data that relate to many hemispheric structures. In view of the wide variety of factors that are involved it is somewhat difficult to evaluate the suggestions of, for example, Vasilevsky (1966). After analysing the background and induced activity of individual "cortical" neurones in the pigeon, he concluded that their reactions resembled those of the archipallium and archicortex in other vertebrates. An overwhelming number showed activity in the presence of stimuli which are of overt "ecological" importance—light, touch and proprioception. An insignificant number responded to sound. As a result he concluded that the pigeon "cortex" may be regarded as the homologue of an ancient "optico-somatic cortex".

Phillips (1964) observed that after total hemisphere extirpation his preparations were still able to body shake, head shake, tail-wag, wing-leg-tail stretch, head scratch, wing-flap and preen as normal. In contrast, escape attempts disappeared throughout the post-operative period as did both spontaneous feeding and spontaneous drinking. In an attempt to investigate whether such preparations would respond to long photoperiods, six extirpated preparations and six control birds were kept under conditions of 14 hours daylight and maintained at comparable dietary levels for 23 days. At the end of this period, which was of sufficient length to enable the game-farm stock which he used to begin laying, the ablated preparations showed no response to the light régime. Phillips added, however, that they had undergone a slow but progressive weight loss and this may have been influential. More specific results are detailed in the subsequent sections.

## II. PREPARATIONS WITH INCOMPLETE TELENCEPHALIC ISOLATION

### A. The Peduncles Behind the Anterior Commissure Remain Intact

The lesions produced in 12 pigeons by Akerman et al. (1962) using high energy proton beams were more rostral than those in the preparations of Chapter 11, Section II. Although the regions of the paleostriatum which lie in front of the anterior commissure, together with the nuclei of the ventromedial forebrain wall and septal region, were all necrotic, the small area of the peduncles lying behind the anterior commissure and the paleostriate material at that level both remained intact. Thus the only residual intact connections between the telencephalon and more posterior regions were a relatively small number of fibres belonging to the occipito-mesencephalic tract. As with the preparations of Chapter 11, Section II which had undergone total telencephalic isolation, both spontaneous eating and spontaneous drinking disappeared although the swallowing reflex was still present.

Overt escape reactions generally vanished at some point between the 6th and 26th day although in one preparation they were never completely extinguished but persisted in a weak form. During the same post-operative period 10 of the 12 preparations ceased to show spontaneous flight although they were able to carry out coordinated flight patterns if thrown into the air (see also Karamyan, 1956). On such occasions they avoided obstacles and landed normally. Both normal blinking and pupillary reflexes were intact, suggesting that they are under the predominant control of posterior nuclei such as the isthmi complex. Also, with the exception of three preparations, all showed obvious reactions to auditory stimuli. In the three exceptions these reactions, present during the early post-operative period, disappeared between 7 and 14 days after lesions were produced.

Electrical stimulation of the pre-optic area, the anterior thalamus or the hypothalamus resulted in enhanced activation. This took the form of both ipsiversive and contraversive movements of the head and trunk together with a general pattern which was analogous to that seen in preparations with total telencephalic ablation.

## B. The Peduncles and Septal Area in Front of the Anterior Commissure Remain Intact

In four preparations the lesions were located 0·5–1·0 mm more rostrally than in those which were described above. As a result the septal area and some additional paleostriatal material remained intact. Such preparations contrasted with the foregoing in the retention of spontaneous eating and drinking until at least 20 days after irradiation. Two preparations retained these abilities throughout the post-operative period. All also retained escape reactions and in two of them these were actually enhanced rather than destroyed (see Section V A below). With one exception they flew spontaneously and in this solitary exception the ability did not disappear until the 26th day. As would be expected in view of the conditions which were described above, the blinking and pupillary reflexes were normal, they responded to sounds, and they were observed indulging in the bowing and calling components of social behaviour.

Taken together with their other results Akerman et al. concluded from this that the telencephalon is indeed concerned with the recognition of stimuli which are related to certain categories of instinctive behaviour. In particular their results suggested involvement in eating, drinking, escape and social behaviour. The presence of definitive escape behaviour in the second group is, however, in conflict with the results of Phillips (1964). The post-telen-cephalic structures seem to comprise control systems for the important reflexes which are involved in locomotory behaviour, and also the basic visual sensibility which permits both unimpeded locomotion and obstacle avoidance. Such conclusions confirm those of many older workers. Their ability to avoid obstacles during flight and walking, and to estimate distances even in the total absence of the hemispheres complements the conclusions of Renzi (1864), Stefani (1881), Schrader (1889), Noll (1915) and Visser (1932). It also clearly contradicts the results of Bouillaud (1830), Flourens (1824) and Munk (1883) who considered that telencephalic extirpation abolished vision and produced blind preparations.

## C. Conditioned Reflexes

It was suggested by Tuge (1957) that the temporary connections which are involved in developing conditioned reflex activity are formed in a "step-like" manner from fish to mammals. Leaving aside the confusions which arise from any extensive discussion of such a premise one may note that it led Tuge and Shima (1959) to investigate the production of conditioned reflexes in avian preparations after forebrain extirpation. Other studies of a comparable but more specific nature are outlined in subsequent sections, here we may consider the data of Tuge and Shima for the importance that they attribute to the paleostriatal components. It is worth noting that any such experiment gives information of two kinds. It shows the deficits or effects that follow ablation in certain areas, but it also demonstrates the degree of competence which is ascribable to the areas that remain intact.

The preparations which are involved in this case fall into two groups. In the first the entire "surface" of the hemispheres was removed with some damage to deeper levels. This appears to have involved ablation of the ventral, dorsal and accessory hyperstriate material, that is, the levels above the hyper-striatic lamina (see Table 59). In the second group the damage was more extensive and involved the ectostriatum, neostriatum, the archistriatum and both the augmentatum and primitivum. Histological examination of the brains at the end of the experiment revealed the relative extent of the damage and this is shown in the Table. In the case of Group I the presentation of a visual conditioning stimulus followed by an unconditioned electric shock established a conditioned respiratory reflex as quickly as in unoperated controls. The establishment of a somatic reflex that involved the neck and body was more difficult than in the controls. Nevertheless both were stable once they had been produced.

A totally different situation was found in Group II. Following 100–160 presentations of the stimulus only rare responses were observed. As a result

# TABLE 59

*A summary of the extent of the telencephalic damage which was sustained by the preparations of Tuge and Shima (1959).*

| Cortical area | | Hyperstriatum | | Neostriatum | | Ectostriatum | | Paleostriatum | | Archistriatum | |
|---|---|---|---|---|---|---|---|---|---|---|---|
| Left | Right | Left | Right | Left | Right | Left | Right | Left | Right | Left | Right |

Group I

Group II

Key: —, intact    +, small lesion    ++, large lesion    +++, more or less total extirpation of this Tuge and Shima concluded that lesions which are so deep that the paleostriatum is involved prevent the establishment of conditioned reflexes. On the other hand extensive damage to the hyperstriatum does not result in the abolition of such reflex forming ability. It is fairly clear from Table 59 that a generalization of this nature goes rather further than the data would justify nevertheless Tuge and Shima were interested in the possibility that temporary connections within the paleostriatum are necessary for such reflexes and their data corroborate those of Akermann *et al.* above, and those of Cohen (1967), Layman (1936) and Zeigler (1963) which are discussed in relation to the Wulst (Section VIII below).

## III. LESIONS OF THE OLFACTORY BULB AND NERVE

The function of the olfactory bulbs had been rather neglected prior to the work of Wenzel and Salzmann (1968). They were interested in the possible significance of primary olfactory systems to behaviour patterns which do not in themselves involve the sense of smell. Four groups of pigeons were trained

### TABLE 60

*Group means for hopper approach, magazine training and response shaping on both colours and both sides. Data for the hopper training are in sessions, those for response shaping in re-inforcements.*

| Group | Hopper training | | Response shaping | |
|---|---|---|---|---|
| | First eating | Criterion | Side 1 | Side 2 |
| Sham operates | 0·6 | 1·3 | 19·4 | 7·6 |
| Hyperstriatum | 0·5 | 1·4 | 20·4 | 4·4 |
| Olfactory nerve | 4·2 | 5·2 | 63·0 | 36·8 |
| Olfactory bulbs | 5·2 | 6·2 | 55·4 | 46·8 |

From Wenzel and Salzmann (1968).

in a visual test situation. One of these groups had undegone bilateral section of the olfactory nerves. A second had had both olfactory bulbs extirpated, a third had had an equivalent amount of damage to the hyperstriatal regions and the fourth had merely been exposed to anaesthesia. In the test situation the final task required pecking the left hand key of a pair when both were illuminated by colour 1, and the right hand key when they were illuminated by colour 2. Training was carried out in a series of stages which involved learning to eat from a hopper in a training box; learning to peck one key while it was lit by

19

the first colour and then learning to transfer pecking to the other key when they were lit by the second colour. Each group of birds was divided into two sub-groups. For one such sub-group in each case the left key was correct when red illumination was used, the right key when green was used. In the other sub-group the reverse was the case.

In terms of group differences the results fell into two separate sections, namely those before and after the key pecking response had been established. All the normal birds and preparations adapted quickly as was shown by the promptness with which they ate from the grain-filled dishes. However, the hopper-training and response shaping with both the first and second colours and sides were significantly slower for the two groups with lesions in their primary olfactory system ($P = 0.005$). Table 60 presents the group mean scores for these components of the training. Once the birds and lesioned preparations were actually pecking the appropriate keys these inter-group differences disappeared, although there was some suggestion that the bulb-ablated group took rather longer to reach criterion.

Wenzel and Salzmann suggested that a lowered arousal level might well be expected following transection of the ascending pathways of any exteroceptive sensory modality. However, they pointed out that the close relationship of the olfactory system and the mammalian limbic system, together with the supposed minor role of the olfactory pathway in birds, placed the regulation of avian discriminative ability in a rather special position. As both nerve transection and bulb ablation led to similar results it is possible to suggest that some input level into the bulb is essential if it is to maintain its general influence, or that the glomerular layer must be intact. Taken together with the known hypothalamic and ovarian effects of olfactory bulb destruction in mammals (see Donovan, 1966), these data may support the suggestion of Herrick (1933) that the olfactory system (cortex) serves as a non-specific activator for all cortical activities in mammals, and also the suggestion of Kluver (1965) that the rhinencephalon as a whole is involved not only in olfactory phenomena but also in emotional and sexual activation. It would clearly be exceedingly interesting to carry out similar experiments with species possessing a more developed olfactory system.

## IV. STIMULATION OF THE PALEOSTRIATE REGIONS

The paleostriatal components can be activated by direct electrical stimulation of a number of other brain areas. Much of such activation is most probably a reflection of the activity within the numerous fibres of passage which lie within it and traverse it when passing to or from other hemisphere nuclei or more posterior levels. They are referred to in a number of other sections. However, direct stimulation of the paleostriatal region occurred in two of the electrode sitings used by Phillips (1966). These resulted in homolateral evoked potentials within both the hyperstriatum and lateral forebrain bundle. Once again the absence of activation in closely neighbouring sites suggested that they were restricted to the components of a fibre system which passes dorso-medially and posteriorly, enters the hyperstriatic lamina and is then distributed in hyperstriate regions.

Following the demonstration that goldfish, rats, cats, dogs, monkeys and dolphins work to administer trains of electrical impulses to certain sites in their median forebrain bundle and hypothalamus, Andrew (1967) investigated the situation in chicks. They were presented with two sets of keys each of which would, when pecked at, set in motion a Grason–Stadler electric clock. This timed the duration of the resulting stimuli. When compared with controls it was found that all six of the experimental birds with implanted electrodes showed a marked increase in key operation during the first 10 min after the keys were made available. As each peck resulted in a train of impulses and the pecks were often given in rapid succession, Andrew suggested that the birds would have worked for, and "preferred", trains of considerably greater length. Analogous results were obtained by Goodman and Brown (1966) using electrodes in the lateral forebrain bundle just in front of the anterior commissure, in the paleostriatum augmentatum, and also further posteriorly in the latero-dorsal tegmental nucleus of the mid-brain. These areas appear to correspond to "pleasure centres" and it is interesting that stimulation of hypothalamic and tegmental foci induces a feeling of euphoria in man.

Furthermore Goodman (1970) using multiple electrodes chronically implanted in pigeons was able to investigate the reinforcement effects of unipolar stimulation at both fore and mid-brain loci. Using a place-preference task *approach* sites were identified in the paleostriatum together with the archistriatum. None were apparent in the di-, or mesencephalon. In contrast *avoidance* sites were identified in telencephalic, diencephalic and mesencephalic sites. These results therefore reflected both positive and negative reinforcement effects. In addition to requiring a conditioned response that differed from those employed by Goodman and Brown (1966) and MacPhail (1967) they overcame possible arguments against the earlier results by showing central reinforcement in an experimental situation not involving prior experience with another positive reinforcer, food, during training. The sites in the archistriatum conform with mammalian data. In relation to MacPhail's suggestion that the lateral forebrain bundle might represent a rostro-caudal circuit for positive reinforcement these more recent data raise the possibility of more diffuse foci as three paleostriatal positive sites were situated medial to it, one archistriatal site was lateral to it and a diencephalic site within the bundle was negatively reinforcing. As in mammals current intensity was an important variable. Current increases above certain critical levels tended to increase avoidance or approach behaviour in the relevant foci.

Turning to a different line of investigation the fact that those preparations studied by Akerman *et al.* which retained some paleostriatal material showed spontaneous feeding, conforms with the results of Harwood and Vowles (1966). They found that electrical stimulation of certain paleostriate foci elicited feeding behaviour. This was also true of foci in the posterior ventral neostriatum and the results are discussed in greater detail in Section VII C. The overall results, which involved an increase in the time spent feeding and preening and a decrease in that spent on various miscellaneous activities are summarized in Table 64 below.

The results of Juhasz and Tienhoven (1964) are also discussed further in relation to the archistriatum. However, in the present context it is significant that stimulation of the primitivum resulted in delayed ovulation. Unlike the delayed ovulation which followed archistriatal stimulation it was not, however, accompanied by follicular atresia (see Table 62). As in the case of the archistriatum there was, however, an optimal time for stimulation if it was to have such delaying effect (Table 63). This appeared to lie between 13 and 16 hours before the expected time of ovulation.

Furthermore, potentials were evoked in both the primitivum and augmentatum by the presentation of peripheral chick stimuli. They were evoked more frequently in the former than the latter and in that structure stimuli at both ipsilateral and contralateral ears produced potentials with an amplitude of $35-110\,\mu V$ after latent periods of $11-17$ msec. The principal auditory representation appears, however, to be in the neostriatum and such responses may represent forward projecting fibres which are passing through paleostriate regions.

# V. STUDIES ON THE ARCHISTRIATUM

## A. Lesion Studies

### (i) *Archistriatal Lesions and "Wildness"*

Following their capture wild birds, in common with other animals, frequently refuse to eat and, by struggling, damage themselves. Phillips (1964) undertook a study of such responses by means of brain lesions. As all workers know, in mallard and other birds total telencephalic extirpation usually leads to quiescence (see Chapter 11, Section II; Karamyan, 1956; Ten Cate, 1936; and Tuge and Shima, 1959). Both nonlaying wild, and laying game-farm specimens were used by Phillips for parallel studies. A marked reduction of escape behaviour occurred in 7 out of 12 birds with lesions in the medial archistriatum and dorso-lateral thalamus. Wild-caught specimens which normally made frantic efforts to escape, became as placid as specimens which had had their hemispheres extirpated. They made no attempt to escape at the approach of people, would perch on the hand and, rather than fly when tipped off, they would grasp and attempt to maintain their balance.

When released into a standard test situation consisting of a $10 \times 20$ ft pen which was visually isolated from other ducks, they consistently spent 10 min bathing, drinking and preening. No attempts at escaping were observed. In contrast intact controls, together with preparations that had undergone lesions at other sites, remained at the end of the pen furthest away from the observer. They also often attempted to fly. Table 61 contains a summary of the activities which were observed during ten 3-min periods for groups of lesion-tamed, lesioned but untamed preparations and intact controls. The data are presented as the percentage of the observation time during which a given activity was recorded. It can be seen that those activities which one would interpret as reflecting mild anxiety, such as head-shaking, tail wagging and tail-wag-shake, occurred most often in the control birds, least in tamed preparations, and at intermediate frequencies in lesioned but untamed preparations. Pacing, pre-flight bobbing and attempts at flight were also totally absent in tamed preparations. Conversely those activities which one might associate with unfrightened birds, such as preening, drinking, lying down and also sleeping, were all more frequent in the "tamed" preparations.

## TABLE 61

*Behaviour patterns observed during ten 3-min observations of pen groups. The results are expressed as the percentage of the total time during which the activities were seen.*

| | Head-shake | Tail wag | Tail-wag-shake | Pacing | Fly | Lie down | Drink | Sleep |
|---|---|---|---|---|---|---|---|---|
| Lesion-tamed preparations | 75 | 92 | 58 | 0 | 0 | 83 | 75 | 75 |
| Intact controls | 100 | 100 | 89 | 89 | 78 | 22 | 0 | 11 |
| Lesioned but not "tamed" preparations | 100 | 80 | 50 | 90 | 30 | 0 | 0 | 0 |
| Imprinted specimens | 82 | 91 | 91 | 61 | 0 | 36 | 0 | 0 |

From Phillips (1964).

In addition to being tamed some preparations lost their flocking tendency and began to act independently of the flock with which they were being kept. Normally all the birds in a given pen tended to eat, drink, preen, sleep or explore as a group. Some of these with lesions kept apart and exhibited generally different activity schedules by comparison with their companions. Although this was at first interpreted as reflecting sickness a closer investigation showed that this was not the case. Some of the most independent preparations were actually amongst the most active specimens in their respective pens.

Histological investigations showed that all of the tamed preparations had sustained extensive bilateral damage to the medial archistriatum, to the fibres of the occipito-mesencephalic tract, and, probably to fibres which ran from the archistriatum to the septal region. The superficial parvocellular nucleus of both sides was always at least partially destroyed. However, "taming" did not seem to involve the anterior commissure as some definitive tamed preparations had sustained no damage to this structure. Also other specimens in which there had been total ablation of this commissure remained wild. The critical fibre paths appeared to distribute to the bed nuclei of the anterior commissure and to baso-lateral septal areas. Damage to fibres that run from the archistriatum to the medial spiriform nucleus and ovoidal nucleus, together with variable amounts of fibre degeneration within both neostriatal and hyperstriatal areas, and extensive cellular and fibre destruction in the dorso-lateral thalamus, gave no reduction in escape activities unless they were also accompanied by interruptions in the occipito-mesencephalic tract. Lesions to areas situated outside the medial archistriatum resulted in tamed preparations on two occasions. These followed damage to the external cellular stratum of the hypothalamus just to the side of the parvocellular paraventricular nucleus. However, in both these cases it appeared that the lesions interrupted fibres which originated in the archistriate taming site. Taking the results of all these preparations it was not particularly easy to derive any direct relationship between the extent of the lesions and the degree of "taming". There were, however, some indications that the greater tameness of certain preparations was associated with more fibre degeneration than that which was visible elsewhere.

## (ii) *The Effect of Taming Lesions on Ovarian Development*

The inhibiting effect of captivity on reproductive activities in the mallard was shown by Phillips and Tienhoven (1960) to result from a lowered level of hypophyseal gonadotropic output. They had concluded that this depressed hypophyseal activity was caused by *Anxiety* acting via the hypothalamus. This was because Ralph (1959) and Ralph and Fraps (1959) had demonstrated that lesions in the mid-line of the fowl hypothalamus both blocked ovulation and led to ovarian regression (see Chapter 11, Section VI E). In an attempt to assess the effect of median archistriate lesions on ovarian development Phillips (1964) carried out analogous experiments to those outlined in (i) above. He used experimental and control specimens in which the ovarian size was comparable at the time of operation. Four to six weeks later the maximum size of the ovarian follicles in "tamed" preparations averaged two to three times that in the controls. They were also equally large when compared with preparations which had undergone lesions that did not result in taming. The weights of both ovary and oviduct exhibited an even more striking difference. There were also rather small drops in the weight of both thyroid and adrenal glands but these were not statistically significant. A careful examination of the data suggested that the effects of such lesions varied with the season. None of the preparations which were tamed by archistriate lesions at times after the end of the normal breeding season had hypertrophied reproductive organs. This presumably reflects the post-reproductive photo-refractory period (*q.v.*).

## B. Electrical Studies of the Archistriatum

### (i) *General Effects of Stimulation*

Sensory projections to the hemispheres are generally thought to involve polysynaptic pathways and a number of these have been described in earlier chapters. Phillips (1966) found that peripheral visual stimuli in the form of flashes of light regularly evoked archistriatal activity although none followed the presentation of peripheral chick stimuli to either ear, movement of the feathers at any point on the body surface or direct stimulation of the sciatic nerve. Similarly no activation potentials followed stimulation of the rotund nucleus or the paleostriatum primitivum. However, to supplement his studies on archistriate lesions Phillips (1964) stimulated 15 sites in the archistriatum and the occipito-mesencephalic tract and observed the results.

The most frequent responses which he obtained by such direct stimulation were rapid bill-movements coupled with motions of the head and neck. These recalled the movements which comprise searching and gabbling and constitute the normal feeding behaviour of mallard. At one locus the result of raising the stimulus intensity from 0·2 to 0·5 mA was to evoke choking movements which were similar to those involved in swallowing dry food. Two electrodes which were situated in the ventral archistriatum and the bed nuclei of the anterior commissure elicited defaecation and urination. This seemed to be "mechanical" and was not associated with either fear or excitement. At median archistriate sites, together with those in the taenia, feeding movements comparable with those described above were also accompanied by crouching, sneaking, sleeking, running, or even flight if the stimulus intensity was increased. These actions all ended abruptly when stimulation ceased. Such apparent fear and escape movements were often associated with vocalizations. These varied from mild tuts and squeaks to loud and distinct quacks and as such represent the type of noises made by normal birds in the presence of some slight disturbance and great alarm respectively. Strong escape behaviour was also evoked in such experiments at all six septal sites and also from three points in the external cellular stratum of the hypothalamus (see page 404).

The most striking feature of evoked responses elsewhere in the brain during such direct archistriatal stimulation was their very extensive ipsilateral distribution (Phillips, 1966). The number of contralateral responses was very restricted but ipsilateral activation occurred throughout the telencephalon, diencephalon and mesencephalon.

### (ii) *Responses with a Short Latency*

To distinguish primary responses and other less direct activation Phillips analysed recordings which had a latency in excess of 2 msec separately from those which had latent periods of this short type. Short latencies were observed in the anterior commissure and its bed nuclei, in the occipito-mesencephalic tract and in the strio-peduncular tract as far back as the posterior commissure and red nucleus. They could also be detected in the septum, in the pre-optic area, in the anterior part of the hypothalamus and the dorsal supraoptic decussation. With the possible exception of the septum all these regions have for a long time been thought to receive fibres from the occipito-mesencephalic tract.

Other responses with a short latency were observed in the thalamic region, within the dorso-medial nucleus and especially the ovoidal nucleus, around the fore part of the rotund nucleus, and within the tectum. Phillips concluded that the distribution of tectal responses indicated a dual input from the archistriatum. One path passed laterally and into the anterio-dorsal tegmental region at the level of rotundal, dorso-lateral and pretectal nuclei. The other, and more ventral route, led through or just medio-ventral to the ovoidal nucleus, through the spiriform nucleus and then, passing sideways to the parvocellular part of the principal isthmi component, entered the tectal grey just above it. The posterior commissure also exhibited responses with a latency of less than 2 msec. All such short latency responses imply more or less direct connections between the archistriatum and these foci.

In the hemispheres themselves analogous activation potentials were observed behind the anterior commissure. They occurred within a narrow band along the ventral border of the caudal neostriatum and also as a rather diffuse projection situated above and behind the archistriatum and more or less encapsulating the neostriatum. The loci concerned correspond to the terminations of diffuse fibres that are visible in Weigert preparations or in unstained preparations when these are viewed through crossed polarizing filters.

Additional potentials which were possibly related to these systems occurred within the hyperstriatic lamina. Some followed with a latent period of less than 1 msec but there was considerable variation and others took 2, or even in excess of 3 msec. Sites in front of the anterior commissure were only investigated in one bird. These revealed short latency effects in the archistriatum itself and also within the fronto-archistriatal tract where direct anatomical connections would clearly be expected. In all these cases the standard stimulus was 10 V but responses could be evoked with voltages down to 1–2 V.

## (iii) *Responses of Longer Latency*

It is rather difficult to interpret the responses which had latent periods in excess of 2 msec. Ipsilateral recording within the dorsal and accessory hyperstriate regions gave small potentials which frequently disappeared as the electrode penetrated the ventral hyperstriatum. The neostriatum also gave potentials but these were of a variable nature. In general the initial response appeared after an interval of 3–10 msec and had an amplitude which varied between 50 and 150 or even 200 $\mu$V. Although the potentials elicited by different stimulating foci were generally similar sometimes a 1 mm difference in the placement of the stimulating electrode gave very different results. One site might elicit no response at all whilst the other produced well-marked potentials. The results suggested that the caudal archistriate regions may possess more caudal projections than do more anterior archistriate foci.

In the paleostriate regions archistriatal stimulation was reflected by responses with latent periods that ranged from 2 to 10 msec. Behind the anterior commissure the amplitudes tended to be 100 $\mu$V or less. The maximum values were recorded within the augmentatum. Here they could achieve 400–800 $\mu$V. However, the shortest latencies occurred at sites in the primitivum. The great magnitude of the potentials in the augmentatum, coupled with the rather long latent periods which elapsed prior to their appearance, were taken by Phillips to indicate massive activation within the region consequent upon archistriatal stimulation.

In the thalamic region slow, biphasic, ipsilateral responses were observed over and above the fast ones mentioned above. These had peaks at 3–5 and 5–10 msec and were detected in most parts of the region. A number of them, and particularly those in the dorso-lateral, ovoidal and lateral spiriform nuclei, achieved amplitudes between 150 and 500 $\mu$V. As in the case of the paleostriatum augmentatum Phillips concluded that these represented widespread and massive activation. Similar responses could be observed as far back as the ventral part of the posterior commissure and down into the tegmentum to the side of the red nucleus. Although the evoked potentials within the hypothalamic and pre-optic nucleus generally had short latent periods, some birds also exhibited others of 50–60 $\mu$V after 5–20 or even 40 msec.

Within the tectum there were maximum amplitudes of 800 $\mu$V, these appeared after 5–10 msec with the majority occurring at 6–8 msec. They were particularly noticeable in the parvocellular part of the principal isthmi component and in the tectal region just above it.

Besides all these long latency ipsilateral potentials some contralateral responses were also observed. At one site close to the hyperstriatic lamina a small potential occurred some 20 msec after stimulation but stimulation of the archistriatum apparently has only a limited effect in the contralateral ventral hyperstriatum. Contralateral neostriatal responses were also only found in 2 out of the 18 monitoring tracks. The failure to find any more presumably reflects meagre fibre projections. Similarly no potentials could be detected at the paleostriatal sites on the opposite side from that stimulated.

However, contralateral septal responses were obtained after latent periods of 0·7–8 or 10 msec. The slowest of these probably represented pickup from adjacent areas in the hyperstriatum as they were limited to the narrow dorsal region. However the archistriatum of the opposite side gave exceedingly clear responses. In some cases stimulation at either of two electrodes situated on either side of the mid-line gave evoked potentials at the site of the other one. Under these conditions the latent periods differed from 2 to 30 msec depending upon the position of the electrodes, the potentials never exceeded 100 $\mu$V and they were frequently very much less than this value, but they emphasize the close functional relationships which exist between the two sides.

## (iv) *Archistriatal Stimulation and Ovulation*

The question of the involvement of the avian telencephalon in endocrine functions, and, more specifically, oviposition and ovulation, was surveyed by Juhasz and Tienhoven (1964). Using White Leghorn hens they stimulated areas in the paleostriatum augmentatum, septal area, archistriatum and neostriatum. It was noted in the section on the primitivum that such stimulation of that area resulted in delayed ovulation. A comparable delay occurred after the archistriatum had been stimulated. In this case it was also accompanied by atresia of the largest follicle in two hens. In the other three specimens, although the largest follicle had actually ovulated, the next largest was atretic. As a consequence of this atresia egg production was interrupted for up to 40 days and during this period the combs also regressed and there was an incomplete moult. The divergent results of Opel (1963a, b, 1967) were discussed in relation to the diencephalon. In view of their investigations Juhasz and Tienhoven concluded that the archistriatum was exerting an effect via the pre-optic area. The fact that the lesion experiments of Phillips showed that damage to the occipito-mesencephalic tract abolished the ovarian inhibition resulting from captivity also intimated that the archistriatal effects may involve this tract. The hypothalamus certainly receives fibres from both the archistriatum and paleostriatum primitivum via this tract and these would suggest themselves as possible pathways.

Whilst noting that the results of Opel conflict somewhat with those of Juhasz and Tienhoven the results of these latter workers are contained in Tables 62 and 63. Fraps (1961) concluded that the actual release of the ovulation-inducing-hormone takes place at about 10 p.m. Juhasz and Tienhoven found that the optimal time for stimulation of both the archistriatum and paleostriatum primitivum was some 5–8 hours prior to this. The mechanism was, however, not clear and they pointed out that, whilst stimulation delayed ovulation, lesions of the relevant areas had no effect at all. They certainly did not, in their experience, advance ovulation. Opel's piqure experiments in the diencephalon did however, but in that case both piqure and stimulation had the same results. Juhasz and Tienhoven suggested that the archistriatum and paleostriatum primitivum may have selective effects on ovulation. Ralph (1959) had previously shown that lesions of the ventral optic region caused oviposition to be delayed for 2–3 days in hens. It therefore seemed possible that the striatal control involved the inhibition of those hypothalamic areas that are concerned with oviposition.

The data of Table 63 show quite clearly that, as for the diencephalon, there is an optimal period during which the stimulation is most likely to achieve a clear result. Stimulation of the same sites at other times had no such effect. Furthermore the optimal time is identical in the case of both archistriatum and primitivum. For the purpose of these experiments they used hens which had previously shown a response. Each was then stimulated 11 hours prior to ovulation and after at least one subsequent normal cycle of ovulation and oviposition she was subjected to a control stimulation. Following two further normal cycles she was stimulated 12 hours prior to ovulation. These sequences of control and experimental stimulations were repeated at 13, 14, 15, 16 and 17 hours before the expected time of ovulation. From Table 64 it is clear that the critical period was between 13 and 16 hours before ovulation.

## (v) *The Results in General*

The distribution of all the evoked potentials which were observed by Phillips parallels the known connections between the regions. More specifically they fit the anatomical destination of projections from the archistriatum. These include direct intra-hemisphere pathways to the neostriatum, paleostriatum and septum, and extra-hemispheric pathways to the ventral diencephalon. Short latency responses at the hypothalamic level are therefore not surprising, and concur with the involvement in functions such as ovulation

## TABLE 62

*The effects of electrical stimulation of the chick telencephalon on both ovulation and oviposition (Juhasz and Tienhoven, 1964). Prematurely laid eggs were actually laid close to the period when they would have been laid in the absence of stimulation.*

| Location of electrode | No. of hens | Normal | | Premature | | Delayed | | | | Atresia no. | Moult no. |
|---|---|---|---|---|---|---|---|---|---|---|---|
| | | Ovul. no. | Ovip. no. | Ovul. no. | Ovip. no. | Ovul. no. | Ovip. no. | Ovul. hr. | Ovip. hr. | | |
| Archistriatum | 20 | 7 | 2 | 0 | 13 | 13 | 5 | 2–3 | 2–7 | 5 | 8 |
| Paleostriatum primitivum | 12 | 6 | 2 | 0 | 0 | 6 | 10 | 2–4 | 2–48 | 0 | 0 |
| Paleostriatum augmentatum | 6 | 6 | 0 | 0 | 0 | 0 | 0 | | | 0 | 0 |
| Neostriatum | 6 | 6 | 0 | 0 | 0 | 0 | 0 | | | 0 | 0 |
| Septal area | 6 | 7 | 0 | 0 | 0 | 0 | 0 | | | 0 | 0 |

and oviposition which the results of Juhasz and Tienhoven imply. This is nevertheless not the case for the dorso-medial and dorso-lateral thalamic nuclei. However, it was pointed out in Chapter 12, Section VI B that both auditory and visual stimuli can evoke ovarian maturation in budgerigars and doves. The sight of the mate in the absence of sound stimuli, and the sound of the male courtship vocalizations in the absence of direct visual stimuli both being effective. In view of the implication of both ovoidal and rotundal nuclei in the auditory and visual pathways respectively, and the occurrence of evoked potentials in the dorsal thalamic region following the presentation of peripheral click stimuli, connections with forebrain areas controlling ovulation would not be unexpected (see Lehrman, 1965).

## TABLE 63

*The effect of the time at which electrical stimulation took place on the time of ovulation and oviposition in hens.*

| Location of electrode | Interval between stimulation-ovulation (hours) | Delayed ovulation | Delayed oviposition |
|---|---|---|---|
| Archistriatum | 11 | 0/13 | 0/5 |
| | 12 | 0/13 | 0/5 |
| | 13 | 2/13 | 0/5 |
| | 14 | 10/13 | 5/5 |
| | 15 | 10/13 | 5/5 |
| | 16 | 11/13 | 4/5 |
| | 17 | 0/13 | 0/5 |
| Paleostriatum primitivum | 11 | 0/6 | 0/10 |
| | 12 | 0/6 | 0/10 |
| | 13 | 0/6 | 0/10 |
| | 14 | 4/6 | 10/10 |
| | 15 | 4/6 | 10/10 |
| | 16 | 4/6 | 10/10 |
| | 17 | 0/6 | 0/10 |

From Juhasz and Tienhoven (1964).

The stimulation studies also corroborate the projections of Phillip's taming lesions and suggest that they reflect a direct pathway. The dorsal supra-optic decussation, together with both the anterior and posterior commissures, can mediate bilateral control. The supraoptic fibres could contribute to contralateral hypothalamic activation, the posterior commissure to bilateral effects on the tectal regions. The marked activation in this area would conform with the large size of the occipito-mesencephalic tract and suggests an important role for the archistriatum together with the neighbouring neostriatal components in mesencephalic-telencephalic integration. In view of the large size of the potentials in the tectal region Phillip's suggested that they were post-synaptic in origin. Edinger *et al.* (1903) and Kalischer (1905) long ago suggested that the apparent blindness consequent upon archistriatal lesions, or lesions in the associated occipito-mesencephalic tract, argued for visual involvement of the archistriatum. Kalischer showed that visual activity persists in most of the contralateral retina following archistriatal extirpation. However, the exact role is disputable. The electro-encephalographic depression which is detected at the occipital pole during initial photic and sonic stimuli and undergoes habituation with repetitive stimulus presentation (Gusel' nikov and Drozhen-nikov, 1959), suggests that archistriatal or posterior neostriatal regions may be involved in arousal. Similar results can, however, also be recorded at frontal foci. Lesions of the occipito-mesencephalic tract lead to a marked reduction of the escape responses to tactile and auditory stimuli. The absence of evoked archistriatal potentials to such stimulation therefore remains un-explained. Phillips (1966) suggested that this is a reflection of the known integrative function of the optic tectum. After extirpation of the hemispheres preparations can avoid obstacles. They also approach light of moderate intensity whilst avoiding excessive darkness or brightness. These patterns of behaviour argue for post-telencephalic integration of those sensory stimuli which are involved, for example, in the fear response. Perhaps the hemisphere involvement only represents an amplifying effect. It is also possible that the neostriate projection areas for tactile and auditory stimuli have a close re-lationship with the archistriatum, particularly in escape reactions but also in ovarian control. The regions of the neostriatum in which potentials can be detected following archistriatal stimulation are certainly close to those activated by peripheral click stimuli and with his taming lesions Phillips interrupted the diencephalic and mesencephalic connections to these regions within the strio-mesencephalic tract. He thus modified escape responses to several sensory modalities which have different areas of representation in the telencephalon.

More diffuse reactions also occur. In his studies on wildness Phillips (1964) observed cardio-vascular changes. These followed the application of stimuli to both the archistriatum and the occipito-mesencephalic tract. Monopolar stimulation was carried out at 86 sites distributed at 1 mm intervals through the neostriatum, archistriatum, paleostriatum, septum and optic tectum. A sharp drop in blood pressure which was accompanied by arousal was elicited by stimulation of both the ventral archistriatum and the tract but not from the neostriatum. No responses were detected following stimulation of the dorsal thalamus but a similar sharp decrease was elicited from the septal area. The tectal sites gave arousal which compared with that obtained from the ventral archistriatum but there were no concomitant vascular changes. This clearly suggests that, as would be expected from its involvement in fright reactions and gonadal control, the archistriatum is an area of extensive autonomic integration. Other autonomic functions such as the control of pupillary activity are discussed in subsequent sections, blood vascular changes and arousal featuring in Chapter 15, Section IV.

## VI. ACTIVATION AND STIMULATION OF THE ECTOSTRIATUM

The telencephalon was systematically explored for slow potentials and evoked unit discharges by Revzin and Karten (1967) during their studies on the rotund nucleus which is the principal diencephalic station in the visual pathway. The largest evoked potentials were detected within the ectostriatum and neighbouring parts of the intermediate neostriatum. Large potentials which differed slightly from these were also recorded in the hyperstriatic lamina, in that part of the ventral hyperstriatum which overlies the ecto-striatum and also in the lateral areas of the paleostriatum. As far as these last are concerned it is most unlikely that they represent a major projection area. Much of the activity was in those fibres that traverse the region of the aug-mentatum directly subjacent to the ectostriatum. These correspond to the lateral part of the thalamo-frontal tract which is the long established rostral projection from the rotundus. Furthermore no post-synaptic evoked potentials occurred in the paleostriatum although there was a certain amount of electrical activity close to the ectostriatum which Revzin and Karten ascribed to electrotonic spread. There was no evidence for the existence of any paleostriate units directly driven by the rotund nucleus. The fibre degeneration which Powell and Cowan (1962) observed presumably reflects lesions in the thalamo-frontal tract itself. Apart from small potentials in the accessory hyperstriatum none were recorded from other hyperstriate regions, nor from either the frontal or caudal neostriatum, the archistriatum or any other telencephalic sites.

The waveform of the ectostriate potentials had two principal components. At first there were di, or triphasic spike-like waves. These were then followed by a larger negative-positive one which was sometimes complex and frequently followed by secondary negative ones. If the frequency of stimulation was in-creased from 1 pulse/sec the secondary negative-positive potentials dis-appeared at a stimulus frequency of about 3/sec. This was before any of the other components were affected. The primary negative-positive wave am-plitude was reduced by about 50% at a stimulus rate of 10 pulses/sec. In contrast the initial spikes exhibited little change up to stimulus frequencies of 100/sec. Revzin and Karten concluded that these spikes represent impulses radiating out within the rotundo-ectostriate pathways. The larger waves comprise later post-synaptic activity in the ectostriate neurones themselves. This was corroborated by the fact that during single-unit studies the ectostriate units discharged in phase with the large negative-positive peaks. The latency of this post-synaptic activity ranged from 6 to 8 msec; that of the initial spikes from 2·5 to 3·5 msec. The largest responses occurred within the posterior half and medial two-thirds of the nucleus. The amplitudes of the initial radiation spikes were maximal in the ventral third and electrode placements in more dorsal regions gave lower values. Complementarily the spike-size increased as the electrodes penetrated the fibres of the thalamo-frontal tract which lay within the underlying augmentatum.

Direct stimulation of thalamic sites above the rotund nucleus also sometimes gave small, variable, long latency responses in the ectostriatum. However, short latency responses only followed rotundal stimulation, or at least the nucleus itself together with those parts of the diencephalon which lie immediately medial to it. Such results are in striking agreement with the visual intensity and pattern discrimination deficits which were reported by Karten and Hodos (1970) and Hodos and Karten (1970). Responses were nevertheless evoked by using the more lateral thalamic regions and these were complex. As Revzin and Karten pointed out this is not particularly surprising in view of the presence of tecto-rotundal fibres passing via the lateral regions on their way to the diencephalon.

Direct electrical stimulation of the ectostriatum itself did not provide any evidence for reciprocal ectostriato-thalamic connections. Although there were responses within the rotundum these took the form of small and inconstant spikes which could well represent antidromic discharges along rotundo-ectostriate fibres. Further investigations also failed to evoke any potentials in the rotund nucleus when the accessory, dorsal or ventral hyperstriatum, or the caudal neostriatum were stimulated. There is therefore no evidence for any reciprocity in structures involved in the anterior ascending projections of the visual pathway. What is more there are no obvious explanations for the elaborate hemisphere projections from the rotund nucleus which are visible anatomically (see Huber and Crosby, 1929; Powell and Cowan, 1962). The evidence for a possible inter-relationship between the ectostriatum and the hyperstriatal regions above it is, however, interesting in view of Zeigler's (1963) study on the impaired visual discrimination learning which is attendant upon hyperstriatal lesions.

# VII. THE NEOSTRIATUM

## A. Telencephalic Projections of the Ovoidal Nucleus

Fibres from the ovoidal nucleus to the hemispheres assemble at its side and give rise to the medial part of the lateral forebrain bundle. The lateral region of this fasciculus carries the rotundo-ectostriate projections mentioned above. When traced by degeneration studies these fibres originating in the ovoidalis pass dorsally, traverse in front of the anterior commissure and turn backwards into the hind part of the telencephalon. On their way the majority pass through the mid-part of the primitivum and augmentatum, cross the dorsal medullary lamina, and enter the neostriatum. Here they contribute to the medial part of the thalamo-frontal tract (= internal occipital capsule) and end in great numbers on its intercalated cells. Although many degenerating fibres can be seen within the lateral part of the caudal neostriatum there is no evidence for any significant number of terminations there. Instead massive axonal and terminal degenerations occur within a clearly demarcated medial area of the caudal medial neostriatum. This sharply defined area of small cells is closely coincident with field L of Rose (1914) (see Table 53). In the front it is separated from the hyperstriatic lamina by a layer of larger cells which are situated on its medio-dorsal side. The anterio-lateral margin is apposed to the medial margin of the caudal ectostriatum. A few terminations were indeed seen in that nucleus and compare with the limited number of rotundal fibre terminals detectable in the neostriatum. Fibres of passage also traverse these ectostriate areas. No degenerations occur in field L after ablation of the rotund nucleus and the two neighbouring regions—ectostriatum and field L—are strongly contrasted in this respect.

The fibres penetrating this zone of the caudal neostriatum are organized in parallel columns. These are orientated along a ventro-lateral to dorso-medial axis. Many fibres pass through the major mass of the caudal neostriatum which is composed of larger and more scattered cells and lies below field L. However, no terminations occur there. Lesions produced in the caudal neostriatum by Karten (1968) resulted in extensive retrograde degenerations together with gliosis in both the ipsilateral ovoidal and semilunar parovoidal nucleus. However, such lesions usually entailed damage to the overlying hyperstriatal and underlying paleostriatal complexes. They are therefore somewhat difficult to interpret and could not in themselves be taken as definitive corroboration of the ovoidal-neostriatal projections.

## B. Evoked Potentials in the Neostriatum

In his electrophysiological studies on the avian brain Erulkar (1955) discovered that light tactile stimulation of even a very small number of feathers at any point on the body surface resulted in evoked potentials within the anterior part of the caudal neostriatum. Similar responses were also produced by direct stimulation of a branch of the sciatic nerve. This suggested that peripheral cutaneous sensitivity was represented in that region. The responses had a latency of 17–19 msec and the system seemed to have a refractory period of 200 msec because evoked potentials usually failed to follow a stimulus rate in excess of 5/sec. Although no definitive topographical pattern emerged Erulkar concluded that in general the front of the body is represented in the upper part of the activated region whereas the posterior part of the body was represented further back and ventrally. In no case was it possible to detect potentials in either the ventral hyperstriatum or the Wulst structures. Most of the activated sites were in fact well away from the surfaces of the neostriatum although a few points situated close to the hyperstriatic lamina also gave positive results.

The presentation of peripheral auditory stimuli produced potentials in the hemispheres which are analogous to those at stations elsewhere along the ascending auditory pathway. It was mentioned in Section IV above that responses varying from 35 to 110 $\mu$V occurred within the paleostriatum primitivum after a latent period of 11–17 msec. Other activation sites occur in the augmentatum. More definitive auditory representation appears to be present in the neostriatum. In the caudal components of this nuclear grouping complex potentials follow sound presentation to the opposite ear. There are first of all initial waves which appear after 13–17 msec (Harman and Phillips, 1967) and hence have latencies equal to the longer ones in the paleostriatum. These initial responses had amplitudes which varied between 25 and 250 $\mu$V. They were subsequently followed by slower waves with latencies of 18–40 msec. When stimuli were presented to both ears simultaneously the resulting potentials followed after 13–15 msec but had rather lower magnitudes and varied between 65–110 $\mu$V. They could persist for up to 25–30 msec and there-fore represented an apparently longer lasting activation than that which resulted from unilateral sound presentation.

Within the frontal neostriatum it was possible to detect responses to con-tralateral sound stimuli. Some of these took the form of negative waves with a latency of 8–11 msec and an amplitude of 10–250 $\mu$V. Others were positive waves with latencies of a comparable duration (8–12 msec) and amplitudes of 100–110 $\mu$V. These were often part of a complex of responses in which slower components were represented after periods of 14–25 msec. These had am-plitudes of 40–110 $\mu$V and negative potentials preceded further positive ones.

The potentials within the caudal neostriatum faded rather faster than those evoked at diencephalic and mesencephalic relay stations such as the ovoidal nucleus and the dorsal part of the lateral mesencephalic nucleus. They also exhibited a marked drop in amplitude when the stimulus frequency was in-creased. Neither ipsilateral nor contralateral sound presentation produced a second response when the interval of time which separated the individual clicks was less than 7 msec. This contrasts with the condition at lower levels of the ascending pathway because an interval of 3·6 msec completely restored the mesencephalic potentials. The neostriatal potentials also showed a linear decrease in amplitude with increasing rates of sound presentation. Activation reached maxima in the range of 240–280 $\mu$V with single stimuli and fell to zero at 10/sec.

## C. Electrical Stimulation of the Neostriatum

Direct electrical stimulation of the neostriatum was reported by Harwood and Vowles (1966) whose results are similar to those which I have observed. They were investigating the feeding behaviour of *Streptopelia rissoria*. No significant effects were found after stimulation of the hyperstriatum, nor following stimulation at certain foci within the anterior or dorsal neostriatum. There was, however, a statistically significant increase in feeding behaviour after stimulation of certain sites within both the paleostriatum and also the more ventral and posterior neostriatum. A quantitative analysis was made of the total amount of time which the birds spent in various categories of behaviour. This was accompanied by data on the actual number of bouts and their length. The principal difference which they observed between the feeding behaviour elicited by hemisphere stimulation and that exhibited by hypothalamic preparations was the much greater variability of the former. This showed itself in long and variable latencies, which often exceeded 1 min, frequent interruptions of feeding bouts whilst the birds preened themselves, and also the prolonged after effects of stimulation. At the end of a period of stimulation there was often a series of unusually long feeding bouts which each lasted 2–3 min. During such a bout pecking built up over 15–30 sec until it reached the rate which is typical of hungry birds. This, of course, contrasts very markedly with the more lackadaisical pecking actions which are shown by satiated specimens. The current strength which was used to elicit such feeding lay within a range close to 30 $\mu$A.

The frequency with which both feeding and preening were evoked, together with the duration of the bouts concerned, was affected by the immediately preceding history of the specimen. In two successive periods of stimulation no feeding might occur and then a third period might produce it. Thereafter successive stimulatory phases during the same session would produce feeding bouts of increasing length and after progressively shorter latent periods. This gave Harwood and Vowles the impression that the initial stimulations primed the feeding behaviour. Such a conclusion clearly conforms to those of von Holst and Saint Paul (1963). The most significant result of such studies was an overall increase in the time which was spent on feeding and preening largely at the expense of various miscellaneous activities. These results are summarized in Table 64. Clearly such responses to hemisphere stimulation are complementary to the loss of spontaneous feeding and drinking which is consequent upon hemisphere ablation.

## TABLE 64

*The behaviour of* Streptopelia rissoria *as a function of electrical stimulation of forebrain foci within the paleostriatum and ventral or posterior neostriatum.*

| Condition | Attention | Walking | Preening | Feeding | Miscel-laneous | Total |
|---|---|---|---|---|---|---|
| | | Percentage of total observation time | | | | |
| Stimulated | 30 | 13 | 24 | 24 | 9 | — |
| Unstimulated | 11 | 11 | 19 | 27 | 32 | — |
| Controls | 32 | 17 | 16 | 12 | 23 | — |
| | | Mean number of bouts in 10 min | | | | |
| Stimulated | 11 | 4 | 8 | 9 | 2 | 34 |
| Unstimulated | 3 | 2 | 3 | 5 | 2 | 15 |
| Controls | 10 | 5 | 5 | 5 | 3 | 28 |
| | | Mean bout length in sec | | | | |
| Stimulated | 15·9 | 20·3 | 18·3 | 18 | — | — |
| Unstimulated | 22·9 | 31·1 | 42·3 | 36·2 | — | — |
| Controls | 22·3 | 19·3 | 19·0 | 12·8 | — | — |

From Harwood and Vowles (1966).

## TABLE 65

*The number and the duration of feeding bouts which were directed to different potential foods.*

| Food | Controls | | During and after stimulation | | |
|---|---|---|---|---|---|
| | N | Duration in sec | N | Duration in sec | t |
| Millet | 23 | 10·9 | 19 | 22·5 | 2·38 |
| Hay | 33 | 11·5 | 25 | 20·3 | 2·81 |
| Faeces | 12 | 16·3 | 8 | 11·8 | 1·40 |
| Grit | 4 | 13·8 | 14 | 18·3 | 0·34 |
| Total | 72 | | 66 | | |

From Harwood and Vowles (1966).

Harwood and Vowles also observed that a higher rate of response occurred if the period of stimulation followed 5 min of normal pecking. Again a corroboration of the conclusions of von Holst and St. Paul, they concluded that it reflected facilitation of activities which were already occurring. As a result of putting the preparations on a deprivation schedule and then comparing the effects of stimulation and no stimulation on a reinforcement programme in which progressively more numerous pecks were necessary to obtain a reward, they concluded that stimulation does not lead to the preparations working to higher reinforcement ratios.

The data of Table 65 are also of interest. These show that when a variety of potential foods was provided the result of stimulating the preparations was to increase their consumption of millet and hay. These are the most common components of the normal diet. It also increased the eating of faeces. However, it would appear that stimulation facilitates most strongly those types of behaviour which have the greatest tendency to occur normally.

# VIII. EXPERIMENTAL STUDIES ON THE WULST COMPONENTS

## A. Peripheral Activation

### (i) *Activation by Peripheral Auditory Stimuli*

Following the presentation of peripheral auditory stimuli Adamo and King (1967) recorded wide-spread activation potentials in the Wulst of 12–20-week-old chicks. Responses to click stimuli were largely surface positive and comprised a single brief positive component followed by a small negative wave. Alternatively there was a series of 2–3 brief positive-negative components. The latencies of the majority lay within a range of 5–8 msec, although longer latent periods occurred in the more caudal regions where they were of the order of 14 msec. Such responses lasted approximately 20 msec except, once again, towards the hind end where those with a long latent period could persist for up to 40 msec. In certain cases sites which responded to peripheral auditory stimuli with surface positive activity also responded to peripheral photic stimuli but then exhibited surface negative activity. Recordings at greater depths from the surface usually resulted in a drop in the amplitude of the observed potentials at a level some 200–300 $\mu$ below the surface. This was also sometimes accompanied by a reversal of the initial polarity. These widespread activation phenomena which were evoked by auditory stimulation contrasted very markedly with the results of cutaneous electrical stimulation. Generally speaking this did not activate the Wulst region and the only electrode which recorded potentials produced in this way was situated at a depth of 3200 $\mu$, and therefore below the Wulst. This agrees with the results of Gogan (1963) who reported responses to tactile stimuli at such levels.

These very definitive results clearly implicate Wulst cells, or cells in the vicinity of the Wulst, in the rostral projections of the ascending auditory pathway. The actual characteristics of the responses, coupled with the fact that they are abolished by surface application of procaine, suggest active sites which are close to the hemisphere surface. As such they must involve loci which are either within the accessory hyperstriatum or else in the superficial corticoid areas. Furthermore the very short latencies which were observed in all regions apart from those at the back, suggest that the ascending pathways which are involved comprise more or less direct paths with relatively few synaptic relay stations. A larger number of junctions was presumably involved in the spread of activation to those areas with latent periods which lasted some 14 msec. In view of the rather long lasting effects within such areas it is probably that they also comprised widespread and complex activation.

## (ii) *Activation by Peripheral Photic Stimuli*

A comparison of the responses which could be detected within the Wulst following auditory and photic stimulation led Adamo and King (1967) to point out that the potentials which resulted from photic stimuli had a longer latency, longer actual duration and larger amplitude. The latent periods varied from 10 to 20 msec and the actual duration of the potentials reached 150 msec. The typical results consisted of initial waves of surface negativity which were then followed by a longer lasting positive-negative complex. Clearly these phenomena were different from those following click stimuli, and within the Wulst they were diffuse and lacked any precise topographical localization. In many respects they are comparable with reptilian responses. Furthermore the results of recording from electrodes at different depths were similar. The potentials were constant and did not show the regular diminution of amplitude or the occasional reversal of polarity which occurred with the sound evoked potentials at a depth of 200–300 $\mu$. The fact that surface procainization had no effect upon them also suggested a fundamental difference in location between the two groups of phenomena. Topological application of anaesthetic abolished the potentials that were evoked by click stimuli and it would seem probable that whilst these involved electrically active foci in a close topographical relationship to the corticoid regions, the effects of photic stimulation were associated with deeper, and possibly more extensive structures. The source was certainly not on the surface of the telencephalic hemispheres, nor in the immediately subjacent laminae.

The gradual changes which seem to occur in such widespread light evoked potentials during embryonic and post-embryonic development were studied by Paulson (1965). Working with Pekin ducks he showed that recordable responses were frequently totally absent in embryos below the age of 26 days. During maturation further changes also took place and the potentials very often diminished in amplitude several days after hatching. A comparison of the latency at various levels within the brain showed, as one would expect, that the potentials within the hemispheres had a markedly longer latency than those at tectal levels and that, whilst the responses in the vicinity of the tectum were usually complex, those at the level of the forebrain were more or less identical over the whole extent of the hemispheres.

## B. Lesions to Wulst Structures

### (i) Generalized Effects

Following extirpation of parts of the Wulst the resultant preparations are both hypo-reactive and hypo-kinetic. Karamyan (1956) recorded that they tend to sit with ruffled feathers and only very intense visual or tactile stimuli evoke escape responses. Cohen (1967) made a large number of observations on such preparations. Using 25 male and female pigeons which varied in age from 2½ to 6 months he observed that after light to moderate damage of the accessory hyperstriatum their pecking ability was accurate and drinking was unimpaired. Indeed the results of monitoring during both the pre-, and post-operative period showed no significant differences in water-intake, body weight or cloacal temperature. Food intake was, however, decreased. He concluded that this did not amount to aphagia but that the daily consumption was somewhat depressed. The heart and respiratory rates were unaffected and fell within the normal limits. Mean heart and respiratory rates for 16 un-operated normal birds were 142·3 beats/min and 32·6 respiratory cyc/min. In eight preparations which had lesions to the Wulst the equivalent values were 142·7 and 34·7/min respectively. Reflexes such as those involving the cornea and pupil were normal, as indeed were the visual, tactile, vestibular and placing reactions. Flexor withdrawal, grasp and extensor thrust were all present. In general the only exceptions which fell outside normal patterns of activity were periods of transient anorexia.

### (ii) Reactivity and Conditioning Performance

In the preparations which were cited above the most extensive damage was to Wulst structures and more particularly the posterior two thirds of what Cohen referred to as the accessory hyperstriatum. The ventral hyperstriatum was less consistently involved. However, it appears that the dorsal hyper-striatum was included by Cohen in his term accessory hyperstriatum and also experienced damage. Furthermore the difficulty of ablating the Wulst components without harming the overlying corticoid structures resulted in damage to both the parahippocampal area and the dorsal part of the hippocampus. Attempts to correlate the subsequent behaviour with the actual extent of the lesions were largely unsuccessful.

The resulting preparations all exhibited deficits in their escape behaviour. This conforms with other reports of hypokinesia following hyperstriatal damage (see Rogers, 1922; Zeigler, 1963). Cohen also undertook an assessment of their reactivity by measuring their overall activity in an activity cage. This was not a very accurate analysis and provided no indications of the actual types of activity. Each specimen was placed in it for 70 min on the 8 days preceding the operation and then again during the 8 post-operative days. Following a 60 min stabilization period 5 light flashes were given. The gross movements were then measured for the following 10 min. As such the data provided information about the reactivity to light. All other extraneous visual stimuli were excluded. A comparison between the control and experimental data showed that a significant impairment followed hyperstriatal ablations. Only one control bird showed a significant post-operative decrease in reactivity whilst there were three such changes in the hyperstriatal group (37·5%).

Both the control and experimental groups were also compared in a light proof conditioning chamber that was provided with a moderate degree of sound attenuation. Red and green lights were placed on the chamber wall and used to produce conditioned reflexes. The unconditioned stimulus was a 60-cycle a.c. foot shock. Both the electrocardiogram and respiratory activity were monitored and there were no qualitative differences between the two groups of birds. Furthermore the bulk of the conditioning occurred during the first session of 20 trials in both cases. The magnitude of the conditioned tachycardia tended to increase over the next four sessions (see also Cohen and Durkovic, 1966). Although there were indications of somewhat greater response levels within the control group the difference was not statistically significant. However, the controls did differ from the lesioned preparations in their respiratory conditioning. The performance of the controls resembled that of normal birds. Most conditioning occurred during the first session. Birds with lesions in the Wulst exhibited a minimal level of conditioning prior to the third session. After this their performance approached that of controls. A day by day comparison also demonstrated that the responses of lesioned preparations were of a significantly lower magnitude than those of the controls on days 1 and 2.

Cohen concluded from this that at least one major attribute of Wulst function is visual. His suggestion that this is associated with the accessory hyperstriatum is clearly unproven in the absence of differentiation between this and the dorsal hyperstriatum. More specifically these results implicated such structures in the mediation of certain somatic responses to light stimuli. The deficits which were observed had certain features in common with those that followed deep lesions involving either rotundal projections to the ectostriatum or ectostriatal input to the accessory hyperstriatum. More detailed studies are clearly required to determine the properties of the visual receptive fields within the hyperstriatum, and also those parameters of visual function that necessitate the accessory hyperstriatal and Wulst region generally for their discrimination.

## (iii) *Discrimination Learning*

The multiplicity of studies which have been carried out on birds (see Thorpe, 1958; Hinde, 1966) reveal a variety of complex behaviour patterns which far exceed the relatively straightforward conditioning outlined above. Zeigler (1963) was one of the first workers in the last decade to attempt to obtain information about visual discrimination learning after telencephalic ablations. He used the pigeon and an operant conditioning situation in a discrimination apparatus that involved high and low levels of illumination and also triangles and circles of equal area. A summary of the brain regions which were extirpated is contained in Table 66. Again the situation is not straightforward because of the absence of specific distinctions between the ventral and dorsal hyperstriatal components.

A statistical comparison of the learning performance in the control animals and preparations with local hyperstriate ablations showed that the lesions produced significant defects. Information which was derived from experiments on the control birds alone showed that the pattern discrimination was the most difficult of the two problems. The data of Table 67 show that, as would be expected in view of this, the lesions had a more profound effect upon this than they did on brightness discrimination. It is also apparent that the primary result is an impairment of the acquisition of the ability rather than its retention.

Comparisons of the discrimination learning ability of individual preparations indicated that there was a direct relationship between the lesion sizes and the observed deficits in learning ability. Those that showed no signs of learning, even when a considerable period had elapsed after the operation, had the most extensive lesions. Conversely the damage sustained by the retentive group was neither so extensive nor so variable as that in the non-acquisitive group. Layman (1936) had previously obtained analogous results to these in his studies on the role of the superficial hemisphere components of chickens. His lesions had resulted in a considerable amount of damage and the few preparations which failed to learn to discriminate to his required level were those with the most extensive lesions. These involved particularly the paleostriatum and hyperstriatal levels.

Taking into consideration the results of Powell and Cowan (1962), who showed that thalamic projections pass to the paleostriatum and the dorsomedial margin of the hemispheres, it is of interest that lesions in both these areas result in a marked deficit of conditioning and visual discrimination.

## TABLE 66

*An analysis of the lesions showing the experiments in which the preparations were used.*

| Brightness acquisition | Brightness retention | Pattern acquisition | Pattern retention | Accessory hyperstriatum | Hyperstriatum | Neostriatum | Paleostriatum | Archistriatum | Cortical regions |
|---|---|---|---|---|---|---|---|---|---|
|  |  |  |  | +(+) | +++ |  | — | — | ++ |
|  |  | + |  | +++ | ++++ | +++ | — | — | — |
|  |  |  |  | +++ | ++(+) | (+) | — | — | +(+) |
|  |  | ++ |  | +(+) | +++ | +(+) | — | — | +++ |
|  |  |  |  | ++ | +++ | — | — | — | (+) |
|  |  |  | ++++ | +++ | ++(+) | (+) | — | — |  |
|  | ++++ |  |  | +++ | +++ | — | — | — | (+) |
|  |  | + |  | ++(+) | +(++) | (+) | — | — | (+) |
| ++++ |  |  |  |  |  |  |  |  |  |

From Zeigler (1963).

Key: —, Untouched    +, Slight lesion    ++, Moderate lesion    +++, Extensive lesion
Crosses in brackets indicate this greater damage to one side only.

Results of this nature following damage to the areas below the dorsal medullary lamina could well reflect extirpation of fibres of passage. Furthermore Zeigler's results conform to those of Cohen which were outlined above and indicated that only minimal respiratory conditioning followed damage to the Wulst. Together with the works of Layman (1936), Tuge and Shima

## TABLE 67

*Statistical comparisons between the discrimination learning ability of pigeons and preparations with hyperstriatal lesions.*

| Discriminating problem | Mean number of days to criterion level in controls | Mean number of days to criterion level in preparations with hyperstriate lesions | "t" | "p" | df |
|---|---|---|---|---|---|
| Brightness discrimination (Acquisition) | 5·45 (N = 11) | 8·25 (N = 4) | 2·74 | 0·01 | 13 |
| Brightness discrimination (Retention) | 2·40 (N = 5) | 5·75 (N = 4) | 2·28 | 0·05 | 7 |
| Pattern discrimination (Acquisition) | 13·0 (N = 10) | 26·0 (N = 4) | 7·22 | 0·001 | 12 |
| Pattern discrimination (Retention) | 6·5 (N = 4) | 19·25 (N = 4) | 9·96 | 0·001 | 6 |

From Zeigler (1963).

(1959) and Shima (1964) these results clearly suggest that avian preparations with damage to the hyperstriatal regions generally, and more especially the components of the Wulst, are deficient when faced with a variety of visual learning situations. Following such lesions the preparations exhibit deficits in their instrumental conditioning ability and these deficits are related to the relative difficulty of the tasks concerned.

## (iv) *Effects on Auditory Perception*

Following the reports of activation potentials in the Wulst, and more especially the superficial regions, which followed the presentation of peripheral sound stimuli (Adamo and King, 1967) it was clearly of interest to investigate the effects of hyperstriatal ablation. Adamo and Bennett (1967)

determined the overt responses of such preparations to a sound stimulus. Using 22 adult birds 11 of them were exposed to white noise for 0·2 sec, and the remainder for 2 sec. The resulting head orientation responses which occurred during ten consecutive trials in each of five test sessions were then observed and rated. After this bilateral hyperstriatal lesions were produced by suction. Each stimulus duration group included one bird which was sham-operated and another which was left unoperated. Seven days after the operation were allowed for post-operative recovery before the pre-operative testing procedure was repeated. A composite picture of the lesions which were produced in the individual birds implicates most parts of the Wulst. Some of the larger areas that were destroyed overlapped the areas of destruction in those preparations with small lesions.

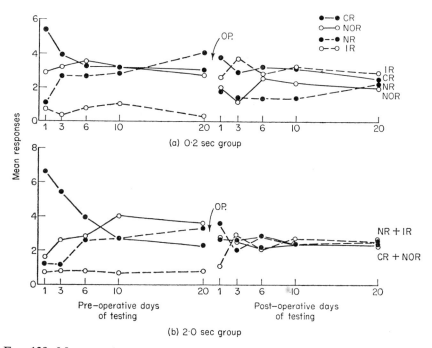

Fig. 128. Mean number of each category of responses for each stimulus duration group. (Adamo and King, 1967).

In rating the responses during the pre-operative period any movement of the head towards the stimulus was considered to be a correct orientation. Turning the head to the side opposite the stimulus rated as incorrect. Failure to respond at all was also noted. Any behavioural responses other than head orientation were recorded as non-orientation responses and included startle reactions or lifting of the head, both of which indicated that the preparation had heard the sound. Habituation of the head orientation response, which was defined as an increase in the tendency for the preparation or bird not to respond, was reflected in a decrease in the number of correct responses together with an increase in the number of nil responses.

In the pre-operative curves for both stimulus duration groups habituation occurred as a function of the days of testing. The curves for correct responses suggest that it occurred primarily between days 1, 3 and 6. Following days 6 and 10 it approached an asymptote after which the level was fairly constant for the remainder of the experimental period. An analysis of variance confirmed that significant inter-session habituation occurred during this pre-operative period ($P < 0.01$). When considered either individually or combined the data from the test sessions showed no significant intra-session habituation as a function of the trials. The data for both groups during the pre-operative period are represented on the left hand side of Fig. 128.

Analyses of the data showed that in the 0·2 sec group there was a significant interruption in the final pre-operative degree of habituation during the interval prior to post-operative testing. This was not the case in the 2·0 sec group. Post-operative inter-session habituation was also found wanting. As far as the actual head orientation responses were concerned the operated preparations made significantly more incorrect responses than they had done prior to the lesions. An attempt to relate the number of incorrect responses to the actual extent of the lesions was unsuccessful. This therefore compares with the comparable lack of success which Cohen (1967) had in correlating the degree of hypo-reactivity with lesion area and volume. Furthermore an analysis of variance showed that the number of incorrect responses did not relate to either the actual lesion position in the rostro-caudal plane, or the depth of the damage below the hemisphere surface. Lesions to the accessory hyperstriatum either with or without concomitant damage to the underlying structures produced similar results. The data which are portrayed in Fig. 128, and the significant difference in the final pre-operative degree of habituation which was produced by the lesions, clearly implicate the Wulst, and more particularly the accessory hyperstriatum, in the habituation of head orientation responses following sound stimuli. They also indicate its probable importance in reflex head orientation to sounds. The suggestions of Batteau (1967) have been discussed in relation to the peripheral acoustic system, the primary relay nuclei in the medulla, and both the ovoidal and dorsal thalamic nuclei. It may well be at the accessory hyperstriatal level that the signed additions and attenuations enable directional discrimination to be undertaken. In view of the results of Adamo and King (1967) it is also likely to be in the more superficial regions which underlie the hemisphere surface.

## C. Electrical Stimulation of the Wulst

The studies of Kalischer (1900, 1901, 1905) led him to suggest the presence in the avian brain of localized hemisphere foci concerned with motor function. However, for 60 years no clear topographical distribution of such foci was demonstrated in any discrete region although there were of course various reports of somatic responses which followed stimulation of a variety of telencephalic regions. As a result of this Cohen and Pitts (1967) undertook an extensive mapping of the Wulst in *Columba livia* with respect to the experimentally induced somatic responses. They observed the results of constant current stimulation in the corticoid regions and also within both the dorsal and accessory hyperstriatum. No attempt was made to differentiate between these two and, in all, 50 electrode sites were located above the superior frontal lamina and within the definitive cell masses of these nuclei. The ten most caudal sites were located dorsal to the ventricle and in the corticoid region. Three others were within the transitional zone between the hyperstriatum and corticoid region. A further six electrodes penetrated below the superior frontal lamina to enter the ventral hyperstriatum and three more were situated laterally in the boundary between the Wulst and the ventral hyperstriatum.

The results of stimulation at the foci in the Wulst involved a number of peripheral responses. In the semi-restrained bird these included beak movements, swallowing, eyelid closure and nictitating membrane closure, feather ruffling in the beak area and pupillary changes, particularly pupillary dilation. These various responses had highly variable thresholds, were elicited from a variety of loci and had no obvious topographical localization. A number of them, and particularly both swallowing and feather ruffling, frequently persisted into the post-stimulation period. During this time the electro-encephalograms from the Wulst regions showed spiking. Eyelid and nictitating membrane closure had low thresholds and sometimes preceded all other somatic movements.

In the restrained birds, although these responses appeared frequently, the predominant and most reproducible effects were head movements. It is worth emphasizing that small opto-kinetic movements in the pigeon and other birds usually involve head movements. In the experimental situation the majority of these were rotations to the contralateral side and often had either upward or downward components as well. Some electrode sites gave pure down responses but pure upward ones were never seen. A few movements with a downward component were also accompanied by ipsilateral abduction of the head and neck so that they were displaced away from the mid-line and towards the side being stimulated. Other motor responses which also occurred infrequently included the arrest of movement on stimulation, rotation to the side in which the electrode was sited, and forward movements of the head in the mid-line. There were also both cardiac and respiratory effects. The cardiac response was invariably a small to moderate tachycardia which could sometimes be associated with arhythmia. The pattern of such arhythmia could not be correlated with either particular stimulation foci or the particular stimulation parameters. It was, however, more frequent at higher currents.

The primary respiratory response was usually an increase in the amplitude of the first full expiration after the onset of the stimulus. This was customarily followed by one or more inspirations of reduced amplitude. These respiratory responses were of short latency and usually occurred without any concomitant cardiac changes. Later in the stimulation the most frequent respiratory response was an acceleration phase which occurred either with or without a change in amplitude. This was usually accompanied by tachycardia and both effects could then persist into the post-stimulatory period. This was particularly true when high currents were used or in cases of long latency. Vedyaev (1964) also reported both inhibitory and excitatory effects on respiration but the area from which Cohen and Pitts obtained respiratory acceleration was not studied by him.

The actual magnitude of the head movements varied as a function of both the relevant electrode site and the stimulation parameters. For example a contralateral rotation could vary from a few degrees away from the mid-line to as much as 180°. In some cases such contralateral movement or rotation was accompanied by slight downward components. Other almost pure downward movements were also accompanied by a slight contralateral displacement. As a result there was a spatial continuum of the possible directions of movement. This gave a very large number of final head positions. Besides the actual direction of movement other properties were also detectable. Two displacements in similar directions could differ in latency, speed and smoothness. On some occasions the final head position was approached in a series of jerky movements while more frequently the change of position was both smooth and continuous. All three properties could be modified by altering the stimulation parameters. Another important variable was the activity at the final head position. The bird could either remain stationary or its head could undergo phasic movements which were centred on that position. At higher currents there was often a considerable amount of non-directed head and neck activity after the actual stimulation had ceased. This was accompanied by spike activity in the electro-encephalogram.

The evoked response was clearly related to the initial position of the head and this was therefore always standardized by being in the mid-line at the onset of stimulation. Initiating stimuli when it was in any other position could result in increased latency and speed of the resulting movement. It did not, however, vary its directional coordinates. The disadvantage of maintaining this standard initial position is that movements towards it are precluded. It is also difficult to identify arrest reactions. Because of this any intense stimulus which failed to elicit movement from the mid-line was then repeated with the head in other initial positions. This usually produced an arrest of all movement, and such stationary head postures were occasionally preceded by a return of the head to the mid-line.

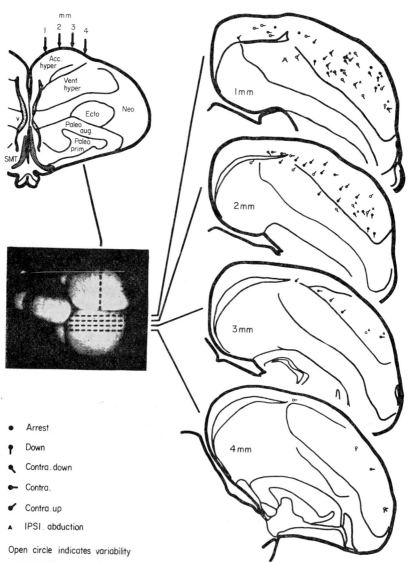

Fig. 129. The direction of head movement following stimulation (Cf. Fig. 130). (Cohen and Pitts, 1967).

Figure 129 shows the various directions of head movement which were evoked at each electrode placement. A distinction is drawn between those points which gave highly repeatable responses and others at which the response was variable. It is important to note that a *response* was considered to be *variable* if it only occurred at high current values ($>400\,\mu A$). This contrasts with the *sites* that gave responses of variable direction. These last are shown by multiple flags in a single location. Because of the occurrence of such variations the 73 electrodes gave 82 responses.

The three electrodes which were located in the transition area between the accessory hyperstriatum and the corticoid regions were considered to be in the accessory hyperstriatum. Those between the Wulst and the ventral hyperstriatum were ascribed to the latter formation. One further site in the neostriatum gave a variable contralateral up response. Within the Wulst and corticoid regions the most medial electrodes in the anterior part of the Wulst tended to give pure down responses. Further away from the midline points which elicited down responses were still within the Wulst but they were situated at more ventral levels. No pure down responses occurred within the most lateral regions. In the case of sites that were situated rather nearer to the occipital pole this down zone merged into a mixed area that gave both contralateral up and down movements. There was also some suggestion of a distinct contralateral down region intervening between the pure down area and the mixed zone.

An area that was predominantly associated with contralateral upward movements occupied most of the caudal and lateral areas. A region of pure contralateral displacements also extended from the posterior parts of the hyperstriatal formations to the corticoid areas. Bremer *et al.* (1939) also obtained analogous movements from that region and showed that an excitable zone extended over the dorso-lateral surface. Sites which gave ipsilateral abduction were confined to the anterior half of the Wulst. Arrest reactions were predominantly obtained from medial sites in both the hyperstriatum and the "corticoid" components. They fell within a superficial anterio-posterior strip on the dorso-medial aspect of the hemisphere; similar arrest responses have also been elicited from neostriatal foci (Machne, 1952).

No comparable correlation between electrode site and evoked reactions was detected by stimulating the ventral hyperstriatum. The stimuli to that region elicited responses which were highly variable in both direction and frequency of occurrence. Expressed as a percentage of the total number of stimulation sites in the appropriate formation a comparison shows that 23% of those in the ventral hyperstriatum gave highly repeatable results. This value can be compared with 74% for the Wulst in general and 64% for the superficial "corticoid" zones.

As far as the necessary threshold levels are concerned it is interesting that

20

both the pure and contralateral downward movements had thresholds which were close to the overall modal value of 200 $\mu$A. Contralateral upward responses consistently showed rather higher thresholds whilst those involving pure contralateral movements were often associated with thresholds which lay below the modal value. Consequently Cohen and Pitts concluded that as the pure contralateral responses were particularly associated with superficial and "corticoid" areas these may be more electrically excitable areas. As the stimulating current was raised above the threshold level the observed latency of the evoked movement decreased and the magnitude of the response itself increased. Very often a rather more complex pattern of movements was also characteristic of the higher currents. At 600–800 $\mu$A there was an outburst of generalized struggling and occasionally some seizures. The fact that a localized destruction of the tissue at the electrode tip increased the observed thresholds very sharply, and could also change the direction of response, suggests that all the movements at low current values were autochthonous and did not reflect a widespread and diffuse spread of the current.

Such results clearly imply a fairly close functional integration between all the regions of the Wulst. This is especially so in view of the fact that the topographical organization seems to be continuous over the whole complex. If it is true that this representation does not extend into the ventral hyperstriatum it is yet further and non-anatomical evidence for the individuality of the Wulst. It also emphasizes the unfortunate historical accidents by which the structures above the hyperstriatic lamina share a common name.

## IX. OCULAR CONTROL

In a study which involved stimulating foci in the accessory hyperstriatum, dorsal hyperstriatum, ventral hyperstriatum, neostriatum and archistriatum in 20 crows and gallinaceous birds Showers (1964, 1965) demonstrated control of pupillary activity. As was emphasized in Chapter 10, Section V above the isthmi complex is closely involved in pupillary control. These telencephalic data therefore suggest a very intimate functional relationship with that part of the mesencephalon. Excitation of the neostriatum at the median side of the occipital pole gave both pupillary constriction and vertical movement of the eyes. Stimuli which were delivered to the lateral side of the occipital pole gave pupillary dilation and horizontal eye movements. Histological preparations showed that fibres from the isthmi complex run to the paleostriatum, neostriatum and hippocampus via the caudal branch of the septo-mesencephalic tract.

Phillips (1966) also reported relevant data. He observed that activation of both the neostriatum and mesencephalic isthmi complex followed archistriatal stimulation. Furthermore, Putkonen (1967) observed pupillary constriction when the archistriatum was stimulated, and also when stimuli were delivered to the paleostriatum, the neostriatum and the septo-mesencephalic tract. Pupillary dilation occurred during stimulation of the archistriatum, paleostriatum, neostriatum and lateral forebrain bundle. Both Putkonen (1967) and Showers and Lyons (1968) noted that the mydriasis from telencephalic stimulation was *insidious* and rather difficult to detect. Putkonen concluded that the long latencies indicated the existence of one or more poly-synaptic pathways comparable with those suggested in relation to the isthmi complex.

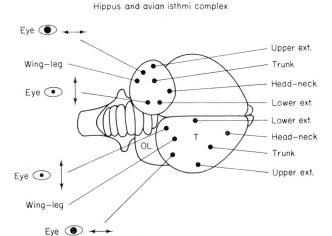

FIG. 130. The correlation between hippus, somatic movements and various points of focal stimulation. (Showers and Lyons, 1968).

Morphological connections between all the hemisphere sites might be represented by the short inter-connecting fibres which are found within the dorsal medullary lamina. A poly-synaptic relay series could involve fibres arising within the isthmi complex and then forming an arc projecting to the forebrain and then back to the isthmi region by way of the septo-mesencephalic tract and the lateral forebrain bundle. Direct connections between the nucleus isthmi and the Edinger–Westphal nucleus provide a potential mechanism for both pupillary dilation and constriction. Impulses from the forebrain, and more particularly the neostriatum, to the isthmi complex may be responsible for the mydriasis and miosis which both Showers (1964, 1965) and Putkonen (1967) reported.

## X. DRUG EFFECTS ON VISUAL RESPONSES

In connection with the visual phenomena that have been noted in the sections dealing with the ectostriatum and Wulst it is of some interest to outline the results of administering certain drugs. It should be emphasized that precise localization of the sites at which such drugs act has not been definitively worked out. Using autoradiographic methods I, initially together with Dr. Salzen, have found that labelled barbiturates have a widespread distribution in the central nervous system after intra-carotid injections.

Studies on the results of drug administration in terms of visual behaviour were first carried out by Dews (1955) and Blough (1956, 1957). The field was reviewed by Brown (1967). Blough used a conventional, automated, grain reinforced, key-pecking, operant conditioning technique. The pecking key was

## TABLE 68

*Differences between control and experimental data on visual discrimination performance at 30 and 120 min after drug administration.*

| Birds | Chlorpromazine | | | | LSD-25 | | | | Meperi-dine | | Caffeine | | Pento-barbital | |
|---|---|---|---|---|---|---|---|---|---|---|---|---|---|---|
| | 10 mg/ kgm | | 30 mg/ kgm | | 0·2 mg/ kgm | | 0·5 mg/ kgm | | 30 mg/ kgm | | 10 mg/ kgm | | 10 mg/ kgm | |
| | % | Σ | % | Σ | % | Σ | % | Σ | % | Σ | % | Σ | % | Σ |
| 1 | − | − | − | − | + | − | + | − | − | − | − | + | − | + |
| 2 | − | − | − | − | + | − | + | − | − | − | − | + | − | + |
| 3 | − | − | − | − | + | − | | | | | − | + | − | + |
| 4 | + | − | − | − | − | − | − | − | − | − | + | − | − | + |
| 5 | + | − | − | − | + | + | + | − | − | − | − | + | − | + |

From Brown (1967).

differentially illuminated so that one or other combination of the central and two lateral regions was contrasted with the remainder of the key. This remained dark. The birds were trained to peck at this residual non-illuminated area, irrespective of its actual position, if the central bar was illuminated. In contrast they had to peck at the illuminated semi-circle, irrespective of its side, if the central region was not lit up.

After training Blough investigated the consequences of administering ethanol (1·6 mg/kg body weight of 25% solution) and pentobarbitol. Both were given orally and had a biphasic effect. There was first of all a decrease and then an increase in the percentage of correct responses. In later reports (Blough, 1957) he extended these data to other drugs. The results are summarized in Table 68. The plus and minus signs indicate the direction of drug induced changes by comparison with the values which were obtained for control birds. The table does not indicate the actual quantity of the change but in all cases apart from caffeine this was very considerable. In view of public interest it is perhaps relevant to direct attention to the consistent facilitatory action of both dosages of LSD-25! It would seem that this produced an improvement in the accuracy of visual discrimination even though the actual level of response was depressed. It is also worth noting that when using comparable doses of chlorpromazine and pentobarbitol it was the latter drug which was the most effective disruptor of behaviour. Clearly, as was implied above, such effects may reflect direct action on the ascending visual pathway itself, action on ectostriate or hyperstriate, particularly accessory hyperstriate loci; or, thirdly, diffuse and gross neural suppression or enhancement. In the particular experimental situation these last would only find expression in visual behaviour as the other sensory modalities are not being monitored.

## XI. THE EFFECTS OF GROSS FOREBRAIN DAMAGE ON MATING BEHAVIOUR

### A. Introduction

The foregoing sections implicate various components of the hemispheres in a variety of basic functions. Hyperstriatal ablation was also reported to abolish mating behaviour by Rogers (1922) but it was subsequently not clear whether this was specifically related to hyperstriatal damage (Beach, 1951). It was indicated in the discussion of diencephalic structures that a number of factors influence mating behaviour. These include auditory and visual stimuli associated with the presence of a mature mate and coupled with the influence of day-length on the reproductive organs; this suggests the involvement of a variety of brain areas. Beach therefore investigated the results of both unilateral and bilateral hemisphere damage on the overall courtship and mating behaviour.

The courtship of the pigeon was described by Whitman (1919) and a detailed analysis appears in Fabricius and Jansson (1963). The effects of castration were investigated by Carpenter (1933a,b). Collias (1944) demonstrated that following the administration of testosterone propionate the bowing and cooing components were restored and this exogenous steroid also induced pair formation in hypophysectomized preparations. Lofts et al. (1968) alluded to further unpublished results on the influence which exogenous steroids have on courtship, and Murton et al. (1969) concluded that a cyclic pattern was involved (see Chapter 11, Section VII B). Erpino (1969) gave a detailed description. He found that both sexual activity and displacement preening were very much reduced after castration but were restored by exogenous testosterone. He also concluded that inhibitory effects exerted by progesterone must occur at the level of the central nervous system as the development of the male ducts, which is elicited by androgens, is not influenced by it. If one assumes that progesterone is involved in inducing incubation behaviour (Lehrman, 1965) and also the nest defence reactions of parent birds (Vowles and Harwood, 1966) the reduction of male sexual behaviour by rising progesterone levels would not be unexpected. Diminished male sexual displays may also encourage incubatory behaviour on the part of the female. All these considerations taken together with the observed ovarian and testicular development which accompanies male courtship song in the budgerigar, and mere sight of a mate in doves, serve to emphasize the complexity of the autochthonous and environmental parameters which, when integrated together, result in courtship.

Beach summarized the components of pigeon courtship behaviour and then concentrated his attention on nine of them. Each can be shown by both individuals of a pair but they are predominantly associated with one sex. They are:

 (i) *Strutting*; in which the male walks back and forth or around the female;
 (ii) *Bowing*; shown by the male as he struts;
 (iii) *Cooing*; occurring with bowing;
 (iv) *Nibbling*; which refers to gentle pecks directed at either the eyes or the top of the head of the partner;
 (v) The *wing signal*; in which the bird holds its wings close to the body and jerks them vertically;
 (vi) *Billing*; in which one bird inserts its beak within that of the other and both bob their heads in unison;
 (vii) *Squatting*; which is typical of the female when she is ready for mating.
 (viii) *Mounting*; when the male steps on to the back of the squatting female and treads alternately with both feet;
 (ix) Copulation.

## B. Unilateral Forebrain Extirpation

The results of extirpating one of the telencephalic hemispheres in 4 males are summarized in Table 69. The copulatory response was eliminated in every case although one male resumed copulating after an injection of 3 mg of testosterone propionate, and another after 5 mg. Although preparations 1 and 3 received 20 and 10 mg of androgen respectively neither undertook any postoperative mating. The restoration of copulatory activity in preparations 4 and 5 appeared to be definitely related to the administered androgen. Both mated 24 hours after they had been injected, and, although they did so again a short time later, they subsequently failed to. There were no obvious or significant differences between the areas which were destroyed in the four preparations which might have explained these different responses to androgen therapy.

# TABLE 69

*The effects of unilateral hemisphere damage on the sexual behaviour of the pigeon* (Beach, 1951). *Frequencies are expressed as averages of the performance during the pre-operative and post-operative periods.*

| Item | Condition of male | Preparation | | | |
|------|-------------------|:---:|:---:|:---:|:---:|
| | | 1 | 3 | 4 | 5 |
| Copulation frequency | Normal | 1·3 | 0·5 | 2·0 | 2·0 |
| | Operated | 0·0 | 0·0 | 0·0 | 0·0 |
| | Injected | 0·0 | 0·0 | 0·3 | 0·2 |
| Billing frequency | Normal | 3·1 | 1·7 | 17·4 | 9·4 |
| | Operated | 7·1 | 0·0 | 0·3 | 0·1 |
| | Injected | 8·6 | 0·2 | 7·0 | 11·2 |
| Cooing frequency | Normal | 4·9 | 2·0 | 9·4 | 4·2 |
| | Operated | 5·6 | 0·4 | 0·8 | 0·3 |
| | Injected | 15·0 | 1·9 | 4·0 | 4·6 |
| Bowing frequency | Normal | 1·9 | 0·1 | 0·7 | 0·3 |
| | Operated | 0·9 | 0·3 | 0·0 | 0·0 |
| | Injected | 1·4 | 2·6 | 1·2 | 1·2 |
| Strutting frequency | Normal | 0·7 | 0·3 | 1·3 | 0·7 |
| | Operated | 6·5 | 0·4 | 0·7 | 0·2 |
| | Injected | 13·4 | 0·7 | 4·7 | 0·4 |
| Nibbling frequency | Normal | 0·0 | 0·8 | 1·7 | 1·5 |
| | Operated | 17·2 | 2·5 | 0·7 | 0·2 |
| | Injected | 6·4 | 2·7 | 0·2 | 3·8 |
| Wing signal frequency | Normal | 0·0 | 1·6 | 1·3 | 1·9 |
| | Operated | 8·8 | 0·4 | 0·0 | 0·0 |
| | Injected | 6·5 | 5·3 | 0·1 | 5·0 |
| Body weight in grams | Before operation | 385 | 340 | 435 | 390 |
| | After operation | 365 | 300 | 410 | 368 |
| | Before injection | 382 | 313 | 435 | 373 |
| | After injection | 372 | 294 | 435 | 375 |
| | Sacrifice | 382 | 300 | 477 | 390 |
| Mg. androgen injected | | 20 | 10 | 3 | 5 |

Both preparations that failed to copulate did, however, mount the female once or twice after the operation. In one case this occurred prior to the administration of any androgen. In the other it was consequent upon this. Bird number 1 differed from the other *hemi-decerebrates* in the frequency of billing during the post-operative period. The others exhibited this less frequently but in that specimen the frequency was increased. Cooing was diminished in three preparations and then subsequently enhanced by hormone therapy. It was again closely tied to this and disappeared 7–14 days after injections had ceased. The effects of both the operative and hormone treatment on the remaining components of courtship activity were inconclusive.

One particularly interesting result was that three of the four preparations occasionally displayed the squatting response which is typical of a receptive female. Although such behaviour does occur in males it had not been seen in these specimens prior to the operation. Very frequently courtship between a pair would proceed normally up to the point at which billing occurred under normal circumstances. Both individuals would then squat. In general billing, cooing, strutting and nibbling all persisted into the post-operative period whilst, in contrast, both bowing and wing-signalling were totally eliminated in two of the preparations and reduced in a third. Beach emphasized the significance of the fact that the various behavioural components either increased or, alternatively, re-appeared at different rates and in different strengths under the influence of androgen administration. One recalls the very specific effects of pre-optic lesions and androgen administration in Barfields studies (see Chapter 11, Section VII). Billing, cooing and strutting increased in all four preparations and both nibbling and wing-signalling in three. The frequency of copulation only rose in two cases but attempts at copulation not only increased in frequency in these two but also reappeared in a third. Furthermore such incomplete and therefore unsuccessful copulatory attempts occurred after fewer injections than the successful coital act.

## C. Bilateral Forebrain Damage

Twelve pairs of birds in which the male had had extirpation in certain areas of both hemispheres were easily divisible into two principal groups on the basis of their overt behaviour. Eight pairs copulated without any prior androgen administration. There was no significant change in the observed frequencies of any behavioural items and qualitative observations gave no indication of any post-operative changes. The four remaining preparations (8; 14; 16; 18) did not mate spontaneously. The actual extent of the individual lesions is surveyed in Table 70.

There are three categories of factor which may contribute to such a loss of mating behaviour after injury to the forebrain. Briefly these are:

(i) The removal of either all or part of the neural mechanisms which are necessary for such behaviour;

(ii) An indirect interference with the production of testicular hormone, presumably via diencephalic structures;

(iii) A general lowering of the overall vigour as the result of chronic weakness and poor health.

Taking the third of these factors first one can note that there was no consistent correlation between a loss of weight and the disappearance of sexual behaviour. At the end of the experimental period the eight preparations which mated had an average weight which was 3% below that of the pre-operative period. Those that failed to mate had an average loss of 2%. The fact that these latter preparations even failed to mate after the administration of exogenous androgens led Beach to also negate the second possibility. This only leaves the first which is, of course, more complex and with more potential sites of involvement than the other two.

Reference to Table 70 will show that in no case was there total ablation of one particular telencephalic formation. Neither was there one in which the damage was restricted to a particular cell mass. It was not possible, for this reason, to assign a specific role in the hierarchical organization of precopulatory behaviour to any single region. Clearly the simple invasion of Wulst or neostriatal regions does not preclude courtship. However, it is equally clear that those preparations which failed to mate had sustained more extensive lesions than those that mated in the absence of hormone therapy. Although a few preparations which did mate had also sustained such damage this was always restricted to one side. In no post-operative copulator was there a large, bilateral loss of the same structure. However, it is worth recalling that in the experiments of Akerman *et al.* (1962) those preparations in which only the septal area in front of the anterior commissure remained intact, but some paleostriatal material persisted, were seen to bow and coo (Section II B above). They therefore had some residual social behaviour but not full copulatory activity.

Although very large lesions could clearly eliminate behavioural components by destroying the neural control systems which are essential for their survival Beach did not think that this was likely in his specimens. In view of the reappearance of the components after hormone therapy his conclusion is understandable. In fact he concluded, and this conforms with the results of olfactory nerve transection together with all other studies on the hemispheres, that the disappearance of certain behavioural traits reflects a general reduction in the overall level of responsiveness to external stimuli. In some preparations the administration of exogenous hormones could compensate for these deficits. The results of hormone therapy are therefore an integrated consequence of

## TABLE 70

*A synopsis of the bilateral injuries to the telencephalic hemispheres in the males of 12 pairs of pigeons.*

| Pre-paration | Wulst | | Ventral hyperstriatum | | Neostriatum | | Notes on behaviour after operation |
|---|---|---|---|---|---|---|---|
| | Right | Left | Right | Left | Right | Left | |
| 6 | Intact | Intact | Intact | Slight invasion | Intact | Intact | Mated without treatment |
| 11 | Intact | Slight invasion | Slight invasion | Slight invasion | Slight invasion | Slight invasion | Mated without treatment |
| 15 | Intact | Slight invasion | Intact | Moderate invasion | Intact | Slight invasion | Mated without treatment |
| 17 | Intact | Intact | Intact | Large invasion | Slight invasion | Slight invasion | Mated without treatment |
| 19 | Intact | Slight invasion | Slight invasion | Large invasion | Slight invasion | Moderate invasion | Mated without treatment |
| 21 | Slight invasion | Slight invasion | Slight invasion | Slight invasion | Slight invasion | Intact | Mated without treatment |
| 22 | Intact | Moderate invasion | Intact | Large invasion | Intact | Moderate invasion | Mated without treatment |
| 23 | Intact | Slight invasion | Slight invasion | Moderate invasion | Intact | Intact | Mated without treatment |
| 8 | Loss complete | Loss nearly complete | Loss nearly complete | Loss nearly complete | Moderate invasion | Moderate invasion | Mated after 9 mg of androgen |
| 14 | Large invasion | Loss complete | Loss nearly complete | Loss nearly complete | Large invasion | Large invasion | Never mated. Received 20 mg of androgen |
| 16 | Loss nearly complete | Loss complete | Very large invasion | Loss complete | Large invasion | Large invasion | Never mated. Received 20 mg of androgen |
| 18 | Slight invasion | Moderate invasion | Large invasion | Large invasion | Very large invasion | Very large invasion | Never mated. Received 20 mg of androgen |

From Beach (1951).

the dosage of hormone used, the amount of forebrain damage and the preoperative vigour of the specimen. Clearly these results corroborate the data on breeding behaviour following hemisphere stimulation which were mentioned in connection with diencephalic evoked breeding behaviour in Chapter 11, Section VIII (Akerman, 1965a,b).

## References

Adamo, A. J. and Bennett, T. L. (1967). The effect of hyperstriatal lesions on head orientation to a sound stimulus in chickens. *Exptl. Neurol.* **19**, 166–175.

Adamo, A. J. and King, R. L. (1967). Evoked responses in the chicken telencephalon to auditory, visual and tactile stimuli. *Exptl. Neurol.* **17,** 498–504.

Akerman, B. (1965a). Behavioural effects of electrical stimulation in the forebrain of the pigeon. I. Reproductive behaviour. *Behaviour* **26,** 328–338.

Akerman, B. (1965b). Behavioural effects of electrical stimulation of the forebrain of the pigeon. II. Protective behaviour. *Behaviour* **26,** 339–350.

Akerman, B., Fabricius, E., Larsson, B. and Steen, L. (1962). Observations on pigeons with prethalamic radiolesions in nervous pathways from the telencephalon. *Acta Physiol. Scand.* **56,** 286–298.

Andrew, R. J. (1967). Intra-cranial self-stimulation in the chicken. *Nature, Lond.* **213,** 847–848.

Batteau, D. W. (1967). The role of the pinna in human sound localization. *Proc. Roy. Soc. B.* **168,** 158–180.

Beach, F. A. (1951). Effects of forebrain injury upon mating behaviour in male pigeons. *Behaviour* **4,** 36–59.

Bickel, A. (1898). Zur vergleichenden Physiologie des Grosshirns. *Pfluger's Arch.* **72,** 190–215.

Blough, D. S. (1956). Technique of studying the effect of drugs on discrimination in the pigeon. *Ann. N.Y. Acad. Sci.* **65,** 334–344.

Blough, D. S. (1957). Some effects of drugs on visual discrimination in the pigeon. *Ann. N.Y. Acad. Sci.* **66,** 733–739 (see also *J. Opt. Soc. Amer.* **47,** 827).

Bouillaud, J. (1830). Récherches experimentales sur les fonctions du cerveau en géneral, et sur les celles de sa portion antérieur en particulier. *J. Physiol. exp. Path.* **10,** 36–98.

Bremer, F., Dow, S. and Moruzzi, G. (1939). Physiological analysis of the general cortex in reptiles and birds. *J. Neurophysiol.* **2,** 473.

Brown, H. (1967). Behavioural studies of animal vision and drug action. *Internat. Rev. Neurobiol.* **10,** 277–322.

Carpenter, C. R. (1933a). Psychobiological studies of social behaviour in Aves. (1) The effect of complete and incomplete gonadectomy on the primary sexual behaviour of the pigeon. *J. comp. Psychol.* **16,** 25–57.

Carpenter, C. R. (1933b). Psychobiological studies of social behaviour in Aves. (2) The effect of complete and incomplete gonadectomy on secondary sexual activity, with histological studies. *J. comp. Psychol.* **16,** 59–98.

Cate, J., Ten (1936). Physiologie des Zentralnervensystems der Vögel. *Ergebn. d. Biol.* **13,** 93–173.

Cate, J., Ten (1965). The nervous system. *In* "Avian Physiology" (Sturkie, P. D., ed.), pp. 697–751. Baillière, Tindall and Cassell.

Cohen, D. H. (1967). The hyperstriatal region of the avian forebrain: A lesion study of possible functions including its role in cardiac and respiratory conditioning. *J. comp. Neurol.* **131,** 559–570.

Cohen, D. H. and Durkovic, R. G. (1966). Cardiac and respiratory conditioning, differentiation and extinction in the pigeon. *J. exp. Anal. Behav.* **9,** 681–688.

Cohen, D. H. and Pitts, L. H. (1967). The hyperstriatal region of the avian forebrain and autonomic responses to electrical stimulation. *J. comp. Neurol.* **131,** 323–336.

Collias, N. E. (1944). Aggressive behaviour among vertebrate animals. *Physiol. Zool.* **17,** 83–123.

Couty, M. (1882). Sur la zone motrice du cerveau des perroquets. *C.R. Soc. Biol. Paris,* **34,** 81–82.

Dews, P. B. (1955). Studies on behaviour. II. The effects of pentobarbitol etc. *J. Pharmacol. Exptl. Therapy* **115,** 380–389.

Donovan, B. T. (1966). The regulation of the secretion of follicular stimulating hormone. *In* "The Pituitary Gland" (Harris, G. W. and Donovan, B. T., eds), vol. 2, pp. 49–98.

Edinger, L., Wallenberg, A. and Holmes, G. (1903). Untersuchungen uber die vergleichende Anatomie des Gehirns. 5. Das Vorderhirn der Vögel. *Abh. Senkenberg. nat. Ges.* pp. 343–426.

Erpino, M. J. (1969). Hormonal control of courtship behaviour in the pigeon (*Columba livia*). *Anim. Behaviour* **17**, 401–405.

Erulkar, D. S. (1955). Tactile and auditory areas of the brain of the pigeon. An experimental study by means of evoked potentials. *J. comp. Neurol.* **103**, 421–458.

Ewald, J. R. (1895). Zur Physiologie des Labyrinths. IV. Die Beziehungen des Groshirns zum Tonuslabyrinth. *Pflugers Arch.* **60**, 492–508.

Fabricius, E. and Jansson, A. M. (1963). Laboratory observations on the reproductive behaviour of the pigeon (*Columba livia*) during the pre-incubation phase of the breeding cycle. *Anim. Behaviour* **11**, 534–547.

Fasula, G. (1889). Effetti scervallazioni parziali e totali negli uccelli in ordine alla visione. *Riv. sper. Freniati.* **15**, 229–265, 317–351.

Ferrier, D. (1879). "Die Funktionen des Gehirnes." Braunschweig.

Flourens, P. (1824). "Récherches experimentales sur les propriétés et les fonctions du système nerveux dans les animaux vertébrés", vol. 1 Paris (vol. 2, 1842).

Fraps, R. M. (1961). Ovulation in the domestic fowl. *In* "Control of ovulation", (C. A. Villee, ed.), Pergamon Press, pp. 133–162.

Gallerani, G. and Lussana, F. (1891). Sensibilité de l'écorce cerebrale a l'excitation chimique. *Arch. ital. Biol.* **15**, 396–403.

Gogan, P. (1963). Projections sensorielles au niveau du telencephale chez le pigeon sans anaesthesie géneral. *J. Physiol. Paris* **55**, 258–269.

Goodman, I. J. (1970). Approach and avoidance effects of central stimulation: an exploration of the pigeon fore, and mid-brain. *Psychon. Sci.* **19**, 39–40.

Goodman, I. J. and Brown, J. L. (1966). Stimulation of the positively and negatively reinforcing sites in the avian brain. *Life Sci.* **5**, 693–704.

Gusel'nikov, V. 1. and Drozennikov, V. A. (1959). Reflection of orienting and conditioned reflex activity in the potentials of the cerebral hemispheres in pigeons. *Pavlov. J. Higher nerve activity* **9**, 844–852.

Harman, A. L. and Phillips, R. E. (1967). Responses in the avian midbrain, thalamus and forebrain evoked by click stimuli. *Exptl. Neurol.* **18**, 276–286.

Harwood, D. and Vowles, D. M. (1966). Forebrain stimulation and feeding behaviour in the ring dove (*Streptopelia rissoria*). *J. comp. physiol. Psychol.* **62**, 388–396.

Herrick, C. J. (1933). The functions of the olfactory parts of the cerebral cortex. *Proc. Natl. Acad. Sci. U.S.* **19**, 7–14.

Hinde, R. A. (1966). "Animal Behaviour; a Synthesis of Ethology and Comparative Psychology, pp. 534. London, McGraw-Hill.

Hodos, W. and Karten, H. J. (1970). Visual intensity and pattern discrimination deficits after lesions of the ectostriatum. *J. comp. Neurol.* **140**, 53–68.

Holst, E., von and Saint Paul, U., von (1963). On the functional organization of drives. *Anim. Behaviour* **11**, 1–20.

Huber, G. C. and Crosby, E. C. (1929). The nuclei and fibre paths of the avian diencephalon, with consideration of telencephalic and certain mesencephalic connections. *J. comp. Neurol.* **48**, 1–225.

Jastrowicz, A. (1876). Uber die Bedeutung des Grosshirns fur die sinneswahrnehmung. *Arch. Psychiatr.* **6**, 612–618.

Jolyet, F. (1902). Presentation d'un pigeon décerebré depuis cinq mois. *C.R. Soc. Biol. Paris* **44,** 878.

Juhasz, L. P. and Tienhoven, A., van (1964). Effect of electrical stimulation of the telencephalon on ovulation and oviposition in the hen. *Amer. J. Physiol.* **207,** 286–290.

Kalischer, O. (1900). Uber Grosshirnexstirpation bei Papageien. *Sitzber. prevs. Akad. Wiss. Physik-math*, pp. 722–726.

Kalischer, O. (1901). Weitere Mittheilung zur Grosshirn-localisation bei den Vogeln. *Sitzber. preus. Akad. Wiss*, pp. 428–439.

Kalischer, O. (1905). Das Grosshirn der Papageien in anatomischer und physiologischer Beziehung. *Abhandl. preus. Akad. Wiss.* **IV,** 1–105.

Karamyan, A. I. (1956). Evolution of the function of the cerebellum and cerebral hemispheres, pp. 160. *Gors. izd. med. lit Leningrad Trans.* Israel programme, 1962.

Karten, H. J. (1968). The ascending auditory pathway in the pigeon (*Columba livia*). II. Telencephalic projections of the nucleus ovoidalis thalami. *Brain Res.* **11,** 134–153.

Karten, H. J. and Hodos, W. (1970). Telencephalic projections of the nucleus rotundus in the pigeon (*Columba livia*). *J. Comp. Neurol.* **140,** 35–52.

Kluver, H. (1965). Neurology of the normal and abnormal perception. *In* "Psychopathology of Perception" (Hoch, P., and Zahn, J., eds), pp. 1–40. Grune and Stratton, New York.

Layman, D. H. (1936). The avian visual system. I. Cerebral functions of the domestic fowl in pattern vision. *Comp. Psychol. Monogr.* **12,** 58p.

Lehrman, D. S. (1965). Interaction between internal and external environments in the regulation of the reproductive cycle of the ring dove (*Streptopelia rissoria*). *In* "Sex and Behaviour" (Beach, F. A., ed.), pp. 355–380. Wiley, New York.

Levinsohn, G. (1904). Uber Lidreflexe. *Graefes Arch. Ophthalmol.* **59,** 381–423.

Lofts, B., Murton, R. K. and Thearle, R. J. P. (1968). The effects of 22, 25, diazocholesterol dihydrochloride on the pigeon testis and on reproductive behaviour. *J. Reprod. Fert.* **15,** 145–148.

Lussana, P. and Lemoigne, C. (1871). "Fisiologia dei centri nervosi encefalici." Padua.

Machne, X. (1952). La funzione inhibitrice del neostriato negli ocelli. *Arch. Sci. Biol.* **36,** 1–9.

Martin, E. G. and Rich, W. H. (1918). The activities of decerebrate and decerebellate chicks. *Amer. J. Physiol.* **46,** 396–411.

MacPhail, E. M. (1967). Positive and negative reinforcement from intracranial stimulation in pigeons. *Nature, Lond.* **213,** 947–948.

McKendrick, J. G. (1873). Observations and experiments on the corpora striata and cerebral hemispheres of pigeons. *Proc. Roy. Soc. Edinburgh* **8,** 47.

Moeli, C. (1879). Versuche an der Grosshirnrinde des Kaninchens. *Virchow's Arch.* **76,** 475–484.

Munk, H. (1883). Uber die zentralen Organe fur das Sehen und das Horen bei den Wirbeltieren. *Stzber. preuss. Akad. Wiss. Physik-Math.* pp. 793–827.

Munzer, E. and Wiener, H. (1898). Beitrage zur Anatomie und Physiologie des Zentralnervensystems der Taube. *Monatschr. Psychiatr.* **3,** 379–406.

Murton, R. K., Thearle, R. J. P. and Lofts, B. (1969). The endocrine basis of breeding behaviour in the feral pigeon (*Columba livia*). I. Effects of exogenous hormones on the pre-incubation behaviour of intact males. *Anim. Behaviour* **17,** 286–306.

Musehold, A. (1878). "Experimentelle Untersuchungen uber das Sehcentrum bei Tauben." Dissertation, Berlin (cited by Ten Cate, 1936).

Noll, A. (1915). Uber das sehvermogen und das Pupillienspiel grosshirnloser Tauben. *Arch. f. Physiol.* pp. 350–372.

Onimus, M. (1871). Récherches experimentales sur les phénomènes consecutifs a l'ablation du cerveau et sur les mouvements de rotation. *J. Anat. Physiol.* **7,** 633.

Opel, H. (1963a). Premature oviposition following operative interference with the brain of the chick. *Endocrinology* **74,** 193–200.

Opel, H. (1963b). Delay in ovulation in the hen following stimulation of the preoptic brain. *Proc. Soc. Exptl. Biol. Med.* **113,** 488–492.

Opel, H. (1967). Effects of brain piqure on the first and second ovipositions of the hens' 2 egg sequence. *Proc. Soc. Exptl. Biol. Med.* **125,** 627–630.

Paulson, G. W. (1965). Maturation of the evoked response to light in the duckling. *Exptl. Neurol.* **11,** 324–333.

Phillips, R. E. (1964). "Wildness" in the mallard duck. Effects of brain lesions and stimulation on escape behaviour and reproduction. *J. comp. Neurol.* **122,** 139–155.

Phillips, R. E. (1966). Evoked potential study of the connections of the avian archistriatum and neostriatum. *J. comp. Neurol.* **127,** 89–100.

Phillips, R. E. and Tienhoven, A., van (1960). Endocrine factors involved in the failure of pintail ducks, *Anas acuta,* to reproduce in captivity. *J. Endocrinol.* **21,** 253–261.

Powell, T. P. S. and Cowan, W. M. (1962). Centrifugal fibres to the retina of the pigeon. *Nature, Lond.* pp. 194–487.

Putkonen, P. T. S. (1967). Electrical stimulation of the avian brain. *Ann. Acad. Sci. Fennica, A. Medical.* **130,** 95.

Ralph, C. H. (1959). Some effects of hypothalamic lesions on gonadotropin release in the hen. *Anat. Rec.* **134,** 411–427.

Ralph, C. H. and Fraps, R. M. (1959). Long term effects of diencephalic lesions on the ovary of the hen. *Amer. J. Physiol.* **197,** 1279–1283.

Renzi, P. (1863–1864). Saggio di fisiologia sperimentale sui centri nervosi della vita psic. *Schmidts. Jahrb.* **123,** 151–161.

Revzin, A. M. and Karten, H. (1967). Rostral projections of the optic tectum and nucleus rotundus in pigeons. *Brain Res.* **3,** 264–275.

Rogers, F. T. (1922). Studies on the brain stem. VI. An experimental study of the corpus striatum of the pigeon as related to various instinctive types of behaviour. *J. comp. Neurol.* **35,** 21–60.

Rose, M. (1914). Uber die cytoarchitektonische Gliederung des Vorderhirns der Vogel. *J. Physiol. Neurol. Leipzig.* **21,** 278–352.

Rosenthal, J. (1868). Uber Bewegungen nach Abtragung der Grosshirn hemispharen. *Zbl. med. Wiss.* **6,** 739–740.

Schrader, M. E. G. (1889). Zur Physiologie des Vögelgehirns. *Pfluger's Arch.* **44,** 175–238.

Schrader, M. E. G. (1892). Uber die Stellung des Groshirns im Reflexmechanismus des zentralen Nervensystems der Wirbeltiere. *Arch. exper. Path.* **29,** 55–118.

Shaklee, A. O. (1921). The relative heights of the eating and drinking arcs in the pigeon's brain, and brain evaluation, *Amer. J. Physiol.* **55,** 65–83.

Shaklee, A. O. (1928a). Decerebrate pigeon fear-signs. *Proc. Soc. exp. Biol. Med.* **25,** 186–188.

Shaklee, A. O. (1928b). The anatomy of fear. *Proc. Soc. exp. Biol. Med.* **25,** 331–333.

Shima, I. (1964). Behavioural consequences of striatal spreading depression in pigeons. *J. comp. physiol. Psychol.* **57,** 37–41.

Showers, M. J. (1964). Stimulation study of the telencephalon and mesencephalon of *Gallus domesticus. Anat. Rec.* **148**, 335.

Showers, M. J. (1965). Telencephalon of the Numidae. *Anat. Rec.* **151**, 416.

Showers, M. J. C. and Lyons, P. (1968). The avian nucleus isthmi and its relation to hippus. *J. comp. Neurol.* **132**, 589–616.

Stefani, A., (1881). Alcuni fatti sperimentali in contribuzione alle fisiologia dell encefalo dei colombi. *Sep-Abdr Ferrara.*

Steiner, J. (1891). Sinnesphären und Bewegungen. *Pfluger's Arch.* **50**, 603–614.

Thorpe, W. H. (1958). "Learning and Instinct in Animals" (2nd ed. 1963). Methuen, London.

Trendelenburg, W. (1906). Weitere Untersuchungen uber die Bewegung der Vögel nach Durchschneidung hinterer Rückenmarkswurzelen. I and II. *Arch. Physiol.* suppl. pp. 231–245.

Tuge, H. (1957). A conception of the formation of conditioned connection, based upon the comparative studies of physiology and neuro-anatomy. *Congr. Nat. Sci. Med. Bucarest* pp. 61–70.

Tuge, H. and Shima, I. (1959). Defensive conditioned reflex after destruction of the forebrain in pigeons. *J. comp. Neurol.* **111**, 427–446.

Vasilevsky, M. M. (1966). Functional characteristics of individual cortical neurones of the cerebral hemisphere of the pigeon. *Bull. exp. Biol. Med.* **6**, 9–13.

Vedyaev, F. P. (1964). The role of striatal and thalamic structures in the central nervous system of birds in the control of respiration. *Fed. Proc.* **23**(5) part II.

Visser, J. A. (1932). "Optische reacties van duiven zonder groote herrenen." Dissertation. Leiden.

Visser, J. A. and Rademaker, G. G. J. (1934). Die optische reaktionen der Grosshirnloser Taube. *Arch. neerl. Physiol.* **19**, 482.

Voit, C. (1868). Beobachtungen nach Abtragung der Hemisphären des Grosshirns bei Tauben. *Sitzber. Akad. preuss. Wiss. Math-Physik.* **2**, 105–108.

Vowles, D. M. and Harwood, D. (1966). The effects of exogenous hormones on aggressive and defensive behaviour in the ring dove (*Streptopelia rissoria*). *J. Endocrinol.* **36**, 35–51.

Vulpian, A. (1866). "Leçons sur la physiologie génerale et comparée du système nerveux." Leçons 28 and 29. Paris.

Wenzel, B. M. and Salzmann, A. (1968). Olfactory bulb ablation or nerve section and behaviour of pigeons in non-olfactory learning. *Exptl. Neurol.* **22**, 472–479.

Whitman, C. O. (1919). The behaviour of pigeons. *In* "Posthumous Works of C. O. Whitman" (Carr, A. H., ed.), vol. 3, pp. 1–161. Publ. Carnegie Inst.

Zeigler, H. P. (1963). Effects of endbrain lesions upon visual discrimination learning in pigeons. *J. comp. Neurol.* **120**, 183–194.

# 15 Electro-encephalographic Activity

## I. NORMAL PATTERNS WHEN AWAKE OR SLEEPING

There is not a large number of EEG studies on Birds (see Klemm, 1969). The normal sleeping and waking patterns were, however, studied by Ookawa and Gotoh (1965) using *Gallus*, and various papers have dealt with certain drug induced, thermally induced and natural aberrations. Ookawa and Gotoh reported that during the waking state the patterns of telencephalic activity had an amplitude of 50 $\mu$V and frequencies of 17–24 cyc/sec during the unexcited state, or 30–60 cyc/sec in the excited state. During the last-named phases the fast waves predominated and those of 50–60 cyc/sec which could be recorded in the region of the parietal bone were particularly marked. During the resting phases the amplitudes were similar but the emphasis shifted towards slower rhythms of 17–24 cyc/sec. (Peters *et al*. 1958, 1965).

It is worth recalling that the anatomical substrates of sleep are no longer regarded as comprising a single unitary, or even diffuse centre, although the brain stem reticular systems are implicated to a greater or less degree (Akert, 1965). At the present time it is more usual to consider sleep as the integrated result of humoral factors, peripheral stimuli, ascending and descending pathways in the reticular formation and forebrain activity. The very special case of hibernation usually involves telencephalic silence. These inter-related systems are envisaged as forming multiple interacting circuits which are themselves influenced by circulating hormones. It is probable that in the mammalian brain a number of specific areas are particularly important. These include the intra-laminary thalamic nuclei, the caudate, the pre-optic area and some medullary foci in the neighbourhood of the solitary nucleus. In view of the coordinative function of similar areas in the avian brain they may have similar attributes. Sleep is a function of the inhibition of the reticular activating system. Akert envisaged the sleep and arousal systems as mirror images of each other. However, there is probably not complete symmetry. Furthermore sleep is *not* a state in which there is necessarily a generalized reduction in neuronal activity. There can, indeed, be an actual increase in the rate of discharge within certain neurones (see also Evarts, 1965).

Ookawa and Gotoh's (1965) study suggested that the onset of sleep in birds was accompanied by three principal stages. These are characterized as follows:

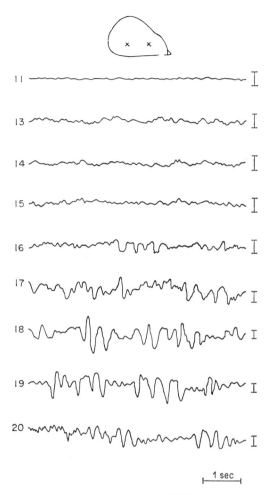

FIG. 131. Beginning and development of the EEG in the chick embryo; 11–20 days of incubation. Standardization: 50 $\mu$V. Time: 1 sec. In the upper part is shown the distribution of the electrodes upon the surface of the forebrain hemisphere: bipolar leads. 11: the small oscillations seen are produced by the apparatus. 13–15: "early" rhythm. 16: start of "late" rhythm. 17–20: further development of the "late" rhythm. (Garcia-Austt, 1954).

*Stage I*: when forced to perch the bird maintained this posture for a few minutes and then assumed an alert attitude with both eyes open and the pupils dilated. During this period there were bursts of waves with a frequency of 6–12 cyc/sec and an amplitude of 200–300 $\mu$V. These were interspersed with both fast waves and irregular low voltage potentials.

*Stage II*: most specimens sat with their legs held under their body. Both

21

eyes were closed and their head sometimes drooped like that of a drowsy man. Slow waves of 6–12 cyc/sec and an amplitude of 200–300 $\mu$V were predominant although there were also some large amplitude waves with a frequency of 3–4 cyc/sec.

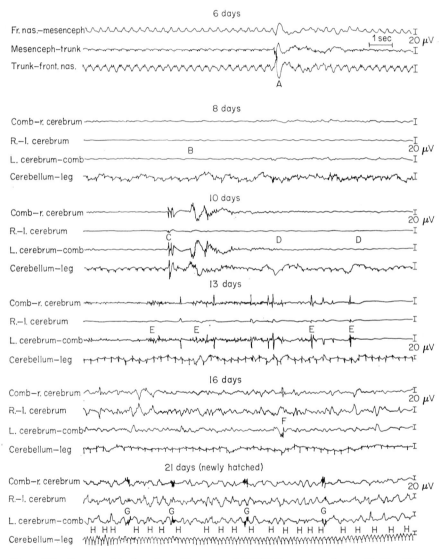

FIG. 132. Comparable data from Peters *et al.* (1956) for comparison with Fig. 131.

*Stage III*: this was marked by two distinct patterns which were designated (a) and (b). Most specimens exhibited a comparable pattern of behaviour, had their head tucked under one wing and both their eyes were closed. However, there was some degree of variation in the patterns which appeared during recording from corresponding fields on different animals. It was these which formed the basis for the two sub-divisions of the stage.

(a) in the first variant slow waves of 3–4 cyc/sec predominated and had a high amplitude which ranged from 200–300 $\mu$V.

(b) the second pattern also involved slow waves of large amplitude but these could be replaced quite suddenly by fast waves with low amplitudes of around 50 $\mu$V. These appeared at intervals among waves of the other type. No spontaneous movements were seen during the prevalence of this state. A comparable picture was obtained by Tradardi (1966) who studied the pigeon. Using electro-encephalograms, electro-myograms and electro-oculograms he observed that during the period of motionless posture there was a slow high voltage electro-encephalographic pattern. This was associated with some activity in the neck muscles. At the time that the head drooped somewhat there was a low voltage but fast pattern.

In a comparative study of sleep in birds and reptiles (Klein, 1963) the encephalogram of chickens and turtles also changed from low voltage fast activity to high voltage slow activity as sleep ensued. In the birds there were also episodes of the low voltage activity during sleep which were interpreted as paradoxical sleep. However, Klemm (1969) emphasized that paradoxical sleep is really only clear in young birds. As to the comparison between sleep and hibernation, no electro-encephalographic studies have been carried out on those species of the Caprimulgiformes which are the only birds known to enter such hypothermal states at regular intervals. However, from the experimental studies on induced hypothermia it is not impossible that there is a state of encephalographic silence, broken by some slow wave activity, and comparable with the electro-cortical silence of hibernating mammals.

## II. DEVELOPMENT OF ELECTRICAL ACTIVITY

The appearance of spontaneous electrical activity at different brain levels has been mentioned in association with cholinesterase activity in a number of the preceding chapters. The electro-encephalogram of development and maturation has also been the basis of a number of studies. These include Garcia-Austt (1954), Peters *et al.* (1956), Tuge *et al.* (1960), Key and Marley (1962), Spooner (1964) and Corner *et al.* (1966). Garcia-Austt combined the data from 64 embryos and concluded that the first detectable traces of electro-encephalographic activity appeared on the 13th day of incubation. These are shown in Fig. 131. The activity is irregular, has a mean frequency of 2·5–3 cyc/sec, and low voltages which range from 5–10 $\mu$V. In broad terms they take the form of short trains of waves which alternate with intervening silent periods that can last several seconds. They have some similarity to movement artifacts in adult animals but, as has been indicated elsewhere, there is a close correspondence between the appearance of this electrical activity and the occurrence of sharp rises in the level of cholinesterase activity.

During the following two days of development there was very little overt increase in either the frequency or the voltage, but the intervening silent periods became progressively shorter and finally disappeared on the 15th day. The next day the whole pattern changed. High voltage trains that vary between 20–100 $\mu$V and have frequencies of 3–4/sec are now superimposed on the earlier activity. At first both brief and sporadic they become progressively longer and more constant during the next 3 days until, at the 20th day, the trace becomes dysrhythmic and the voltage decreases.

During all these developmental phases there was a considerable amount of topographical variation. The rhythms of the two hemispheres were not synchronous and the voltages were different. The greatest degree of constancy was associated with the dorsal region and the most marked voltage was rostral. Exposure of the brain to dry air for 15–30 min evoked epileptoid seizure activity which is comparable to that occurring in adult mammals under similar conditions. Random high voltage peaks appear, become more frequent, achieve a maximum level and then decrease gradually. Such epileptoid phases could persist for as long as 24 hours.

Peters *et al.* (1956) also studied the pattern of development during the period from the 6th day of incubation until hatching. They concluded that there was a definite chronological sequence in the time of maturation of the various brain regions. This went in the order medulla, spinal cord, mesencephalon, diencephalon and finally telencephalon. Recordings from the hemispheres at stage 41 (see Chapter 4) showed periods of electrical silence interrupted by an occasional sequence of slow waves. These had a frequency of 4–7/sec, were of moderate amplitude and, like those of Garcia-Austt, were not synchronized in the two hemispheres. At stage 42 (about 16 days) this pattern was replaced by sustained waves with a dominant frequency of 8–12 /sec on which were superimposed waves of lower amplitude but higher frequencies. These varied between 8–12/sec. The results are shown in Fig. 132.

In the period which immediately precedes hatching, and during the process of hatching itself, this spontaneous activity changes. When the newly hatched chick was motionless the hemispheres had high amplitude waves at the rate of 1–4/sec and superimposed upon these there were other waves with a frequency of 16–22/sec. When the chick became active this pattern was replaced by one in which waves of high amplitude occurred at the rate of 30–40/sec. Should the activity increase even further, and the movements become very vigorous, then the frequency rises even higher and reached 60–70/sec. (See also Corner, *et al*, 1967, and Vos *et al*, 1967.)

Observations on the spontaneous electrical activity occurring on the hemisphere surface of chicks aged 1–14 days during various behavioural situations showed that the patterns which are characteristic of the most mature birds were already present in 1-day-olds (Tuge *et al.*, 1960). There was a wide range of frequencies and amplitudes for the slow wave component and a weaker, faster signal was continuously superimposed upon it. The duration of those periods which are devoid of slow waves increases significantly in this time. This suggests that there is a progressive increase in the attention span as the chick matures (Corner *et al.*, 1966). After hatching there may be a slight increase in the frequency of the fast low voltage component (Key and Marley, 1962) but this is over by the 5th post-hatching day. The two extremes, which comprise total absence of all waves and maximally large slow waves, were associated with behavioural attention and deep sleep respectively. The transition from the first to the second state was both gradual and discontinuous and accompanied decreasing signs of behavioural alertness. Stimulation always shifted the pattern towards desynchronization regardless of whether or not an observable motor response occurred. The behavioural patterns in the testing situation were limited to a few types, all of which were seen sooner or later in almost all the experiments. Nevertheless, there was some degree of individuality in the occurrence of a particular type of response.

# III. THE FACTORS INFLUENCING EEG ACTIVITY

## A. Patterns at Various Foci

The amplitude of the electrical activity differs according to the site at which the record is made. In general the amplitudes which were observed in the parietal region by Ookawa and Gotoh (1965) were smaller than those recorded in the frontal areas. Slow waves with frequencies of 6–12 cyc/sec or 3–4 cyc/sec, and amplitudes of 200–300 $\mu$V were recorded in leads from the medial and outer sides of the caudal part of the frontal bone. Records of the potentials which were recorded simultaneously from both hemispheres were asymmetrical but those from one hemisphere showed widespread synchrony. Relatively similar patterns of slow waves were only observed in some of the phases of sleep and drowsiness at probes located in that part of the parietal region which covers the cerebellum. High voltage slow waves appeared occasionally in one hemisphere independently of the other, and under anaesthesia spikes were induced bilaterally and asymmetrically. Pauses in the trains of spikes which followed deeper anaesthesia could be synchronous on the two sides.

## B. The Effects of Photic Stimuli

The on-off responses of both peripheral and tectal regions were described in Chapter 10. When the eyes of a bird were covered with black vinyl tape the preponderant rhythms were of moderately high amplitude and consisted of slow waves with frequencies of 6–12 cyc/sec. Fast waves of low amplitude could be superimposed upon these and they were sometimes interrupted by very slow waves of 3–4 cyc/sec and a high amplitude. These last resembled those occurring during the drowsy state. When placed into total darkness the fast waves were suddenly replaced by slow waves of high amplitude. The fast waves only returned when the specimens were again illuminated but the actual light threshold at which they reappeared was so low that its "dimness" could not be determined accurately.

## C. The Effects of Sound Stimuli

When acoustic stimuli were presented during the drowsy state they resulted in arousal. If they were presented when the bird was in the deeper phases of sleep it was not aroused at once, but the EEG pattern tended to be reduced in amplitude. An arousal pattern, which also included artefacts from muscle potentials, appeared in the parietal region as a response to a second stimulus. Similar responses were only observed elsewhere on the inner side of the caudo-frontal region. However, the very slow waves of 3–4 cyc/sec and high amplitude were clearly diminished. (See also Garcia Austt 1963).

## D. The Effects of Anaesthesia

During anaesthesia the "Spindle burst" or "barbiturate burst" was not observed, and nor was the fast wave. Both of these are characteristic features of the electro-encephalogram of mammals after barbiturate injections. By comparison with mammals the avian EEG is dramatically changed by the administration of nembutal to give a characteristic spike activity. With a dose of 5 mg/kg body weight of nembutal there were bursts of 6–12 cyc/sec and 200 $\mu$V amplitude. These were interspersed through the arousal pattern although both eyes were kept open. These and the effects of 10 and 30 mg/kg body weight are shown in Fig. 133. The responses to such graded doses varied. The moderately deep anaesthesia which was induced by 10–15 mg/kg consisted of a series of fairly continuous spiky waves which were again at frequencies of 6–12 cyc/sec but of low amplitude. During deep anaesthesia the patterns did not differ substantially from those observed under light anaesthesia. However, in very deep anaesthesia, which resulted from dosages of 30 mg/kg body weight, there was intermittent cessation of spikes.

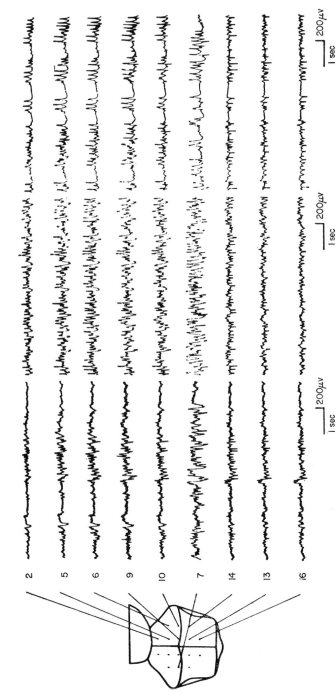

FIG. 133. The effects of nembutal anaesthesia. Recording was made 3 min after the beginning of intra-venous injection. Note the diphasic components on the laterocaudal part of the frontal bone (leads 7 and 10) and monophasic downward waves in the parietal region (leads 13, 14 and 16) (Ookawa and Gotoh, 1965).

During anaesthesia induced by the nasal application of ether the changes in the EEG were divisible into four principal phases on the basis of responses to comb pinching on the one hand, and the wave form on the other. When ether was first applied the animals became very excited. At this time there were fast waves of low amplitude. A few minutes later bursts of waves of 6–12 cyc/sec and 200 $\mu$V amplitude were interspersed between the fast waves of low amplitude. This comprises stage 1. During stage 2 the waves with a frequency of 6 cyc/sec become relatively continuous, overlap with others of 17–24 cyc/sec and are sometimes intermingled with yet others of high amplitude and a frequency of 3–4 cyc/sec. If the comb was pinched during this phase the bird awoke very suddenly. Gradually as the degree of anaesthesia increased this overt response to external stimuli was abolished although the EEG pattern still differed very little from that of stage 2. There were, however, downward spiky components in the recordings from both the outer side of the frontal region and from the parietal region. When stage 4 replaced this third stage the wave disappeared intermittently amongst the spikes.

## IV. THE EFFECTS OF SYMPATHOMIMETIC DRUGS

A very useful analysis of the effects of sympathomimetic amines was provided by Key and Marley (1962). Earlier work on smooth muscle had shown that there is a triple division of activity which depends upon the molecular structure. This division was used to relate the structure to the effects which they observed on behaviour and *electro-cortical* activity. The structural formulae of the amines used are presented in Table 71.

(1a) *Amines with hydroxyl groups in the 3, 4 position on the benzene ring and with another on the β carbon of the side chain*

In the alert 1–28-day-old chick the injection of 1 $\mu$g of adrenalin consistently induced sleep and cortical slow wave activity. Following such injections bursts of 3–6 cyc/sec and with an amplitude of 200–250 $\mu$V appeared either immediately or in a few seconds. The slow wave activity became continuous, then after a minute it was either punctuated with bursts of low voltage fast activity, or, alternatively, changed abruptly to fast activity which was accompanied by behavioural arousal. Intravenous injection of adrenalin to the sleeping chick gave an increase in the amplitude of the slow wave of the *electro-corticogram*.

The *electro-cortical* sleep-like state which was induced by adrenalin was accompanied by a crouching or squatting stance but the bird was easily aroused. Similar phenomena followed the administration of noradrenalin, isoprenaline or cobefrin. It is interesting that substances acting on alpha or beta receptors, or both, should have similar effects. However, by the end of the 4th week of life the birds responses change and this was termed the transitional period. There was faster activity in the normal records of the alert chick and even the slow wave activity of sleep tended to be faster. After injections with the above named amines this stage was characterized by bursts

## TABLE 71

*The structural formulae and group classification of sympathomimetic amines and allied substances used on the chick.*

| Compound | Group | Structure | | | | | | |
|---|---|---|---|---|---|---|---|---|
| Tuamine | 3 | CH₃(CH₂)₄ — CH—CH—NH, CH₃ H | | | | | | |
| Phenmetrazine | | H | H | H | H | O—CH₃ | CH₂ | CH₂ (ring) |
| Methyl phenidate | | H | H | H | H | COOCH₃ | CH₂ | CH₂—CH₂ |
| Amphetamine | | H | H | H | H | H | CH₃ | H |
| β-phenylethylamine | | H | H | H | H | H | H | H |
| Tyramine | 2b | H | OH | H | H | H | H | H |
| Pholedrine | | H | OH | H | H | H | CH₃ | CH₃ |
| Pipadrol | 2a | H | H | H | H | OH and C₆H₁₁ | CH₂ | CH₂—CH₂ |
| Ephedrine | | H | H | H | H | OH | CH₃ | CH₃ |
| Methoxamine | | CH₃O | H | H | CH₃O | OH | CH₃ | H |
| Phenylephrine | Ic | H | H | OH | H | OH | H | CH₃ |
| Oxedrine | | H | OH | H | H | OH | H | CH₃ |
| Metanephrine | | H | OH | CH₃O | H | OH | H | CH₃ |
| Normetanephrine | | H | OH | CH₃O | H | OH | H | H |
| 3-methoxy-4-hydroxy mandelic acid | | H | OH | CH₃O | H | OH | COOH | |
| Epinine | Ib | H | OH | OH | H | H | H | CH₃ |
| Dopamine | | H | OH | OH | H | H | H | H |
| Adrenaline | Ia | H | OH | OH | H | OH | H | CH₃ |
| Noradrenaline | | H | OH | OH | H | OH | H | H |
| Isoprenaline | | H | OH | OH | H | OH | H | CH(CH₃)₂ |
| Cobefrine | | H | OH | OH | H | OH | CH₃ | H |

From Key and Marley (1962).

of low voltage fast electrical activity alternating with bursts of 6–7 cyc/sec and an amplitude of 350 $\mu$V. This transitional phase lasted until about the 7th week after which the injection of such amines produced only behavioural and electro-encephalographic arousal.

21*

(1b) *Amines with hydroxyl groups in the 3, 4 position on the benzene ring but lacking any on the side-chain*

Injecting 10–30 μg of dopamine into the young chick also produced drowsiness and an electro-encephalographic slow wave. As with the amines of (1a) there was again a transitional period between the 4th and 7th week and this was followed by behavioural and electrical arousal in the adult.

However, in contrast, the chemically related molecule of epinine was without effect at any stage when given in the appropriate physiological dosage. That this was the case appeared surprising as it only differs from dopamine in having a $CH_2$ substituent on the terminal amino group. This might indeed have been expected to favour activity in view of the fact that N-methylation of nor-adrenalin to give adrenalin increases the potency of its effects on the central nervous system. However, whilst dopamine is a normal constituent of brains epinine is not.

(1c) *Amines with an hydroxyl group in either the 3 or 4 position on the benzene ring, together with another on the β carbon of the side-chain*

The *electro-cortical* changes which are produced by chemicals such as these were investigated using phenylephrine, oxedrine, the 3-methoxy analogues of both adrenalin and nor-adrenalin, namely metanephrine and nor-metanephrine, as well as 3-methoxy 4-hydroxy mandelic acid. With the first of these, phenylephrine, the same three stages of response with reference to maturation were observed as with groups (1a) and (1b) above. However, there was a slight difference in that the mixed slow and fast activity of the transitional stage were only detectable in the alert rather than the sleeping animal. In the adult both behavioural and electrical arousal occurred, while slow wave activity and sleep appeared instantaneously after intra-venous injection in the previously alert 1–28-day-old chick. As with cobefrine the effects were long-lasting, ataxia often developed, and the irregular high voltage 3–4 cyc/sec activity might be replaced by regular 2/sec cycles and the birds were difficult to arouse even on handling. However, oxedrine, the para-isomer of phenylephrine, only had equivocal effects on both behaviour and electro-encephalographic activity.

Metanephrine and nor-metanephrine seemed to have relatively little pharmacological activity when tested on muscle. However, with large (200 μg) doses sleep was induced in the 1–28-day-old chick although, in the adult bird, doses as high as 300–400 μg of these metabolites of adrenalin and nor-adrenalin produced no observable effects.

(2) *Amines with no hydroxyl group on the aromatic nucleus but possibly possessing one on the β carbon atom of the ethylamine side-chain*

Of the substances within this group pipadrol was a special case as it has a benzene ring as well as a hydroxyl group on the β carbon of the side-chain, whilst the $\alpha$ carbon and the terminal amine group are linked through a ring structure (see Table 71). When injected it consistently produced both behavioural and electro-encephalographic alerting in young and old birds (5·0 mg). In contrast, ephedrine, when given intravenously in doses of 0·25 mg, produced no response in some birds but in two other specimens it induced drowsiness with slow wave activity of 2–3 cyc/sec and an amplitude of 250–300 $\mu$V. In a previously drowsy chick the 4–6 cyc/sec activity disappeared and was replaced by 2 cyc/sec. In general, however, substances of this class produced very little effect on the birds.

With both tyramine and pholedrine, which lack the hydroxyl group on the sidechain, there was no effect when using 1–28-day-old chicks but they induced arousal in adult birds. Although no satisfactory explanation was at hand for these results Key and Marley suggested that the maturation of the receptor sites for group 2 amines is slower than for other groups, although at least in mammals there are differing effects within the central and peripheral nervous systems.

### (3) *Phenylethylamines with no hydroxyl groups on either the aromatic nucleus or the side chain*

From the first day of life amphetamine, phenmetrazine and methyl phenidate all produced alerting. The behavioural responses to amphetamine were actually first described by Selle. There is an almost immediate erection of the feathers and increased motor activity, together with neck and wing extension. Following a period of "aimless" running, ataxia developed and the birds remained stationary but emitted a continuous high-pitched twittering which could continue for 24 hours. Similar results emerged from the use of both phenmetrazine and methyl phenidate although the symptoms were slightly less consistent. With maturation the whole pattern changed slightly because neither ataxia nor twittering occurred in the adult. Slow wave electro-encephalographic activity and sleep, which had been induced by the administration of cobefrine, was replaced within minutes of amphetamine injections. The converse was also true and Key and Marley concluded that the two types of substances acted on different receptors.

All these sympathomimetic amines do then, seem to fall into three classes. Those such as adrenalin produce sleep in the previously alert animal; those such as tyramine appear to have no effect on either the sleeping or alert animal; whilst, as might have been expected, those like amphetamine induce both behavioural and electro-encephalographic arousal. Corroborative data were also obtained by using nicotine which releases adrenal medullary hormones. There was more slow wave electrical activity after nicotine administration in the young bird and Key and Marley concluded that this reflected sleep which had been induced by the catecholamines that had been released from the adrenal gland. The less marked effect in the adult bird was interpreted as reflecting the gradual development of decreasing permeability to catecholamines with the functional appearance of the blood brain barrier.

Spooner and Winters (1966a, b, 1967) also demonstrated the direct effect of many amines on the brain using baby chicks with an immature blood-brain barrier. They too found that nor-adrenalin and adrenalin produced hypertension along with both encephalographic and behavioural signs of sleep. Similar encephalographic patterns were evoked by isoproterenol and carbachol, both of which lowered the blood pressure. 5-hydroxytryptamine consistently produced depression but pressure changes varied from decreases to increases. Alert patterns were produced by DOPA, dopamine and amphetamine although the pressures increased. The lack of a direct relationship between blood pressure and the electro-encephalogram suggested that the effects were direct and did not reflect vascular phenomena (Klemm, 1969).

# V. THE EFFECTS OF TEMPERATURE CHANGES
## A. The Effects of Hyperthermia

Studies on the effects of cooling on the electrical activity of the developing chicken brain showed that there were physiological components which had different degrees of resistance to extreme temperature changes (Peters et al., 1961; Peters et al., 1964). The results of induced hyperthermia were observed using a constant temperature of $44\pm1°C$ (Peters et al., 1964). Under these circumstances the body temperature of 1–21-day-old chicks rose to, and remained at, 45–46°C. During this hyperthermia the patterns of electrical activity in the fore-brain underwent an orderly series of changes that culminated in electrical silence. The actual pattern was somewhat different in newly hatched chicks and others that were a few days old.

In the newly hatched chick slow waves of large amplitude dominate the record as the temperature reaches 44·5°C and fast waves of small amplitude are inconspicuous. With sustained hyperthermia these simple slow waves are modified by the irregular appearance of three characteristic units, a slow spike, a dome and then a plateau with super-imposed low-amplitude fast waves. This slow spike and wave pattern may achieve a frequency of several times per second, persists for 1–10 min, and can be evoked by photic stimulation. As hyperthermia continues both spike and slow waves disappear and the forebrain lapses into silence.

In chicks that are 20 days old the body temperature of 44·5°C induces forebrain activity with a larger than normal percentage of 18–22/sec waves of small amplitude. During periods of activity the results of head waving and contractions of the nictitating membrane superimpose a number of additional artifacts on this. If the high body temperature is maintained the fast waves tend to disappear and well-marked sequences appear which consist of smooth slow waves which have a large amplitude. These are occasionally punctuated by spike-like discharges. At this point about 3% of the older chicks enter a physiological state which only lasts a few minutes and in which the telencephalon emits the slow spike and wave pattern. With continuing hyperthermia the slow waves of large amplitude decrease in frequency and periods of electrical silence occur. In some cases total silence is preceded by a 5–10 sec sustained discharge of very fast waves of small amplitude. In all cases the disappearance of the electro-encephalogram occurs some 2 to 3 min before the final disappearance of the electro-cardiogram.

The characteristic pattern also changes at the level of the optic lobes. At normal body temperatures all chicks show spontaneous fast waves of moderate to small amplitude. In the older chicks both frequency and amplitude are greater. Visual stimulation, using 10 flashes/sec, evokes a corresponding electrical wave at the optic tectum. During hyperthermia, when the forebrain shows mainly slow waves of large amplitude, the optic lobes exhibit photic driving. With prolonged high temperature these photic responses lose amplitude and finally disappear a few minutes prior to the cessation of forebrain activity.

The slow waves in the telencephalon, which have an age dependent amplitude, seem to depend for their existence upon afferent input from elsewhere in the brain. For example, both the onset and the cessation of photic stimulation evokes slow waves from an otherwise silent brain (see Burns, 1958). The appearance of low-voltage fast waves during the initial phases of hyperthermia was taken as indicating that the brain stem reticular formation was functional and that excessive heat was the arousal stimulus. The absence of such low-voltage fast waves in the newly hatched chicks, even when they were exposed to hyperthermia, was thought to reflect the relative immaturity of the reticular formation.

## B. The Effects of Hypothermia

Following their earlier studies on the electrical activity of avian brain, eye and muscles, Peters *et al.* (1961) investigated the effects of lowered temperature. Using New Hampshire chick embryos, stainless steel needles were inserted to a depth of 2 mm and spanning the left eyeball, right hemisphere and the belly portion of the shank muscles. A Grass model III D, 4-channel electro-encephalograph was then used to monitor all three sites at once. The actual contemporary body temperature was measured by 18-gauge thermistor needles in the body cavity. During a roughly 3-hour period the body temperature was gradually lowered from the normal level of 41°C and 38°C for

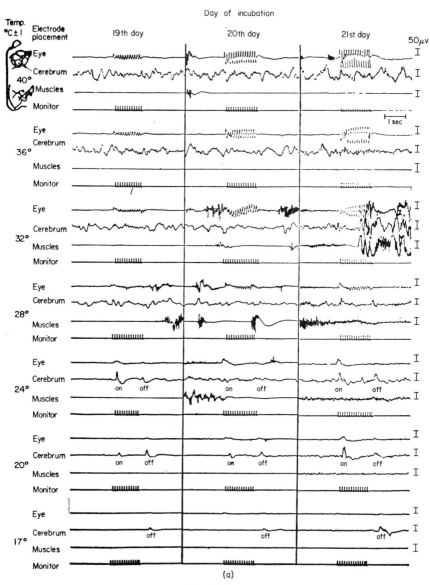

FIG. 134.

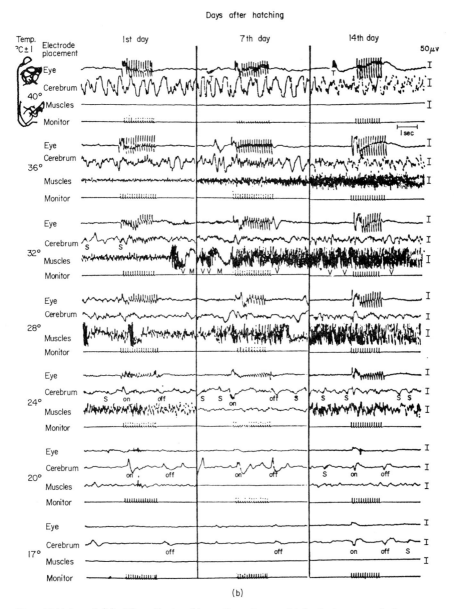

FIG. 134(a) and (b). The effects of hypothermia on chick electro-encephalograms. (Peters *et al.*, 1961).

hatched chicks and embryos respectively, to about 17°C. This was accomplished by placing ice in an Isolette baby incubator, and a plastic bag containing ice on the trunk, or around the egg shell, of the chick. The electrical recordings were made continuously during this 3-hour period of hypothermia and responses to incident light were tested, whenever the body temperature had dropped one degree, by using 10 flashes of light/sec for 1 or 2 sec.

At body temperatures which can be regarded as normal for a particular stage (Sturkie, 1965) they had found that the electro-encephalographic record is broadly similar for chicks between the 19th day of incubation and 2 weeks after hatching. The dominant frequency of 2–8 waves/sec with a superimposed series of waves of smaller amplitude and a frequency of 18–24/sec does, however, gradually give way to a pattern in which the 18–24/sec waves are more prominent. During hypothermia the changes in this observed sequence vary with the age of the chick. At the 19th day stage of incubation a gradual lowering of body temperature is accompanied by a comparable diminution of electrical activity. As 20°C is approached, isolated slow waves appear with a frequency of 10/min, and at 17°C there is electrical silence in the hemispheres.

The gradual drop in temperature in older chicks is accompanied by a comparable decrease in the amplitude of the electro-encephalographic record followed by hemisphere silence (Fig. 134). There are, however, a few exceptions to this generalization. From normal values to about 32°C the predominant feature is a decrease in the amplitude of the 2–8/sec waves but all types of waves eventually exhibit a similar drop. At temperatures lower than 32°C there are brief sequences of waves with a frequency of 30–40/sec in which the components are of very small amplitude. Peters *et al.* suggested that these might be associated with bouts of shivering. Uncommon in the 19-day-old embryos, they were abundant in the older chicks, which showed overt muscular symptoms at the lower temperatures, and could occur superimposed upon the on-off responses to visual stimulation. Actually, the electrical recordings showed that in the developing chick the first definite muscular response to cold was not shivering *per se*, but a periodic tensing of the muscles. With the greater maturity after hatching a lowered body temperature evokes both shivering and this muscular tensing. As such a phenomenon had also been observed prior to shivering in adult pigeons (Steen and Enger, 1957), Peters *et al.* (1961) concluded that it is the initial heat producing process.

## VI. SPREADING DEPRESSION

### A. Introduction

Spreading cortical depression has been quite extensively investigated in mammals. It was first described by Leao (1944, 1947) who discovered that strong repetitive stimulation of any point in the mammalian cortex caused the slow spread of a wave of complete electrical inactivity. Such waves radiate outwards in an ever-widening circle from the original focus of stimulation at a velocity of 0·5 mm/sec. The activity of all cortical neurons in their path is abolished for about 1 min. Along with this wave of transient inactivity there travels a wave of surface negativity which has a peak magnitude of 10 mV. Such depression waves can also be initiated by surface negative polarization, by the topical application of KCL and veratrine, or by brief cortical distortions. In contrast they do not result from the application of eserine, acetylcholine or histamine.

After the initial discovery of this phenomenon many attempts were made to provide an explanation for it and a number of additional characteristics have been established. It has occasionally been attributed to the morbid characteristics of unhealthy brains, indeed Marshall *et al.* (1950), and Marshall and Essig (1951) suggested that it was the result of cortical dehydration. This was, however, subsequently thought to be inherently improbable as it can be demonstrated in any neurologically isolated mammalian cortex. Its somewhat elusive nature in some preparations may result from natural variations in electrical activity.

Originally Leao claimed that the spread of the wave was accompanied by transient vasodilation. Subsequently, Harreveld and Stamm (1952) suggested that, on the contrary, the first phase of the vascular response was a period of constriction and the observed vasodilation was possibly a response to this. Indeed, they actually favoured the possibility that the depression was itself the result of the vaso-constriction.

Grafstein (1956) used cuts of various histologically controlled depths to demonstrate that there seemed to be no particular cytoarchitectonic layer whose integrity was essential for the further spread of the waves. Records which were obtained from electrodes that had been placed at varying depths within the cortex showed that violent unit activity, which is reminiscent of an injury discharge, accompanies the leading edge of the circle of electrical depression. The magnitude of the potential charges suggests that exceedingly large numbers of neurons are involved. Grafstein suggested that the factors which are essential to spreading depression are: (1) an intense local activity of a large number of cortical neurons. This leads to (2) the leakage of $K^+$ from their fine terminal processes out into the interstitial spaces in quantities which are sufficient to depolarize the adjacent neurons so that these are in turn (3) forced into an intense local activity which leads to further local leakage of $K^+$ and so to the excitation of the next peripheral band of neurons in the path of the centrifugally moving wave.

Despite the controversial nature of any explanations Burns (1958) concluded that the effects are at least analogous to the pathological changes which occur in concussion and, as mentioned above, an easy method of inducing it is by brief mechanical deformation of the brain surface. Although incisions rarely produce it, a pin dropped from a few inches is almost always effective. Burns regarded as at least suggestive the diminution or total disappearance of electro-encephalographic potentials during concussive states (Meyer and Denny-Brown, 1955), and following electro-convulsive therapy (Kalinowsky and Hock (1952). Although Denny-Brown (1945) thought that the fundamental neural factor in concussion was the formation of intra-cellular microvacuoles, which are similar to those which are induced by ultra-sound in cases of extreme traumatic concussion, the entire cortex becomes negative, relative to the white matter, by 3–9 mV. The EEG is also reduced for some 80 sec. Differences between such concussive effects and spreading depression may lie in the manner of excitation. When spreading depression is induced mechanically it is the result of local focal deformation, whilst concussion results from a widespread mechanical disturbance which is transmitted almost simultaneously to the whole brain.

## B. Spreading Depression in the Pigeon Brain

The possibility of evoking spreading depression in the avian brain was actually mentioned by Leao himself (1944). A detailed study of the subject was carried out by Bureš *et al.* (1960). They concluded that it was indeed possible to evoke a wave of depression, which is closely comparable with that observed in the mammalian neocortex, over the entire accessible dorsal surface of the pigeon hemispheres. In their account they included points within the accessory hyperstriatum and the dorso-lateral corticoid regions. They also demonstrated that the wave could reach points situated at 5 mm below the brain surface and could attain amplitudes of 22 mV.

Using pigeons under dial anaesthesia Shima *et al.* (1963) investigated the limits of such waves of depression within a single para-sagittal plane. The dosage of dial was 40 mg/kg of body weight. Data were obtained from 37 hemispheres and 21 individual birds. The total number of waves recorded was 214 and these resulted from KCL stimulation at 108 sites. In contrast to the situation in mammals the effects were most easily evoked at sites 1·5–2·5 mm below the surface, not by topical surface application. Whatever the actual site the waves in all nuclear layers were approximately similar in amplitude, duration and shape, despite the differences of cell size and arrangement. In most cases positive results were, however, evoked by stimulation of the accessory hyperstriatum, although eight waves were produced by mesencephalic stimulation and a further 31 from 12 sites in the paleostriatum primitivum and augmentatum. At these latter foci there is usually no slow potential change but, occasionally, 2 mV positive or negative deflections were recorded simultaneously with the spreading waves at the more superficial electrodes.

## TABLE 72

*The percentage of transition of slow potential waves within and between different structures in the pigeon forebrain.*

| Near point | Distant point | Occurrence of positive responses in the first SD wave % | Occurrence calculated from average values for each electrode % | Occurrence calculated from total number of SD waves % | Percentage of points entered by at least one SD wave % |
|---|---|---|---|---|---|
| Wulst | Ventral hyperstriatum | 100 (n = 3) | 100 | 100 (n = 8) | 100 |
| Ventral hyperstriatum | Neostriatum | 54·5 (n = 11) | 66·9 | 59 (n = 39) | 90·9 |
| Accessory hyperstriatum | Accessory hyperstriatum | 100 (n = 4) | 100 | 100 (n = 9) | 100 |
| Ventral hyperstriatum | Ventral hyperstriatum | 88·2 (n = 17) | 95·1 | 96·2 (n = 53) | 100 |
| Neostriatum | Neostriatum | 87·5 (n = 8) | 87·5 | 87·5 (n = 16) | 87·5 |
| Neostriatum | Paleostriatum | 0 (n = 6) | 0 | 0 (n = 14) | 0 |
| Hyperstriatum | Paleostriatum | 0 (n = 9) | 0 | 0 (n = 28) | 0 |

From Shima, Fifkova and Bureš (1963).

The results, which are summarized in Tables 72 and 73, suggest that the spreading depression evoked by application of KCL to hyperstriatal foci fails to reach the paleostriatum. It can, however, enter the neostriatum fairly easily. In some cases the first and third waves reached this level whilst the second failed to. Shima *et al.* suggested that this was indicative of a rather long refractory period within the connections between hyperstriatal and neostriatal structures. Alternatively the long refractory period may be within the neostriatum itself. It certainly seems probable that the various laminae have a more influential effect than do differences in cytoarchitecture. This would mean that the dorsal medullary lamina blocks the transition from the neostriatum to the paleostriatum; the hyperstriatic lamina from the ventral hyperstriatum to the neostriatum. Shima *et al.* did not indicate the effects of the superior or supreme frontal laminae. The former, at least, would be expected to exert a similar affect. They did, however, show that the slow spreading rate from the hyperstriatum to the neostriatum reflects circuitous connections between them and not slow spread within the neostriatum by simultaneous measurements of the rate of spread across the nearest boundary between the two, and round more circuitous routes.

## TABLE 73

*The rate of spread of slow potential waves in the pigeon hemispheres.*

| Structure | Spreading rate in mm/minute |
|---|---|
| Hyperstriatal regions | $3.4 \pm 0.4$ (n = 43) |
| Hyperstriatal regions—Neostriatum | $2.0 \pm 0.3$ (n = 21) |
| Neostriatum | $3.0 \pm 0.5$ (n = 11) |

From Shima *et al.* (1963).

Shima *et al.* (1963) concluded their work with a discussion of some behavioural considerations. They discovered that the spreading depression which can be evoked within the accessory hyperstriatum impaired Skinner-type conditioned reflexes which involve the alimentary canal (Shima, 1962, 1964). There was also a close correlation between the duration of this impairment and that of the electro-encephalographic evidence for spreading depression. From this they concluded that in pigeons hyperstriatal and neostriatal dysfunction parallels the behavioural results of functional decortication in rats.

This does, however, apparently conflict with the suggestion of Beritoff (1926), which was confirmed by Tuge and Shima (1959), that alimentary conditioning in pigeons is possible providing the paleostriatum and ectostriatum are intact. Using conditioned changes of cardiac and respiratory activity these authors demonstrated that even extensive ablation of the hyperstriatum and neostriatum had no effect on the ability to induce such reflexes. Shima *et al.* suggested that such apparent discrepancies may reflect the differing neuro-physiological mechanisms which are disrupted after functional ablation on the one hand, and surgical ablation on the other. The behavioural effects are usually observed to be more severe following functional ablation because the effects of spreading depression are detected immediately, and not masked or enhanced by post-operative compensatory processes which can improve the performance of a preparation lacking hyperstriatal components.

## VII. THE ENCEPHALOGRAM OF ENCEPHALOMALACIA

The administration of dietary régimes which are deficient in vitamin E induces certain well known syndromes in chicks. These include encephalomalacia, exudation diathesis, and muscular degeneration. When it became possible to differentiate the individual nutritional factors which induced these various conditions Sheff and Tureen (1962) investigated the electrical activity which is associated with the first. Selenium protects against exudation diathesis, and linoleic or arachidonic acid is important in the development of encephalomalacia. Selenium and anti-oxidants protect from muscle degeneration on a low sulphur diet. Nichol 108 cockerels were placed on such encephalomalacia inducing diets and then recordings were made from birds aged from 4 to 20 days. Following a 20-min recording session under standard conditions the effects of both photostimulation and pentylenetetrazol injections were investigated.

The birds showed bursts of high voltage activity whilst excessive slowing, to 3–4 cyc/sec for periods of 1–3 sec, appeared erratically. During the intervening period the traces resembled those of normal birds. There were low voltages and a preponderance of 5–7 cyc/sec. Like normal chicks they were sensitive to photo-stimulation and the tracings showed activation at a similar threshold dose of pentylenetetrazol. The optimum level for this in both normal and deprived birds was 70–80 mg/kg body weight. In the experimental birds there was, however, a very prolonged latent period prior to activation which was twice that observed in normal birds.

Histological examinations of comparable birds, but not the actual specimens, showed that encephalomalacia was reflected by lesions in the white fibre tracts of the cerebellum. The absence of any lesions in either the forebrain or the optic lobes renders the responses to light quite reasonable. However, there were lesions in some mesencephalic fibre tracts and Sheff and Tureen suggested that the prolonged latent period which followed the injection of pentylenetetrazol reflected the importance of mesencephalic structures in general, and the reticular activating system in particular, in the induction of convulsions.

## VIII. PARADOXICAL SLEEP

During behavioural sleep there occurs a state which is characterized by fast cortical activity that is reminiscent of the waking state. At the same time there is a complete disappearance of much muscle tone but rapid eye movements. Jouvet (1965) has discussed whether both classical sleep, with its slow waves, and paradoxical sleep are differing expressions of a single hypnogenic mechanism. Polygraphic records show that slow sleep is readily recognizable in birds (Klein *et al*, 1964). As stated above, it expresses itself in hens and pigeons as slow waves which are associated with immobility, eye closure and both respiratory and cardiac slowing. In pigeons, hens and chicks very short bursts of paradoxical sleep also occur at intervals. These last for 6–15 sec. They are characterized by the appearance of rapid hyperstriatal activity, rapid eye movements, a reduction in the nuchal electro-myogram, considerable bradycardia and also postural relaxation. This is expressed as wing drooping. These phases of rudimentary paradoxical sleep comprise some 0·15–0·2% of the total period of behavioural sleep. This is a low value which contrasts very markedly with the far higher values of 6–30% characterizing this phenomenon in mammals. The most comprehensive reviews of such phenomena are Jouvet (1965), Koella (1967) and Klemm (1969).

## IX. HYPNOTIC STATES

In 1646 Kircher described how a trance-like state of immobility could be induced in a chicken by seizing it and holding the head and trunk motionless in an unnatural position while a chalk line was drawn outwards from its beak. This was the so-called *experimentum mirabile*. The chalk line is actually quite unnecessary. Such states of hypnosis in animals are reviewed by Oswald (1962). The response in birds can be produced by overwhelming fear, monotonous stimulation and imposed immobility either alone or in combination with each other. The initial condition is immobility together with extreme alertness and increased muscular tension. This passes into a state of light sleep which can last for some minutes. Byrne (1942) came to the conclusion that it was homologous with the first stage of normal sleep. We have often seen it in canaries and Hinde (1966) records it in wild caught Paridae. If these are held in the hand they sometimes lie on the outstretched palm for a minute or longer with their eyes open but their bodies limp. They will fly at once if exposed to a sharp noise or thrown into the air. Armstrong (1947) and Ratner and Thompson (1960) ascribed the reaction to frustration or conflict. The relation of such responses to the freezing postures used when escaping from predators has not been investigated. Pavlov considered the initial phase as a cataleptic state of partial sleep that passed into general sleep with relaxation should the stimulation be both intense and long-lasting. Gilman *et al.* (1950) found that it gradually disappeared if hens were exposed to forcible bodily restraint and manipulation daily.

A number of electro-encephalographic studies have been made of the state. Silva *et al.* (1959) reported that sleep signs appear and the regular 6 cyc/sec waves are lost. The application of sensory stimuli provoked brief electrical and behavioural arousal signs. However, repetition of the stimulus rapidly led to habituation of first of all the overt, and then the electrical signs, of arousal. The typical signs of the hypnotic state were increases of voltage and reduction of frequency in the corticoid region, hemispheres generally, optic tectum and cerebellum.

# References

Akert, M. (1965). The anatomical substrate of sleep. *Progress Brain Res.* **18**, 9–19.

Armstrong, E. A. (1947). "Bird Display and Behaviour: an Introduction to the Study of Bird Psychology", pp. 431. Lindsay Drummond, London.

Beritoff, I. S. (1926). Uber die individuelle erworbene Tätigkeit des Zentralnervensystems bei Tauben. *Pfluger's Arch.* **213**, 370–406.

Burns, B. D. (1958). "The Mammalian Cerebral Cortex", pp. 119. Edward Arnold.

Bureš, J., Fifkova, E. and Maršala, J. (1960). Leao's spreading depression in pigeons. *J. comp. Neurol.* **114**, 1–10.

Byrne, J. G. (1942). "Studies on the physiology of the eye, still reaction, sleep, dreams, hibernation, repression, hypnosis, narcosis, coma, and allied reactions, 2nd reissue with supplements. H. K. Lewis, London.

Corner, M. A., Peters, J. J. and Rutgers van der Loeff, P. (1966). Electrical activity patterns in the cerebral hemisphere of the chick during maturation, correlated with behaviour in a test situation. *Brain Res.* **2**, 274–292.

Corner, M. A., Schadé, J. P., Sedlaček, J., Stoeckart, R. and Bot, A. P. C. (1967). Developmental patterns in the central nervous system of birds. I. Electrical activity in the cerebral hemisphere, optic lobe and cerebellum. *Progr. Brain. Res.* **26**, 145–192.

Denny-Brown, D. (1945). Cerebral concussion. *Physiol. Rev.* **25**, 296–325.

Evarts, E. V. (1965). Relation of cell size to effects of sleep in pyramidal tract neurones. *Progress Brain Res.* **18**, 81–89.

Garcia-Austt, E., Jnr. (1954). Development of electrical activity in the cerebral hemispheres of the chick embryo. *Proc. Soc. exp. Biol. Med.* **86**, 348–352.

Garcia-Austt, E., Jnr. (1963). Influence of states of awareness upon sensory evoked potentials. *Electroenc. clin. Neurophys.* suppl. **24**, 76–89.

Gilman, T. T., Marcuse, F. L. and Moore, A. V. (1950). Animal hypnosis: a study in the induction of tonic immobility in chickens. *J. comp. physiol. Psychol.* **43**, 99.

Grafstein, B. (1956). Mechanisms of spreading cortical depression. *J. Neurophysiol.* **19**, 154–171.

Harreveld, van A. and Stamm, J. S. (1952). Vascular concomitants of spreading cortical depression. *J. Neurophysiol.* **15**, 487–496.

Hinde, R. A. (1966). "Animal Behaviour: a Synthesis of Ethology and Comparative Psychology", pp. 534. McGraw-Hill, London.

Jouvet, M. (1965). Paradoxical sleep. *Progress Brain Res.* **18**, 20–62.

Kalinowsky, L. B. and Hock, P. H. (1952). "Shock Treatments, Psycho-surgery and other Somatic Treatments in Psychiatry", pp. 396. Grune and Stratton, New York.

Key, B. J. and Marley, E. (1962). The effect of sympathomimetic amines on behaviour and electro-cortical activity of the chicken. *Electroenc. clin. Neurophysiol.* **14**, 90–105.

Klein, M. (1963). "Étude polygraphique et phylogenetique des différent états de sommeil." Thesis, Lyon. Bosc. edit.

Klein, M., Michel, F. and Jouvet, M. (1964). Polygraph study of sleep in birds. *C.R. Soc. Biol. Paris* **158**, 99–103.

Klemm, W. R. (1969). "Animal Electro-encephalography", pp. 292. Academic Press, London and New York.

Koella, W. P. (1967). "Sleep." Thomas, Springfield, Illinois.

Leao, A. A. P. (1944). Spreading depression of activity in the cerebral cortex *J. Neurophys.* **7**, 359–390.

Leao, A. A. P. (1947). Further observations on the spreading depression of activity in the cerebral cortex. *J. Neurophys.* **10**, 409–414.

Marshall, W. H. and Essig, C. F. (1951). Relation of air exposure of the cortex to the spreading depression of Leao. *J. Neurophys.* **14**, 265–273.

Marshall, W. H., Hanna, C. and Barnard, G. (1950). Relation of dehydration of the brain to certain abnormal phenomena of the cortex. *Electroenc. clin. Neurophys.* **2**, 177–185.

Meyer, J. S. and Denny-Brown, D. (1955). Studies of cerebral circulation in brain injury. II. Cerebral concussion. *Electroenc. clin. Neurophys.* **7**, 529–544.

Ookawa, T. and Gotoh, J. (1965). Electro-encephalogram of the chicken recorded from the skull under various conditions. *J. comp. Neurol.* **124**, 1–14.

Oswald, I. (1962). "Sleeping and Waking; Physiology and Psychology", pp. 232. Elsevier, London.

Peters, J. J., Cusick, C. V. and Vonderahe, A. R., (1961). Electrical studies of hypothermic effects on the eye, cerebrum, and skeletal muscle of the developing chick. *J. Exp. Zool.* **148**, 31–40.

Peters, J. J., Vonderahe, A. R. and McDonough, J. J. (1964). Electrical changes in the brain and eye of developing chick during hyperthermia. *Amer. J. Physiol.* **207**, 260–264.

Peters, J. J., Vonderahe, A. R. and Powers, T. H. (1956). The functional chronology in the developing chick nervous system. *J. Exp. Zool.* **133**, 505–518.

Peters, J. J., Vonderahe, A. R. and Powers, T. H. (1958). Electrical studies of functional development of the eye and optic lobes in the chick embryo. *J. Exp. Zool.* **139**, 459–468.

Peters, J. J., Vonderahe, A. R. and Schmid, D. (1965). Onset of cerebral electrical activity associated with behavioural sleep and attention in the developing chick. *J. Exp. Zool.* **160,** 255–262.

Ratner, S. C. and Thompson, R. W. (1960). Immobility reactions (fear) of domestic fowl as a function of age and prior experience. *Anim. Behaviour.* **8,** 186–191.

Sheff, A. G. and Tureen, L. L. (1962). EEG studies of normal and encephalomalacia chicks. *Proc. Soc. Exp. Biol. Med.* **111,** 407–409.

Shima, I. (1962). Narušeni podmíněných reflexu striatovou šiřici se depressi u holubů. *Cs. Physiol.* **11,** 478.

Shima, I. (1964). Behavioural consequences of striatal spreading depression in pigeons. *J. comp. physiol. Psychol.* **57,** 37–41.

Shima, I. and Fifkova, E. (1963). Remote effects of striatal depression in pigeon brain. *Jap. J. Physiol.* **13,** 630–640.

Shima, I., Fifkova, E. and Bureš, J. (1963). Limits of the spreading depression in pigeon striatum. *J. comp. Neurol.* **121,** 485–492.

Silva, E. E., Estable, C. and Segundo, J. P. (1959). Further observations of animal hypnosis. *Arch. ital. Biol.* **97,** 167.

Spooner, C. E. (1964). "Observations on the use of the chick in the pharmacological investigation of the central nervous system. Ph.D. Thesis, California University. University microfilms 64–6387. Ann. Arbor, pp. 222.

Spooner, C. E. and Winters, W. D. (1966a). Distribution of monoamines and regional uptake of D. L. norepinephrin, 7-H³, and dopamine, 1-H³, in the avian brain. *Pharmacologist* **8,** 189.

Spooner, C. E. and Winters, W. D. (1966b). Intra-arterial blood pressure recording in the unrestrained chick during wakefulness and sleep. *Arch. Internat. Pharmacodyn.* **161,** 1–6.

Spooner, C. E. and Winters, W. D. (1967). The influence of centrally active amine induced blood pressure changes on the electro-encephalogram and behaviour. *Intern. J. Neuropharmacol.* **6,** 109–118.

Steen, J. and Enger, P. S. (1957). Muscular heat production in pigeons during exposure to cold. *Amer. J. Physiol.* **191,** 157–158.

Sturkie, P. D. (1965). "Avian Physiology", pp. 766. Baillière, Tindall and Cassell, London.

Tradardi, V. (1966). Sleep in the pigeon. *Arch. ital. Biol.* **104,** 516–521.

Tuge, H., Kanayame, Y. and Yue, C. H. (1960). Comparative studies on the development of the electro-encephalogram. *Jap. J. Physiol.* **10,** 211–220.

Tuge, H., and Shima, I. (1959). Defensive conditioned reflex after destruction of the forebrain in pigeons. *J. comp. Neurol.* **111,** 427–446.

Vos, J., Schadé, J. P. and van der Helm, H. H. (1967). Developmental patterns in the central nervous system of birds. II. Some biochemical parameters of embryonic and post-embryonic maturation. *Progr. Brain Res.* **26,** 193–213.

# 16 The Relative Development of Different Brain Regions

## I. HISTORICAL INTRODUCTION

In everyday life a classification on the basis of intelligence is widespread and frequently relates, for example, to the conduct and docility of animals. In the past a number of formulae have been adduced in attempts to express such behavioural characteristics in arithmetical terms. More particularly to express them in terms of the relative size or complexity of different brain regions. Many of these mathematical functions refer to differences between genera of mammals. Krompecher and Lipak (1966) summarized these in terms of *cerebralization*. However, there are many difficulties. A direct comparison of the total bulk of different brains is of little significance. Indeed it leads to absurd results since both the elephant and the whale have brains whose weight far exceeds that of man. As the result of this it was soon concluded that the weights of both the brain and the body should be taken into consideration. Cuvier introduced such considerations by which the degree of intelligence ($I$) is expressed in terms of brain weight ($E$ = encephalon) over body weight ($C$ = corpus) in grams.

$$I = \frac{E}{C}$$

A correction of this nature makes a drastic change in the relative prominence of the brains of whales and man. Subsequently Dubois (1897) drew attention to the fact that brain size is related to an exponential function of body weight and formulated it as:

$$E = KC^r$$

where $K$ is a cephalization index and $r$ is the relative exponent. He produced this formula in an empirical manner by selecting two pairs of animals on the assumption that their intelligence was equal and took the quotient of their brain weights

$$\frac{\text{brain weight of (a)}}{\text{brain weight of (b)}} = \frac{\text{body weight of (a)}}{\text{body weight of (b)}}r$$

From this be obtained a mean value of 0·56 for $r$. From these exponents he then calculated the index of cephalization:

$$K = \frac{E}{0\cdot56C}$$

The original concept has been modified in a number of ways by more recent workers who all tried to obtain the relative exponent and the cephalization index with greater accuracy. However, the method has deficiencies. Fat content alone can give widely varying values for body weight, but the greatest drawback is the essentially empirical nature of the original formula, and the cephalization index results in a number of anomalies when it is applied to mammalian brains. Prior to the work of Dubois the paper by Manouvrier (1885) had differentiated between two components of the brain. He suggested that a somatic part $(E_s)$ is proportional to the bulk of the body, its surface and its function. In contrast he postulated that a "mental" part relates to what he called the "mind" $(E_a)$. This can clearly be expressed as

$$E = E_s + E_a$$

and the somatic part is related to body weight,

$$E_s = K'C^r$$

The Dubois formula was therefore amended to

$$E_a + K' = KC^r$$

and from this

$$K = \frac{E_a + K'C^r}{C^r} = \frac{E_a}{C^r} + K'$$

Although Manouvrier's conclusions are more useful than those of Dubois they still have very considerable limitations. Lapicque (1944) drew attention to the fact that although

$$E = E_a + K'C^r$$

is the equation of a straight line the measured parameters actually give a curved distribution.

The most highly developed and more frequently used alternative method of attacking this problem is that which was developed by Portmann (1948) and Wirz (1950). It involves the so-called *intra-cerebral* indices and is a development of Kuenzi's work (1918). In the simplest case the overall mass of the brain stem centres is compared with that of *higher* integrating centres. The higher the values for the resulting quotient, that is to say the greater the relative bulk of the higher centres, the greater the degree of *cerebralization*. A modified and rather more effective result is obtained by using the so-called *transcendental* indices. Here the parts of the brain of the species being considered are compared with the size of the brain stem of another species from the same taxonomic group and of comparable body weight but in which their development is minimal. The brain stem value which the species in question would have, in relation to its body weight, if it belonged to the less developed species is called the basic number. Animals of similar size possess similar basic numbers. The individual indices are derived by dividing the brain part in question by the basic number. Thus:

$$\text{Complete brain index } (T_i) = \frac{\text{complete brain}}{\text{basic number}}$$

$$\text{Hemisphere index } (M_i) = \frac{\text{hemisphere}}{\text{basic number}}$$

The developmental rank of a series of species can then be indicated on the basis of $(T_i)$ which in mammals is then subjected to a correction by the analogous neopallial index. The bigger the neopallial index, and the smaller the brain stem index, in terms of $T_i\%$, the greater is the degree of "cerebralization". However, from a practical point of view actually isolating the brain stem is not easy and if one attempts to isolate telencephalic centres which are concerned with "mental" and "somatic" activity there are clearly insuperable difficulties. Furthermore the actual calculation of the basic number has always been performed on the basis of the Dubois formula with its own consequential inadequacies.

## II. ADULT AVIAN BRAINS

It was actually over 50 years ago that an outstanding analysis of avian brains and their component parts was carried out by the Swiss ornithologist Kuenzi (1918). He derived an index $(E)$ for the size of each brain and calculated from this the appropriate index for each region using the relationship:

$$\frac{\text{Surface of the region} \times 100}{E}$$

He presented the results of such calculations on the telencephalon, diencephalon, mesencephalon, cerebellum and medulla in 107 species. Some of his conclusions are contained in Tables 74 and 75. It is interesting that although these data were mentioned by Drooglever and Kappers (1921) they are rarely referred to elsewhere with the exception of Portmann (1946).

A comparable series of values for alciform genera was provided independently by Nikitenko (1965) and these are contained in Table 76. Portmann himself undertook an extensive series of analogous calculations. These are contained in the papers which were published 25 years ago (1946, 1947) and reviewed by him more recently (Portmann and Stingelin, 1961). In his first approximations he found that exponents of the Dubois type varied from 0·45 to 0·82 and he concluded that no single value had any wide validity. It could not be applied to all avian groups. When plotted logarithmically a group of animals in which both body weight and brain weight increase with a constant

relationship should give a straight line relationship between these two parameters. The line should be at 45° to the horizontal. No exact relationship of
this nature was ever observed because the brain weights always increased less
than those of the body. Nevertheless in certain groups the data did approach
such a straight line. In view of these variations he continued his investigations
by using a more standard yard-stick—the so-called brain stem rest, which is
rather smaller than the brain stem as defined by neurologists. It comprises the
diencephalon, tegmentum and medullary bulb, a volume that is shown black
in Fig. 135. The values for this are again not constant and can vary from 1 to

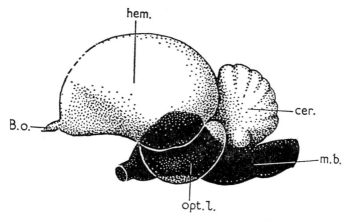

FIG. 135. The brain stem rest of Portmann is shown in black in this lateral view
of the avian brain. (Portmann and Stingelin, 1961).
    Abbreviations: B.o. olfactory bulb; cer, cerebellum; hem, hemisphere; m.b.,
medullary bulb; opt.l., optic lobe.

2·7 in birds of the same body weight. Portmann (1947, Portmann and
Stingelin, 1961) therefore determined which order had the lowest values. This
proved to be the Galliformes and as a result he took as his basic unit the
values of this brain stem rest in a gallinaceous bird of a given body weight.
The values which he obtained by calculations using this basis are contained
in Tables 77–81. These data enabled him to establish average values for each
of the orders and for the families of song-birds. These are shown in Figs 136
and 137. He concluded that amongst the alectoromorph orders only the
Alciformes achieve hemisphere values greatly in excess of 5. Larger values
occur amongst his pelargomorphs but it is in his coraciomorphs that really
high values occur. One can note that an increase in total body weight
from 30 to 1200 g results in an increase in the hemisphere index of Galliformes
from 2·51 to 3·18. In striking contrast the comparable values for Passeriformes
are 7–18·70 and in the case of the Psittaciformes 7·4–27·6. From Fig. 136 one
can also see that there are three main groupings for the relative dominance of

## TABLE 74

*The brain indices for non-passerine birds according to* Kuenzi (1918). *Those species which were considered by* Portmann (1946–47) *are omitted.*

| Species | Fore-brain | Dien-cephalon | Mesen-cephalon | Cere-bellum | Hind-brain |
|---|---|---|---|---|---|
| Tinamiformes | | | | | |
| *Calopezus elegans* | 62 | 15 | 24 | 28 | 22 |
| Galliformes | | | | | |
| *Caccabis saxatilis* | 64 | 14 | 22 | 24 | 21 |
| *Gennaeus melanonotus* | 65 | 14 | 26 | 30 | 19 |
| *Tragopan caboti* | 66 | 15 | 25 | 31 | 20 |
| *Lophortyx californicus* | 67 | 15 | 26 | 24 | 20 |
| Columbiformes | | | | | |
| *Carpophaga rosacea* | 68 | 14 | 24 | 31 | 19 |
| *Osmotreron vernans* | 71 | 15 | 28 | 33 | 21 |
| *Melopelia leucoptera* | 67 | 16 | 32 | 33 | 20 |
| *Geopelia striata* | 69 | 16 | 33 | 35 | 20 |
| *Chalcophaps chrysochlora* | 71 | 16 | 29 | 33 | 21 |
| *Turtur turtur* | 72 | 15 | 25 | 31 | 20 |
| Ralliformes | | | | | |
| *Gallinula angulata* | 67 | 14 | 22 | 25 | 21 |
| *Aramides cayanea* | 69 | 16 | 22 | 30 | 20 |
| *Porphyrio calvus* | 70 | 13 | 18 | 26 | 20 |
| Podicipitiformes | | | | | |
| *Podiceps fluviatilis* | 73 | 13 | 27 | 35 | 21 |
| Alciformes | | | | | |
| *Alca torda* | 66 | 12 | 20 | 36 | 20 |
| Charadriiformes | | | | | |
| *Charadrius apricarius* | 72 | 11 | 29 | 27 | 18 |
| *Tringa subarquata* | 75 | 12 | 23 | 27 | 21 |
| Ardeiformes | | | | | |
| *Ibis molucca* | 79 | 10 | 14 | 27 | 21 |
| *Florida coerulea* | 68 | 15 | 23 | 32 | 18 |
| Pelecaniformes | | | | | |
| *Sula bassana* | 65 | 10 | 14 | 26 | 18 |
| *Pelecanus crispus* | 70 | 10 | 11 | 22 | 19 |
| Anseriformes | | | | | |
| *Chloephaga magellanica* | 71 | 13 | 18 | 28 | 23 |
| *Anser brachyrhynchus* | 71 | 13 | 14 | 28 | 21 |
| *Mergus albellus* | 73 | 14 | 20 | 25 | 22 |
| *Tadorna tadorna* | 74 | 13 | 19 | 28 | 23 |
| *Netta rufina* | 75 | 13 | 20 | 26 | 22 |
| *Oedemia nigra* | 75 | 13 | 18 | 32 | 22 |
| *Chaulelasmus streperus* | 79 | 13 | 19 | 28 | 23 |
| Falconiformes | | | | | |
| *Pernis apivorus* | 69 | 12 | 20 | 28 | 17 |
| *Cerchneis tinnunculus* | 73 | 9 | 20 | 28 | 15 |
| Psittaciformes | | | | | |
| *Lorius flavopalliatus* | 83 | 10 | 16 | 19 | 18 |
| *Cacatua moluccensis* | 77 | 9 | 13 | 16 | 15 |
| *Cacatua sulfurea* | 80 | 10 | 14 | 19 | 14 |

TABLE 74 (*contd.*)

| | | | | |
|---|---|---|---|---|
| *Eclectus pectoralia* | 79 | 10 | 15 | 18 | 18 |
| *Tanygnathus muelleri* | 80 | 10 | 15 | 18 | 17 |
| *Conurus cactorum* | 96 | 10 | 19 | 21 | 18 |
| Coraciiformes | | | | | |
| *Alcedo ispida* | 69 | 11 | 25 | 30 | 17 |
| *Anthracoceros convexus* | 70 | 10 | 19 | 26 | 18 |
| Cuculiformes | | | | | |
| *Eudynamis honorata* | 66 | 14 | 22 | 30 | 18 |
| Piciformes | | | | | |
| *Rhamphastus discolorus* | 70 | 12 | 18 | 25 | 16 |
| *Selenidera maculirostris* | 72 | 13 | 21 | 30 | 17 |

## TABLE 75

*The brain indices for some passerine species.*

| Species | Fore-brain | Dien-cephalon | Mesen-cephalon | Cere-bellum | Hind-brain |
|---|---|---|---|---|---|
| Tyrannidae | | | | | |
| *Megarhynchus pitangua* | 71 | 11 | 24 | 28 | 18 |
| Turdidae | | | | | |
| *Turdus musicus* | 74 | 11 | 21 | 24 | 15 |
| *Phoenicurus phoenicurus* | 76 | 10 | 26 | 30 | 17 |
| *Copsychus saularis* | 78 | 10 | 23 | 36 | 19 |
| *Saxicola oenanthe* | 81 | 10 | 26 | 33 | 19 |
| Laniidae | | | | | |
| *Lanius excubitor* | 82 | 10 | 20 | 26 | 15 |
| Motacillidae | | | | | |
| *Motacilla alba* | 83 | 9 | 24 | 30 | 17 |
| Fringillidae | | | | | |
| *Cyanocompsa cyanea* | 82 | 10 | 25 | 30 | 18 |
| *Fringilla coelebs* | 82 | 9 | 22 | 23 | 17 |
| *Emberiza citrinella* | 82 | 8 | 27 | 26 | 18 |
| *Chloris chloris* | 83 | 8 | 23 | 23 | 17 |
| *Passer montanus* | 86 | 9 | 25 | 24 | 17 |
| *Paroaria larvata* | 87 | 9 | 23 | 31 | 17 |
| *Miliaria miliaria* | 87 | 8 | 19 | 27 | 18 |
| Tanagridae | | | | | |
| *Rhamphocoelus brasilius* | 82 | 10 | 22 | 28 | 17 |
| *Tachyphonus rufus* | 88 | 10 | 22 | 29 | 16 |
| Ploceidae | | | | | |
| *Munia orizivora* | 85 | 9 | 23 | 29 | 19 |
| *Fondia madagascariensis* | 91 | 8 | 20 | 30 | 19 |
| Sturnidae | | | | | |
| *Sturnus vulgaris* | 89 | 9 | 22 | 29 | 15 |
| Corvidae | | | | | |
| *Pica pica* | 76 | 9 | 17 | 21 | 14 |
| *Urocissa occipitalis* | 82 | 9 | 18 | 21 | 14 |
| *Coloeus monedula* | 84 | 7 | 16 | 18 | 13 |

After Kuenzi (1918).

## TABLE 76

*The relative contribution of different brain regions to the brains of alciform species according to* Nikitenko (1965).

| Species | Sex | Olfactory lobe | Hemi-spheres | Dien-cephalon | Mesen-cephalon | Cere-bellum | Medulla |
|---|---|---|---|---|---|---|---|
| *Alca torda* | ♂ | 0·25 | 54·9 | 7·7 | 8·5 | 20·2 | 8·5 |
| | ♀ | 0·25 | 53·8 | 8·2 | 9·4 | 20·0 | 8·4 |
| *Uria aalge* | ♂ | 0·27 | 54·1 | 8·6 | 9·1 | 22·1 | 5·83 |
| | ♀ | 0·29 | 54·2 | 8·5 | 8·9 | 21·7 | 6·4 |
| *Uria lomvia* | ♂ | 0·28 | 52·5 | 7·9 | 8·9 | 21·3 | 9·2 |
| | ♀ | 0·24 | 52·4 | 7·3 | 9·1 | 21·7 | 9·3 |
| *Cephus grylle* | ♂ | 0·21 | 54·7 | 8·6 | 10·3 | 16·6 | 9·5 |
| | ♀ | 0·26 | 54·2 | 8·8 | 10·7 | 16·8 | 9·2 |
| *Fratercula arctica* | ♂ | 0·38 | 56·8 | 5·9 | 9·6 | 18·5 | 8·8 |
| | ♀ | 0·35 | 56·8 | 5·6 | 9·5 | 18·6 | 9·2 |

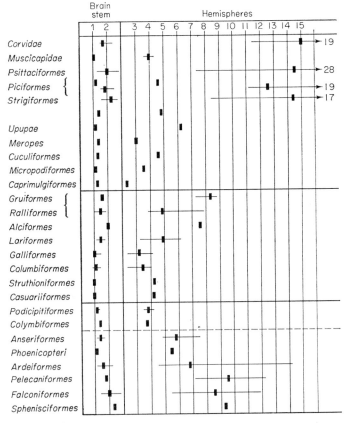

FIG. 136. A summary of Portmann's conclusions on various orders of birds (After Portmann 1946–47).

the hemispheres. The very large values occur in the Passeriformes, Psittaciformes, Piciformes and Strigiformes. In contrast the majority of the taxa have
values of less than 5. Nevertheless it is interesting that a number of aquatic
orders, together with the Falconiformes, have values that are somewhat in
excess of 5. These include the Ralliformes, Alciformes, Lariformes, Anseriformes, Ardeiformes, Pelecaniformes and Sphenisciformes. Although these
are all subject to considerable variation it does appear to indicate a possible
relationship between their hemisphere size and habits. In view of the immense
relative volume of the Passeriformes hemispheres, and their small olfactory

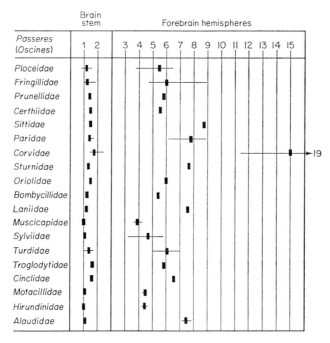

FIG. 137. A summary of the results of Portmann's calculations on some song bird
brains. (After Portmann 1946–47).

bulb size, it would be unwise to draw too definitive conclusions. One can,
however, note that it is amongst such aquatic genera that olfaction is important (see Chapter 13, Section II).

In spite of this wealth of data Krompecher and Lipak (1966) were not
satisfied with Portmann's results. They were interested in a measure of
"intelligence" and claimed that another method avoided many inherent
difficulties whilst it united the advantages of the classical studies. Although
they only dealt with a few avian species their results are of interest and have

22

## TABLE 77

*The brain indices for some so-called Alectoromorphae.*

| Species | Body weight in grams | Basal number | Stem rest | Optic lobes | Cere-bellum | Hemi-spheres |
|---|---|---|---|---|---|---|
| **Galliformes** | | | | | | |
| *Excalfactoria sinensis* | 31 | 0·097 | 1 | 0·618 | 0·512 | 2·51 |
| *Coturnix coturnix* | 85 | 0·17 | 1 | 0·705 | 0·541 | 2·36 |
| *Perdix perdix* | 370 | 0·37 | 1 | 0·667 | 0·564 | 2·49 |
| *Chrysolophus pictus* | 550 | 0·437 | 1·33 | 0·93 | 0·95 | 4·10 |
| *Gallus gallus* | 550 | 0·437 | 1·24 | 0·840 | 0·774 | 3·27 |
| *Phasianus colchicus torquatus* | 1,200 | 0·71 | 1 | 0·684 | 0·66 | 3·18 |
| *Gennaeus nycthemerus* | 1,250 | 0·669 | 1·21 | 0·91 | 0·77 | 4·10 |
| *Lyrurus tetrix* | 1,250 | 0·74 | 1 | 0·662 | 0·825 | 2·77 |
| *Tetrao urogallus* | 2,750 | 1·10 | 1 | 0·58 | 0·78 | 2·82 |
| *Pavo cristatus* | 3,500 | 1·16 | 1 | 0·63 | 0·87 | 3·74 |
| **Columbiformes** | | | | | | |
| *Geopelia cuneata* | 30 | 0·095 | 1·13 | 0·901 | 0·743 | 2·31 |
| *Streptopelia risoria* | 143 | 0·216 | 0·91 | 0·796 | 0·74 | 3·22 |
| *Columba livia* | 300 | 0·318 | 1·38 | 0·85 | 1·04 | 4·0 |
| *Columba palumbus* | 450 | 0·394 | 1·12 | 0·76 | 0·888 | 3·32 |
| *Goura coronota* | 1,500 | 0·738 | 1·14 | 0·855 | 1·11 | 4·05 |
| **Charadriiformes** | | | | | | |
| *Actitis hypoleucos* | 47 | 0·121 | 1·21 | 1·07 | 0·844 | 3·26 |
| *Limnocryptes minimus* | 60 | 0·137 | 1·68 | 0·657 | 0·804 | 3·65 |
| *Capella gallinago* | 100 | 0·179 | 1·71 | 0·52 | 0·75 | 4·08 |
| *Philomachus pugnax* | 180 | 0·244 | 1·11 | 0·74 | 0·98 | 3·92 |
| *Vanellus vanellus* | 200 | 0·257 | 1·40 | 1·20 | 1·26 | 4·73 |
| *Scolopax rusticola* | 290 | 0·312 | 1·38 | 0·67 | 1·04 | 5·22 |
| *Burhinus oedicnemus* | 440 | 0·389 | 1·57 | 0·95 | 1·23 | 5·35 |
| *Haematopus ostralegus* | 500 | 0·416 | 1·79 | 0·493 | 0·925 | 5·45 |
| *Numenius arquatus* | 650 | 0·476 | 1·34 | 0·777 | 1·34 | 5·13 |
| **Lariformes** | | | | | | |
| *Sterna albifrons* | 40 | 0·111 | 1·41 | 1·33 | 1·89 | 3·95 |
| *Sterna hirundo* | 120 | 0·20 | 1·13 | 0·96 | 1·42 | 3·92 |
| *Larus ridibundus* | 250 | 0·289 | 1·42 | 1·25 | 1·70 | 5·36 |
| *Larus argentatus* | 1,000 | 0·598 | 1·12 | 0·89 | 1·15 | 4·31 |
| *Larus marinus* | 1,670 | 0·781 | 1·27 | 0·935 | 1·62 | 6·13 |
| **Alciformes** | | | | | | |
| *Fratercula arctica grabae* | 330 | 0·335 | 2·03 | 1·15 | 2·48 | 7·57 |

TABLE 77 (contd.)

| | | | | | | |
|---|---|---|---|---|---|---|
| **Ralliformes** | | | | | | |
| *Porzana porzana* | 80 | 0·159 | 1·51 | 0·944 | 0·95 | 3·88 |
| *Crex crex* | 110 | 0·189 | 1·005 | 1·08 | 0·85 | 3·97 |
| *Rallus aquaticus* | 120 | 0·197 | 1·52 | 0·80 | 1·25 | 5·38 |
| *Gallinula chloropus* | 230 | 0·276 | 1·22 | 0·87 | 1·01 | 4·37 |
| *Fulica atra* | 410 | 0·375 | 1·41 | 0·80 | 1·09 | 5·57 |
| *Porphyrio porphyrio* | 500 | 0·416 | 1·85 | 0·99 | 1·635 | 7·86 |
| **Gruiformes** | | | | | | |
| *Anthropoides virgo* | 2,000 | 0·857 | 1·61 | 1·03 | 1·81 | 7·24 |
| *Balearica pavonina* | 3,250 | 1·11 | 1·585 | 0·84 | 1·76 | 8·80 |
| *Antigone antigone* | 7,500 | 1·71 | 1·59 | 0·73 | 1·08 | 8·31 |
| **Casuariiformes** | | | | | | |
| *Dromiceius novae-hollandiae* | 40,500 | 4·18 | 0·92 | 0·342 | 0·91 | 4·18 |
| **Struthioniformes** | | | | | | |
| *Struthio camelus* | 90,000 | 6·34 | 0·969 | 0·282 | 0·955 | 4·27 |

From Portmann (1946–47).

## TABLE 78
*Brain indices for some so-called Pelargomorphae.*

| Species | Body weight in grams | Basic number | Brain stem | Optic lobes | Cere-bellum | Hemi-spheres |
|---|---|---|---|---|---|---|
| **Podicipitiformes** | | | | | | |
| *Podiceps ruficollis* | 160 | 0·229 | 1·31 | 0·85 | 1·18 | 4·15 |
| *Podiceps cristatus* | 1,050 | 0·614 | 1·04 | 0·52 | 1·11 | 3·46 |
| **Colymbiformes** | | | | | | |
| *Colymbus stellatus* | 1,200 | 0·659 | 1·370 | 0·679 | 1·63 | 3·69 |
| **Sphenisciformes** | | | | | | |
| *Spheniscus demersus* | 2,700 | 1·01 | 2·28 | 0·777 | 1·94 | 9·31 |
| **Anseriformes** | | | | | | |
| *Anas crecca* | 300 | 0·318 | 1·64 | 0·645 | 1·15 | 5·47 |
| *Anas penelope* | 700 | 0·497 | 1·34 | 0·563 | 0·915 | 5·33 |
| *Mergus serrator* | 900 | 0·566 | 1·62 | 0·574 | 1·56 | 5·14 |
| *Anas platyrhynchos* | 1,200 | 0·658 | 1·41 | 0·608 | 0·84 | 6·08 |
| *Somateria mollissima* | 2,050 | 0·869 | 1·51 | 0·437 | 1·14 | 6·95 |
| *Anser anser* | 3,250 | 1·11 | 1·38 | 0·575 | 1·11 | 7·13 |
| *Cygnus olor* | 11,000 | 2·10 | 1·16 | 0·269 | 1·03 | 5·16 |
| **Phoenocopteriformes** | | | | | | |
| *Phoenicopterus ruber* | | | | | | |
| *roseus* | 3,000 | 1·35 | 1·08 | 0·296 | 1·355 | 5·46 |
| **Ardeiformes** | | | | | | |
| *Ixobrychus minutus* | 136 | 0·21 | 1·34 | 0·753 | 1·08 | 4·52 |
| *Egretta garzetta* | 500 | 0·416 | 1·27 | 0·937 | 1·15 | 4·98 |
| *Botaurus stellatus* | 900 | 0·566 | 1·24 | 0·86 | 1·28 | 5·82 |
| *Egretta alba* | 1,000 | 0·600 | 1·10 | 1·13 | 1·28 | 5·32 |
| *Ardea cinerea* | 1,500 | 0·738 | 1·565 | 1·015 | 1·35 | 6·73 |
| *Ciconia ciconia* | 3,500 | 1·16 | 1·836 | 0·986 | 1·92 | 7·91 |
| *Leptoptilus crumeniferus* | 6,200 | 1·55 | 2·16 | 0·986 | 2·84 | 14·22 |
| **Falconiformes** | | | | | | |
| *Falco tinnunculus* | 230 | 0·276 | 1·88 | 1·51 | 1·67 | 8·24 |
| *Accipiter nisus* | 260 | 0·295 | 1·865 | 1·49 | 1·69 | 5·40 |
| *Buteo buteo* | 900 | 0·566 | 1·96 | 1·59 | 2·14 | 9·78 |
| *Accipiter gentilis* | 1,100 | 0·629 | 1·80 | 1·33 | 1·73 | 7·34 |
| *Pandion haliaetus* | 1,500 | 0·737 | 2·04 | 1·80 | 1·89 | 9·74 |
| *Aquila chrysaetos* | 5,300 | 1·43 | 2·13 | 0·93 | 1·37 | 8·64 |
| *Aegypius monachus* | 9,000 | 1·90 | 1·60 | 0·537 | 1·48 | 9·44 |
| **Pelecaniformes** | | | | | | |
| *Phalacrocorax carbo* | 2,200 | 0·895 | 1·64 | 0·66 | 1·65 | 7·08 |
| *Pelecanus onocrotalus* | 9,000 | 0·89 | 1·85 | 0·529 | 2·20 | 12·07 |

After Portmann (1946–47).

## TABLE 79

*The brain indices of some Coraciomorphae.*

| Species | Body weight in grams | Basic number | Brain stem | Optic lobes | Cere-bellum | Hemi-spheres |
|---|---|---|---|---|---|---|
| Caprimulgiformes | | | | | | |
| *Caprimulgus europaeus* | 70 | 0·146 | 1·26 | 0·73 | 0·93 | 2·35 |
| Coraciiformes | | | | | | |
| *Merops apiaster* | 60 | 0·137 | 1·38 | 0·87 | 1·13 | 3·03 |
| *Alcedo atthis* | 35 | 0·104 | 1·40 | 1·36 | 1·125 | 4·86 |
| *Upupa epops* | 55 | 0·131 | 1·145 | 0·721 | 1·28 | 6·27 |
| Cuculiformes | | | | | | |
| *Cuculus canorus* | 100 | 0·179 | 1·31 | 1·26 | 0·99 | 4·61 |
| Micropodiformes | | | | | | |
| *Micropus apus* | 38 | 4·108 | 1·06 | 0·65 | 1·02 | 3·43 |
| *Micropus melba* | 90 | 0·169 | 1·18 | 0·66 | 1·04 | 3·67 |
| Strigiformes | | | | | | |
| *Otus scops* | 92 | 0·171 | 2·10 | 0·964 | 1·40 | 8·45 |
| *Athene noctua* | 165 | 0·234 | 1·58 | 1·003 | 1·28 | 12·9 |
| *Asio otus* | 250 | 0·289 | 2·48 | 0·865 | 1·51 | 14·22 |
| *Tyto alba* | 290 | 0·313 | 2·70 | 0·687 | 1·47 | 14·53 |
| *Strix aluco* | 450 | 0·394 | 2·48 | 1·155 | 1·75 | 17·0 |
| *Bubo bubo* | 2000 | 0·858 | 2·15 | 0·880 | 1·59 | 15·07 |
| Psittaciformes | | | | | | |
| *Melopsittacus undulatus* | 36·7 | 0·105 | 1·285 | 0·76 | 1·14 | 7·40 |
| *Agapornis fischeri* | 42 | 0·114 | 2·32 | 1·05 | 1·06 | 13·09 |
| *Calopsitta novae hollandiae* | 85 | 0·165 | 1·64 | 0·82 | 1·27 | 11·9 |
| *Palaeornis eupatrius* | 96 | 0·174 | 2·76 | 1·32 | 1·67 | 17·76 |
| *Trichoglossus novae hollandiae* | 136 | 0·210 | 2·19 | 1·19 | 1·81 | 7·90 |
| *Amazona versicolor* | 400 | 0·370 | 2·39 | 0·959 | 1·68 | 16·1 |
| *Lophochroa sulphurea* | 450 | 0·394 | 2·89 | 0·94 | 0·813 | 17·49 |
| *Psittacus erythacus* | 450 | 0·394 | 2·28 | 0·957 | 1·78 | 19·1 |
| *Ara ararauna* | 850 | 0·549 | 2·66 | 0·99 | 2·38 | 28·02 |
| *Ara chloroptera* | 1430 | 0·720 | 2·48 | 1·05 | 2·67 | 27·61 |
| Piciformes | | | | | | |
| *Jynx torquilla* | 37 | 0·107 | 1·21 | 0·82 | 0·86 | 4·625 |
| *Dryobates medius* | 58 | 0·135 | 1·56 | 1·04 | 1·63 | 11·04 |
| *Dryobates major* | 80 | 0·159 | 1·85 | 1·04 | 1·76 | 12·35 |
| *Picus canus* | 122 | 0·199 | 1·53 | 1·21 | 1·76 | 12·91 |
| *Picus viridis* | 200 | 0·257 | 1·87 | 1·03 | 1·63 | 12·53 |
| *Dryocopus martius* | 300 | 0·318 | 2·48 | 1·26 | 2·00 | 19·35 |

After Portmann (1946–47).

## TABLE 80

*The brain indices for some Passeriformes.*

| Species | Body weight in grams | Basic number | Brain stem | Optic lobes | Cere- bellum | Hemi- spheres |
|---|---|---|---|---|---|---|
| **Alaudidae** | | | | | | |
| *Alauda arvensis* | 39 | 0·110 | 1·0 | 0·145 | 0·79 | 7·86 |
| *Melanocorypha calandra* | 55 | 0·131 | 1·20 | 0·93 | 1·08 | 7·18 |
| **Hirundinidae** | | | | | | |
| *Delichon urbica* | 15 | 0·066 | 1·05 | 0·674 | 0·689 | 4·28 |
| *Hirundo rustica* | 18·5 | 0·074 | 0·97 | 0·74 | 1·11 | 4·62 |
| **Motacillidae** | | | | | | |
| *Anthus pratensis* | 16 | 0·068 | 1·16 | 1·09 | 0·945 | 4·65 |
| *Motacilla alba* | 23 | 0·083 | 1·02 | 1·02 | 0·99 | 4·51 |
| **Cinclidae** | | | | | | |
| *Cinclus cinclus* | 60 | 0·137 | 1·58 | 1·01 | 1·42 | 6·60 |
| **Troglodytidae** | | | | | | |
| *Troglodytes troglodytes* | 9·5 | 0·052 | 1·66 | 1·12 | 1·05 | 5·87 |
| **Turdidae** | | | | | | |
| *Erithacus rubecula* | 16·2 | 0·069 | 1·50 | 1·27 | 1·11 | 5·01 |
| *Turdus ericetorum* | 67 | 0·146 | 1·40 | 1·30 | 1·16 | 6·16 |
| *Turdus merula* | 95 | 0·174 | 1·425 | 1·18 | 1·115 | 6·67 |
| **Regulidae** | | | | | | |
| *Regulus regulus* | 5·4 | 0·039 | 1·11 | 1·19 | 1·09 | 5·77 |
| **Sylviidae** | | | | | | |
| *Acrocephalus scirpaceus* | 14 | 0·064 | 1·01 | 1·09 | 1·03 | 4·41 |
| *Sylvia borin* | 19 | 0·075 | 1·19 | 0·99 | 0·93 | 4·70 |
| **Muscicapidae** | | | | | | |
| *Muscicapa striata* | 16 | 0·069 | 1·13 | 0·841 | 0·971 | 4·28 |
| **Laniidae** | | | | | | |
| *Lanius collurio* | 29·7 | 0·095 | 1·21 | 1·12 | 1·18 | 7·57 |
| **Bombycillidae** | | | | | | |
| *Bombycilla garrulus* | 55·5 | 0·132 | 1·26 | 0·91 | 1·10 | 5·38 |
| **Sturnidae** | | | | | | |
| *Pastor roseus* | 55·21 | 0·131 | 1·51 | 1·13 | 1·3 | 7·33 |
| *Sturnus vulgaris* | 80 | 0·160 | 1·37 | 1·01 | 1·16 | 7·63 |

TABLE 80 (*contd.*)

| | | | | | | |
|---|---|---|---|---|---|---|
| Oriolidae | | | | | | |
| *Oriolus oriolus* | 72 | 0·151 | 1·49 | 1·16 | 1·27 | 6·00 |
| | | | | | | |
| Corvidae | | | | | | |
| *Garrulus glandarius* | 160 | 0·229 | 1·92 | 1·62 | 1·66 | 12·71 |
| *Coloeus monedula* | 200 | 0·257 | 1·595 | 1·40 | 1·54 | 13·98 |
| *Pica pica* | 220 | 0·270 | 1·82 | 1·41 | 1·63 | 15·81 |
| *Pyrrhocorax pyrrhocorax* | 356 | 0·347 | 1·67 | 1·02 | 1·41 | 14·60 |
| *Corvus frugilegus* | 430 | 0·383 | 1·75 | 1·24 | 1·67 | 15·68 |
| *Corvus corone* | 520 | 0·418 | 1·79 | 1·27 | 1·615 | 15·38 |
| *Corvus corax* | 1100 | 0·629 | 1·53 | 1·17 | 1·85 | 18·70 |
| | | | | | | |
| Paridae | | | | | | |
| *Aegithalos caudatus* | 7·5 | 0·046 | 1·725 | 1·29 | 0·755 | 6·15 |
| *Parus coeruleus* | 11·0 | 0·056 | 1·43 | 1·06 | 0·92 | 8·77 |
| *Parus major* | 17·5 | 0·072 | 1·44 | 1·17 | 1·09 | 8·92 |
| | | | | | | |
| Sittidae | | | | | | |
| *Sitta europaea* | 23 | 0·083 | 1·50 | 1·02 | 1·50 | 8·75 |
| | | | | | | |
| Certhiidae | | | | | | |
| *Certhia familiaris* | 8·7 | 0·050 | 1·48 | 0·94 | 0·998 | 5·53 |
| | | | | | | |
| Prunellidae | | | | | | |
| *Prunella modularis* | 18·0 | 0·075 | 1·44 | 1·29 | 1·08 | 5·86 |
| | | | | | | |
| Fringillidae | | | | | | |
| *Serinus canaria* | 8 | 0·048 | 1·46 | 0·85 | 1·14 | 5·95 |
| *Carduelis spinus* | 11·5 | 0·057 | 1·21 | 0·95 | 1·21 | 6·21 |
| *Carduelis cannabina* | 18 | 0·073 | 1·09 | 0·86 | 0·89 | 5·87 |
| *Carduelis carduelis* | 14·5 | 0·065 | 1·22 | 0·76 | 1·07 | 6·47 |
| *Fringilla coelebs* | 21·6 | 0·080 | 1·18 | 1·14 | 0·99 | 5·83 |
| *Passer domesticus* | 28 | 0·092 | 1·30 | 0·975 | 1·03 | 7·22 |
| *Loxia curvirostra* | 38 | 0·108 | 1·51 | 1·02 | 1·39 | 8·94 |
| *Montifringilla nivalis* | 45 | 0·118 | 1·255 | 1·025 | 1·01 | 5·97 |
| *Coccothraustes* | | | | | | |
| *coccothraustes* | 52 | 0·127 | 1·50 | 1·26 | 1·26 | 8·78 |

After Portmann (1946–47).

been hailed as a great advance on Portmann's results by some workers. In-corporating the weight of the spinal cord they considered that this was the best reflection of the bulk of the body together with its surface and functions. The brain, leaving aside the relatively small "somatic" part that is related to the spinal cord, was then used, together with the cord, to produce a measure of intelligence. This is tabulated in the final column of Table 82, and gives

## TABLE 81

*Brain indices for some song-bird families (the acromyodine or oscine families of the Passeriformes).*

| Family | Brain stem rest | Optic lobes | Cerebellum | Hemispheres |
|---|---|---|---|---|
| Ploceidae | 1·15 | 0·83 | 0·94 | 5·46 |
| Fringillidae | 1·21 | 0·96 | 1·01 | 5·97 |
| Prunellidae | 1·44 | 1·29 | 1·08 | 6·86 |
| Certhiidae | 1·48 | 0·94 | 1·00 | 5·53 |
| Sittidae | 1·50 | 1·02 | 1·50 | 8·75 |
| Paridae | 1·43 | 1·05 | 1·01 | 7·82 |
| Corvidae | 1·72 | 1·20 | 1·64 | 14·99 |
| Sturnidae | 1·37 | 1·01 | 1·16 | 7·63 |
| Oriolidae | 1·49 | 1·16 | 1·27 | 6·0 |
| Bombycillidae | 1·26 | 0·91 | 1·10 | 5·38 |
| Laniidae | 1·21 | 1·12 | 1·18 | 7·57 |
| Muscicapidae | 1·05 | 1·00 | 0·93 | 3·97 |
| Sylviidae | 1·14 | 1·06 | 1·00 | 4·67 |
| Turdidae | 1·39 | 1·15 | 1·10 | 6·04 |
| Troglodytidae | 1·66 | 1·12 | 1·05 | 5·87 |
| Cinclidae | 1·58 | 1·01 | 1·42 | 6·60 |
| Motacillidae | 1·08 | 1·02 | 0·97 | 4·51 |
| Hirundinidae | 1·02 | 0·70 | 0·89 | 4·45 |
| Alaudidae | 1·10 | 0·98 | 0·93 | 7·52 |

From Portmann (1947).

comparable results to those of Fig. 136 with three passeriform and the single Psittaciform species having values which are greatly in excess of those for hens and pigeons.

## III. DEVELOPMENTAL CONSIDERATIONS

Any consideration of avian brains must take into account the very marked differences between nidifugous and nidicolous habits. There are a very large number of data which relate to the hatching times and development of young birds. These have been collected and collated by the thousands of amateur and professional ornithologists who have carried out detailed investigations during the last century or more. However, a concise anatomical and neurological comparison between the major taxa again owes much to the works of Portmann (1935, 1937, 1942, 1950). He emphasized the general relationships which exist between the assumed taxonomic position of a particular species, its ontogenetic development, and the relative predominance of the hemispheres in its brain.

Broadly speaking nidifugous development is reminiscent of reptilian development. It is therefore not surprising that it occurs in the Struthioniformes, Casuariiformes, Rheiformes and Tinamiformes on the one hand, and amongst the Galliformes, Turnicidae, Pteroclididae, Gruidae, Mesoenatidae, Jacanidae, Otididae, Colymbiformes, Podicipidiformes and Anseriformes on the other. In contrast to these the principal examples of nidicolous development

## TABLE 82

*The brain and spinal cord weights of seven avian species, together with the derived intelligence quotient.*

| Species | Body weight | Brain weight (E) | Spinal cord weight (Ms) | $\dfrac{E}{Ms.}$ |
|---|---|---|---|---|
| *Gallus domesticus* | 1800 | 3·68 | 2·175 | 1·7 |
| *Gallus domesticus* | 1340 | 3·525 | 1·554 | 2·2 |
| *Columba livia* | 326 | 1·79 | 0·575 | 3·1 |
| *Hirundo rustica* | 20 | 0·492 | 0·075 | 6·6 |
| *Corvus cornix* | — | — | — | 11·4 |
| *Fringilla domestica* | — | — | — | 10·9 |
| *Passer domesticus* | 27 | 1·0 | 0·085 | 11·8 |
| *Agapornis personata* | 40 | 1·92 | 0·12 | 16·0 |

From Krompecher and Lipak (1966).

occur in the Pelecaniformes, Procellariiformes, Columbiformes other than the Pteroclididae, Ardeiformes, Cariamidae, *Opisthocomus*, Falconiformes, Strigiformes, Cuculiformes, Coraciiformes, Trogoniformes, Piciformes, and Passeriformes. An intermediate situation marks the Apterygiformes, Phoenicopteridae and Eurypygidae, and although the young of the Laridae and Spheshisciformes can leave the nest at an early stage they remain dependent on their parents for food.

Portmann (1950) was at pains to emphasize that there is no direct relationship between the length of the incubation period and the occurrence of nidifugous or nidicolous young. The presence of nidifugous habits is not a reflection of an excessively long incubation period, nor are nidicolous young the result of a short incubation period. Spector (1956) summarized a large number of data relating to both incubation period and nestling period. Amongst nidicolous species incubation periods of 51–55 days occur in *Aptenodytes*, *Gypaetus barbatus* and vultures. The procellariiform genus *Diomedea*

22*

has a period which exceeds 60 days. These values equal or exceed those for nidifugous genera which include 42–48 days for the ostrich, 52 days for the emu, 59–63 days for megapodid Galliformes and 34 days in the Gruidae. In contrast one can clearly cite at the other extreme nidifugous species such as *Coturnix delagorguei* in which the period is not in excess of 14 days and, as such, approximates to that of those small passerines which have 11–14 day periods.

In summary *Zosterops* has one of the shortest incubation periods with 10 days. *Columba* and *Streptopelia* have slightly longer periods, occupying some 13–17 days, and the gallinaceous birds take longer still involving some 20 days. Amongst the Piciformes, even in those of large size, incubation is of a relatively short duration and takes 12–13 days in *Dendrocopos major*, or 12–14 in *D. martius*. This is also true amongst certain falcons. The Micropodiformes, Strigiformes, the majority of the Falconiformes, the Pelecaniformes, Procellariiformes, Sphenisciformes and Psittaciformes all have relatively long incubation periods by comparison. The Trochilidae incubate for 16–21 days; *Apus caffer* for 20–26 days; *Otus scops* for 24–25 days; *Bubo bubo* for 35 days; and the megapodid *Leipoa ocellata* for 59–63 days. Amongst the Procellariiformes *Hydrobates pelagicus* take 38–40 days; *Puffinus puffinus* 52–54 days, and *Diomedea exulans* from 2–3 months. It is significant that amongst closely related species overt differences in egg size appear to exert no great effect upon the incubation period. Both *Phylloscopus collybita* and *P. sibilatrix* take 13 days, whilst *Corvus corone* and *C. monedula* take 17–18 days.

Portmann (1950) concluded that the particular type of embryological development which is exhibited by an avian species is largely a product of, and characteristic of, its taxonomic position. As such it is related to factors other than the duration of embryonic life. By contrast with the egg of reptiles, which usually requires some environmental moisture in order to achieve satisfactory development, that of birds contains everything which is necessary for the developing embryo. Not only is there enough water, but there is, indeed, actually an excess of it which compensates for losses during incubation. Ontogenetic development in reptiles lasts for at least a month and, except for their small size, the resultant young are well-developed. As they are independent of their parents at birth they are, in avian terms, nidifugous. Anything like such an extreme degree of nidifugous development is rare in birds but is approached most closely amongst the megapodid Galliformes of the Australo-Malaysian region. In *Megapodius* and in *Megacephalum maleo* the eggs are laid, like those of many reptiles, in the sand or humus floor of the forest. Development then ensues by virtue of the environmental temperature and, at hatching, the megapode chick can be as independent of its parents as are the young of reptiles. Indeed it is said to be capable of undertaking the brief flight which is necessary if it is to gain the nocturnal perches in the branches. Such birds are, however, unique in the present-day world avifauna and comprise only a relatively insignificant fraction of it. Even closely related species differ from them in their breeding behaviour and incubate their eggs.

In view of such considerations Portmann (1947, 1950) drew attention to the very extensive differences which exist between the relative development and weights of different body parts of nidifugous and nidicolous genera. A comparison of the weights of brain, eyes, heart, alimentary canal and liver is provided by Table 83. In this table the weights of these structures are expressed as a percentage of the total bodyweight at the time of hatching but the very considerable weight of the gut contents etc., is excluded. Table 84 contains a briefer list of five nidicolous and five nidifugous species. The very considerable differences of the various brain weights are abundantly clear. It is also not unduly surprising that it is amongst the nidifugous genera and passerines that the highest percentage values for relative eye size occur.

Once they are hatched nidicolous young grow at a prodigious rate which is not exceeded by members of any other vertebrate class. For example, at the time of hatching nestling herons weigh 42 g. Some 40 days later, two weeks prior to flying, they equal the weight of their parents and tip the scales at 1,600 g. This represents some thirty-eight times their hatching weight. Amongst the Cuculiformes a young cuckoo weighs 2 g at hatching and 100 g 3 weeks later. This represents an increase of 50 times their newly hatched weight. Table 85 shows that during the period of post-embryonic development the nestling's weight frequently actually exceeds that of its parents, a characteristic which reaches its apogee amongst members of the Procellariiformes and Pelecaniformes. A comparison of the developmental weights of young *Cygnus* and young *Pelecanus* shows that whilst the swan takes more than 6 months to reach 11 kg, young pelicans exceed the parental weight some 50 days after hatching. It is therefore exceedingly interesting that, whilst the hemispheric index for brain development in the swan is 5·16, that of *Pelecanus* reaches 12·07.

The process of myelination also confirms the fact that post-embryonic development is a function of taxonomic position rather than a reflection of the period of incubation. By a study of transverse sections of the brain at the level of the stato-acoustic nuclei in the genera *Picus* and *Sturnus*, and the galliform genera *Gallus* and *Coturnix* (Schifferli, 1948) it is quite clear that, in spite of the marked differences in the actual incubation time within the pairs, the degree of nervous development which is achieved at the time of hatching is comparable in a given order. Conversely the degree of development in different orders differs, in spite of similarities in the duration of the incubation period.

Portmann (1947) tried to summarize the post-embryonic development of the brain. As a first approximation to a description of the changes which have

# TABLE 83

*A comparison of the relative weight of the brain, eyes, heart, alimentary canal and liver expressed as a percentage of total weight, less gut contents, in newborn birds.*

| Species | Brain | Eyes | Heart | Alimentary canal | Liver | Total weight |
|---|---|---|---|---|---|---|
| **Galliformes** | | | | | | |
| *Pavo cristatus* | 2·74 | 2·99 | 0·73 | 7·37 | 3·29 | 50·08 |
| *Meleagris gallopavo* | 2·95 | 3·88 | 0·81 | 9·14 | 3·45 | 43·22 |
| *Gallus gallus* (wyandotte) | 2·58 | 2·41 | 0·80 | 9·55 | 2·92 | 32·73 |
| *Numida meleagris* | 4·03 | 3·36 | 0·90 | 7·55 | 2·96 | 23·05 |
| *Phasianus colchicus* | | | | | | |
| *torquatus* | 4·16 | 4·19 | 0·69 | 6·49 | 2·71 | 19·31 |
| *Chrysolophus pictus* | 5·31 | 5·10 | 0·91 | 8·77 | 3·82 | 14·08 |
| *Perdix perdix* | 6·14 | 4·69 | | | | 8·22 |
| *Coturnix coturnix* | 6·19 | 5·48 | 0·80 | 9·66 | 3·07 | 4·58 |
| **Lariformes** | | | | | | |
| *Larus ridibundus* | 4·85 | 6·36 | 1·08 | 14·9 | 4·38 | 19·28 |
| *Sterna hirundo* | 4·56 | 5·46 | 0·74 | 8·62 | 2·81 | 12·82 |
| **Charadriiformes** | | | | | | |
| *Vanellus vanellus* | 6·05 | 9·72 | 0·77 | 6·13 | 2·95 | 13·00 |
| *Haematopus haematopus* | 5·44 | 8·68 | 0·51 | | 3·53 | 10·39 |
| *Glareola pratincola* | 4·66 | 8·69 | 0·64 | 9·20 | 3·02 | 5·75 |
| *Charadrius alexandrinus* | 7·21 | 10·67 | 0·61 | 6·62 | 3·29 | 4·59 |
| **Alciformes** | | | | | | |
| *Fratercula arctica grabae* | 4·16 | 4·84 | 0·81 | 5·47 | 4·71 | 31·6 |
| **Ralliformes** | | | | | | |
| *Fulica atra* | 4·49 | 2·88 | 0·59 | 11·90 | 2·84 | 22·36 |
| *Rallus aquaticus* | 6·16 | 4·53 | 1·07 | 10·51 | 3·16 | 8·85 |
| **Columbiformes** | | | | | | |
| *Columba* | 2·92 | 4·94 | 1·27 | 10·33 | 3·48 | 11·49 |
| *Streptopelia risoria* | 3·43 | 7·87 | 0·96 | 10·29 | 2·29 | 5·24 |
| **Strigiformes** | | | | | | |
| *Strix aluco* | 6·71 | 6·66 | 0·74 | 4·85 | 2·80 | 21·44 |
| *Tyto alba* | 5·14 | — | 0·95 | 8·71 | 2·92 | 11·69 |
| **Micropodiformes** | | | | | | |
| *Micropus melba* | 3·11 | 6·12 | 0·83 | 14·64 | 3·05 | 4·19 |
| *Micropus apus* | 3·20 | 6·19 | 1·12 | 13·04 | 3·85 | 3·13 |

TABLE 83 (*contd.*)

| | | | | | | |
|---|---|---|---|---|---|---|
| Piciformes | | | | | | |
| *Jynx torquilla* | 3·42 | 4·09 | 1·09 | 11·08 | 2·40 | 1·91 |
| Psittaciformes | | | | | | |
| *Melopsittacus undulatus* | 7·58 | 3·69 | 1·04 | 9·14 | 2·11 | 1·42 |
| Podicipitiformes | | | | | | |
| *Podiceps cristatus* | 3·23 | 2·89 | 0·79 | 10·34 | 2·20 | 22·72 |
| Anseriformes | | | | | | |
| *Anser* (var. dom) | 2·60 | 1·61 | 0·80 | 10·00 | 3·64 | 81·04 |
| *Cairina moschata* | 4·91 | 3·13 | 0·65 | 7·27 | 3·30 | 32·18 |
| *Anas crecca* | 6·69 | 2·03 | | | | 15·3 |
| Falconiformes | | | | | | |
| *Falco tinnunculus* | 5·06 | 9·48 | 0·71 | 6·47 | 2·60 | 13·88 |
| Ardeiformes | | | | | | |
| *Ciconia ciconia* | 3·11 | 5·21 | | | | 54·86 |
| *Ardea cinerea* | 3·61 | 5·30 | 1·2 | 9·03 | 5·24 | 36·22 |
| *Threskiornis aethiopica* | 4·20 | 3·30 | | | | 28·95 |
| Passeriformes | | | | | | |
| *Corvus corone* | 3·02 | 5·06 | 0·68 | 13·10 | 2·80 | 13·56 |
| *Coloeus monedula* | 3·57 | 5·48 | 0·69 | 12·15 | 2·51 | 7·44 |
| *Pica pica* | 5·13 | 7·32 | 0·64 | 11·00 | 2·14 | 6·31 |
| *Turdus merula* | 4·16 | 6·93 | 0·86 | 13·44 | 2·98 | 4·69 |
| *Sturnus vulgaris* | 3·21 | 4·03 | 0·97 | 14·09 | 2·85 | 4·66 |
| *Acrocephalus arundinaceus* | 4·60 | 7·58 | 0·76 | 11·97 | 2·54 | 2·28 |
| *Lanius collurio* | 6·87 | 9·90 | 0·86 | 12·93 | 2·54 | 2·08 |
| *Passer domesticus* | 4·53 | 4·78 | 0·94 | 16·69 | 3·08 | 1·92 |
| *Emberiza citrinella* | 5·85 | 7·29 | 0·93 | 13·64 | 2·70 | 1·81 |
| *Chloris chloris* | 6·25 | 5·68 | 0·94 | 13·06 | 3·00 | 1·60 |
| *Sylvia atricapilla* | 6·69 | 7·64 | 1·06 | 13·43 | 2·74 | 1·39 |
| *Fringilla coelebs* | 8·17 | | 1·18 | 17·73 | 3·54 | 1·27 |
| *Phoenicurus phoenicurus* | 7·06 | 8·90 | 0·92 | 11·51 | 3·02 | 1·19 |
| *Parus major* | 7·63 | 7·62 | 0·82 | 10·27 | 2·49 | 1·15 |
| *Carduelis cannabina* | 7·38 | 5·13 | 1·08 | 14·77 | 2·61 | 1·11 |
| *Hirundo rustica* | 6·57 | 7·85 | 0·88 | 13·70 | 3·10 | 1·12 |
| *Troglodytes troglodytes* | 8·45 | 8·30 | 0·96 | 10·80 | 2·52 | 1·05 |
| *Serinus canaria* | 7·30 | 5·50 | 0·67 | 11·36 | 3·14 | 0·89 |
| *Parus coeruleus* | 7·39 | 5·68 | 0·93 | 9·04 | 2·42 | 0·81 |
| *Phylloscopus collybita* | 8·84 | 9·36 | 0·77 | 11·02 | 2·30 | 0·78 |

From Portmann (1946–47).

to occur during this period in order to attain the adult condition he used the quotient of the definitive brain weight, or alternatively that of its component regions, alongside the weight at hatching. The greater the degree of development at the moment of hatching then the less growth and fewer changes which

## TABLE 84

*A comparison of the relative weight of the brains, eyes, alimentary canal and liver in the new born young of five nidifugous and five nidicolous species.*

| Species | Total weight | Weight as a percentage of total body weight | | | |
|---|---|---|---|---|---|
| | | Brain | Eyes | Alimentary canal | Liver |
| **Nidifuges** | | | | | |
| *Vanellus vanellus* | 13·0 | 6·05 | 9·72 | 6·13 | 2·95 |
| *Chrysolophus pictus* | 14·08 | 5·31 | 5·10 | 8·77 | 3·07 |
| *Rallus aquaticus* | 8·85 | 6·16 | 4·53 | 10·51 | 3·16 |
| *Coturnix coturnix* | 4·58 | 6·19 | 5·48 | 9·66 | 3·07 |
| *Charadrius alexandrinus* | 4·59 | 7·21 | 10·67 | 6·62 | 3·29 |
| **Nidicoles** | | | | | |
| *Columba livia* | 11·49 | 2·92 | 4·94 | 10·33 | 3·48 |
| *Corvus corone* | 13·56 | 3·02 | 5·06 | 13·10 | 2·8 |
| *Coloeus monedula* | 7·44 | 3·57 | 5·48 | 13·10 | 2·51 |
| *Sturnus vulgaris* | 4·66 | 3·21 | 4·03 | 14·09 | 2·85 |
| *Micropus melba* | 4·19 | 3·11 | 6·12 | 14·64 | 4·19 |

From Portmann (1946–47).

## TABLE 85

*The day of attainment of maximum weight and a comparison between this weight and that of the adult.*

| Species | Duration of post-embryonic period | Day of maximum weight | Maximum weight in gm | Adult weight in gm |
|---|---|---|---|---|
| *Pelecanus onocrotalus* | 100 | 63 | 13,850 | 10,000 |
| *Phaeton americanus* | 62 | 41 | 570 | 410 |
| *Oceanodroma leucorrhoa* | 70 | 40 | 69·5 | 43 |
| *Tyto alba* | 55 | 32–52 | 375 | 330 |
| *Hirundo rustica* | 21 | 15 | 23 | 18 |
| *Micropus melba* | 54 | 35 | 110 | 90 |

From Portmann (1950).

are necessary in order to attain the adult condition. The results of such calculations are contained in the developmental indices of Table 86.

## TABLE 86

*The developmental indices for the telencephalic hemispheres and brain stem.*

| Species | Hemispheres (a) | Brain stem rest (b) | a : b |
|---|---|---|---|
| Galliformes | | | |
| *Meleagris gallopavo* | 4·13 | 4·13 | 1·00 |
| *Phasianus colchicus torquatus* | 5·19 | 4·66 | 1·11 |
| *Chrysolophus amherstiae* | 5·15 | 3·35 | 1·54 |
| *Chrysolophus pictus* | 4·97 | — | — |
| *Perdix perdix* | 4·31 | 3·20 | 1·34 |
| *Coturnix coturnix* | 3·17 | 2·43 | 1·30 |
| Charadriiformes | | | |
| *Vanellus vanellus* | 3·15 | 2·28 | 1·38 |
| *Charadrius alexandrinus* | 2·35 | 1·29 | 1·82 |
| Alciformes | | | |
| *Fratercula arctica grabae* | 3·84 | 2·49 | 1·54 |
| Lariformes | | | |
| *Larus ridibundus* | 3·24 | 2·28 | 1·42 |
| Ralliformes | | | |
| *Rallus aquaticus* | 3·53 | — | — |
| *Fulica atra* | 3·65 | 2·66 | 1·37 |
| Columbiformes | | | |
| *Streptopelia risoria* | 10·53 | 3·52 | 2·99 |
| *Columba* | 8·48 | 3·77 | 2·25 |
| Podicipitiformes | | | |
| *Podiceps cristatus* | 5·61 | 3·70 | 1·51 |
| Anseriformes | | | |
| *Anas platyrhynchos* | 4·93 | 3·29 | 1·49 |
| *Anas crecca* | 2·91 | 2·74 | 1·06 |
| Falconiformes | | | |
| *Falco tinnunculus* | 6·66 | 3·33 | 2·00 |
| Ardeiformes | | | |
| *Ardea cinerea* | 7·86 | 3·85 | 2·04 |
| *Threskiornis aethiopica* | 11·86 | 5·04 | 2·35 |
| *Ciconia ciconia* | 11·91 | 5·23 | 2·27 |

Table 86 (*contd.*)

| Species | Hemispheres (a) | Brain stem rest (b) | a : b |
|---|---|---|---|
| Micropodiformes | | | |
| *Apus apus* | 9·25 | 3·97 | 2·33 |
| Psittaciformes | | | |
| *Melopsittacus undulatus* | 19·40 | 4·35 | 4·46 |
| Piciformes | | | |
| *Jynx torquilla* | 19·04 | 5·42 | 3·51 |
| Strigiformes | | | |
| *Strix aluco* | 7·71 | 3·58 | 2·15 |
| *Bubo bubo* | 14·61 | 3·57 | 4·09 |
| Passeriformes | | | |
| *Corvus corone* | 32·59 | 6·73 | 4·84 |
| *Coloeus monedula* | 28·95 | 6·31 | 4·59 |
| *Pica pica* | 27·20 | 6·28 | 4·33 |
| *Chloris chloris* | 14·75 | 3·57 | 4·13 |
| *Fringilla coelebs* | 13·82 | 4·75 | 2·91 |
| *Passer domesticus* | 21·48 | — | — |
| *Passer montanus* | 16·77 | 4·14 | 4·05 |
| *Emberiza citrinella* | 12·07 | 3·55 | 3·40 |
| *Carduelis cannabina* | 12·65 | 3·48 | 3·63 |
| *Serinus canaria* | 11·67 | 3·50 | 3·33 |
| *Sturnus vulgaris* | 17·71 | 4·76 | 3·72 |
| *Parus coeruleus* | 19·04 | 5·06 | 3·76 |
| *Parus major* | 19·55 | 4·73 | 4·13 |
| *Sitta europaea* | 15·85 | — | — |
| *Phoenicurus phoenicurus* | 10·63 | 2·80 | 3·79 |
| *Motacilla alba* | 11·36 | 3·15 | 3·61 |
| *Turdus ericetorum* | 11·78 | — | — |
| *Turdus merula* | 13·98 | 4·59 | 3·04 |
| *Troglodytes troglodytes* | 9·06 | 3·78 | 2·39 |
| *Lanius collurio* | 12·00 | 2·95 | 4·07 |
| *Phylloscopus collybita* | 9·08 | 2·89 | 3·14 |
| *Acrocephalus arundinaceus* | 12·93 | 3·33 | 3·88 |
| *Acrocephalus scirpaceus* | 10·48 | 2·95 | 3·55 |
| *Hirundo rustica* | 12·21 | 3·00 | 4·07 |
| *Sylvia atricapilla* | 11·41 | 3·28 | 3·48 |

From Portmann (1946–47).

The lowest values for the "brain stem rest" (1·29–2·43) are only found amongst nidifuges such as the Charadriiformes and Galliformes. Conversely the highest values of 5·0–6·3 occur in the extreme nidicoles such as some Ardeiformes, Piciformes and passerines. Between these two extremes the values ranging from 2·4–4·66 are found equally amongst both nidifugous and nidicolous genera. The developmental indices for the hemispheres demonstrate the gradual augmentation of these regions, an augmentation which is comparable with that taking place during embryonic life (see Haefelfinger, 1958 and Chapter 4). They vary from 2·35 for *Charadrius alexandrinus*, to 32·59 in the case of *Corvus corone*. Clearly these values reflect the relative degree of *cerebralization* in these two very contrasting genera (see Chapter 13, Section XIII).

A comparison between this post-embryonic developmental history of the brain stem component and that of the forebrain hemispheres enabled Portmann to define the differences which are associated with various ontogenetic types. The third column of Table 86 gives this further quotient which he derived by simply dividing the values of the other two columns. There is clearly a considerable amount of variation. On the one hand species such as *Meleagris gallopavo* have a roughly equal development of the two regions. On the other hand in species like *Corvus corone* the quotient attains values of 4·84. In all such cases the greatest difference occurs in the most pronounced nidicoles. A value in excess of 3 occurs in the so-called coraciomorph genera such as those in the Piciformes, Psittaciformes and Passeriformes. It is essential to point out that such data are lacking for the Ratites, Sphenisciformes, Procellariiformes and Pelecaniformes, many of which lay one or more large eggs and only the alciform genus *Fratercula* approximates to this condition.

# References

Drooglever, F. and Kappers, C. U. A. (1921). "Anatomie des Nervensystem der Wirbeltiere." 3 vols.

Dubois, E. (1897). Sur le rapport de l'encéphale avec le grandeur du corps. *Bull. Soc. Anthropol. Paris* **8**, 337–376.

Haefelfinger, H. R. (1958). "Beiträge zur vergleichenden Ontogenese der Vorderhirns bei Vögeln", pp. 99. Helbing and Lichtenhahn.

Krompecher, St. and Lipak, J. (1966). Simple method for determining cerebralization. *J. comp. Neurol.* **127**, 113–120.

Kuenzi, W. (1918). Versuch einer systematischen Morphologie des Gehirns der Vögel. *Rev. Suisse. Zool.* **26**, 17–112.

Lapicque, L. (1944). Considerations quantitatives sur l'anatomie comparée du système nerveux. *Rev. Neurol.* **76**, 117–134.

Manouvrier, L. (1885). Sur l'interpretation de la quantité dans l'encephale et dans le cerveau en particulier. *Mem. Soc. Anthropol. Paris* **3**, 137–326.

Nikitenko, M. F. (1965). Principal features of adaptation to the aquatic way of life and the structure of the brain in auks. *Zh. Obshchei. Biologii.* **26**, 464–474.

Portmann, A. (1935). Die Ontogenese der Vögel als Evolutionsproblem. *Acta Biotheoretica*, A, **1**, 59–91.

Portmann, A. (1937). Beobachtungen uber die post-embryonale Entwicklung des Rosenpelikans. *Rev. Suisse Zool.* **44**, 363–370.

Portmann, A. (1942). Die Ontogenese und das Problem der morphologischen Entwicklungszustande von verschiedener Wertigkeit bei Vögeln und Saugern. *Rev. Suisse Zool.* **49**, 169–185.

Portmann, A. (1946–47). Études sur la cérébralisation chez les oiseaux, I, II, III. *Alauda*, **14**, 2–20; **15**, 1–15, 161–171.

Portmann, A. (1948). "Einfuhrung in die vergleichende Morphologie der Wirbeltiere. B. Schwabe, Basle.

Portmann, A. (1950). Les oiseaux. "Traité de Zoologie", vol. 15, pp. 1164. Masson, Paris.

Portmann, A. and Stingelin, W. (1961). The central nervous system. *In* "The Biology and Comparative Physiology of Birds" (Marshall, A. J. ed.), vol. 2, pp. 1–36. Academic Press, London and New York.

Schifferli, A. (1948). Uber Markscheidungbildung im Gehirn von Huhn und Star. *Rev. Suisse Zool.* **55**, 117–212.

Spector, W. S. (1956). "Handbook of Biological Data", pp. 584. Saunders, London.

Wirz, K. (1950). Studien uber die Cerebralisation zur quantitativen Bestimmung bei Saugetieren. *Acta Anat.* **9**, 134–196.

# ENVOI

It will, I hope, have become clear from the foregoing chapters that much remains to be elucidated about the avian brain. Not only is there more to investigate about the function of many discrete nuclear entities than we already know, but also there remain the innumerable problems about the degree of difference or similarity which exists between avian brain structures and the analogous or homologous structures in other vertebrates. It is also impossible to over-estimate the contribution which the growing number of semi-automatic and automatic control processes involved in telegraphy and traffic control are likely to make to our knowledge of those brain structures serving analogous functions. However, as these are analogues they merely serve to reinforce the zoologist's inherent scepticism of trite conclusions which unhesitatingly homologize structures in groups that are widely separated taxonomically.

It was implicit in several of the observations made in the Introduction that the fact that birds are homoiotherms, and as such comparable with mammals, is both a strength and a disadvantage in terms of brain studies. As such they have been the subject of some research simply because they are convenient laboratory animals providing an alternative to small rodents and Primates. On the other hand it has on occasions resulted in certain homologies being drawn between avian and mammalian conditions in the absence of in-equivocal evidence. It is certainly worth emphasizing that the two homoio-thermal stocks have had a widely divergent phylogenetic history at least since the Permian period, some 280 million years ago, and possibly since the Carboniferous, some 300–340 million years ago. The anatomy and sparse fossil history of birds shows that they are indisputably highly modified derivatives of the diapsid reptiles. As such they are related to the archosaurs and their descendant dinosaurs. The synapsid stock, from which the definitive mammals evolved during Permo-Triassic times, was a quite distinct and un-related phylogenetic line by Permian times. Indeed they were the dominant reptiles at that time and it was only later, and during the Mesozoic Era, that the diapsid stock achieved the supreme importance which is reflected in the term "Age of Dinosaurs". In terms of such phylogenetic considerations any trite analyses of avian brain function in terms of supposed nuclear homologies in the mammalian brain are clearly dangerous. However, it is extremely interesting that many of the recent data confirm that numerous homologies and analogies do exist between the demonstrable functions and biochemical composition of units situated at medullary, cerebellar, mesencephalic, di-encephalic and paleostriatal levels. The differences in the basic organization

## TABLE 87

The volume of some medullary structures in $\mu^3 \times 10^7$.

| | Body weight | V sensory | V motor | VI | Laminar | Angular | Magno-cellular | Triang-ular | Superior olive | Reticular formation |
|---|---|---|---|---|---|---|---|---|---|---|
| Melopsittacus | 37 | 39·02 | 15·46 | 3·88 | 9·72 | 17·24 | 13·26 | 10·99 | 1·62 | 906·0 |
| Agapornis | 42 | 121·4 | 21·69 | 5·47 | 13·26 | 14·88 | 9·62 | 14·93 | 3·88 | 1435·0 |
| Merops | 60 | 4·98 | 5·18 | 6·81 | 3·26 | 5·31 | 5·43 | 12·77 | 4·09 | 598·1 |
| Capella | 100 | 146·5 | 7·84 | 3·58 | 9·13 | 10·01 | 8·63 | 21·46 | 6·88 | 840·3 |
| Armides | — | 49·16 | 23·16 | 11·79 | 13·12 | — | 14·72 | 56·94 | 14·00 | — |
| Strix | 450 | 50·32 | 28·81 | 11·58 | 169·44 | 82·77 | 98·56 | 72·31 | 54·25 | 4504·0 |
| Corvus | 520 | 32·44 | 32·85 | 22·40 | 12·50 | 15·82 | 14·22 | 46·71 | 12·19 | 1813·0 |
| Aix | 600 | 118·5 | 27·65 | 9·56 | 18·29 | 13·55 | 14·31 | 35·09 | 16·83 | 1615·0 |
| Ibis | ~1200 | 1236·0 | 81·25 | 32·63 | 25·98 | 39·55 | 40·79 | 79·53 | 55·74 | — |

From Stingelin (1965).

## TABLE 88

The index values for the medullary structures of Table 87.

| | V. sensory | V. motor | VI. | Laminar | Angular | Magno-cellular | Triang-ular | Superior olive | Reticular formation |
|---|---|---|---|---|---|---|---|---|---|
| *Melopsittacus* | 802·0 | 318·0 | 85·0 | 203·0 | 355·0 | 273·0 | 226·0 | 33·3 | 1864 |
| *Agapornis* | 2295·0 | 410·0 | 107·5 | 252·0 | 286·0 | 183·0 | 283·0 | 73·7 | 2720 |
| *Merops* | 78·3 | 81·7 | 107·5 | 51·5 | 87·2 | 85·7 | 201·0 | 64·6 | 947 |
| *Capella* | 1765·0 | 92·5 | 43·0 | 110·0 | 120·0 | 104·0 | 295·0 | 80·6 | 1019 |
| *Strix* | 276·0 | 158·0 | 63·6 | 930·0 | 451·0 | 541·0 | 396·0 | 297·0 | 2470 |
| *Corvus* | 167·0 | 170·0 | 115·0 | 64·5 | 81·7 | 73·4 | 241·0 | 63·0 | 935 |
| *Aix* | 562·0 | 131·0 | 45·5 | 86·7 | 64·2 | 67·8 | 166·5 | 79·7 | 765 |
| *Ibis* | 4044·0 | 265·9 | 106·8 | 85·05 | 129·4 | 133·5 | 260·3 | 170·9 | — |

From Stingelin (1965).

of other hemisphere regions render any detailed comparison somewhat point-less except in terms of generalized control or integrative functions.

Discursive considerations of the various available data are of course con-tained in earlier chapters and a repetitious citation would be out of place here. However, it is necessary to reiterate a few points to emphasize certain aspects. In particular it is essential to direct attention to the very considerable variations which occur both between the different orders and also between genera or species within a particular order. Clearly gross ecological differences such as a predominantly nocturnal or diurnal way of life lead to the differential prominence of the auditory and visual pathways amongst the primary sensory modalities. Intra-generic differences can also reflect differing body size as is the case in the cerebellum. As far as the student of behaviour is concerned the various responses to a particular *set* of stimuli are influenced by long and short term circadian rhythms and their biochemical and hormonal con-comitants.

The medullary region is frequently thought of as a relatively well-defined group of structures of similar basic form in all avian orders. Indeed inter-ordinal and inter-generic differences are frequently ignored. Extensive analyses of medullary organization, such as that provided by Stingelin (1965), show that here too variations of nuclear size parallel differences of body size, behaviour and habitat. These include, of course, the rather better known differences in the structure of the auditory nuclei. Using the volume of the individual nuclei together with indices derived from them Stingelin (1965) observed that the size of the sensory nucleus of V exceeded that of other nuclei in *Ibis* and *Melopsittacus*, and was more or less equivalent to the motor nucleus in *Corvus*. As would be expected the laminar nucleus is largest in *Strix* (Table 87) and both the angular and magnocellular far exceed the size of their homologues in *Capella*, *Ibis*, *Melopsittacus*, *Merops* and *Corvus*. In these other genera the triangular nucleus tends to be larger than the auditory nuclei which are all of comparable size. It is interesting to note that the superior olive is of similar size in *Strix* and *Ibis* and that in *Merops*, apart from the olive, the various nuclei all tend to be of similar size.

Another feature of considerable interest, although perhaps not in itself too surprising, is the existence of somatotopic representation at the central levels of various sensory modalities. For example one can cite the organization of afferent foci within the sensory nucleus of the trigeminal, the cochlear nuclei, the optic tectum and the cerebellum. To these one can add the point to point inter-relationships which exist between pontine and olivary sites on the one hand and the cerebellum on the other.

In the sensory nucleus of the trigeminal Zeigler and Witkovsky (1968) found no discrete modality segregation but a clear sequence of peripheral representation in a dorso-ventral direction. Although no such somatotopic organization was apparent in either the medio-lateral or anterio-posterior planes there was a progressive representation of units responding to mandibular, maxillary and ophthalmic stimuli, and also a successive representation of distal and then more proximal foci as the monitoring electrode advanced from above. Despite the fact that the series of mandibulo-maxillary sites sometimes underwent a reversal those units which responded to orbital stimuli were always ventral.

Table 34 summarizes the afferent projections to the cerebellum. Whitlock (1952) concluded that the activation of Larsell's folia III, IV, V, VIa, VIb and VIc by stimulation of the ipsilateral peripheral nerves demonstrated a definite somatotopic pattern of representation. The caudal part of the body projected rostrally, the leg, wing and facial areas followed in sequence behind. Furthermore he drew attention to the fact that folia VIc, VII and VIII could be designated an audio-visual area. Similarly both de Long and Coulombre (1965) and Hamdi and Whitteridge (1954) established the existence of a definite topographical representation in the retinal projections to the optic tectum. Their tectal maps showed that the entire ventro-lateral and postero-ventral tectal surfaces receive projections from upper retinal areas.

Many of the recent discussions of general integrative function have been in terms of *labelled line coding*. This refers to the commonplace observation that those signals generated along the visual pathway yield vision, those along the auditory pathway elicit acoustic experiences, along the olfactory pathway— smell, etc. In birds such overall pathways have been established in the case of vision and hearing. However, it is as well to remember that the source of many central neurophysiological phenomena is peripheral. Even where central processes are involved they can generally be traced to a peripheral origin. Thus in vision events at the retina include adaptation, flicker fusion and masking (Leibovic and Sabah, 1969). Rushton (1968) suggested the presence of variable *gain control* as a result of his studies on adaptation. Dowling (1967) proposed that the amacrine-bipolar cell network is the site of at least some adaptation and Rodieck (1968) considered that some occurs prior to the bipolar cell layer. Leibovic and Sabah suggest that the amacrine-bipolar net would qualify well for a variable *gain control box* since the output could be modulated by the general level of activity in the amacrine process. If the bleached fraction of a visual pigment contributes a signal to an *adaptation pool* then this may also act on the horizontal and amacrine cells and thereby be involved in adaptation.

The actual ascending pathway of the visual system appears, from the electro-physiological data that are cited in earlier chapters, to involve forward projections from the tectum to the rotund nucleus in the thalamic region. From there the input to more anterior levels passes, via the axons of the thalamo-frontal tract, to the ectostriatum and also, possibly, the neighbouring regions of the intermediate neostriatum. Diffuse polysynaptic projections then connect with centres within the Wulst. In particular the accessory hyperstriatum of the dorso-medial margin of the hemispheres has been mentioned. Within the Wulst the responses to photic stimuli lack any apparent topographical organization. Unlike those which follow auditory stimuli they have a constant form when measured at various depths and the absence of any sensitivity to surface anaesthetics implies that their source is below this level. By comparison with the potentials that are elicited by click stimuli the potentials are of longer latency and duration and reach a greater magnitude.

Impulses from the cochlea are transmitted to the angular and magnocellular nuclei, both of which contribute fibres to the lateral lemniscus. This ends in the contralateral dorsal part of the lateral mesencephalic nucleus which projects to the ovoidal nucleus. From here fibres pass to Rose's field L within the caudo-medial neostriatum. A smaller projection also reaches the contralateral ovoidal nucleus via the dorsal supra-optic decussation. This is projected to the neostriatum of that side (see Fig. 138).

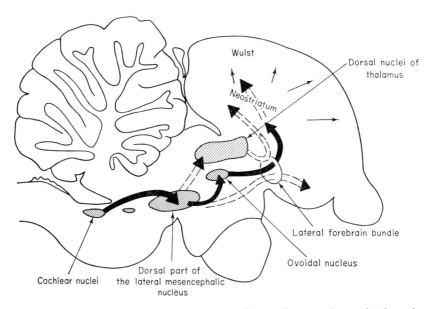

FIG. 138. The principal features of the ascending auditory pathways in the avian brain as elucidated by electrophysiological investigations. (After Phillips).

A useful comparison of the avian and mammalian medullary auditory stations in terms of the situation in the rat led Boord (1968) to suggest that the true cochlear components of the magnocellular nucleus correspond to the anterior division of the ventral cochlear nucleus. The medial part of the angular nucleus would be the avian equivalent of the posterior division of the ventral cochlear nucleus. The involvement of such stations on the ascending auditory pathway in motor function is suggested by the data of Brown (1965a, b) who established that both the torus and tectum were motivation centres for sound production in the alarm call.

In the Wulst the responses to acoustic stimuli are typically surface positive. Although some sites studied by Adamo and King (1967) responded to both acoustic and visual stimuli the polarities in the two situations were different. Furthermore surface procainization abolished the auditory responses and there was a drop in the potential magnitude at a depth of 200–300 $\mu$ from the surface. Clearly such data implicate Wulst cells in the anterior projections of the auditory system, and also suggest a differential representation of vision and hearing within the Wulst. However, any easy comparison of the various forms of hemisphere organization with the predominance of one or other primary sensory modality is not possible at the present time although it is interesting that both owls and song birds, in which hearing is important, have the well-developed Wulst situated anteriorly. One is also tempted to direct attention to the prominence of the neostriatum and ventral hyperstriatum in certain other orders. Nevertheless vision is important to the oscines and hearing to, for example, the Psittaciformes.

It is also not possible to allocate specific filtratory, stimulatory or inhibitory functions to particular nodes of forebrain activity other than those cited in the earlier chapters. An overall picture of sensory input and motor outflow within the avian brain is summarized by Fig. 139. It is clear from this that many possible sites exist at which filtering may occur. The avian brain has therefore the theoretical requirements needed for systems controlling *attention* as postulated by psychologists such as Broadbent and Treisman. Broadbent's (1958) formulation, upon which more recent psychological workers rely, involves information entering the system by way of a number of channels. Such channels may not only include such direct pathways as the auditory and optic nerves of each side, but also a wave envelope of a particular fundamental frequency, or similar functional channels. In such theories these are assumed to have a distinct neural representation permitting messages to be selected on the basis of say, pitch, loudness and spatial characteristics. Situated further along the system there are limited capacity channels in the information theory sense. As their capacity is less than that of the initial input lines loss of information must occur between parallel inputs and more anterior or distant projections. Alternatively some form of recoding is necessary if the bottleneck is to be overcome. Broadbent postulated a short term memory store at the inner end of the parallel input lines. This was followed by a filter able to select one of the input lines and permit its information direct access to the limited capacity channel. Lines not selected in this way can retain messages

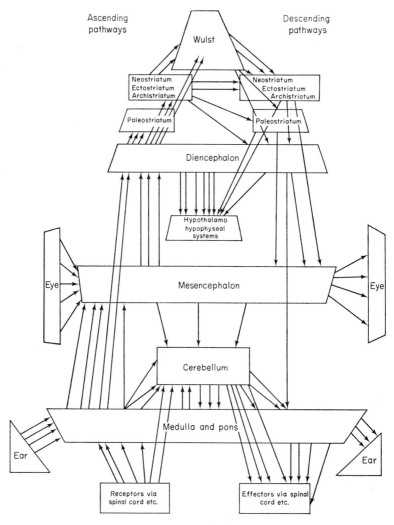

FIG. 139. An overall flow diagram summarizing the established input and output of the avian brain. This can be compared with the anatomy represented in Figure 140 and with the theoretical filter systems of experimental psychologists.

in the short-term memory store for a limited period of a few seconds. At the downstream end of the channel there are the motor systems.

Reference to the conclusions of von Holst and Saint Paul which were referred to in the introduction will make it clear that both these and the theories of psychologists such as Broadbent, although produced under very different circumstances, have a number of factors in common. This is also true of the rather more elaborate suggestions of, for example, Treisman (1966), who gave more explicit information about the rules which govern the filter. In terms of von Holst and Saint Paul's examples the particular locomotory pattern elicited by a given set of stimuli will be a function of the characters selected by the filter at a given time—at stimulation. That various chemicals such as reserpine, dartal, luminal and nembutal can have a direct effect on the thresholds for centrally evoked clucking and locomotor responses (von Holst and Schleidt, 1964; von Saint Paul, 1965) provides further evidence for the way that autochthonous molecules can effect behaviour, and emphasizes the basic importance of molecular characteristics such as electronic structure, and also the circadian rhythms of production, in eliciting specific behaviour patterns. It is in this way that the multiple feedback loops influence the overall pattern of nervous activity which is reflected by impulses passing along particular afferent or efferent pathways. In the case of those behaviour patterns which have been exhaustively analysed (see Hinde, 1966, for example) it is quite clear that we can now elucidate some of the brain centres which must necessarily be involved when certain organs such as wing, tail or eye are concerned. Future work will elucidate the significance of those brain stem structures whose function is as yet unknown to us.

## References

Adamo, A. J. and King, R. L. (1967). Evoked responses in the chicken telencephalon to auditory, visual and tactile stimuli. *Exptl. Neurol.* **17,** 498–504.

Boord, R. L. (1968). Ascending projections of the primary cochlear nuclei and nuclei laminaris in the pigeon. *J. comp. Neurol.* **133,** 523–542.

Broadbent, D. (1958). "Perception and Communication." Pergamon, London.

Brown, J. L. (1965a). Vocalization evoked from the optic lobe of a song bird. *Science* **149,** 1002–1003.

Brown, J. L. (1965b). Loss of vocalization caused by lesions in the nucleus mesencephalicus lateralis of the red-winged blackbird. *Amer. Zool.* **5,** 693.

de Long, G. R. and Coulombre, A. J. (1965). Development of the retino-tectal topographic projection in the pigeon. *Exptl. Neurol.* **13,** 351–363.

Dowling, J. E. (1967). The site of visual adaptation. *Science* **155,** 273–279.

Hamdi, F. A. and Whitteridge, D. (1954). The representation of the retina on the optic tectum of the pigeon. *J. exp. Physiol.* 111–118.

Hinde, R. A. (1966). "Animal Behaviour; a Synthesis of Ethology and Comparative Psychology", pp. 594. McGraw-Hill, London.

Holst, E. von and Saint Paul, U. von (1963). On the functional organization of drives. *Anim. Behav.* **11,** 1–20.

Holst, E. von and Schleidt, W. M. (1964). Wirkungen von Psychopharmaka auf instinktives Verhalten. *Neuropsychopharm.* **3,** 22–29.

Kappers, C. U. A. (1947). "Anatomie comparée du système nerveux". Masson, Paris, pp. 754.

Leibovic, K. N. and Sabah, N. H. (1969). On synaptic transmission, neural signals and psycho-physiological phenomena. *In* "Information Processing in the Nervous system" (Leibovic, K. N., ed.), pp. 273–296. Springer, Berlin.

Rodieck, R. W. (1968). Quoted by the fore-going.

Rushton, W. A. H. (1968). The Ferrier lecture; visual adaptation. *Proc. Roy. Soc. Lond.*, B **162**, 20–46.

Saint Paul, U. von (1965). Einfluss von Pharmaka auf die Auslobarkeit von Verhaltweisen durch elektrische Reizung. *Z. vergleich. Physiol.* **50**, 415–446.

Stingelin, W. (1965). Qualitative und quantitative Untersuchungen an der Kerngebieten der Medulla oblongata bei Vogeln. *Bibliotheca. Anat.* **66**, 1–116.

Treisman, A. (1966). Our limited attention. *Adv. Sci.* pp. 600–611.

Whitlock, D. G. (1952). A neuro-histological and neuro-physiological study of the afferent fibre tracts and receptive areas of the avian cerebellum. *J. comp. Neurol.* 567–635.

Zeigler, H. P. and Witkovsky, P. (1968). The main sensory trigeminal nucleus in the pigeon. A single unit analysis. *J. comp. Neurol.* **134**, 255–264.

# AUTHOR INDEX

Numbers in *italics* indicate the pages on which the references are listed.

23

# SUBJECT INDEX

Page numbers which are in italics indicate illustrations.

## A

Acanthodii, 319
Accessory hyperstriatum, 7, 118–119, 122, 343, *399*, 449–451, 461, 463, 465–468, 477–482, 484, 486, 488, 491, 502–504, 515–516, 538–551, 583
*Accipiter*, 20, 26, 27, 106, 490
  *cooperi*, 23, 25
  *gentilis*, 600
  *nisus*, *112*, *142*, 295, 487–489, 600
  *velox*, 487–489
Accipitridae, 299
Acetylcholine, 42, 63–69, 147, 148, 257, 504
  esterase, 63–69, 147–148, 166, 229, *257*, 312–313, *331*, 364, 504, 568
Acetylthiocholine iodide, 166
*Acrocephalus arundinaceus*, 609, 612
  *schoenobaenus*, 299
  *scirpaceus*, 602, 612
*Actitis hypoleucos*, 243, 250, 598
Adeno-hypophysis, *see* hypophysis
Adenosine, monophosphate, 57
  5 phosphate aminohydrolase, 50
Adrenalin, 42, 59–63, 406, 504, 572–573
*Aegithalos caudatus*, *292*
*Aegolius acadicus*, 345–346
*Aegypius monarchus*, 600
*Aethia*, 135
  *cristatella*, 136–137
  *pygmaeus*, 136–137
*Agapornis*, 196, 218, 616–618
  *fischeri*, 601
  *personata*, *245*, 250, 605
  *roseicollis*, 396, 502
*Agelaius phoeniceus*, 300
*Aix*, 196, 218
  *galericulata*, *184*, 616–618
Alanine, 57
*Alauda*, 252
  *arvensis*, 299, 602

Alaudidae, 299, 597, 601, 604
Albatross, *see Diomedea*
Albumin, 48, 77
*Alca*, 135, 236, 451
  *torda*, 136–137, 221, 297, 594, 596
Alcedinidae, 296, 300, 452
*Alcedo*, 5
  *atthis*, *290*, 300, 601
  *ispida*, 192, *194*, 595
Alcidae, 297, 452
Alciformes, 6, 135, 280, 487, 593–594, 596–599, 608, 611
Aldolase, 50
*Alectorix*, 289
  *rufa*, 294–295
Ali-esterases, 166
Alimentary canal motility, 216–217
*Alligator*, 460, 482
Amacrine-amacrine synapses, 308–311
Amacrine-bipolar synapses, 308–311
*Amazona*, 465, 473, 476–478
  *ochrocephala*, *472*, 474
  *versicolor*, 601
Amino-acids, 55–59
Amino-butyric acid, 55–56
Amphetamine, 573, 575
Ampulla, 173–181, 184
Anaesthesia, 570–572, 582
*Anas*, 22, 105, 106, 113, 165, 175, 214, 238, 260, 326, 334, 349–350, 421, 424, 455, 465, 481
  *boschas*, 174, *292*, 418
  *crecca*, 600, 609, 611
  *domesticus*, 107, 487–489
  *penelope*, 600
  *platyrhynchos*, 140, *142*, *176*, 221, *247*, 261, 297, 345, 457, 472–474, 479, 600, 611
Anastomosis
  anterior cerebral, 27
  dorso-ventral, 29
  cerebro-extracranial, 29
  inter-carotid, 15–22, 30, 415